Designing Value-Creating Supply Chain Networks

Alain Martel · Walid Klibi

Designing Value-Creating Supply Chain Networks

Alain Martel
Université Laval
Quebec City
Canada

Walid Klibi
KEDGE-Business School
Bordeaux
France

ISBN 978-3-319-28144-5 ISBN 978-3-319-28146-9 (eBook)
DOI 10.1007/978-3-319-28146-9

Library of Congress Control Number: 2016932348

Printed on acid-free paper

This Springer imprint is published by Springer Nature
The registered company is Springer International Publishing AG Switzerland

To Lise and Feriel

Preface

Companies must periodically adapt their supply chain networks (SCNs) to changes in their business environment to create value for their stakeholders. They must question their market offers; the number, location, technology, and mission of their production-distribution facilities; as well as their choice of suppliers, subcontractors, and logistics service providers. Furthermore, their SCN must be sufficiently robust and resilient to cope with any plausible future. With the inherent uncertainty of the current business environment and the breadth of contemporary supply chains (SCs), strategic SCN reengineering problems have become crucial. They are, however, extremely difficult. Supply chain management professionals and students need to understand the factors affecting SC performances as well as available SCN design approaches and techniques to be able to help their company use their SCN to develop a competitive advantage.

Designing Value-Creating Supply Chain Networks presents an innovative reengineering methodology to significantly improve the bottom line of SC-driven businesses in an uncertain world. The book starts by discussing current business challenges and strategies affecting SCN design and it examines the nature of production–distribution centers and facilities, transportation means, internalization–externalization decisions, and planning and control paradigms. It then gradually introduces a modeling framework that can be used to optimize network structures, starting with basic design problems. Capacity planning issues, multinational business factors, and sustainable development requirements are also studied. SCN design problems under uncertainty are then examined. Finally, a synthesis of the SCN design methodology progressively developed in the book is presented and practical SCN reengineering project planning and control issues are discussed.

The book grew out of lecture notes developed over several years for graduate students in supply chain engineering and management at Laval University's business school in Quebec City, Canada. These lecture notes were also used to support industrial engineering and executive education courses in various countries and in particular at the ISLI master programs of KEDGE Business School in Bordeaux, France. During that period, the notes were enriched with material from several SCN reengineering projects completed by the authors in commercial and industrial firms and the military. Much of the modeling methodology presented was developed during research projects on the design of effective and robust SCNs at the Interuniversity Research Centre on Enterprise Networks, Logistics and Transportation (CIRRELT) in Canada.

Using the same material to teach in business, engineering, and executive programs is quite challenging because each group has different expectations. Business students want to develop a good understanding of the issues they will have to confront in practice and the strategic decisions they will have to make, but they tend to shy away from the mathematical tools required to develop superior SCN designs. Engineering students are motivated by the formulation and solution of mathematical models, but they tend to content themselves with a superficial understanding of industry structures and business problems. It is our firm conviction that both facets are important and necessary to develop robust value-creating SCNs. It is therefore worth pushing business students into examining mathematical formulations. This provides a much deeper understanding of problems and the tradeoffs to be made, and it develops good decision-making reflexes. Also, managers are reluctant to adopt a model-based solution proposed by a consultant if they do not understand how good candidate SCN designs can be generated using models. Conversely, engineering students must develop their soft skills and learn to interact with real organizations and problem contexts. Generic predefined problems are useful teaching tools but SCN designers must be able to analyze and understand real problems and make recommendations in terms that top executives can understand. This book provides the material required to support a balanced teaching approach that is adequate for both groups.

The book covers classical SCN design methods and models but it goes much further. Strategic issues related to product-market offers, activity internalization–externalization decisions, facility platform choices, logistics service provider selection, international commerce Incoterms adoption, sustainable development strategies elaboration, and so forth are addressed in the proposed modeling framework. The studied concepts and methods are pedagogically introduced using examples and exercises. The book combines in a simple way key management, engineering, optimization, statistics, and risk analysis notions required to design robust value-creating SCNs. The proposed reengineering approach is based on a visual SC activity representation formalism that facilitates the analysis and modeling of SCN design problems. Several of the example problems presented are based on real cases, and we show how standard Excel statistical analysis and optimization add-ins can be used to solve them. The chapters also include review questions as well as several exercises that can be solved with Excel tools or standard statistical and optimization software packages.

The book is intended for an academic and professional audience. It is primarily targeted at postgraduate and final-year undergraduate students in industrial engineering and operations and SC management. Students in operations research and management science can also use it to discover a stimulating application area. Practitioners in industry and consulting involved in SCN reengineering projects will find that the book provides several avenues to enhance their practice. It will also help SC managers understand how design models can be used to develop better SC strategies. Researchers in SCN design will find that it offers a detailed review of the current state of the art helpful for the investigation of new research avenues.

We trust that the students and practitioners using the book will find it helpful. We would like to thank our many graduate and executive students who, over the years, have stimulated our thinking and helped develop the material in the book. We would also like to thank the executives who, during SCN reengineering projects completed in their company, challenged our methodology and raised insightful questions that are the source of several of the innovations presented in the book.

Alain Martel
Walid Klibi

Contents

Chapter 1
Supply Chains: Issues and Opportunities

Today's shoppers generally expect to find large product assortments in the stores they visit. Yet product availability depends on a whole chain of sourcing, transportation, warehousing, and manufacturing activities. Many of the products we buy are made on the other side of the world and travel great distances before reaching our store shelves. Some components may have been made in China and others in Mexico, with the final product itself being assembled in Poland and stored in a distribution center in Germany before final shipment to a European store. The places a product was made and warehoused, and the itinerary it took to get to the store shelf, define the supply chain (SC) that ultimately determines its price. Without SCs, we would be unable to feed or clothe ourselves, stay healthy, or take part in leisure activities. Supplychains are the cardiovascular system of our economy and the basic fabric of the industrial and commercial organizations ensuring our survival and prosperity. In the industrialized world, they account for a large percentage of GDP and employment.

The competencies needed to design and manage SCs have become critical success factors and key competitiveness drivers in the business world. At a given point in time, a company may be simultaneously involved in a number of SCs within its sector. It can be responsible for running several activities in these chains, such as collaborating with partners operating upstream or downstream, and even outsourcing some of its activities. The business model that companies adopt leads to the implementation of internal SC systems that are typically deployed in several production, distribution, or sales centers. In order to develop a sustainable competitive advantage, these SC systems must be continuously adapted to comply with expansion, merger, or acquisition decisions; benefit from new technological or outsourcing opportunities; or simply follow changes in the company's commercial environment. This requires the adoption of SC strategies and system reengineering processes that are capable of preserving and improving the company's ability to create value. This chapter provides an overview of the SC concepts associated with this approach and lays the foundations for the following chapters. After identifying

A. Martel and W. Klibi, *Designing Value-Creating Supply Chain Networks*,
DOI 10.1007/978-3-319-28146-9_1

the main socioeconomic trends having an impact on SCs, the chapter studies the nature of SC systems' structures, processes, and missions. It also takes a look at strategic SC instruments that companies can use to ensure their survival and long-term prosperity.

1.1 Business Context and Trends

Before examining the structure and internal processes of SC systems, we analyze some contemporary trends and issues that are shaping modern economies' competitive environments, and have a strong impact on SC design and management. As illustrated in Fig. 1.1, these trends and issues are grouped into nine categories, each of which will be examined briefly.

1.1.1 Globalization of Markets and Companies

Undoubtedly, global competition has been redefined by the opening of markets worldwide and subsequent emergence of new global competitors. In 2007, UNCTAD (United Nations Conference on Trade and Development—www.unctad.org) counted 79,000 transnational corporations in the world as well as a total of 790,000 foreign affiliates. In a similar vein, global exports have clearly risen much faster in recent years than world production (CST-Québec 2010). This globalization of business has been encouraged by bodies such as the WTO (World Trade

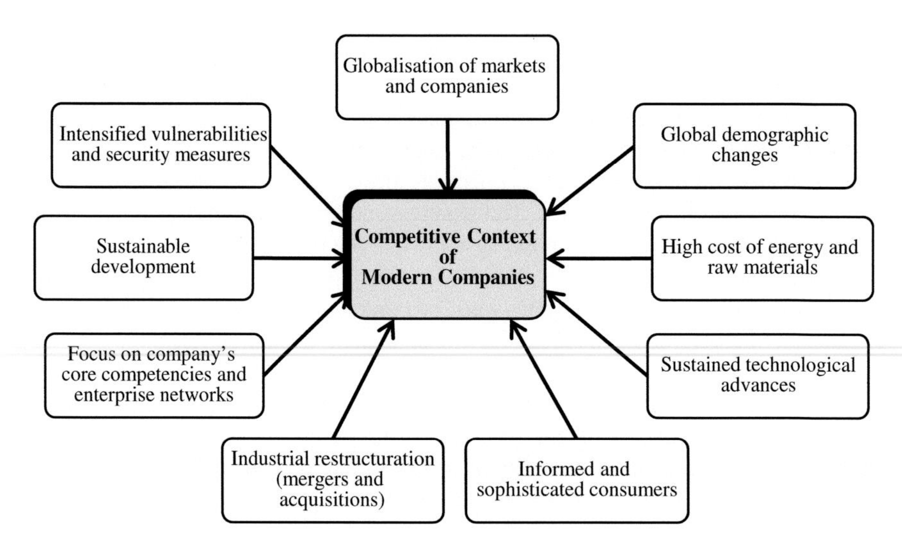

Fig. 1.1 Trends affecting business supply chain systems

Organization—www.wto.org), and by the emergence of trading blocs like the European Union (28 countries in 2014), NAFTA (United States, Canada and Mexico), Mercosur (Argentina, Brazil, Paraguay, Uruguay, and Venezuela), ASEAN +3 (Southeast Asia, China, Japan and South Korea), and APEC (the Asia Pacific Economic Cooperation forum). According to the WTO, in 2010 almost 300 preferential trade agreements were in force and, on average, a WTO member was part of 13 agreements (WTO 2011). As stressed by the anti-globalization movement, the phenomenon is not always fair or orderly. At the same time, it seems irreversible.

For years, the globalization of business was largely motivated by companies' search abroad for more advantageous factors of production. One specific driver has been their desire to reduce production costs, thanks to the abundant and inexpensive workforce found in the developing world. This often led to the delocalization of business activities, moves made even more attractive by the subsidies, and tax advantages offered by local governments. More recently, however, it has become clear that this is not always a profitable strategy, especially in highly automated industries. Indeed, delocalization has been damaging many multinationals' image, if only because of the social and environmental misconduct with which it is sometimes associated. It can also be very expensive to move from one country to another. Similarly, some countries' inadequate SC infrastructures, as well as the distances involved, can lead to significant indirect costs. Lastly, certain markets' comparative labor advantages have evolved quickly in recent years. All of this explains the recent trend towards relocalization. For example, several labor-intensive Western European companies' delocalization to Eastern Europe has been followed by a relocalization, following certain unfortunate experiences abroad and a growing desire to work in the proximity to customers.

The gradual rise in developing countries' output has increased their purchasing power and created new markets. Increased overseas demand has also contributed to companies' internationalization. Domestic demand in the developed world has had a tendency to stagnate while developing country demand has shot up. Yet the average income remains higher in the developed world. The end result is a two-speed global economy (BCG 2010): lower demand for branded products but high service requirements in older industrialized countries and higher demand for basic products in the developing world. In this context, emerging economies, which had previously been mainly used for production purposes, are becoming increasingly important actors in global R&D, distribution, and sales. China, for instance, moved ahead of United States in 2009 to become world's leading market for new motor vehicle sales (BCG 2010). Besides from the emerging powerhouses that are collectively known as the BRICs (Brazil, Russia, India, and China), other countries including South Africa, Indonesia, Thailand, and Malaysia now present exciting opportunities. Strengthened by their successes, developing economies now account for 39 % of global foreign direct investment flows, compared with only 12 % at the beginning of the 2000s (UNCTAD 2014). Hence the emergence of new strategies such as "Blue Ocean" (www.blueoceanstrategy.com) that focus directly on globalization. The focus here is no longer on increasing companies' share of existing markets but instead on encouraging innovation and originality, enabling access to global market sectors that are totally free from competition. Companies such as Apple and Cirque du Soleil apply this strategy with great success.

1.1.2 Global Demographic Changes

The effects of globalization have been accentuated by major demographic changes. The world's population has crashed through the 7 billion barrier, with 8 billion expected by the year 2025 (NIC 2008). It is in China and India that population has grown the fastest in recent years. China has taken steps to slow demographic expansion but the trend remains strong in India, which is projected to be the world's most populous country by 2025. Population growth is much slower in the developed world, and particularly in Western Europe. The fastest growth rates are found in certain underdeveloped countries of sub-Saharan Africa and the Middle East. Population growth trends in the emerging world, coupled with economic growth in the same countries, have created increasingly attractive markets in certain regions. China is expected to become the world's largest economy by 2025, and India's consumer spending is also increasing at a fast pace. Pressures caused by rising population in Asia and other fast-growing regions will all have a major effect on urbanization and cross-border migration. With many cities also expanding in size, distributing merchandise into city centers is bound to become more complex. In addition, growing and shifting world population patterns will increase the pressure on the availability of energy, natural resources, and foodstuffs.

There has also been a significant increase in life expectancy, particularly in developed and developing countries. Life expectancy for G7 countries currently exceeds 78, creating the conditions for a new category of consumers with some very specific needs. Baby boomers are retiring and many are financially comfortable. Life expectancy is also increasing rapidly in the emerging world and currently stands near 74 in both China and Brazil. Soon this will generate new consumer product needs. At the same time, however, a very high percentage of the population in the world's underdeveloped countries remains under the age of 30.

1.1.3 High Cost of Energy and Raw Materials

Sustained commercial activity in the developed world, combined with strong growth in emerging countries, has contributed to the depletion of rare natural resources and nonrenewable energy sources. Rising demand has led to a sharp rise in raw material, fuel, and energy prices, a trend that should accentuate over the next few years. One example is European Union oil imports, which reached €210 billion in 2010 (EC 2011). Initiatives aimed at preserving rare resources and developing alternative energy solutions have become necessary. Shell, for instance, has devised a number of scenarios for the year 2050 (Shell 2008). These assess the planet's energy needs and resources based on global political, technological, and industrial developments, and on decision-makers' behavior. There are two future scenarios in this vision: *Scramble*, dominated by an irresponsible and inefficient use of resources; and *Blueprints*, characterized by responsible development and energy efficiency. The study shows that the two scenarios lead to very different outcomes in terms of global resources reserves and energy consumption: 13 % less energy is consumed in *Blueprints* than in *Scramble*, which would be very difficult to sustain. Given the uncertainty about the future cost and availability of energy and raw materials, strategic SC planning has become increasingly challenging. A global study carried out by IBM shows that rapid and continuous changes in important inputs such as oil, raw materials, or labor are overwhelming SC managers' ability to adapt (IBM 2009). Because of this volatility, controlling SC costs will be a serious challenge for many years to come.

1.1.4 Sustained Technological Advances

Unlike resource-based products, knowledge-based products have experienced rapid expansion in recent years. Mobile communication and data transfer technologies (using text, image, and voice) are now very widespread. It is clear that Internet technologies have revolutionized our way of living and doing business. Whether this involves business-to-business (B2B) trading or business-to-consumer (B2C) sales, users' imagination appears to be the only limit on the opportunities that the Internet offers. Companies have also taken advantage of rising enthusiasm for social media to create new promotional platforms. One example is Wal-Mart's recent initiative on Facebook called *My local Walmart*, which enables targeted customer offers covering new products and promotions.

New information technologies (ITs) such as surface computing, cloud computing, software as a service (SaaS), virtual reality, intelligent software agents, real-time positioning and detection tools, the Internet of Things (IoT), and so on are continuing to arrive at a fast pace. Yet as much as these technologies facilitate the dissemination and processing of data, they do little per se to improve SC decisions. Given the size and complexity of modern SC networks, global companies seeking to manage their material flows must set up sophisticated information, communications, and decision support systems. This may lead to information overloads, however, making it hard for managers to detect what is really important. Business intelligence technologies have become an increasingly popular response to this problem, using analytical, forecasting, and optimization applications to transform *big data* into knowledge in a way that facilitates judicious decision-making.

There have also been technological changes in areas other than IT, including production and distribution. Several SC innovations have enabled improvements in transportation activities and reduced international trade costs. Transportation infrastructures are being transformed with the advent of new multimodal platforms, such as Delta3 (www.delta-3.com) in north France and Tanger Med (www.tmsa.ma) on the Moroccan side of the straits of Gibraltar. Gigantic post-Panamax container ships will be able to use the Panama Canal once its expansion is completed in the near future (www.pancanal.com), completely modifying SC flows between Asia and America. Drones are also starting to be used for the transportation of products. Otherwise, new flexible production/handling technologies, such as 3D printing, are making product customization easier. The *Physical Internet* manifesto (Montreuil 2011) offers a vision of global material flows based on connective technologies such as RFID (Radio Frequency IDentification), GPS (Global Positioning Systems) and the Internet, along with a number of SC innovations. Production and export patterns for high technology products (that is, R&D-intensive goods) are also changing. Over the past decade high tech products' share of manufacturing exports have fallen sharply in the United States and Japan but have risen rapidly in China and India (with Europe remaining relatively stable). All these technological changes could alter SC flow patterns and reduce design-to-market times significantly in a number of industries.

1.1.5 Informed and Sophisticated Consumers

In recent years, the cost of living has tended to rise more quickly than wages. Alongside, according to the World Bank, the share of women employed in non-agricultural jobs has reached 48 % in North America, Europe, and Central Asia. The result is that in many households all of the adults are now working. In turn, this means that they have little time to shop, and therefore require immediate access to products. They also demand greater quality and service than in the past. Increased quality and health standards have created a need for greater product SC traceability. Buying patterns are also changing due to the wide use of the Internet, mobile phone apps, and e-commerce. Many consumers today compare products and buy online instead of going to the shops. Some even compare product offers in stores, using their mobile phone, before they buy. As a consequence, retailers are now adopting *omni-channel* selling strategies (Carroll and Guzman 2013). This has led to a significant increase in the demand for small package transportation.

The Internet also lets customers become more active in product design. A recent study shows that consumers are now a major source of innovation (Von Hippel et al. 2011). This is why Proctor & Gamble created the Tremor Web community (www.tremor.com), mainly composed of teenagers who help develop new product ideas. Lego (www.lego.com) has been encouraging its customers to download Lego Digital Designer to make their own designs, and inspire the company's new product development. The end result has been open innovation and crowdsourcing (www.openinnovators.net), which involves using the creativity and know-how of people to facilitate new product designs and launches. The total effect of competition, technological progress, and increased customer expectations has been to shorten product life cycles and nurture demand for customized products. For manufacturing firms, this imposes a fragmentation of production and a compression of cycle times. Companies must shift from mass production to mass customization, a transition that can be very challenging due to increased flexibility requirements (Poulin et al. 2006).

1.1.6 Industrial Restructuration

The emergence of new competitors and markets, pressure from rising costs, more demanding consumers, new technological opportunities and, of course, the 2008 financial crisis, have provoked a major restructuration of many industrial sectors. Many of today's SC configurations result from successive mergers and acquisitions. Where SCs used to be organized on a country-by-country basis, as in Europe, recent industry restructuration has led to current factories and distribution centers now serving several EU countries. One consequence has been the congestion of regional transportation networks—costing an estimated €80 billion annually in Europe, or 1 % of the EU's total GDP (EC 2011). Also, the retail sector is now dominated by a few major players like Wal-Mart and Carrefour. These companies can impose, because of their enormous purchasing power, price and service conditions on their SCs. This led to the disappearance of several intermediaries since large retailers now usually trade directly with manufacturers. Globalization has also led to a wave of mergers and acquisitions in the logistic services sector, culminating in the birth of enormous global third-party logistics (3PL) providers such as DHL (www.dhl.com), UPS (www.ups.com) and Geodis (www.geodis.fr).

This new global economic landscape does offer a number of attractive opportunities. However, some sectors/regions suffer from an ongoing shortage of qualified human resources, intimating further restructuring efforts. The SC networks resulting from corporate mergers necessarily contain much duplication. To take advantage of a merger, SC networks must be re-optimized. Also, succeeding in the current two-speed economy requires operational flexibility and responsiveness that is not necessarily provided by traditional SC networks. These also lead to SC re-optimization projects. Companies increasingly participate in integrated global SC networks. Coordinating material flows in such interconnected SCs has become extremely complex (longer and more variable delays; larger number of suppliers, transporters, customers, countries, etc.). To be competitive in this context, companies must continuously adjust the structure of their SC networks, and the processes of their planning and execution systems, to the changes in the business environment. This necessitates the adoption of efficient reengineering methods.

Another industrial restructuring trend that has gained momentum is the pooling of transportation, storage, and/or sourcing activities. This usually involves activity-sharing agreements between companies to reduce SC costs. One example is the French food industry, which today widely shares common transportation and logistics means. The Physical Internet manifesto (Montreuil 2011) stresses this propensity to share by suggesting that global transportation and warehousing networks be converted into a portfolio of infrastructures and resources that are open to all industrial actors.

1.1.7 Core Competencies and Enterprise Networks

Few companies have the know-how they need to be competitive individually in the global marketplace. Future success depends, for instance, on the ability to design and manage global research, production, and distribution center networks. Most world-class companies have undergone a number of restructurings and are now focused on core competencies and distinctive know-how, relying on third parties to execute nonstrategic activities. Competition today no longer involves one company against another but rather a network of companies against other networks. The competitiveness of today's companies reflects their success in establishing efficient partnerships, notably with product-service suppliers, customers, or technological centers. Of course, these may be spread worldwide. In this context, a company must learn to share its information and expertise with partners, who in turn need to learn how to share the profits and risks resulting from the collaboration. In a similar vein, it is often useful today for dynamic SMEs to join powerful industrial networks. Since logistics is often not a core activity for modern industrial companies, it tends to be outsourced. Indeed, combined with globalization, the demand for logistics services has driven the emergence of aforementioned global 3PLs.

1.1.8 Sustainable Development

Everyone is now aware of the impact that economic activity can have on environmental health and social equity. For instance, merchandise transportation-related energy consumption rose in the Canadian province of Québec by 51 % between 1990 and 2005, with greenhouse gas (GHG) emissions jumping by 92 % (CST-Québec 2010). A growing number of organizations and governments are waging a high-profile battle aimed at reducing energy consumption, GHG production, congestion, and the degradation of natural resources. Many measures have been implemented to protect the environment, including green taxes and severe regulations targeting production and transportation companies to get them to diminish their GHG emissions. Movements for social justice and human rights (social sustainability) have also stressed the need for ethical business practices. As a result, according to a 2011 global executive survey (Haanaes et al. 2012), most managers now believe a sustainability strategy is a competitive necessity. These changes have led to the development of green transportation and manufacturing technologies, ecologically friendly production–distribution centers (LEED certification—Leadership in Energy and EnvironmentalDesign), green purchasing, reverse logistics, product reuse, improved labor conditions, and so on.

1.1.9 Intensified Vulnerabilities and Security Measures

Supply chains' globalization and interconnections have increasingly exposed them to all kinds of disturbances and dangers. Several components of the global business environment, including demand, price, interest rates, and exchange rates, are by their very essence quite variable. Other disruptions affecting companies include natural disasters (whose number, according to EM-DAT (www.emdat.be) statistics, has practically tripled over the past 30 years), economic and political crises, industrial accidents, failed partnerships, and long-term strikes—all of which seriously compromise SC effectiveness. Corporate performance has suffered additional operating costs, expensive recourse actions and income losses due to these phenomena. Between 1989 and 2000, the stock market performance of 835 companies which experienced SC disruptions was 33–40 % worse than the rest of their sector (Van Opstal 2007). According to the World Bank (2011), the March 2011 Japanese tsunami caused a 1.1 % decline in worldwide industrial production the following month.

Aside from these catastrophic events, companies are facing daily incidents of theft, fraud, and counterfeiting, forcing them to make their SC systems ever more secure. Similarly, a number of countries have tightened border security and imposed preventative measures on business. These measures are necessary but can cause significant SC delays. Recent examples include security procedures responding to demands from customs authorities in the USA (C-TPAT), Canada (PEP), and Europe (OEA) in line with the World Customs Organization's SAFE charter (www.wcoomd.org). All of these factors demonstrate the growing vulnerability of today's SCs. Significant adjustments are needed to make them more robust and resilient.

This initial section has allowed us to highlight those socioeconomic trends and issues that are likely to have a major effect on tomorrow's SCs. It is clear that if companies do not adapt to changes in the business environment, their competitive position will deteriorate. Hence the following sections' closer analysis of the nature of SC systems, and study of what companies must do to adapt and create value.

1.2 Supply Chain Systems

Several stages are required to transform raw materials into consumer products requiring subsequent shipment to markets. These stages can occur in different companies and involve many different design, procurement, production, distribution, and sales activities. The industrial and commercial networks created in this way are complex lattices but they are customarily referred to as *supply chains*, despite the reductionist nature of this image. A company will only get involved in a subset of all the activities of the SCs to which it belongs. The resources and processes it uses to fulfill its role in one or several industrial networks define its internal *supply chain network* (SCN). Internal SCN nodes correspond to the company's facilities (factories, warehouses, transshipment terminals, points-of-sale, etc.) and to all the processes and resources that they encompass. Network arcs correspond to the transportation, communication, and payment means used to move products, information, and funds from one node to another. To understand a company's SCN, it is thus necessary to study the factors affecting its nodes and arcs.

Figure 1.2 illustrates the roles that companies can play in their SCs. A company does not necessarily maintain a direct contact with end users since it might sell its goods to downstream manufacturers, or to intermediaries such as distributors, retailers, brokers, or industrial agents. Similarly, it does not necessarily do business with upstream sources of raw material. It may even act as a subcontractor for other companies operating upstream and downstream from its own position. In the clothing sector, for instance, it frequently occurs that a company is responsible for product design, fabrics sourcing, and distribution, with the actual manufacturing being delegated to a specialist apparel company.

Five *primary* activities drive companies' internal material flows: source; make; deliver; sales and service; and recovery. Clearly, not all of these activities necessarily exist in all sectors. Indeed, in some cases what is being processed is information rather than materials. The basic concept remains the same, however. On top of these primary activities comes a set of *support* activities that are needed to manage and improve primary activities. These activities make it possible to acquire, maintain, and use the company's resources and knowledge; plan its material, information, and monetary flows; and facilitate interactions with its business

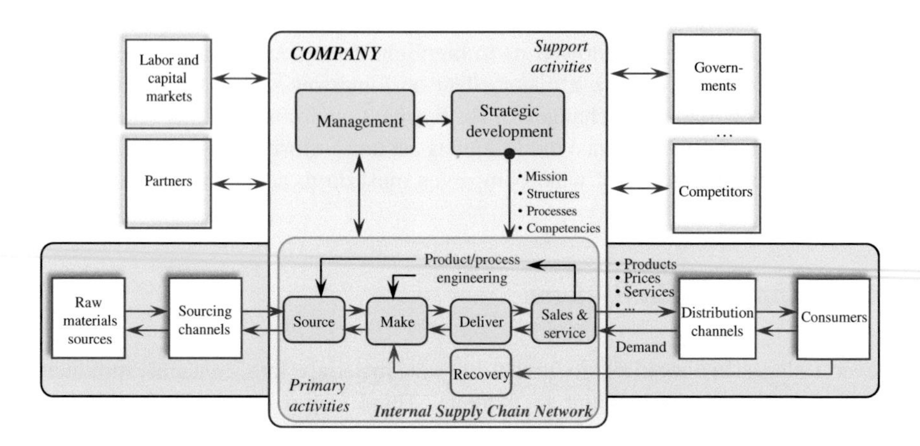

Fig. 1.2 A company in a supply chain

environment (competitors, partners, labor unions, capital markets, governments, etc.). They also contribute to the company's long-term competitiveness by developing the structures, processes, and core competencies needed to win customer orders. Consumer goods manufacturers often view the design of products and production processes as a support activity. However, for companies manufacturing unique products to order, such as Bombardier's (www.bombardier.com) public transportation system division, product and process engineering is viewed as a primary activity. Indeed, with current trends toward product customization and concurrent engineering, this vision is shared by more and more companies.

A company's SCN revolves around its primary activities. When all the support activities needed to improve and manage primary activities are also considered, the organizational construct delineated is the company's *supply chain system* (SCS). In other words, above and beyond the SCN, the SCS incorporates all the support activities required to plan and control the acquisition, development, deployment, maintenance, use, and withdrawal of the company's human, financial, material, and informational resources.

For a given *mission*, it is a company's acquisition, deployment, and withdrawal decisions that shape interfaces with its business *environment* (relationships with customers, suppliers, and value-added third parties) as well as its physical (production, distribution, and sales facilities network; facilities layout; bills of material; telecommunications networks; data bases; etc.) and organizational (organization chart, reward system, etc.) *structures*. Resource utilization decisions shape corporate engineering, operational (e.g., customer-order processing, handling, production, shipment, accounting) and management (e.g., demand shaping, sales forecasting, material procurement planning, production sequencing) *processes*. Figure 1.3 uses two planes to distinguish between environment, structures, processes, and mission. The functional plane highlights the relationship between primary SCN activities and their planning and control (P&C) system. It also shows that an SCS's mission is defined by the product-markets it serves and order-winning criteria (price, quality, availability, lead times, etc.) it provides. On the structural plane, it shows that SCSs incorporate numerous human, material, and informational resources. The figure also illustrates the fact that SCSs are not inert objects but living organizational systems that evolve over time within their environment.

Structures and processes are inseparable like two sides of a coin. Indeed, it is the way that they relate to one another within their environment that shapes and explains organizations' productivity and competitiveness. The development of value-creating SCSs therefore requires a deep understanding of these interrelationships and their effects on a company's key success factors. Given their importance, it is worth examining SCSs' structures, processes, and mission more closely. As for their social and informational structures, these are also important but have a greater relevance to organizational and information system theory than to SC management. This book will therefore only discuss them indirectly.

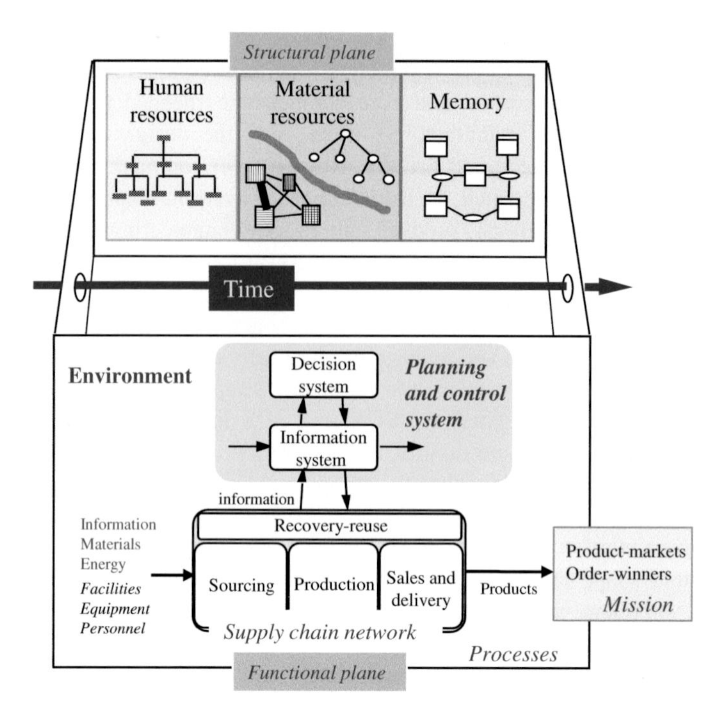

Fig. 1.3 Processes, structures and mission of a supply chain system

1.2.1 Material Structure of a Supply Chain Network

As Fig. 1.4 illustrates, the primary activities of a company's SCS can be depicted as a network of nodes and arcs deployed across the geographic territory that a company covers. Network nodes include supply sources (suppliers), factories, distribution centers, import/export terminals, points-of-sale, and demand zones (customers). Primary activities occurring in these nodes include transformation (production), handling, warehousing, information processing, and customs clearance. Between the nodes, activities such as transportation (by road, rail, air, sea, or intermodal), information interchange and funds transfer are taking place. Such activities can involve a slew of partners: subcontractors, transport specialists, public warehouses, 3PLs, distributors, retailers, banks, industrial agents, import/export firms, international freight forwarders, and customs brokers.

When a network's production–distribution nodes are examined more closely, it becomes apparent that, as illustrated in Fig. 1.5, their physical layout defines subnetworks with nodes associated to production (equipment, assembly lines, etc.) or storage (for raw materials, semi-processed goods, or final products) systems, and arcs linked to handling activities. SCNs can assume several forms and one of the aims of this book is to demonstrate that, in a given competitive environment, a company's ability to provide the kinds of prices, response times and services that

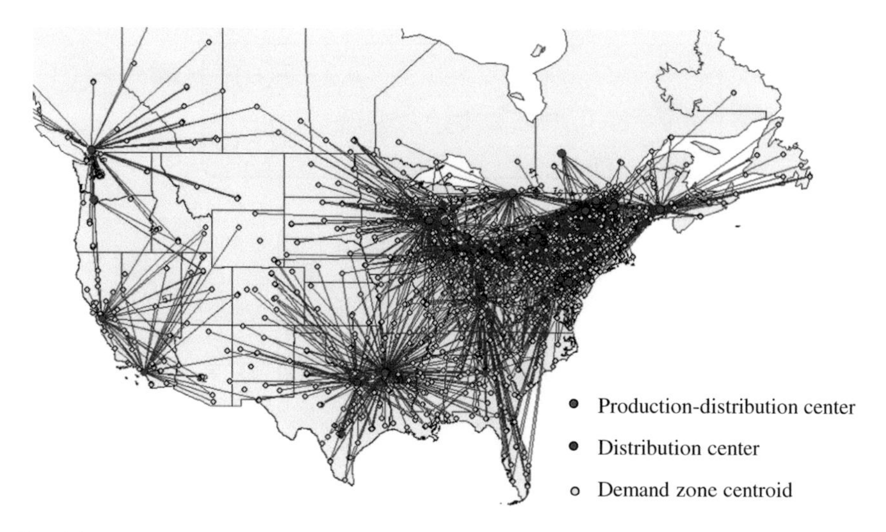

Fig. 1.4 North American supply chain network

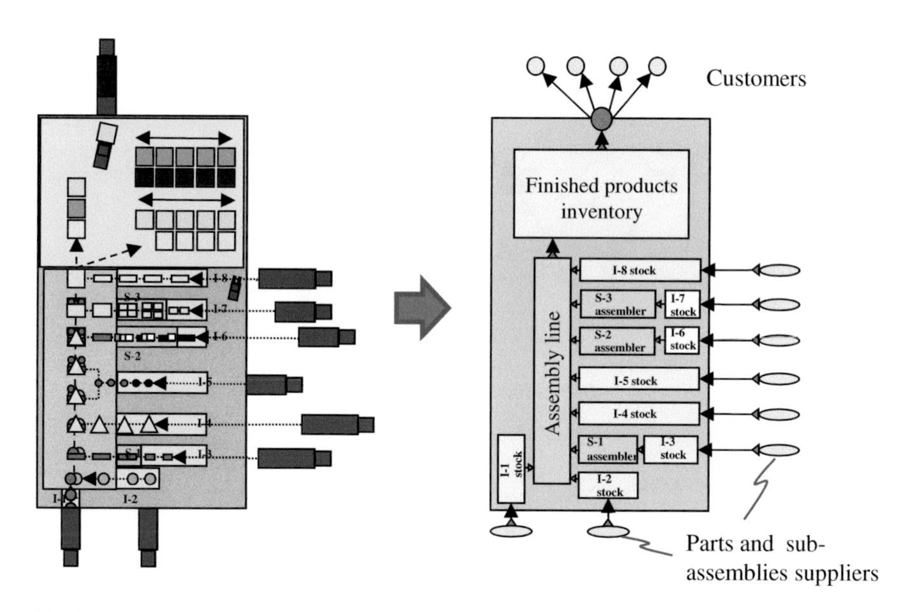

Fig. 1.5 Production–distribution center's internal network—*Source* Montreuil (2006)

will increase its market share and profitability is largely predicated on the spatial organization of its SCN.

Figure 1.6 portrays three typical SCNs by illustrating schematically how resources and primary activities (fabrication, assembly, distribution, sales, transportation, handling and storage of raw materials, work in progress, and finished products) might be deployed in a company's facilities (plants, distribution centers,

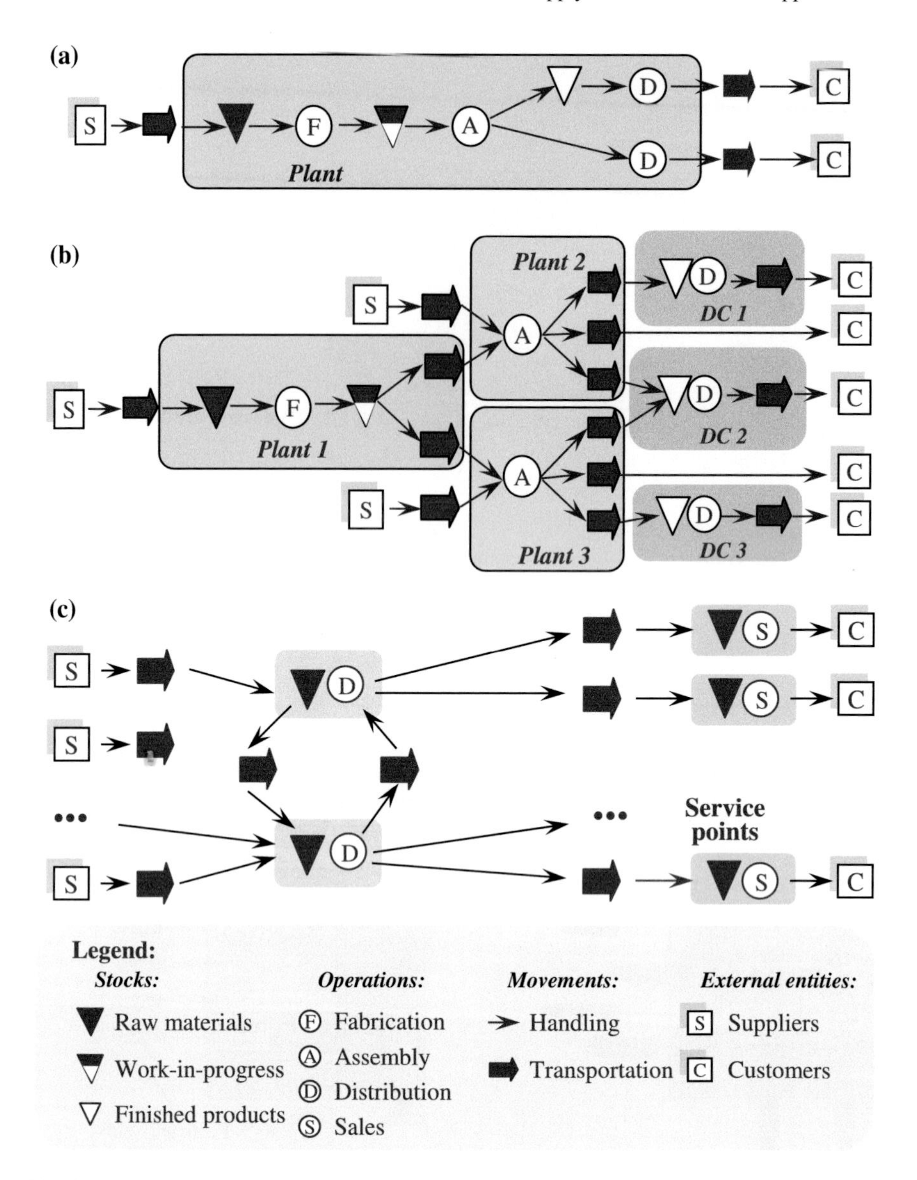

Fig. 1.6 Typical supply chain networks

and sales point). Network (a) corresponds to the situation for many SMEs manufacturing a family of similar products for a local market. All products use the same fabrication and assembly technology and some are made to stock. All of the company's activities happen in its factory and it uses public transportation. When a manufacturer expands and starts distributing its products in several markets (see descriptions of the Domtar and PSA Peugeot Citroën networks in the box below), it must adopt a structure that looks more like network (b). When the company diversifies and starts to manufacture different families of ancillary products, more deployment options are possible since it may be advantageous to group activities that are similar for all product families.

Several Companies Have Highly Developed SC Networks

Domtar (www.domtar.com)

With annual sales of around US$6 billion, Domtar is North America's leading producer and distributor of fine paper. Its SCN includes sawmills, pulp mills, paper mills, conversion plants, distribution centers, and wholesalers. Some of its plants, like the one in Windsor (Quebec), incorporate systems for making wood chips, pulp, and paper; for converting intermediate products into finished goods, and to distribute them. Others focus on a subset of these activities. For instance, the Port Huron plant in Michigan only makes paper. Finished product conversion activities are often delegated to subcontractors. Domtar runs 22 plants in Canada and the United States. Its production network is supported by a vast network of distribution centers serving about 15,000 ship-to-points.

PSA Peugeot Citroën (www.psa-peugeot-citroen.com)

As Europe's second largest automaker (after Volkswagen), PSA Peugeot Citroën runs a highly developed SCN. Vehicle production is carried out in factories that generally revolve around four major activities: metals stamping, body assembly, painting and final assembly. Six are located in France with others found in Spain, Portugal, Italy, Slovakia, the Czech Republic, Russia, Argentina, Israel, and China. The mechanical parts (engines, gearboxes and suspension systems) and forged parts that the factories use are made in dedicated centers, 12 of which are located in France, one in Brazil, one in Argentina, and one in China. An assembly site in Turkey also manufactures under license vehicles that are delivered in kit form. PSA Peugeot Citroën collaborates with other carmakers like Fiat, Mitsubishi, and Toyota to manufacture components. Its automobiles are shipped to points-of-sale in 160 countries worldwide.

Carrefour (www.carrefour.com)

As Europe's leading distributor, Carrefour has become a leader in global mass distribution. Today, the company counts more than 10,800 stores in 33 countries. It offers four formats: hypermarkets, supermarkets, cash & carry and local stores selling dry, fresh, and frozen products from hundreds of suppliers. Some products are delivered directly to the stores by suppliers. Most go through distribution centers that stock and consolidate supplier shipments. Carrefour manages 30 distribution centers in France alone. Currently, the company is trying to convert its dedicated dry product platforms into cross-docking centers to run leaner operations and cut inventories.

McDonald's (www.mcdonalds.com)
With more than 36,000 restaurants in 118 countries worldwide, McDonald's expanded at a lightning pace over the past 20 years. This growth has changed the way it sources merchandise. During the 1970s, procurement generally relied on local suppliers. In the 1980s, a more regional approach was adopted. Today, McDonald's gets its supplies from certified sources located worldwide, relying on consolidation zones to receive shipments from its international suppliers. The company tracks the food's transfer from place of origin (often farms) to restaurant. This includes a stopover at the company's distribution centers, where it ensures that all products satisfy its high quality standards and that the SC is as environmentally friendly as possible. McDonald's runs around 180 distribution centers globally. Each serves about 200 restaurants that are generally supplied three times a week, and even more frequently in dense urban zones.

Industrial companies are not the only ones facing challenging activity and resource deployment decisions. Network c) shows a supply network used to support a range of sales point or service. This includes retail chains (see Carrefour in the box), restaurant chains (see McDonald's example), telephony services, equipment rental and repairs, utilities networks (see Hydro-Quebec (www.hydroquebec.com) SCN in Fig. 1.7), and so forth. It is also possible to find combinations of type (b) and (c) networks in large companies.

Business competitiveness depends directly on the deployment of a company's resources and primary activities. In turn, this depends on the product-markets that it wants to serve. All in all, senior managers have a number of strategic questions to answer, including:

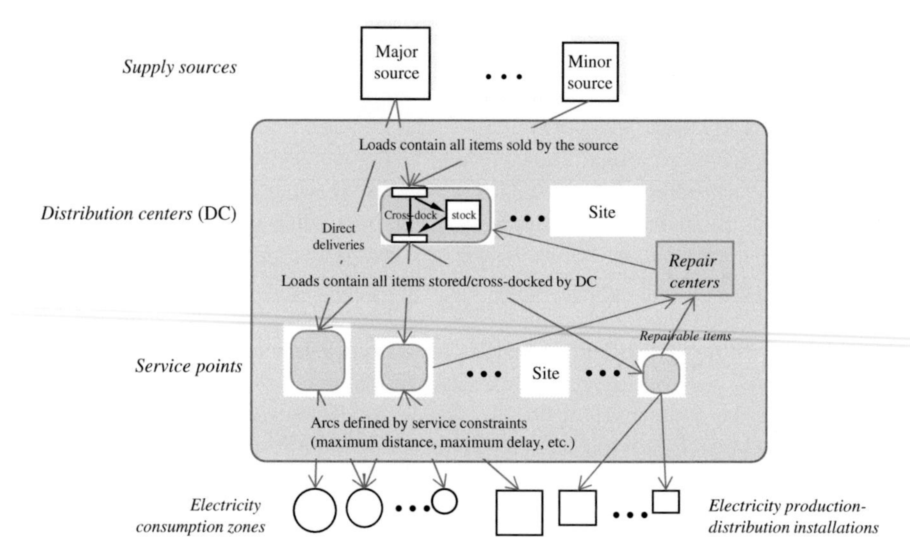

Fig. 1.7 Hydro-Quebec's supply network

- Which product-markets should we target?
- Which activities should we outsource?
- Which partners should we choose?
- How many production and distribution centers should the network have?
- What should be their mission and where should they be located?
- Which production and handling technologies should we adopt and what should their capacity be?
- How should the technologies adopted be exploited (facilities layout)?
- Which customers should be supplied from each production–distribution center?
- Which products should be supplied to each DC by each factory?
- Which supply sources should be used by each factory?
- Which embarkation/debarkation points should be used for imports/exports?
- What means of transportation should be used and should we run an in-house fleet?
- What would be the effect of our introducing new products?

All industrial/commercial companies have SCNs. In many cases, the network developed by itself in response to day-to-day events. Given the speed at which companies and their environment change in today's economy—be it through the addition of new product-markets, the advent of new technologies, mergers and acquisitions, the emergence of new outsourcing opportunities, or the appearance of new competitors—an ad hoc approach no longer suffices to build a sustainable competitive advantage. A reengineering approach systematically addressing these issues must rather be adopted. It is this type of SCN design methodology that the book presents.

1.2.2 Supply Chain Planning and Control Processes

The efficient coordination of all SCS activities is as crucial to a company's competitiveness as the optimization of its SCN structure. As aforementioned, it is by planning and controlling the acquisition, deployment, utilization, and withdrawal of human, material, financial, and informational resources that a company can offer the products and services that its customers want. After being transformed and moved through the SCN, *processed* resources such as raw materials, energy, and information become the products and services that the company sells on the market. On the other hand, *durable* resources like manpower, equipment, land, and buildings are needed to enable this transformation. The configuration of the SCN, as discussed earlier, involves the strategic planning of these durable resources to provide transformation, transportation, and storage capacity for day-to-day operations. The flow of processed resources in the network (material flows) must continually be planned and controlled to comply with available capacities, which must also be adjusted in the short term, using recourses such as subcontracting or overtime, to

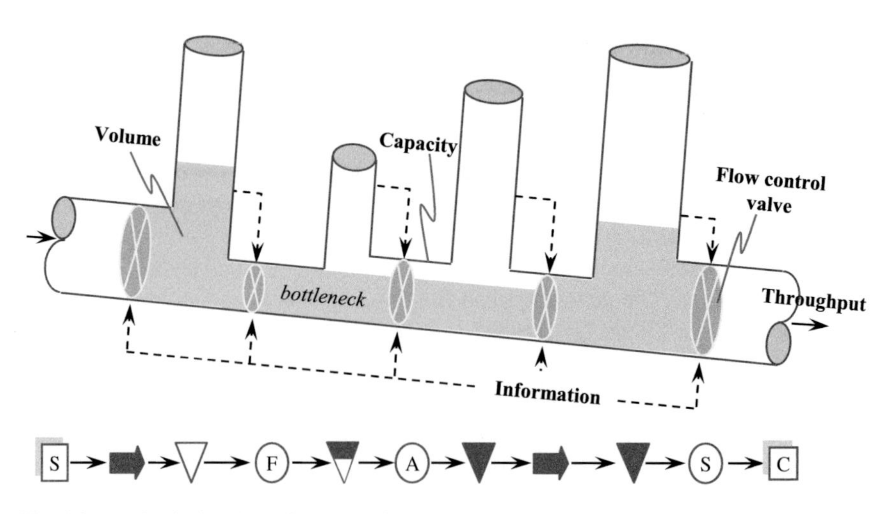

Fig. 1.8 Mechanical analog of material flows in a supply chain network

respond efficiently to customer demands. Hence the present section's focus on material flow P&C issues.

Figure 1.8 offers a mechanical analogy for an SCN in an attempt to foster intuitive understanding of how concepts such as capacity, volume, and throughput influence material flow management. This construct is comprised of a series of pipes moving a liquid from an upstream source (a supplier) to downstream consumers, along with stacks (stocking points) that can absorb flow fluctuations. The size of the pipes and stacks limit their capacity. Each pipe (representing transportation, manufacturing, assembly, or distribution equipment) has a valve that can be adjusted to *control* the flow of liquid. Clearly, the downstream rate of flow (throughput) is constrained by the size of the pipes and by the volume of liquid (stocks) available upstream. When all the valves are opened and the stacks are empty, the throughput cannot exceed the capacity of the smallest pipe (bottleneck). When demand is stable, to get the mechanical system to function effectively each valve's rate of flow must be the same as the customer demand, meaning that the valves' control must be coordinated. This is only possible if there is an exchange of information between control points. However, if customer demand is irregular, if there are bottlenecks, if supply is uncertain or if the transmission of information is delayed, it is practically impossible to equalize the rate of flow of all of the valves all of the time. In this case, the flow can be coordinated effectively only if a certain volume of liquid (stocks) remains in the stacks. Considering pipe capacities, the volume of liquid in the pipes and stacks, the delays involved and the supply inflows, for the system to function properly downstream throughput requirements (demand) must be *forecasted* periodically and volumes *planned* to ensure the satisfaction of the demand at all times.

This simple analogy shows that managing material flows in an SCN requires several P&C activities. It also indicates that capacities, volumes, and throughputs are fundamental variables in P&C systems, with information being their raw material. In practice, several P&C strategies can be elaborated for any given SC depending on how information is transmitted throughout the system and how volume and throughput decisions are made and executed. To satisfy customer demands at the operational level, the two main P&C instruments that people use are *demand-orders* and *execution-orders*. A demand-order is a request for materials (e.g., customer order, requisition, purchase order) or services, sent by a downstream activity to its upstream counterpart. Such orders are not executed automatically: they are acted upon only if the manager of the upstream activity decides to follow through, leading to the issuance of an execution-order (e.g., shipping order, manufacturing order). As an example, items that are unavailable cannot be shipped. Some P&C strategies avoid using demand-orders but activities cannot be controlled without execution-orders.

To illustrate this, consider the type of distribution centers found in most retail SCs. Two kinds of fundamental events tend to spark most material flows in this type of SCS:

- Customer orders (often for product packs) that—insofar as on-hand stocks allows this—lead to picking, staging, and shipping orders, and eventually to picking racks refilling orders;
- Purchase orders (often for pallets of products) that trigger suppliers' delivery activities and receiving orders at the DC.

These two types of events are independent: the DC P&C system triggers purchases; but sales result from customer's orders. As illustrated in Fig. 1.9, the stocks maintained are what allow the DC to function harmoniously despite these two types of events' independence from one another. If customer demands and supplier delivery times are known with certainty, management can ensure that merchandise arrives at the warehouse just-in-time to satisfy customer demand. In this context,

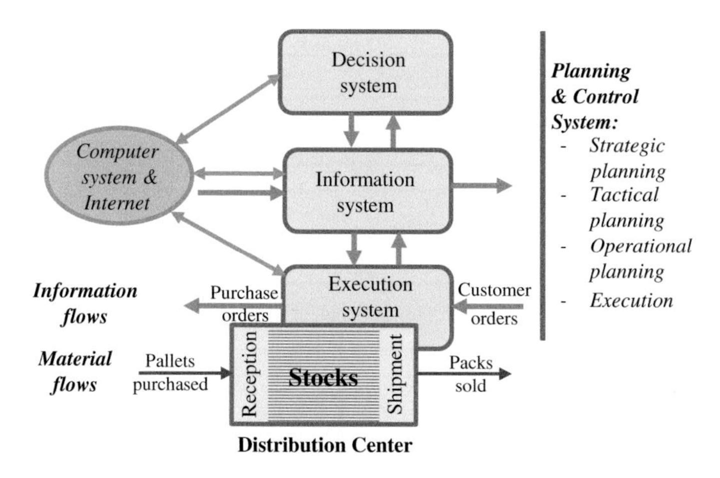

Fig. 1.9 Flow P&C system for a distribution center

there is no need to hold stocks, meaning that the distribution center's role is solely to break and sort the loads received from suppliers to enable the shipment of customer's orders. Minimizing stocks therefore requires minimizing the uncertainty surrounding purchasing and sales processes. By exploiting information and communications technologies, it is possible today to transmit sales information directly to suppliers, thereby enabling a continuous replenishment process adapted to actual needs, while minimizing the stocks needed to ensure an adequate synchronization between supply and demand.

Unfortunately, procurement lead times and demands are rarely known with certainty in the real world. To satisfy demand in this context, safety stocks must be maintained, inventories controlled and replenishments managed. This requires up-to-date information and suitable decision-making processes. It also leads to the development of *P&C systems* integrating decision, information, and execution subsystems, as shown in Fig. 1.9. The information system is an amalgam of all processes that input, process, memorize, and communicate primary information relating to demand-orders, execution-orders, and the state of the DC (e.g., inventory levels, storage locations); as well as any aggregate information (e.g., sales time series), external information (e.g., supplier prices), plans (e.g., manpower requirements, sourcing plans, delivery schedules), or policies (e.g., service levels) needed to manage the distribution center.

The decision system encompasses all the processes needed to develop plans over the short (operational planning), the medium (tactical planning), or the long-run (strategic planning). It generally includes several interrelated subsystems that create a hierarchy of policies and plans used to manage the SCN. Each subsystem generates plans satisfying the constraints and objectives resulting from higher level planning processes, while considering data transmitted by lower or same level processes. Figure 1.10 illustrates a typical distributed planning and an execution framework. It shows how decisions are broken down according to the duration of

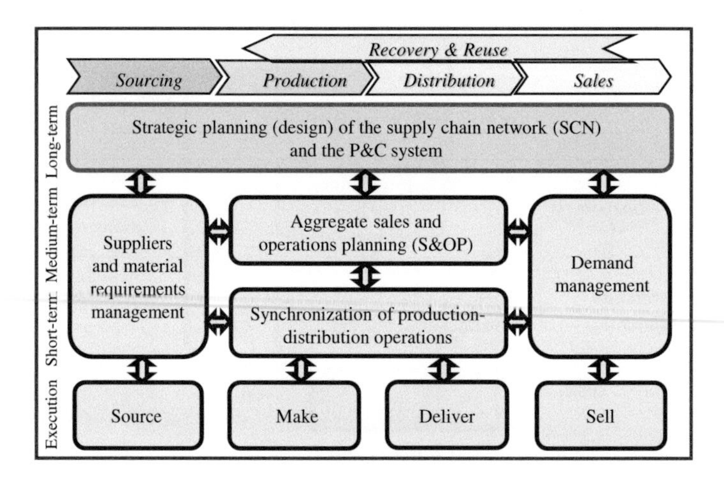

Fig. 1.10 Distributed planning and execution framework

their impact and the SC activities covered. Note that the strategic level encompasses the long-term decisions that shape the future of the SCS, including the structural resource acquisition and deployment decisions discussed in the previous section, along with decisions relating to the design of the P&C system itself. It is for this reason that strategic design decision processes are often referred to as "meta-planning."

P&C can be either formal or informal. Nowadays, formal processes are often computerized. The information system then relies on extensive databases, transactional procedures, data analysis algorithms, and Internet-based communication protocols. The decision system, on the other hand, often relies on decision support models and tools to optimize operational, tactical, and strategic choices. This typically leads to the implementation of *Enterprise Resource Planning* (ERP) systems and *Advanced Planning Systems* (APS). These P&C applications must be adapted to the company's needs, SC strategy and culture, and designed to create a competitive advantage. Chapter 3 studies these issues, in particular the development of P&C strategies and systems. Methods available for the tactical and operational planning of sourcing, production, distribution, and demand management decisions are not covered in detail in this text. A number of excellent books, such as Nahmias (2005) or Vollmann et al. (2005), are available on tactical and operational decisions techniques and systems.

1.2.3 Mission of the Supply Chain System

The mission of a company's SCS is not restricted to ensure that the products purchased, manufactured, distributed, and sold are available at the right place at the right time in the right quantity and quality at the lowest possible cost. Even if this reductionist view of SC's role often prevails in practice, it is much too restrictive. It positions SCSs as cost centers needed to produce and deliver products to customers. It completely neglects the fact that SCSs have become one of the main instruments that a company can use to develop a sustainable competitive advantage. The mission of an SCS is, therefore fundamentally, to create value for the company and its stakeholders.

In practical terms, an organizational system creates value if it contributes to the sustainable improvement of its company's market value. It can be shown that because this performance measure requires complete information, it provides a suitable mechanism for arbitrating between the conflicting objectives of a firm's various stakeholders. In finance, value is defined as the sum of all of a company's future *residual cash flows* (RCF), discounted at the firm weighted average cost of capital (WACC). Schematically, a company's RCF is defined by the expression

$$\text{RCF} = (\text{Revenues} - \text{Operating expenses})(1 - \text{Tax rate}) - \text{Capital expenditures}$$

Since it is not possible to sum over an infinite series of cash flows, companies often base their performance assessment on static measures such as *economic value added* (EVA) or *return on capital employed* (ROCE). Unfortunately, these measures tend to focus on the short term. The notion of sustainable return implicit in our definition of value is of particular importance at this level. It evokes the necessity of designing SCSs that are sufficiently robust to cope with any plausible future. It also forces us to account for a company's social responsibility. If, for instance, a company does not behave in an environmentally sound manner, it may be capable of improving some of its static performance indicators in the short run. If it persists in this vein, however, its long-term survival may be compromised.

As defined above, RCF implies that three broad categories of value drivers must be taken into account in SCSs' engineering and management: operating costs; capital expenditures; and revenues. To improve RCF, the cost of inputs and SC operations (sourcing, production, warehousing, handling, transportation, maintenance, recycling, etc.) must clearly be minimized. Capital expenditures related to the acquisition of durable resources and ownership of current and fixed assets must also be minimized. Among other things, this encourages the rationalization of capacity investments and raw materials, work in progress, and finished products inventories. Minimizing income tax is another necessity that becomes particularly important when the SCN is deployed in several countries. These spending minimization objectives have traditionally been the leading performance drivers for logistic managers. They continue to be very important.

The definition of RCF also encourages the maximization of sales revenues. Clearly, to increase revenues, a company must constantly be better at winning orders than its competitors. This raises a fundamental question about what motivates customers to buy a particular company's products and services, and which order-winning criteria are most important to customers. Table 1.1 enumerates the main value criteria prevailing in customers' purchasing decisions. The notion of value invoked here is the same as above, but from a customer's perspective. To simplify the discussion, value criteria are grouped into order winners. The criteria and groupings listed do not necessarily all apply to any given industrial sector, but

Table 1.1 Order winners valued by world-class customers

Order winners	Value criteria	Order winners	Value criteria
Products	Attractiveness Innovation Range	Flexibility	Design Customization Volume
Product quality	Conformance with specifications Reliability	Service	Coverage Availability Sales assistance Technical support
Prices	Price structure Payment methods	Sustainability	Carbon footprint Recovery/recycling Social equity …
Response times	Speed Reliability		

they can be used as the basis for developing an array of order winners and criteria adapted to a specific company's context.

The first order winner involves the *products* that a company sells. The predominant value criterion from this perspective is the product's attractiveness, which depends on several factors. For products like toothpaste, what dominates is the appearance and in some cases the packaging, as well as performance (ability to prevent cavities). For consumer goods like televisions or cars, important factors include user-friendliness, what is covered by the warranty, ease of maintenance, operating costs, and resale value. All of these elements also apply to specialized or industrial goods. With respect to the latter, purchasing decisions are often based on a lifecycle costing approach involving many other factors, especially where expensive equipment is involved.

Some value criteria apply to the product assortment that a company sells instead of any specific product. One of these criteria is the ability to market innovative products that make competitors' offers seem old-fashioned. For instance, by doing business with suppliers manufacturing innovative electronic equipment, an industrial firm can ensure that it is buying leading-edge technology and reduce capacity upgrade problems. To create this kind of value, the company must develop and preserve its perceived technological leadership. Another criterion affecting all products sold by a company is the ability to offer a relatively broad range. This is particularly important in retail since it allows consumers to satisfy as many of their needs at one and the same point-of-sale. It is one of the explanations for the success of retail chains like Home Depot (www.homedepot.com) in the USA and Brico Dépôt (www.bricodepot.fr) in France.

The value criteria we associate with *product quality* could be part of the preceding grouping but they are listed separately because of their significant impact on purchasing and production decisions. The main quality criteria of a product are its compliance with specifications and reliability. A product must be free of defects, that is, its performance must match customers' expectations and meet standards enforced by relevant authorities. The product must also be reliable, with a very low probability of failure and a long life expectancy.

The ability to offer low *prices* is an order winner that is becoming increasingly important in certain markets, notably during periods of recession and for products in the maturity or decline phase of their life cycle. Pricing issues go well beyond the price levels specified for different market segments. They also include discount policies, sales conditions, (e.g., incoterms) and payment terms. All three are useful ways of getting customers to adopt behaviors that are likely to lead to lower costs for the company. Pricing policies are important per se but their strategic implications largely lie in the fact that in order to offer low prices, companies must also have low costs. For industrial products low prices have become a crucial order winner. For instance, automakers often sign multi-year contracts with their suppliers including a requirement to lower prices over time. To remain profitable under such conditions, a company must know and understand its costs, and be able to capture learning effects.

Quality and price have long been perceived as critical success factors. Nowadays, however, *response times* are just as important. In their seminal work, Stalk and Hout (1990) showed that speed plays a crucial role on several levels. First, the ability to develop new products and launch them quickly is a necessary precondition for harvesting the fruits of innovation. Second, companies that lower production times can cut inventories and increase productivity markedly. By squeezing cycle times, they can also reduce uncertainty throughout their SCs. From the customer's point of view, the value of response times lies mainly in criteria such as delivery speed and reliability (punctuality). Response times include all of the time that elapses between the moment a customer's need is expressed and when it is satisfied. Clearly, it is an advantage to have shorter response times than competitors. Yet in and of itself, this is not enough. Customers increasingly require a rigorous satisfaction of their demands, notably in just-in-time contexts. Deliveries must be on the right day at the right time, that is, neither early nor late. They must also match the order exactly. Partial deliveries or substitute products are seen as noncompliant.

One important factor for reducing response times is *flexibility*. However, the value of flexibility goes much further than timing issues. Not all customers have the same needs and it is generally impossible to produce a range of standard products or services that is sufficiently broad to satisfy all needs. Companies capable of customizing products to meet customers' specific demands have a clear advantage. In some cases, it is not enough to adapt an existing product. Instead, the company must be capable of manufacturing nonstandard products to customers' specifications, that is, it must adapt quickly to new designs. On the other hand, many products (like skates or swimsuits) are subject to irregular demand throughout the year and in some cases model sales shares can be hard to forecast. Product demand can also rise suddenly during crisis like droughts, wars, epidemics, or hurricanes. The company that can adapt to the ensuing changes in volume can earn customers' long-term loyalty. As an example, when Hurricane Katrina hit New Orleans in 2005, Wal-Mart delivered water, fuel, and toilet paper to thousands of victims before the authorities were able to provide any assistance. Companies that are both flexible and rapid demonstrate *agility*.

Customer service is another multidimensional order winner that is highly appreciated. Speed and flexibility are value drivers that tend to be associated with service, but there are others as well. Market coverage, product availability at the point-of-sales, sales assistance, technical support, and after-sales service are also important value criteria. Products must be easy to find and purchase. When customers go shopping, the goods they want must be in stock, especially when commodities are involved. Customer accounts must be opened and managed smoothly. Customers must also be able to easily obtain any and all commercial or technical information that they might need. Where necessary, products must be easy to install and any potential user problems resolved promptly. Spare parts must be available and maintenance and repair facilities should be accessible.

Lastly, consumers and companies have become increasingly focused on the ecological impact and social equity of their purchasing/manufacturing decisions.

This brings an advantage for any vendor adopting sound *sustainability* practices, that is, behaving in an environmentally and socially sound manner, from manufacturing all the way through delivery (and possibly product reuse). Conversely, companies neglecting their SC's carbon footprint and engaging in nonethical business practices risk losing market share and even being ejected from certain markets. Nowadays, people expect companies to run energy-efficient facilities, adopt reusable packaging and containers, use eco-efficient means of transportation, accept product returns, take charge of goods' end-of-life recovery and recycling, respect human rights and safety standards, and so forth. FedEx (www.fedex.com), with its fleet of 700 airplanes and 44,000 motor vehicles consuming 20 million liters of fuel daily, is a good example here. To improve energy efficiency, FedEx has replaced its entire fleet of airplanes with Boeing 757 s consuming 36 % less fuel. It has also added solar energy systems to its distribution terminals and runs hybrid trucks that are 42 % more energy-efficient.

As aforementioned, the criteria under discussion here do not apply to every company nor are they exclusive. Having said that, companies should make an effort to identify the prevailing value criteria in their product-markets since this is the starting point for any strategic process building competitive advantage. In addition, it is worth noting that company's customers are not necessarily external. When a firm englobes different operational units (factories, distribution centers, points-of-sale, etc.), downstream units become customers of their upstream counterparts, each applying their own value criteria. This defines the company's in-house *value chain*.

Generally, not all value criteria favored in a given market segment have the same importance. Moreover, the relative prominence of each can change as the segment's needs evolve. Also, some criteria are critical to win orders whereas others are needed for customers to view the company as a potential supplier (*qualifiers*). When a new product is introduced, it is often criteria like attractiveness and innovation that help to generate sales, with price and service being less important. This is what happened, for example, when the first smart phones were offered in the market. On the other hand, when a product reaches maturity, competition is more based on prices and service. Products with insufficient performance or quality tend to disappear. Even if some criteria (like price) are not crucial to order-winning, any exaggeration at this level could lead to orders being lost, turning this into an *order-losing criterion*. Hill (1999) discusses all these nuances in some detail, offering an approach to weight the importance of different value criteria.

Lastly, it is worth noting that to be successful in a market, it is not necessary for all order-winning criteria to be satisfied. Dominant companies often succeed by concentrating on operational excellence, customer intimacy, and/or product leadership. Companies like Dell Computer (www.dell.com) and Wal-Mart (www.walmart.com) emphasize operational excellence by opting for criteria like price, quality and response times, whereas companies like Kraft Foods (www.kraftfoodscompany.com) favor customer intimacy by focusing on products' attractiveness, wide ranges, customer service, and customization. Otherwise, companies like Nike (www.nike.com) and Johnson & Johnson (www.jnj.com) seek product leadership by innovating and launching new products rapidly. A few companies have been able to master two or three of these disciplines. Toyota (www.toyota.com), for instance, allies operational excellence and product leadership; and Apple (www.apple.com), the company with the world's largest stock market capitalization ($353 billion), excels on all these levels. Such companies are rare, however, since each strategy tends to nurture a different culture and business model.

This discussion of customer value drivers reveals the important role that SCSs play in increasing sales revenues. Indeed, market shares largely depend on the SCS's capabilities. If a company's value-for-money offer—in terms of products, quality, response times, flexibility, service, and sustainability—is superior to the competition, it will attract more customers and revenues will rise. Creating value therefore requires acting on several factors simultaneously. This involves the specification of adequate tactical and operational planning sub-objectives, as well as a hierarchy of performance metrics. Since all value criteria are interdependent, suitable compromises must be found. Recall also that value derives from discounting the company's RCF during its useful lifespan, thereby intimating a need for sustainable performance. This involves anticipating the future. Yet the future is ineludibly uncertain.

A SCS must not only be capable of increasing RCF in the business environment that prevails at a given point in time but also be sufficiently *robust* and *resilient* to create value irrespective of which plausible future is likely to emerge in coming years. It must therefore be capable of minimizing exposure to all the different kinds of risks that the system might face. Many of these risks are related to random day-to-day business variables such as prices, demands and exchange rates but they can also involve sporadic disturbances (e.g., natural catastrophes, politico-economic failures, industrial accidents, and terrorist attacks) that might have a major effect on the system's performance. Looking backwards, it is easy to use historic data to calculate performance indicators. However, it is much more difficult to anticipate the future performance. To do this, one must elaborate plausible future scenarios and estimate performance deviation metrics—like value-at-risk (VAR)—in addition to expected value metrics. In short, risk mitigation is a crucial aspect of SCS engineering and management that must not be neglected. Issues pertaining to value creation and the choice of appropriate performance metrics are examined in detail in Chap. 2.

1.2.4 Companies in Their Supply Chains

As illustrated by Fig. 1.2, few companies incorporate all of the facilities and activities found in their SCs. Generally, SC engineering and management is studied from a company's perspective. This involves examining the part of the consumer product SCs in which the company is participating, rather than the whole SC associated with any one product. Some companies are *vertically integrated* and cover a large proportion of the activities found in associated end-products' SCs. For instance, several companies in the forest products sector, like Domtar (www.domtar.com), run activities ranging from cutting down trees in forests to selling consumer products, including primary transformation in pulp and paper mills, secondary transformation in conversion plants and distribution of finished products through specialized merchants. At the other end of the spectrum, some companies have become *network orchestrators* known for their trademarks but outsourcing almost all their activities. Most companies operate between these two extremes, concentrating on core competencies while networking with a number of partners. In Brazil's automotive sector, for instance, certain production facilities are built nowadays in a condominium form. The automaker focuses on final assembly with the main component suppliers manufacturing their products in one section of the plant before transferring them to nearby assembly lines. A company can therefore be operating at a significant distance from its raw material sources or final consumer markets, and not necessarily have to consider the whole of its SCs when making design or planning decisions. The APICS Supply Chain Council (www.apics.org) suggests that analysis focus on the *extended enterprise*, a construct ranging from the suppliers of a company's suppliers to the customers of its customers.

Flows between the facilities of companies involved in the SCs of a given industrial sector create what is known as business networks. Companies are the nodes of these networks, with their arcs symbolizing inter-firm buyer–supplier relationships. This can involve the purchase or sale of products or services, one example of which is transportation capacity provided by a 3PL. The relationship can be either ad hoc or strategic in nature, the latter being exemplified by a formal partnership based on a long-term agreement. Figure 1.11 illustrates the business network of the US automotive industry. A particular car model may include

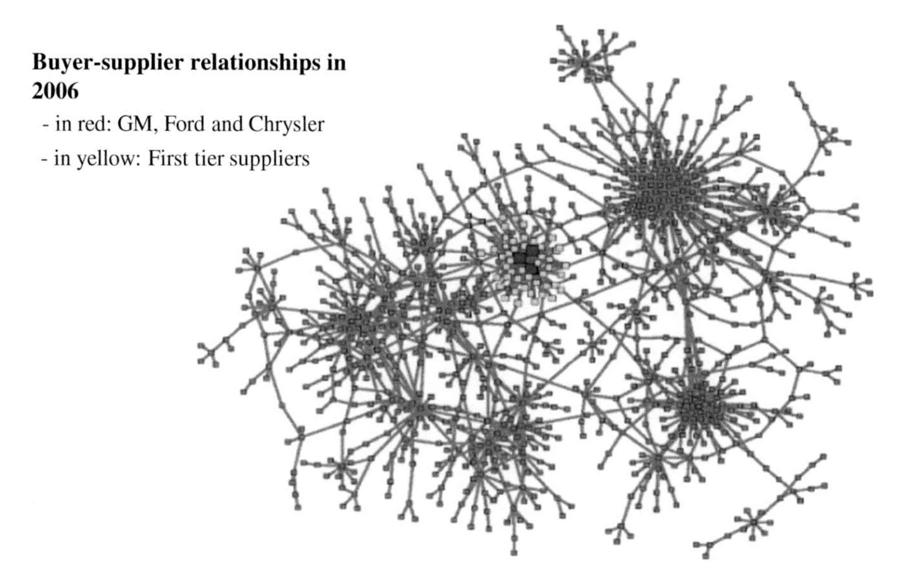

Fig. 1.11 Business network associated with US automotive SCs (Atalay et al. 2011)

hundreds of components purchased from first-tier suppliers. The figure shows the network of major buyer–supplier relationships obtained in 2006 from the Compustat database of American companies (Atalay et al. 2011). The red points correspond to the US automakers (GM, Ford, and Chrysler), and the yellow points designate their first-tier suppliers. These suppliers must source their inputs from other companies, thereby spawning what is ultimately a very complex network. Note within this network that some companies are characterized by a very large number of incident arcs and occupy a central position in a sub-network, whereas others operate on the periphery. In 2006, the *focal company* with the largest number of suppliers in the United States was Wal-Mart.

1.3 Supply Chain Strategy and Reengineering

1.3.1 Corporate SC Strategies

In the preceding section, constructs such as SCS's structure, processes, and mission were explained using Fig. 1.3. It was shown that an SCS can fulfill its mission by exploiting its resources and P&C processes. Figure 1.3's temporal axis also illustrated that SCSs are living organizational systems that must adapt to changes in their environment. Section 1 showed that today's economic environment features numerous challenges that are likely to accentuate in the near future. Senior management is responsible for assessing the opportunities and threats associated with these changes, establishing future targets to improve profitability and competitiveness, and developing strategic SC directions to reach targets. Developing an SC strategy is part of a company's strategic management process and the strategy elaborated must be congruent with other functional strategies (marketing, finance, human resources, etc.) and general policies. An SC strategy is articulated on three interdependent levels. First, a customer *offer* capable of enhancing the company's market share must be concocted. Second, to nurture this offer, an SC *doctrine* that builds on distinctive competencies and is hard to copy must be adopted. Lastly, the doctrine must be implemented via concrete SCS processes and structures development *means*. Figure 1.12 summarizes the main elements of an SC strategy.

The offer is the value proposal that the company makes to its customers, in connection with aforementioned order winners (Table 1.1). We have seen that to get orders, qualifiers must be satisfied and the company must dominate competitors with respect to a certain number of order winners. Yet it is not enough to simply want to improve the factors that customers value to achieve the expected results. To beat the competition, there must also be a fit between targeted order winners, the company's competencies (know-how), and the supply chain management (SCM) paradigm that the top management advocates. Various approaches can be adopted to design and manage SCSs. Numerous schools of thought have emerged in recent years towards this end (TQM—Total Quality Management; 6σ;

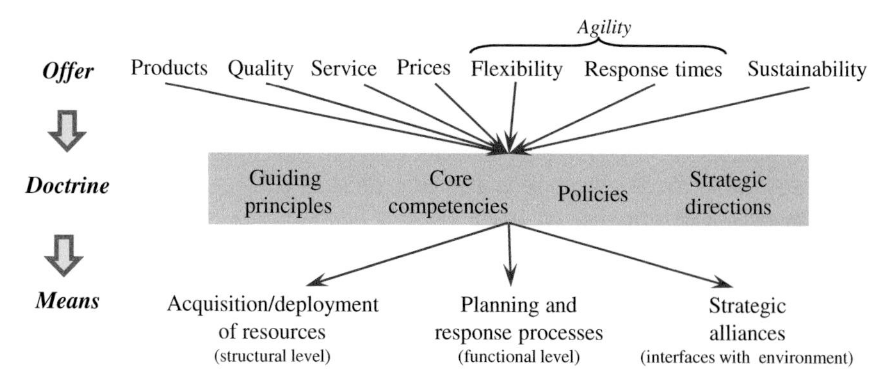

Fig. 1.12 Elements of a supply chain strategy

JIT—Just-In-Time; TOC—Theory of Constraints; Lean Manufacturing; Agile Manufacturing; QR—Quick Response; ECR—Efficient Consumer Response; Mass Customization, Sense and Respond, etc.). These movements overlap, with each generally offering no more than a partial vision. Companies can draw inspiration from them to develop the guiding principles based on which they can build their SC strategy. Here are a few examples of the kinds of fundamental behavior and principles that can be mobilized:

- Capacity-stock-delay interdependencies;
- Concentrating activities to encourage economies of scale;
- Eliminating product and process defects;
- Reducing cycle and setup times;
- Eliminating bottlenecks;
- Process and structure modularity to nurture flexibility;
- Reducing variability to avoid congestion;
- Buying insurance to protect against risks;
- Sharing information across the SC;
- Adapting the SCS to environmental changes;
- Aligning with partners in terms of profit and risk sharing.

A number of these principles are examined in detail in future chapters. For the moment, it is worth emphasizing the importance of understanding the interactions between capacity, stocks, and delays. The lean approach (Womack et al. 1991) inspired by Toyota's practices seeks a simultaneous reduction in stocks, capacity, and cycle times. It provides excellent results in a stable environment but requires the stabilization of upstream and downstream flows, which leads to certain rigidities. This being the case, it may be difficult to adapt to short-term changes in supply or demand, to adjust to external upheavals, or to take advantage of new opportunities. Any disturbance may translate into major delays. A good illustration is the tsunami that hit Japan in March 2011, delaying the delivery of several automobile models for several months. Conversely, locating component stocks or capacity buffers strategically in an SCN can create the kind of agility that is needed to react quickly to disturbances or opportunities (Lee 2004).

The guiding principles expounded by a company to deliver its product or service offer help to identify the distinctive competencies it must develop, the operational guidance policies it must formulate and the strategic directions it must follow to improve its SCS. An SC *doctrine* is thus prescribed to guide a company's operational, tactical, and strategic activities. Policies and strategic directions are concerned, among other things, with

- Products and manufacturing processes (products modularity, postponement, eco-efficient processes, etc.);
- Technology selection (flexible technologies, cross-docking, green vehicles, etc.);
- Insourcing, outsourcing, or co-sourcing of activities;
- Sourcing policies (e-sourcing, price, quality, delays, etc.);
- Distribution channels (resale network, company stores, online sales, etc.);
- Decoupling points (MTS—make-to-stock, ATO—assemble-to-order, MTO—make-to-order, etc.);
- P&C strategy (push-pull, information exchange modalities, forecasting and decision-making methods, etc.);
- Risk mitigation (insurance, contingency plans, response policies, etc.).

Decoupling points refer to the positioning of the first inventory within an SC: customer orders penetrate as far as this initial stocking point. The choice of decoupling points therefore has a direct impact on response times and SC costs. A given company can have different decoupling points for different product-markets. This is illustrated by the fine paper SC portrayed in Fig. 1.13. Furthermore, competitors from the same industry do not necessarily use the same decoupling points. For instance, when Dell (www.dell.com) introduced its assembles-to-order strategy, it had a decoupling point further upstream than most of its competitors. Conversely, by selling furniture in kit in its stores, IKEA (www.ikea.com) has a decoupling point that is more downstream than most actors in its business. Its kit sales strategy cuts production, transportation, and inventory costs while guaranteeing extremely rapid delivery times. Somewhat differently, the success of Zara (www.zara.com) can be largely attributed to its ability to respond quickly to customer demand by adjusting weekly plans. The company's SC strategy is based on postponement, which allows it to take advantage of mass production while remaining flexible. Otherwise, with its SmartVille concept, Mercedes-Benz (www.mercedes-benz.com) in the automotive sector has been successfully using modular production to provide an efficient response to complex compatibility and scalability needs. A number of companies believed that they could use Internet technologies to build stock-free SCs in which production and sourcing operations would be entirely triggered by customer orders. Unfortunately, they soon realized that by purchasing/producing everything to order, it is very hard to offer short delivery times or achieve economies of scale. On the other end, carrying everything

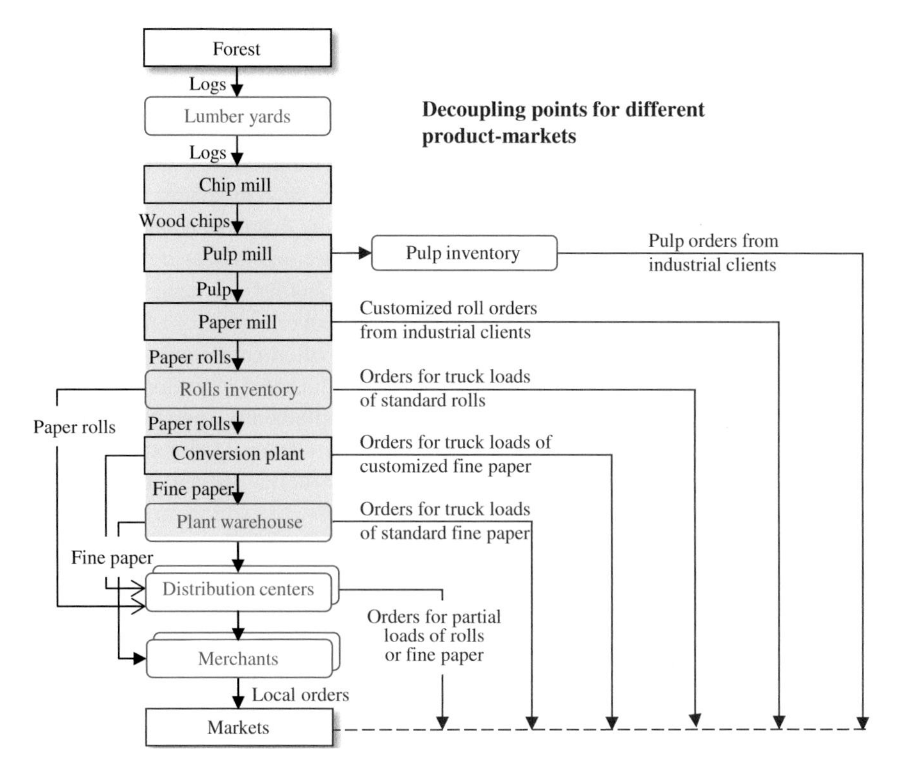

Fig. 1.13 Decoupling points for a fine paper manufacturer

in inventory is no more suitable. Finding a good compromise between these two extremes is a vital strategic challenge.

The doctrine that a company adopts must translate into concrete action plans improving its SCS. The means leveraged to achieve this goal revolve around the acquisition, withdrawal, or redeployment of resources on the structural plane, the improvement of P&C processes on the functional plane, and the negotiation of value-added partnerships. Clearly, these means are not the only ways of influencing a company's competitiveness. Several other elements, such as stakeholders' vision, marketing strategy, financial strategy, and organizational structure, are also essential. The present book does not deal explicitly with these topics, which relate to strategic management, marketing, and finance more than to SCN engineering and management.

Strategic decisions about the product-markets to conquer, the activities to outsource, and the activities to deploy in the geographically dispersed facilities of the firm specify the mission of each facility in terms of products to manufacture, distribute and sell, and of customers (internal or external) to serve. Decisions must also be made about the technologies (resources) each facility should use to carry out its mission efficiently, that is, to create value for its customers. The insertion of new technologies into an SCS must result from an effort to reengineer its processes and structures. This entails much more than simply choosing and purchasing equipment. A technological system can only produce the expected outcomes if the configuration and methods of its host facilities are adapted, and if it is operated by motivated people with the necessary know-how. The primary technologies to consider are those necessary to transform, move, or store materials, such as robots, auto-guided vehicles, flexible machine tools, or automated conveyors that can be configured into assembly lines or cells. Other technologies that have become increasingly important enable the processing, communication, storage, and analysis of data. These technologies are important to modern P&C systems. Moreover, nowadays the design of production and operational management processes is often combined with product design by means of a concurrent engineering approach rooted in customers' needs.

In short and as illustrated in Fig. 1.14, several complementary perspectives must be examined to develop an SC strategy. First and foremost, a superior strategy must be based on the qualifier and order-winning criteria valued by targeted product-markets, the capabilities of the company and its competitors to satisfy these criteria, and the technological opportunities and partnerships offered on the industrial scene. From this examination of the SCS's internal and external environment—and from the strategic vision that the company's senior management advocates—it is possible to shape the offer to make to different target markets, and the SC doctrine to inculcate in-house. This offer and this doctrine must then be transformed into concrete means, using an SCS reengineering methodology that the next section outlines. When the process is applied in a continuous manner, it creates a sustainable competitive advantage due to the fact that competitors find it increasingly difficult to overcome the gap that is gradually arising between the new expectations that customers are induced to develop and the new competencies that are needed to satisfy them. To become and remain a leader, companies must stay very vigilant. After all, an acquired advantage can only be maintained as long as no other actors develop the kinds of know-how that make them more attractive to customers. The process described in Fig. 1.14 is a part of a more global approach to companies' strategic development. The box below shows how Wal-Mart has been able to establish a winning SC strategy that goes some way towards explaining its global success.

Wal-Mart: Retail and Distribution Leadership

Wal-Mart is the world's leading retailer with total revenues estimated at $473 billion (versus $60 billion for Costco and €100 billion for Carrefour, its main competitors) and a 25 % return on equity. The company owes its position to regular growth sustained by an appropriate SC strategy. With more than 11,000 stores worldwide, its average sales per square meter, inventory rotation rate, and operational profits are among the best in the industry. Despite more than 100 million weekly customer visits in the United States, however, the domestic market has been stagnating for several years and the company has had to seek growth abroad. Wal-Mart now manages stores in 26 other

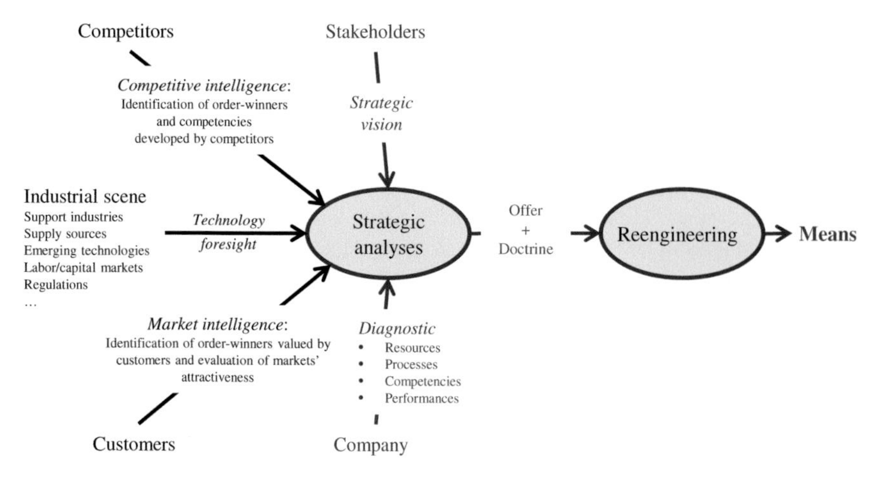

Fig. 1.14 Process for developing a supply chain strategy

countries, accounting for more than 25 % of its total sales. This expansion has been partially achieved through alliances (Bharti in India, CITIC Pacific in China) and acquisitions (MassMart in South Africa).

Wal-Mart quickly realized that customers in its sector of activity greatly value order winners such as low prices, an extensive product range, and quality service. The company therefore decided that these order winners would become a strategic goal. It has been able to offer them by developing a number of distinctive competencies, including a very efficient cross-docking approach for the consolidation of cargo in its distribution centers (DCs). In and of itself, customers do not attach much value to this competency, yet the savings enables that Wal-Mart has been able to offer everyday low prices across a wide range of products. Clearly, it was not sufficient to decide to use cross-docking to successfully implement it. Cross-docking is a complex practice requiring a variety of means, including investments in sophisticated handling, transportation and communications technologies, remodeled DC layouts, redefined working methods, and revised employee reward systems. These are the elements that the company has worked on to develop its order-winning offer.

Wal-Mart has become a world leader in distribution, which accounts for less than 2 % of its revenues versus 4 % for its main rivals (Carrefour, Tesco, Metro AG, and Costco). Its United States SCN counts 158 DCs (for a total surface of 119 M sq. ft.) working 24/7. 80 % of its DCs are company-owned and each employs about 800 staff members. The company supplies each store daily (versus once every five days for most rivals) using a fleet of 7,200 trucks and 53,000 trailers. Its DCs ship 78 % of the merchandise sold, with company policy dictating that they should cover the needs of about 100 stores within a range of 200 miles. International operations are supported by 134

DCs situated worldwide, including two centers consolidating exports from the US. Of these, 34 are owned and run by Wal-Mart, 38 are rented and run by the company, and 62 are run by third parties.

Despite this success, Wal-Mart is proactively working to develop a sustainable competitive advantage. With its new slogan "Save more. Live better," the company launched a *Sustainability 360* initiative focused on building a new generation of environmentally friendly stores and developing a range of recycling activities. It still faces the challenge of finding a suitable compromise between sustainability and an international sourcing policy requiring an enormous amount of transportation resources. Facing an increasingly risky environment, the company opened its own crisis management center in Arkansas, called Walmart Alarm Central, which receives an average of one emergency call a day for incidents occurring at one of its stores. In 2005, thanks to this center and its efficient contingency plans, Wal-Mart was able to deliver water, fuel, and toilet paper to thousands of victims of Hurricane Katrina in New Orleans, even before state authorities were in a position to provide assistance.

1.3.2 Reengineering SC Systems

In most cases, a company's SCS is the product of its merger/acquisition or expansion decisions. This means that its natural development is not necessarily congruous with changes in the commercial environment or with the SC strategy of the company. To continue to fulfill its mission, an SCS must adapt continually to changes in the competitive environment, such as the arrival of new competitors, the opening of new markets, the modification of international trade rules, changes in consumer needs, the retreat of a big player or the advent of new technologies. It must also adapt to changes in a company's products or services offer, or in its SC doctrine. The gradual improvement of the system is generally sufficient to adapt to minor changes, but a more drastic reengineering becomes necessary when major opportunities or threats arise. The reengineering of the SCS can be viewed as cyclical improvement endeavor where some of the system's processes and structures are periodically updated.

Figure 1.15 recaps the methodology that this book suggests for the reengineering of a company's SCS. It illustrates the basic activities to accomplish (in boxes), as well as associated deliverables (on arrows). The approach starts with a detailed system analysis of the SCS in place at the beginning of the study and of its environment. This activity is directly related to the development of an SC strategy, as discussed in the previous section. It requires the compilation and analysis of a large amount of historical data and more or less relies on the process described in Fig. 1.14 to develop the company's product/service offer as well as its SC doctrine.

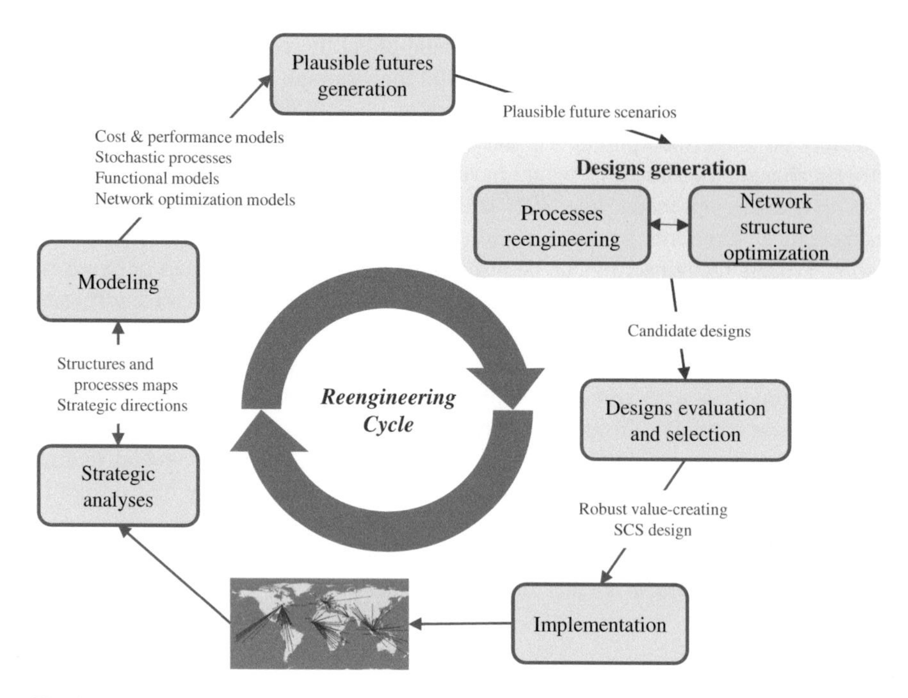

Fig. 1.15 Supply chain system reengineering methodology

Based on these analyses, modeling activities can be launched. This mainly involves modeling the risks to which the SCN is exposed; elaborating improved methods for some of the planning and execution processes identified in Fig. 1.10; and formulating an optimization model to support the reengineering of the SCN structure. The later incorporates strategic decisions relating to the product-markets to serve, the internalization or externalization of future activities, the selection of technologies, capacity investments, and the deployment of activities/resources across geographically dispersed facilities. It also specifies each facility's mission in terms of the products to manufacture, distribute, or sell, as well as the customers to serve. The following activity involves the generation of plausible future scenarios, mainly using Monte Carlo methods grounded in previously defined stochastic processes. Plausible future scenario samples are generated to support the reengineering of processes and structures, and also to enable an integrated value, risk, and sustainability assessment of the SCS designs being developed.

This is followed by parallel process and structure optimization activities. The arrow between these activities indicates that they are interdependent, meaning that when we act upon one facet of the system, we alter the basic data of the design problem associated with the other. For instance, changes in the nature of P&C processes—such as those resulting from using the Internet to facilitate communications between SC partners—would affect the costs associated with the material flows across the SCN arcs, and with their storage in network nodes. As a result, a network structure that had been optimal before the use of Internet would probably no longer be optimal. To take full advantage of the consequences of this process change, the structures would have to be adjusted, and vice versa. For given P&C processes, optimizing the SCN structure requires the solution of large-scale mathematical programs for several independent samples of plausible future scenarios. Process reengineering can be inspired by industry best practices or based on customized developments exploiting new technologies, new commercial software, or original operations management models. P&C system methods are chosen by comparing results obtained from simulations for a given network structure. This approach produces a set of candidate designs for the future SCS. In the following activity, these candidate designs are compared to the status quo, and assessed using performance measures based on the economic, environmental, and social returns they produce for a broad sample of plausible future scenarios. The best design is then selected and implemented. This ongoing SCS improvement process can be reinitiated as required with the eventual launch of a new reengineering cycle. Chapter 13 will return in greater detail to the activities associated with this SCS design methodology.

Above all, SCS design involves the search for superior structural and functional compromises to face a necessarily uncertain future. The success of SCS reengineering projects is predicated on an in-depth knowledge of the nature of SCs, along with competencies in statistical analysis, mathematical optimization, and simulation. The purpose of this book is to help readers develop these competencies. Chapter 2 includes an analysis of relevant SC costs and revenues, and it recommends a performance measurement approach suitable for designing and managing SCs. Chapter 3 focuses on the design of material flows P&C systems. As aforementioned, excellent books have already been written on operations management, which explains why the chapter primarily focuses on SC planning strategy choices. Chapters 4–12 deal with different aspects of SCN design. Chapters 4 and 5 study the nature and decisions relating respectively to network nodes (production–distribution centers) and arcs (means of transportation). Chapter 6 examines internalization-externalization issues, as well as the eventual selection of SC partners. Network optimization is explicitly addressed in the next chapters. Chapter 7 studies the nature of the SCN design problem and presents basic optimization models. Chapter 8 suggests extensions for technology selection and capacity planning. Chapter 9 looks at optimizing multinational SCNs. Chapter 10 examines SC risks and studies the generation of plausible future scenarios. Chapter 11 shows how the modeling concepts previously studied can be extended to design robust and resilient SCN under uncertainty. Chapter 12 shows how to design closed-loop SCNs including recovery and revalorization activities for sustainable development. Chapter 13 concludes by deepening the design methodology outlined in Fig. 1.15.

1.4 The Food Industry Supply Chains

Humans must eat and most people buy their own food. This absence of self-sufficiency intimates the need for diversified SCs to ship food to consumer households. Making food available requires a number of transformation and distribution activities stretching from agricultural production to end-user delivery. Food sector SC structures depend on the nature of the products sold. Some transit directly from producer to consumer, whereas others are the outcomes of a sophisticated transformation system. Figure 1.16 offers a classification of food sector products such as they are seen by grocers. Three product types can be distinguished: fresh products, frozen products, and dry products. Fresh products include foods like fruit and vegetable, meat, and dairy goods. Frozen products include pre-prepared meals and ice cream. Dry products include all of the other items that a food retailer sells.

Each product has its own SC, which can feature a number of different raw materials and intermediate products. Some are very simple, like when consumers

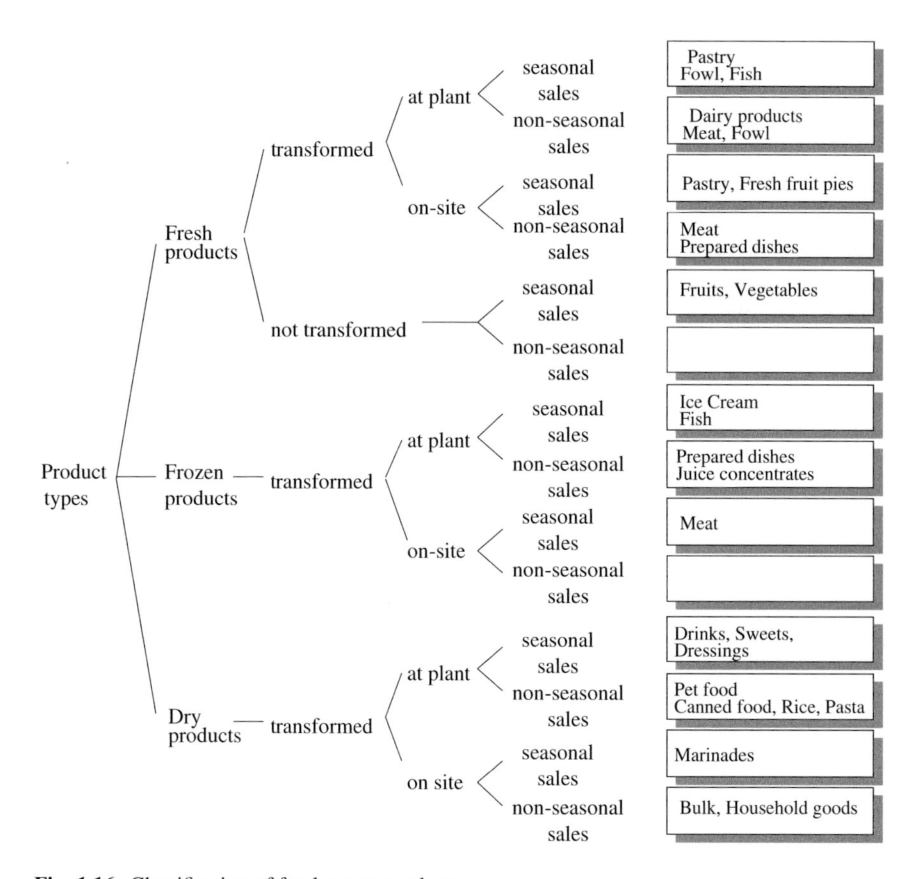

Fig. 1.16 Classification of food sector products

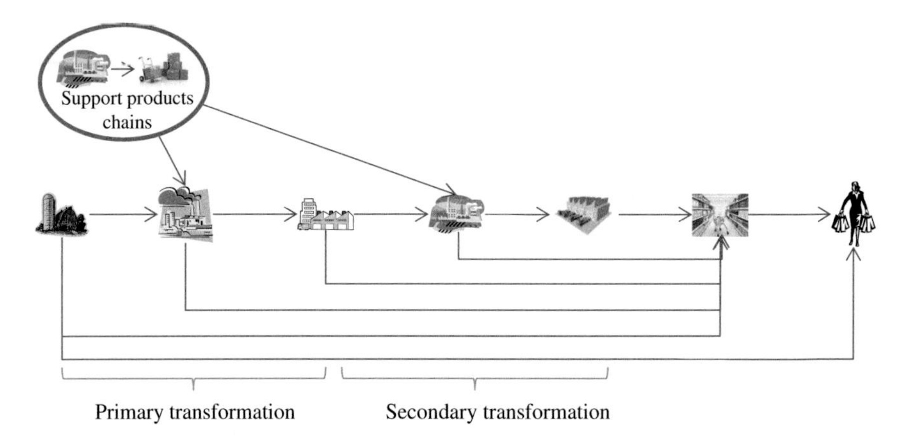

Fig. 1.17 Generic supply chains in the food industry

visit an apple orchard to pick fruit themselves. Most of the time, the transaction requires an intermediate point-of-sale. In this context, there needs to be some kind of sourcing activity as well as transportation between the producer and the point-of-sale. Also, raw materials are often transformed and shipped to consumers in another form. For instance, a dairy farm will collect milk and warehouse it until a tanker takes it to a plant for transformation into packaged finished products (milk cartons). The finished product is then stored or delivered immediately by truck to retail outlets or consumers. Some products comprising different ingredients are subject to much more sophisticated transformations. Figure 1.17 displays some of the SCs found in the food sector, showing how raw materials can flow from producer to end-user and how SCs can incorporate different transformation and distribution stages. It also illustrates the role of support products in the production and distribution of food products. For instance, food product SCs incorporate supply sources for packaging products. Note finally that all of these stages could be performed by different companies.

We have seen that merchandise flows within a particular DC tend to be triggered by two events, namely purchasing and sales. Figure 1.18 illustrates the role that DCs play in a grocery chain that source goods from external suppliers. The figure shows that deliveries from the DCs result from the stores' purchasing behavior, which depends in turn on consumers' purchasing behavior. It also indicates the main tactical and strategic management instruments available to suppliers, DCs, and stores. Clearly this is a simplified representation, but it does illustrate some of the SC challenges that the food retailing industry faces. Among others, SC flows are deeply influenced by the many promotional mechanisms that the industry uses, and more specifically by the preponderance of temporary low prices (offered to consumers in weekly flyers) at all echelon within the SC. Temporary low prices cause great volatility in demand up and down the SC and greatly complicate operations. Although the downside of these practices has been fully explored by industry

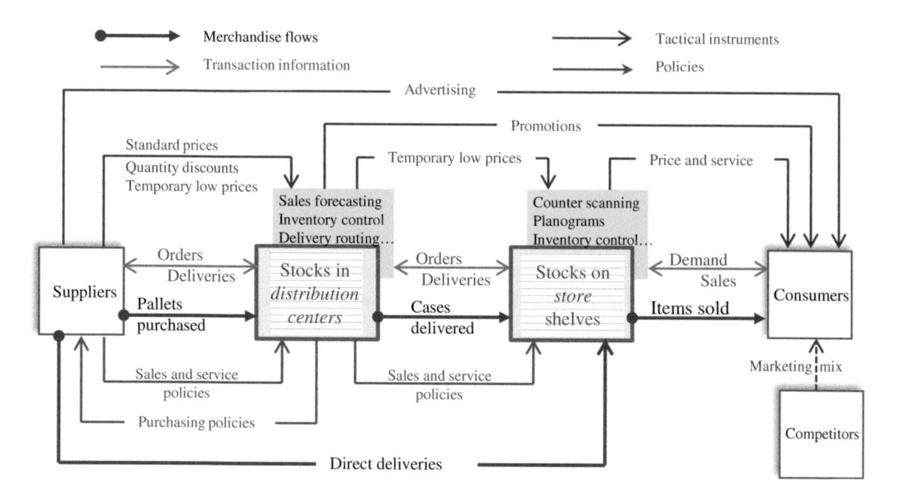

Fig. 1.18 Instruments shaping behaviors in the grocery industry

initiatives such as the *efficient consumer response* (ECR) movement (http://ecr-all.org), they are deeply engrained and there is little chance of their disappearing in the near future.

Let us now turn to the role that RCFs play in a food retailing industry, and study the relationship between revenue, expenses, capital employed, and material flows in the DCs of the SCN. Some revenues and costs depend directly on the nature of the processes being triggered by demand and purchase orders, whereas others relate to annual purchasing and sales volumes. The former can be categorized as *operating* costs and revenues in the sense that they are controlled or influenced by operating actions or decisions associated with specific purchases or deliveries. The latter are overhead costs and revenues since they do not depend on operating actions but instead on the rules of the game (purchasing and sales policies) and on infrastructures (offices, warehouses, IT systems, etc.).

When a purchase is made, the operating costs incurred include the amount paid to the supplier plus other procurement costs. The former depends on the supplier's price structure and the size of the order. The latter includes transportation and receiving costs. Both contain a variable component, which depends on the size of the order. They also feature a fixed component. The operating costs for all of the purchases made during a year can be characterized as follows:

$$\text{Operating costs} = \sum_{\text{orders}} (\text{pallets purchased}[\text{pallet price} + \text{transportation cost} \\ + \text{receiving cost}] + \text{order cost})$$

The operating revenues associated with any one delivery include the amount of the sale minus any distribution costs. The sales amount depends on the current selling price and the size of the customer order. Distribution costs include preparation and staging costs plus any delivery costs. All three are a function of the size of the order. Operating revenues for all of the deliveries being made during a given year can be transcribed as follows:

$$\text{Operating revenues} = \sum_{\text{deliveries}} \text{cases delivered}[\text{sales price} \\ -\text{staging costs} - \text{delivery costs}]$$

Some of the overhead costs incurred, such as administrative charges or IT costs, depend mainly on the total number of transactions needed to purchase and deliver the merchandise. Others, such as DC space and maintenance costs, depend on the volume of merchandise received, shipped, or kept in stock. Lastly some of these costs, like customer discounts or rebates, depend mainly on total sales volumes. Overhead revenues such as supplier volume discounts largely depend on the total volume of a company's purchases.

Capital employed can be broken down into three main components:

$$\text{Capital employed} = \text{inventory} + \text{fixed assets} + \text{working capital}$$

Working capital encompasses everything needed to facilitate the company's financial transactions and cover elements like accounts payable or receivable. Fixed assets include all of the capital frozen in equipment and facilities that are not being rented externally or managed by an independent corporate entity. Distribution centers capacity requirements largely depend on throughputs and merchandise inventory levels. The value of inventories constitutes a significant part of the capital employed.

To create value, RCFs must be maximized. Based on the aforementioned revenue, cost and capital elements, this business objective can be broken down into a hierarchy of sub-objectives for the management of material flows in the SCS. Table 1.2 highlights the relative importance of a few of these value drivers for the SCNs of a few

Table 1.2 Marginal contributions for consumer goods distributors

Return factors	% of sales	Marginal contribution (%)
Total sales	*100*	*2–12*
Overhead revenues	*5–22*	*0.4–0.7*
Total purchases	84–93	1.5–11
Total receptions costs	0.3–0.75	0.01–0.04
Total shipment costs	1–4	0.05–0.2
Total delivery costs	1–2	0.04–0.15
Overhead costs	8–24	0.4–1
Working capital	0–22	0–0.4
Merchandise inventory	7–17	0.15–0.3
Fixed assets	0.4–15	0.0–0.3

Quebec consumer goods distributors. The last column shows the marginal contribution of a factor, that is, the increase in ROCE ([revenues-costs]/capital employed) generated by a 1 % increase (decrease) in this factor. Fixed assets are very low in some cases because space or IT equipment costs are charged as rent.

These figures provide hints about which activity to focus on to improve returns. However, such analyses must also take into account the leeway provided by each factor. The following improvement opportunities illustrate this:

- Any contribution that the SCS makes to an increase in sales will have a major impact. Improving factors such as the quality of customer service, the breadth of the product assortment or products' in-store availability, and enabling the company to lower its prices due to lower SC costs (in other words, responding better to the order winners valued by customers) contributes to increase the company's market share, hence sales revenues.
- Purchasing costs have a major impact, meaning that a high priority must be given to the search for ways of cutting procurement prices and transportation costs (coordinated replenishments, internalization of inbound transportation, speculation on temporarily low prices, vendor managed inventory based on point-of-sale transaction information, etc.).
- Overhead costs are very high. It is crucial to ensure that they add value.
- The marginal contributions of shipment and delivery costs, and of the capital invested in inventories, are of the same order of magnitude. However, the leeway for the first two factors is much smaller than for the third. This is because, as we saw above, when uncertainty surrounding purchasing and sales processes can be reduced, it is possible to operate with very low inventory levels. On the other hand, the preparation and transportation of orders to stores is inevitable, meaning that there is a threshold beyond which it is physically impossible to lower shipment and delivery costs.

These observations intimate that there is much to be gained by focusing on improving sourcing and sales processes. This is the aim of industry initiatives such as ECR.

Review Questions

1.1. What are the main socioeconomic trends today that have a noticeable effect on SCM?
1.2. How would you distinguish between concepts such as industrial networks, SCs, extended enterprises, SCSs, and SCNs?
1.3. What is the difference between an SCS's structure and processes?
1.4. What constitutes flows P&C within an SCN?
1.5. What is the mission of an SCS?
1.6. What are the main SC decisions that a company must make?
1.7. Why do companies often seek to maximize value creation?
1.8. Why is risk analysis, measurement, and mitigation such an important SC issue?

1.9. Specify the nature of the SCS in your company (or in a company you know) in both functional and structural terms.
1.10. Describe a typical SC in the industrial sector to which your company (or a company you know) belongs.
1.11. Based on Figs. 1.6 and 1.17, for each of the following services, try to specify what a typical SC might look like:

- Toaster
- Hardware store product (like a hammer)
- Tailor-made clothing
- Clothing bought from a large department store
- Box of cereal for a known brand
- Fresh fruit or vegetable

1.12. What role do qualifier and order-winning criteria play in value creation?
1.13. What is a doctrine from a strategic perspective?
1.14. What is the link between a company SC offer, doctrine, and means?
1.15. Are SCs important in agribusiness? Why?
1.16. Based on the food industry study presented in Sect. 1.4, perform an analysis of the SC of another industry in which ECR plays a key role (cosmetics, pharmaceutical, etc.).
1.17. What is the impact of decoupling points and MTS/MTO strategies on SCN deployment decisions?

Bibliography

Atalaya E, Hortaçsua A, Roberts J, Syversonc C (2011) Network structure of production. PNAS 108–13:5199–5202
Ballou R (1992) Business logistics management, 3rd edn. Prentice-Hall, Upper Saddle River
BCG (2010) Rethinking operations for a two-speed world, Special Report, Boston Consulting Group
Bechtel C, Jayaram J (1997) Supply chain management: a strategic perspective. Int J Logistics Manag 8–1:15–34
Bradley J, Suwinski J, Thomas J, Thomas D (2001) What has not changed in supply chains because of e-business. The ASCET project. http://mthink.com/article/what-has-not-changed-supply-chains-because-e-business/
Carroll D, Guzman I (2013) The new omni-channel approach to serving customers. Accenture
Copacino W (1997) Supply chain management. St. Lucie Press
CST-Québec (2010) L'innovation dans la chaîne logistique des marchandises, Conseil de la science et de la technologie, Gouvernement du Québec
EC (2011) Roadmap to a single european transport area. White Paper, European Commission
Fisher M (1997) What is the right supply chain for your product? Harvard Bus Rev 75–2:105–117
Haanaes K, Reeves M, Strengvelken I, Audretsch M, Kiron D, Kruschwitz N (2012) Sustainability nears a tipping point. Research report, MIT sloan management review
Hill T (1999) Manufacturing strategy: text and cases, 3rd edn. McGraw-Hill/Irwin
Hopeman R (1969) Systems analysis and operations management, Merrill, Indianapolis
IBM (2009) The smarter supply chain of the future. IBM Corporation

Jacobs K, Jordan P, Plujim R, Vethman A, Ritter S (2006) 2016: The future value chain. Executive Outlook 4:46–63

Lapide L (1976) The essence of excellence. Supply Chain Manag Rev 18–24

Lee H (2004) The triple-A supply chain. Harvard Bus Rev 102–112

Martel A, Klibi W (2011) A reengineering methodology for supply chain networks operating under disruptions. In: Gurnani MR (eds) Managing supply disruptions. Springer, New York

Martel A, Oral M (eds) (1995) Les défis de la compétitivité: Vision et stratégies. Publi-Relais

Montreuil B (2006) Facilities network design: a recursive modular protomodel based approach. In: Meller R (ed) Progress in material handling research. Material Handling Industry of America (MHIA), pp 287–315

Montreuil B (2011) Towards a physical internet: meeting the global supply chain sustainability grand challenge. Supply Chain Res 3:71–87

Nahmias S (2005) Production and operations analysis, 5th edn. Irwin/McGraw-Hill

Naylor B, Naim M, Berry D (1999) Leagility: integrating the lean and agile manufacturing paradigms in the total supply chain. Int J Prod Econ 62:107–118

NIC (2008) Global trends 2015: a transformed world. National Intelligence Council, US Government

Pine II J (1993) Mass customization: the new frontier in business competition. HBS Press, Massachusetts

Poulin M, Montreuil B, Martel A (2006) Implications of personalization offers on demand and supply network design: a case from the golf club industry. EJOR 169:996–1009

Ross D (1998) Competing through supply chain management. Chapman & Hall, UK

Shell (2008) Shell energy scenarios to 2050. Shell International BV

Stalk G, Hout T (1990) Competing against time: how time-based competition is reshaping global markets. Free Press, New York

Stalk G, Evans P, Shulman L (1992) Competing on capabilities: the new rules of corporate strategy. Harvard Bus Rev 57–69

Treacy M, Wiersema F (1993) Customer intimacy and other value disciplines. Harvard Bus Rev 84–93

Treillon R, Lecomte C (1996) Gestion industrielle des entreprises alimentaires. Technique et documentation, Paris

UNCTAD (2014) World investment report 2014. United Nations

Van Opstal D (2007) The resilient economy:i competitiveness and security. Council on Competitiveness

Vollmann T, Berry W, Whybark C, Jacobs R (2005) Manufacturing planning and control systems for supply chain management, 5th edn. McGraw-Hill, New York

Von Hippel E, Ogawa S, De Jong J (2011) The age of the consumer-innovator. MIT Sloan Manag Rev 53:27–35

Vonderembsea M, Uppalb M, Huangc S, Dismukes J (2006) Designing supply chains: towards theory development. Int J Prod Econ 100:223–238

Womack, J, Jones D, Roos D (1991) The machine that changed the world. Harper Perennial, New York

World Bank (2011) Global industrial production declined 1.1 % in April in the wake of the tsunami and earthquake in Japan, Propects Weekly. The World Bank, 21 June 2011

WTO (2011) July 2011 Word trade report, World Trade Organization

Yucesan E (2007) Competitive supply chains: a value-based management perspective. Palgrave Macmilan, UK

Chapter 2
Performance Evaluation and Value Chains

A supply chain system can only achieve top performance if it finds a way of fulfilling its mission while ensuring that managerial actions make a positive contribution at every step of the way. Success depends on the mission being broken down into a set of objectives and performance indicators that will help to determine priorities and promote good decision-making. A strategy is of no use if it does not come together with performance indicators concretizing its aims. On the other hand, what a responsibility center measures is often the only thing deemed important. Consequently, a misalignment between missions and performance indicators does not only lead to a waste of resources, but it may endanger the company's overall success.

Chapter 1 stressed that the fundamental mission of a supply chain system is continuous value creation for the company and its partners. The performance indicators that a company adopts must therefore contribute to value creation, the latter being defined as the discounted sum of all future revenues, operating expenses, and capital expenditures. The fulfillment of this ultimate aim can be verified using financial performance indicators such as EVA and ROCE. These systemic indicators are crucial for top management but they are not appropriate for supply chain operations managers. Ultimately, performance depends on operations, not on strategies. Hence, the need to identify value drivers leading to suitable performance indicators for the responsibility centers of the supply chain system. These indicators do not have to be financial in nature and can be expressed, for instance, in terms of time (cycle times, setup times, delivery times, etc.), process productivity or reliability (products/hour, defect rate, failure rate, fill rate, etc.), and/or resource use (capacity utilization rate…).

This may lead to the definition of large numbers of indicators. The proliferation of measures can become problematic, because either it requires too much effort or it masks fundamental underlying behaviors. Designing a performance evaluation system that is balanced and congruent with a company's mission can be a challenge. After discussing several facets of this problem, the chapter looks at ways of evaluating a supply chain system's economic performance. More specifically, it studies the modeling of the revenues and costs necessary to design value-creating

A. Martel and W. Klibi, *Designing Value-Creating Supply Chain Networks*,
DOI 10.1007/978-3-319-28146-9_2

networks. Since these networks must be designed to provide superior performance during several years, future results are necessarily uncertain, and the measurement of value-at-risk over the planning horizon considered is also discussed.

2.1 Performance Evaluation in SC Management

A performance indicator is a verifiable measurement associated with a performance object. A measurement is *verifiable* if it is based on an estimation process specifying the data and calculation methods required to obtain the indicator. The *performance object* refers to what is being evaluated. It depends on where we are looking (forward or backward), on what we are trying to do (control, improve or communicate), and on the level of the responsibility centers that it affects. An indicator must elucidate what allows the responsibility center to create value for the company and its customers. As such, it should ideally be tied to value drivers, that is, to factors affecting the value added by the center.

When the purpose is to control operations or communicate outcomes, it is necessary to look backwards, with a focus on the outcomes achieved with the actions taken over a given time period. Once relevant observations have been made, the calculation method specified is used to evaluate the indicator. To control operations, this evaluation is compared with pre-set targets. These can be absolute standards like zero defects or the complete elimination of downtime; internal aims like budgets, standard times, and improvement targets; or industrial benchmarks like environmental standards or matching top-performing companies in the industry. Benchmarks of this kind are published by specialist organizations such as the Supply Chain Council (now part of APICS—www.apics.org), which has, for instance, collaborated with APQC (www.apqc.org) to develop a SCORmark^{SM} benchmark for its members. The performance indicators provided in the SCOR (*Supply Chain Operations Reference*) model allow a comparison with the best practice in the field of supply chain management. Furthermore, the publication of periodic financial statements is a perfect example of how one can proceed to communicate results.

If, on the other hand, the purpose is to improve supply chain system performance, it is necessary to look forward. Figure 2.1 shows how to proceed to make improvement decisions. The first step is to set an objective, then comes the choice of performance indicators and the evaluation of these indicators, decisions are made, and lastly, an action is taken. Decisions can entail short-term execution, operational or tactical planning, and even supply chain network reengineering. The ensuing actions should normally contribute to achieve the objective. Ex post, this is verifiable through the chosen performance indicators. Ex ante, these indicators become an essential element of the decision-making process, which relies on a formal or informal optimization model grounded in the perceived relationship between the decisions that must be made and expected performance. Where a formal normative model is used, performance indicators are associated with certain

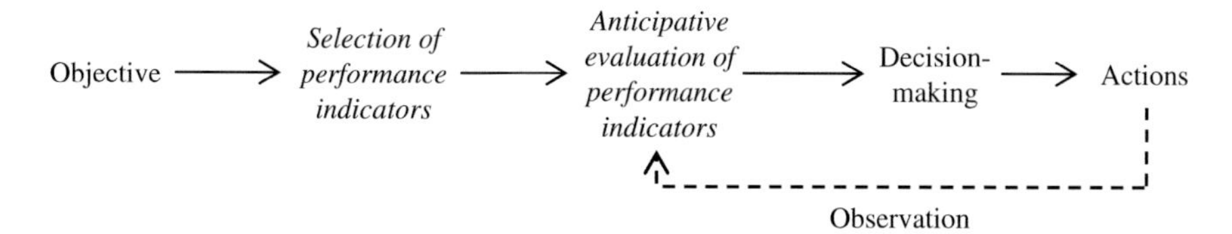

Fig. 2.1 Performance indicators' role in decision-making

model parameters. If the objective is to minimize relevant costs, the model's economic function could, for instance, associate operating costs with activity levels, and acquisition costs with added resources. Its capacity constraints would require parameters such as the resources consumed by unit of activity, the inventory turnover ratio, and so forth. If future decisions relate to short-term operations management, it is reasonable to assume that latest ex post indicator estimates suffice to forecast future behavior. On the other hand, when making long-term decisions, this assumption is generally unrealistic and consideration must be given to evolutionary trends associated with plausible futures. It is then no longer sufficient to rely on static performance indicators calculated using historic observations. Instead, performance functions depending on a number of contextual variables must be used. The estimation of these functions involves regression analysis as much as management accounting techniques.

In addition, the method used to estimate performance indicators/functions from operational data (quantities, loads and delays associated with purchases, deliveries, production, inventory, and sales, as well as with the resources employed) largely depend on what we want to do, and more broadly on the underlying performance model. This shows the importance of congruence between decision-making and performance models. Take the example of a product unit cost. Suppose a responsibility center uses space, equipment, and staff to transform raw materials into finished products. To set an adequate sales price for these products, it is important that the unit production cost calculated covers all the resources used. Using activity-based costing (ABC), the cost of the resources used during a particular period can be allocated to products, enabling the calculation of unit production costs that are good enough to set sales prices. Assume now that different equipments with different costs and throughputs (which affects staffing needs) can be used to make the products, and that the quantity of each product to make on each equipment must be planned to minimize production costs over the next week in light of the capacity available for each equipment and the volume of customer orders received. To be able to elaborate an optimal production plan, a distinct production cost must be calculated for each product on each piece of equipment. The ABC costs calculated cannot be used to do this because they are associated with historic production plans (looking backwards instead of forward) and they do not distinguish between equipment. This is why accounting system-based costs often cannot be used to optimize supply chain network planning and design decisions.

The objective pursued and the associated performance objects depend on the level of the responsibility center considered. A responsibility center is a generic group of resources pursuing a particular objective, and it must define performance indicators to plan and control its activities. Since a supply chain generally cuts across several companies that may include several strategic business units (SBUs) organized into departments performing production or service activities, a hierarchy of performance objects must be specified. At the top level, a responsibility center could encompass the whole of a supply chain, or else correspond to an actual company. Conversely, at the local level, it can be limited to a specific process. At the top level, performance indicators that focus on value creation (EVA, ROCE, etc.) and performance failure risks (value-at-risk, downside risk, etc.) are favored. Locally, the focus might be on the total cost of a subsystem or process, or on measures like response times, defect rates or productivity indexes. As aforementioned, indicators are useful if they are linked to value drivers. For instance, a relevant indicator for a distribution center may be the "average number of pallets loaded per truck" since this creates an incentive to reduce the resources required for shipments. An indicator like "percentage of on-time deliveries" is also useful because it relates directly to an order winner valued by customers. By relying on the definition of residual cash flows (RCF) and on the discussions featured in Chap. 1, four main categories of value drivers can be distinguished:

- Drivers that help to increase revenues by affecting order winners valued by customers;
- Drivers that help to reduce operating costs and current assets (inventory and receivables) by improving processes;
- Drivers that help to reduce fixed assets by improving network structures and interfaces with the environment (strategic alliances);
- Drivers that help to reduce risk.

The most widely used tool for tracking performance is a scorecard. To help fulfill a company's mission and to implement its business strategy, Kaplan and Norton (1992, 1996) have suggested using balanced scorecards such as the one illustrated in Fig. 2.2. The basic idea is to allow senior managers to monitor a company from four complementary perspectives by observing a limited number of performance indicators revolving around mission and strategy. The indicators chosen address four basic issues:

- How do customers see us? (Customer perspective)
- What must we excel at? (Internal business perspective)
- Can we continue to improve and create value? (Innovation and learning perspective)
- How do we look to shareholders? (Financial perspective).

Since this textbook focuses on reengineering supply chain networks, it will not emphasize performance monitoring but instead the prospective evaluation of alternative networks' performance over a multi-year planning horizon. At the strategic level, since value creation is the main objective, the SCN designs

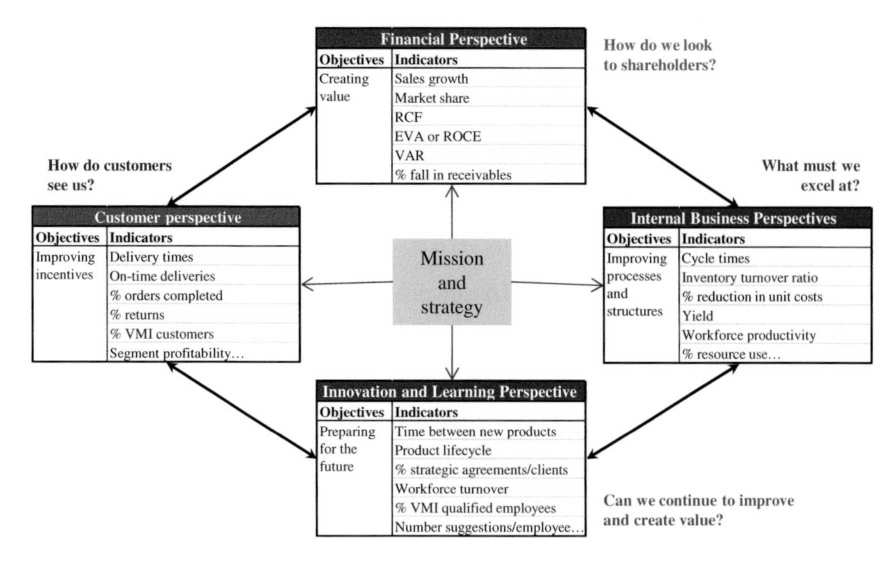

Fig. 2.2 The balanced scorecard. Adapted from Kaplan and Norton (1992, 1996)

considered are evaluated essentially from an economic point of view, that is, using expected value and value-at-risk measures. Two approaches have been widely used for this kind of strategic decision-making. Management accounting generally compares a number of alternatives developed by managers to the status quo (the default solution if nothing is done) by preparing differential investment budgets. These investment budgets identify the operating revenues and expenditures, as well as the capital expenditures, modifications induced by an alternative over the status quo for a given planning horizon. However, there is no guarantee that the alternatives considered would perform particularly well. Hence the second approach, which involves formulating a mathematical programming model with an objective function enclosing all relevant expected monetary flows, and possibly a risk evaluation. Where the mathematical model captures the essence of the problem, this approach leads to solutions that are close to the optimum.

Independently of the approach, to obtain valid results, the explicit or the implicit comparison of designs must be made over the same planning horizon. In addition, consideration should be given to all relevant monetary flows across the supply chain. The fact that several partners (customers, subcontractors, 3PLs, suppliers, etc.) might be involved can cause serious difficulties. When the supply chain system is in a stationary state, it may be sufficient to compare monetary flows over one year. This is not trivial, however, because operational revenues and costs are incurred on a daily basis, but capital expenditures provide resources whose productive life lasts several years. To perform such evaluations, it is therefore necessary to aggregate daily material and monetary flows, and to calculate an investments' annual "lease." If the system is not in a stationary state—due, for instance, to companies' desire to conquer new markets or increase market share—a multi-year planning horizon must be considered. This raises questions as to the value of money over time and of fixed assets' value at the end of the planning horizon. These concerns are examined in detail below.

2.2 Economic Performance Measurement

Corporate supply chain costs are often difficult to determine due to the inability of financial and management accounting systems to isolate them. Accounts defined for financial monitoring purposes are generally associated with resource categories (labor, equipment, buildings, etc.) and they are elaborated to be able to report past financial results. The cost functions required for prospective supply chain system evaluations may be derived from corporate accounting information, but their estimation usually requires a reorganization of available accounting data. In this section, before examining the revenues and costs relevant to strategic supply chain decision-making, we review the fundamental nature of costs, as well as issues related with investment projects evaluation.

2.2.1 Cost Accounting and Cost Functions

Companies estimate different types of costs. Not all are relevant to decision-making, however. It is important to understand the fundamental nature of costs to determine which to consider in a particular decision-making context. Estimating costs is the purpose of management accounting but, as aforementioned, the approach required to estimate a cost depends on what we want to do, that is, on the decision-making process requiring it. When looking forwards, a cost is not a fact but a prospective anticipation. Costs can be fixed or variable or a mixture of the two, and in some cases, they are affected by economies of scale or scope. A distinction must also be made between unit and marginal costs, historic and opportunity costs, relevant and sunk costs, and between operating and capital expenditures. The next paragraphs explain all of these concepts.

2.2.1.1 Fixed and Variable Costs

Fixed costs do not depend on the volume of activity during a given time period. Activity volumes are typically expressed as quantities produced, stocked, shipped, and so on. Wages paid to a company supervisory or support staff members are fixed costs since they do not vary depending on the volume of activity. Other expenses such as building leases, heating, or electricity are fixed but considered "indirect" because it is difficult to associate them with a specific product. Conversely, costs

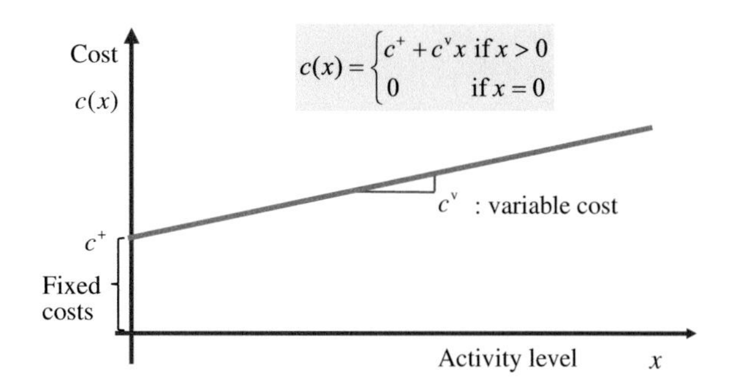

Fig. 2.3 Representation of a mixed (fixed plus variable) cost function

that evolve depending on activity levels are called *variable*. The cost of the raw materials used to make a product is variable because the quantities needed depend on the number of products being manufactured. Similarly, the costs of the workers employed directly to transform materials into products are variable.

Most activities have costs that are partially fixed and variable. The fixed portion involves costs incurred to set up or use a process, and the variable part depends on an activity variable such as the run time or the quantity produced. The cost for using a workstation combining equipment and human operators to manufacture specific products is generally mixed since it contains a fixed setup cost as well as a variable production cost. Figure 2.3 illustrates the cost function thus defined. The cost $c(x)$ incurred to make x products is the sum of fixed cost c^+ and a variable cost c^V for each unit produced.

2.2.1.2 Economies of Scale and Scope

The cost function $c(x)$ represented in Fig. 2.3 is linear, because it assumes that the variable cost is directly proportional to the volume of activity. In fact, given the possibility of achieving economies of scale or scope, this linearity assumption does not always apply. Economies of scale happen when marginal costs fall as activity levels increase. Thus, a product costs less to make as production quantities rise. Economies of scale stem from a number of factors such as learning effects, the use of higher performance technologies, a better utilization of available capacities, quantity discounts offered by suppliers and transportation companies. Let $c(x)$ be the total cost incurred to produce x items. Economies of scale exist when $c(x_1) + c(x_2) > c(x_1 + x_2)$. This behavior is illustrated on the left-hand side of Fig. 2.4.

The opposite of economies of scale can also occur and are called *diseconomies* of scale. Here, the marginal cost increases with the volume of production. Generally, diseconomies of scale start to arise when congestion effects occur, that is, when there is no longer sufficient capacity to use available resources efficiently. This phenomenon is reflected in the rising curve on the right-hand side of Fig. 2.4. Economies and diseconomies of scale are often modeled using a power function

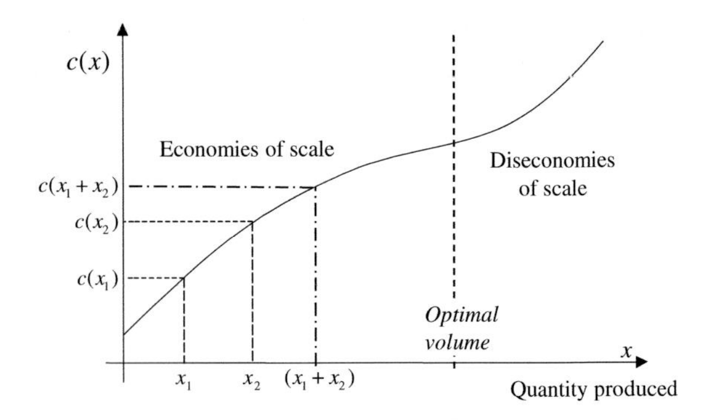

Fig. 2.4 Economies and diseconomies of scale

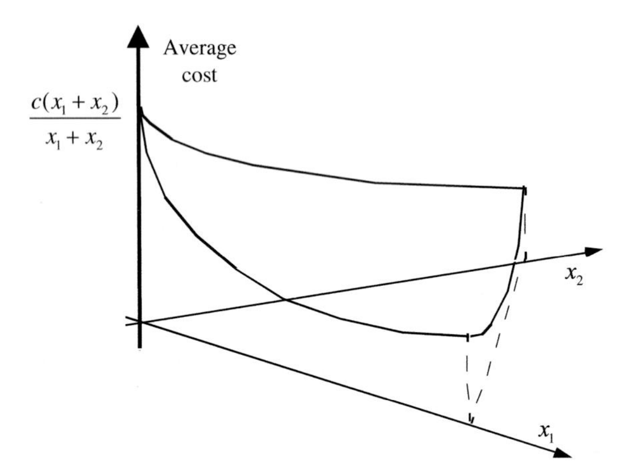

Fig. 2.5 Economies of scope

$c(x) = c^{o}x^{b}$, where c^{o} is a basic unit cost and the exponent b reflects the magnitude of the economies or diseconomies of scale. If $b = 1$, then $c(x) = c^{o}x$ and we fall back to the aforementioned linear variable costs. If $b < 1$, economies of scale exist; but if $b > 1$, there are diseconomies of scale. The parameters c^{o} and b are usually estimated using regression analysis.

Economies of scope are generated when several products share the same resources (supervisors, equipment, IT systems, etc.). Let $c(x_1, x_2)$ be the total cost for producing x_1 and x_2 quantities of two different products. Economies of scope exist when $c(x_1, 0) + c(0, x_2) > c(x_1, x_2)$. This behavior is illustrated in Fig. 2.5. Flexible modern technologies like robots are often a source of economies of scope.

2.2.1.3 Marginal and Unit Costs

As indicated above, accountants generally include a share of a company's indirect costs in their estimates of a product or activity's *unit cost*. Fixed costs are spread over all outputs produced. Suppose that the real activity costs can be described by

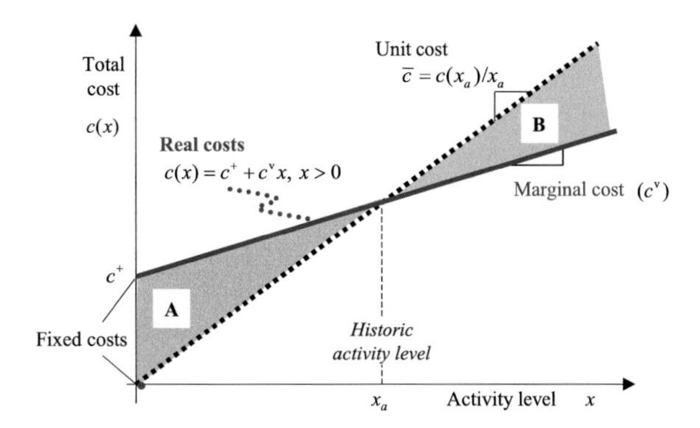

Fig. 2.6 Marginal and unit costs

the function $c(x) = c^+ + c^v x$ if $x > 0$ (with $c(0) = 0$) and that the level of activity observed over the last fiscal year was x_a, as illustrated in Fig. 2.6. In this case, the unit cost is obtained by dividing the total cost $c(x_a)$ incurred during the fiscal year by the observed level of activity, giving $\bar{c} = c(x_a)/x_a$. *Marginal costs*, on the other hand, reflect the real cost incurred for an additional unit of activity. In the example, real costs are linear, meaning that the marginal cost corresponds to the variable cost, that is, to the slope c^v of the straight cost line. As can be seen in the figure, on the left-hand side of the activity level x_a, the unit cost as calculated underestimates the real total cost (region A). On the other hand, to the right of this point, the unit cost overestimates the real total cost (region B). In a planning process, a correct evaluation of total costs must be based on a cost function defined in terms of the decision variables involved (e.g., planned level of activity x), while accounting for economies of scale or scope where this is relevant. Using unit costs to make decisions can lead to some major errors. For instance, a unit cost-based analysis might indicate that a product is not profitable and causes it to be abandoned. Yet if the calculations were based on marginal costs, it might become apparent that the product can increase the company's RCF. Having abandoned it, the indirect costs would be spread across fewer products, thus reducing the value created by the company!

2.2.1.4 Historic and Opportunity Costs

Accounting's factual focus demands the use of current costs or of an allocation of historic costs, such as depreciation for equipment. It remains that in a decision-making context, it is not the past that counts but the current and future impact of a decision. Hence, the need to rely on current or opportunity costs. The latter is a sacrifice associated with a particular decision. An example is the return that a company would have made on the sum invested in raw materials, semi-processed goods, and final products inventories had this capital not been tied up in stocks. Another is the revenues that could be obtained from leasing or selling a warehouse if the company was not using it.

2.2.1.5 Relevant and Sunk Costs

Not all of the costs that a company incurs are relevant for supply chain decision-making. A cost is considered relevant if it influences a decision in one direction or another. A simple example is deciding where to purchase a car model available from three dealerships. Assume the manufacturer sets the same sales price for all dealerships (say $40,000), then this becomes irrelevant to the purchaser's decision—despite the large sum being spent. Relevant costs will then probably involve things like the distance that needs to be traveled to maintain the vehicle, plus any other service costs. On the other hand, if the dealers propose a different purchasing price, this becomes an important relevant factor in buyers' decisions. Thus, the idea of a relevant cost has nothing to do with the payment of a certain amount but reflects instead the way this might affect decision-making.

Sunk costs are an important category of irrelevant costs and involve expenses that have already been incurred by the company and do not have a direct effect on decision-making. For instance, a company may wonder whether it should keep or replace a piece of equipment. The costs that are relevant to this decision include the equipment's maintenance and operations expenditures; their current commercial value; and any tax aspects. On the other hand, the acquisition price paid several years ago for the current equipment is no longer relevant. Although this is often hard to accept, current assets' book value is seldom relevant to decision-making.

2.2.1.6 Operating Revenues and Expenses, and Capital Expenditures

Operating revenues and expenses are monetary flows associated with day-to-day activities. Operating expenses involve things like paying suppliers, employees, or energy bills as part of a company's primary or support activities. Operating revenues are linked to sales. Capital expenditures relate to a company's acquisition of long-term resources (facilities, equipment, vehicles, etc.). Although these strategic acquisition decisions involve an initial investment, owning long-term resources can also generate monetary inflows (tax repayments, residual value) as well as outflows (property taxes, insurance, etc.) throughout the assets' useful life.

In supply chain decision-making, it is often necessary to evaluate solutions (designs or plans) involving these three types of monetary flows across a single-period or multi-period planning horizon. For comparative purposes, revenues/expenditures must be linked to the decision variables specified for the time periods considered (e.g., merchandise flows between a factory and a DC during a year, opening a new factory at the beginning of a period, etc.). Relevant cash flows must be assigned to period costs/revenues in such a way to maintain their additivity over all the periods of the planning horizon. In general, the operating costs/revenues defined are derived from intra-period monetary flows. On the other hand, periodic capital costs are obtained by dividing supra-periodic commitments over several periods (typically years), which is usually not obvious. The breakdown must reflect the resources' commercial value over time, which may be quite different from the book value obtained using normal accounting methods. The next section takes a closer look at this issue.

A good discussion of cost estimation assumptions and methods is found in Giard (2003). To conclude this section, it is worth emphasizing that supply chain decisions must be based on a calculation of relevant operating revenues/expenses and capital expenditures, and not on accounting conventions. The three following principles (Magee et al. 1985) should be kept in mind when analyzing strategic supply chain options:

- Costs must represent real out-of-pocket expenditures or profit opportunities that have been sacrificed.
- Only revenues/expenditures and opportunity costs affected by the particular supply chain decisions made should be taken into account.
- It is more important to clearly capture the nature of all relevant revenues/expenditures than to measure them precisely.

It is also crucial that double counting be avoided.

2.2.2 *Discounting Cash Flows*

When evaluating a strategic project, consideration must be given to all associated fund inflows and outflows during its useful life. Strategic investments provide assets lasting several years, which raises questions as to the value of money over time. Clearly, a sum invested today will produce interest and therefore be worth more in a few years. The opposite is also true. More specifically, when annual interest income is reinvested (compounded), \$1 invested at an annual interest rate of r will grow as shown in Fig. 2.7a). The growth term $(1+r)^t$ obtained after t years is called the *capitalization factor*. Similarly, the amount to invest to receive \$1 after t years can be determined retroactively as shown in Fig. 2.7b). The annual term $(1+r)^{-1}$ used in this calculation is called the *discount factor*. As a result, the *present value* V of a sum S received (spent) at the end of year t is

$$V = \alpha^t S, \quad \alpha = (1+r)^{-1}$$

If the interest is compounded f times a year, the discount factor α becomes

$$\alpha = \left(\frac{1}{1+(r/f)}\right)^f$$

When the interest is compounded continuously, the discount factor becomes $\alpha = e^{-r}$, where e is the Neperian constant $e = 2.71828\ldots$

To evaluate a strategic project, the *net present value* (NPV) of all associated cash flows must be calculated, using the opportunity cost of capital instead of the annual

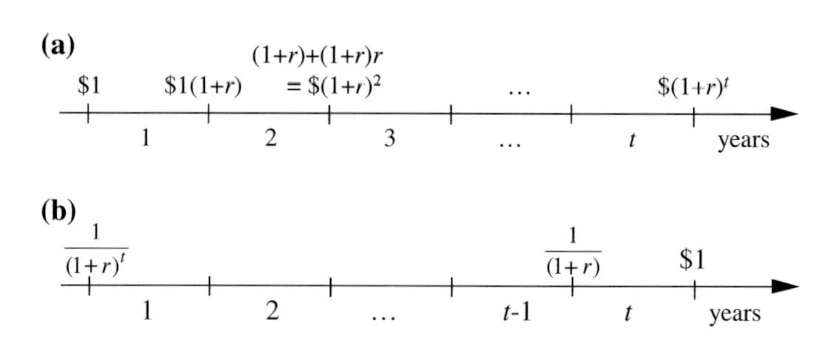

Fig. 2.7 Value of \$1 in time under an annual interest rate r

interest rate. This opportunity cost is referred to as the *discount rate*. The choice of an appropriate rate is a difficult problem that receives a great deal of attention in finance. Several issues may have to be considered to calculate this rate, including the company's capital structure, taxation, inflation, borrowing rates, insurance on borrowings, and financial risks. Studies in this area typically rely on simplistic assumptions that are crucial to whatever approach is being recommended. It is widely recognized, however, that two main categories of stakeholders should share a company's operating profits after tax, namely shareholders and creditors. In the following let:

r^{S} Return rate that shareholders expect to receive for comparable (notably in terms of risk) investments on the capital market
r^{C} Risk-free interest rate paid on the debt
τ Corporate tax rate

The general recommendation is to use the weighted average cost of capital (WACC) $r = \theta r^{\mathrm{S}} + (1 - \theta)(1 - \tau)r^{\mathrm{C}}$, where θ represents the proportion of capital employed that comes from shareholders. Since debt interest can be subtracted from a company's taxable income, the interest rate must be adjusted. Most companies have specific policies regarding the discount rate to use when choosing strategic projects, which is the assumption we make in the rest of this text.

If the initial investment (outflow of funds) associated with the project is S_0, and an inflow (outflow) of funds S_t occurs at the end of year t (an inflow being a positive monetary flow and an outflow a negative one), the NPV for the series of net annual flows $S_0, S_1, \ldots, S_T$ observed over a planning horizon of T years is

$$NPV = \sum_{t=0}^{T} \alpha^t S_t \tag{2.1}$$

A positive *NPV* signifies enrichment since the project provides a return superior to the cost of capital. Conversely, a negative *NPV* involves an impoverishment. Clearly, companies will only opt for projects producing positive NPVs.

Consider the case of a company deciding on whether to purchase a new equipment with an initial cost of $100,000. It is forecasted that the net inflows of funds, over the next 3 years, resulting from this purchase will be $30,000, $35,000 and $40,000. The equipment's resale value after 3 years is estimated at $40,000. With the company using a cost of capital of $r = 12.5\ \%$, what is the NPV of the cash flows involved? This requires calculating NPV with (2.1) using $S_0 = -100{,}000$, $S_1 = 30{,}000$, $S_2 = 35{,}000$ and $S_3 = 40{,}000 + 40{,}000$, with a discount factor $\alpha = 1/(1 + 0.125) = 0.888$, giving

$$NPV(\text{in } \$\ 1{,}000) = -100 + 30(0.888)^1 + 35(0.888)^2 + 80(0.888)^3 = 10.257.$$

In other words, it is expected that the company will be enriched by $10,257 if it purchases this piece of equipment.

A recurrent problem in practice is the choice between alternative investment projects with very different cash flows, useful lifespans, and residual values (the asset's value at the end of its useful life). It is important in these problems to compare projects on the same basis. Suppose that a company must invest in production equipment and that it is considering two options $i = 1, 2$. These options do not affect the products' sales price or actual sales, meaning that the only relevant cash flows are those relating to the equipment's purchase, operation, and maintenance. The data required to make a decision is thus the following:

S_{i0} Initial investment required for equipment i
n_i Useful life of equipment i (in number of years)
RV_i Residual value of equipment i at the end of its useful life (can be negative)
S_{it} Total operating and maintenance expenses for equipment i during year t

The net present cost NPC_i of each of these options is given by

$$NPC_i = (S_{i0} - \alpha^{n_i} RV_i) + \sum_{t=1}^{n_i} \alpha^t S_{it} \qquad (2.2)$$

Suppose that $n_1 < n_2$. Basing our choice on the NPC's would not be fair due to the fact that NPC_2 covers a longer period of time. To make the right decision, the options must be compared for the period of the shortest equipment's useful life (n_1 in this case), and the cost calculations for option 2 must be revised. This can be illustrated with the data found in Table 2.1. This simplified example assumes that the expenditures $S_{it}, t = 1, \ldots, n_i$, are the same for each year of the equipment's useful life, and that the company uses a discount rate $r = 6\ \%$ (and thus a discount factor $\alpha = 1/(1 + 0.06) = 0.943$). The column $(S_{i0} - \alpha^{n_i} RV_i)$ determines the equipment's present value in light of its residual value at the end of its useful life. The NPCs of the two options are calculated using relation (2.2). These values cannot be relied upon to make a decision due to the fact that NPC_1 covers relevant

Table 2.1 Example of choice between two pieces of equipment

i	n_i	S_{i0}	RV_i	$S_{i0} - \alpha^{n_i} RV_i$	$S_{it}, \forall t$	NPC_i
1	4	$100,000	$20,000	$84,158.13	$40,000	$222,762.35
2	8	$180,000	$50,000	$148,629.38	$30,000	$334,923.20

cash flows during the first 4 years, whereas NPC_2 also includes expenditures in year $t = 5, \ldots, 8$.

To make a valid comparison, the relevant annual cost of option 2 during the years $t = 1, \ldots, 4$ must first be evaluated. This is done by calculating the annuity required to recover, with interest, the initial value of equipment 2 during the 8 years of its useful life. Using the annuity calculation formula[1] (Crundwell 2008) leads to

$$AN_2 = (S_{20} - \alpha^{n_2} RV_2)\left[\frac{r(1+r)^{n_2}}{(1+r)^{n_2} - 1}\right] = 148{,}629.38\left[\frac{0.06(1.06)^8}{(1.06)^8 - 1}\right] = \$23{,}934.67$$

Consequently, the annual investment and usage costs for equipment 2 during its first 4 years are $AN_2 + S_{2t} = 23{,}934.67 + 30{,}000 = \$53{,}934.67, \quad t = 1, \ldots, 4$. The net present cost of the relevant cash flows is therefore

$$NPC_2(4\ \text{years}) = \sum_{t=1}^{4} \alpha^t (AN_2 + S_{2t}) = 53{,}934.67 \sum_{t=1}^{4} (0.943)^t = \$186{,}889.34$$

Since $NPC_2(4\ \text{years}) < NPC_1$, equipment 2 is the one that should be purchased. However, this analysis implicitly assumes that at the end of year 4, equipment 1 will be replaced by an identical machine. If this is not the case, an eight-year analysis considering the replacement equipment explicitly must be performed. As we shall see in Chap. 8, when several options are available over a given planning horizon, it is preferable to formulate the problem as a mathematical program.

A simpler approach to select an option would be to calculate the annuity needed to recover the capital invested, with interest, and the annual usage costs during the useful life of each piece of equipment. This is tantamount to calculating an annual *lease* L_i that should be charged for each option. These calculations are made using the following formula:

$$L_i = \left(\sum_{t=0}^{n_i} \alpha^t S_{it} - \alpha^{n_i} RV_i\right)\left[\frac{r(1+r)^{n_i}}{(1+r)^{n_i} - 1}\right], \quad i = 1, 2 \tag{2.3}$$

The results obtained for the example are $L_1 = \$64{,}287.32$ and $L_2 = \$53{,}934.67$. Since $L_2 < L_1$, the decision is the same as above. Note, however, that this approach

[1]With *Excel*, this calculation is done using the financial function PMT(r; n_2; initial value).

implicitly assumes that a more economical alternative than equipment 2 would not be available at the end of equipment 1's useful life. As Chap. 7 will show, when the company operates in a stable business environment, supply chain network design decisions are often taken using an optimization model defined for a typical year. Possible options are then assessed using this approach.

The preceding discussion neglects the effects of tax rebates. Equipment purchases can be amortized over several periods by applying the company's depreciation methods (linear, digressive, etc.) and applicable tax rules. As a result, it is possible with each option to recover taxes paid during the equipment's useful life. These cash flows are relevant to evaluate the options and they must be included in the analysis. Clearly, anticipated cash flows should also take inflation into account. Assuming, as we did, that the usage expenditures $(S_{it}, t = 1, \ldots, n_i)$ will not change over the next 8 years is not very realistic. Each company operates within a specific context and investment decisions can be very diverse. Valid evaluations require substantial expertise in financial analysis.

Lastly, note that selecting investment projects based on NPV is not without flaws, in part because projected cash flows are random variables and not known quantities. To some extent, the discount rate can be selected to account for risk, but this ignores the fact that managers can modify their projects midstream, if necessary, to adapt to unexpected events. All sorts of *real options* (Trigeorgis 1996) such as delaying, abandoning, extending, or limiting the project exist in practice so that people can adapt to unforeseen events. The analysis should consider this. Several projects assessed as unprofitable when evaluated using NPV become profitable when real options are considered in the analysis.

2.2.3 SC Costs and Revenues

This section discusses the different costs and revenues that must be evaluated when trying to improve a company's supply chain system. All of these costs/revenues are inter-related in the sense that a decision taken to optimize one of them might increase or diminish others. Hence the importance of adopting a systemic approach, that is, one that seeks to maximize the value added by the supply chain. It has been explained that, conceptually, a company's value depends on the discounted sum of its RCF over its lifetime. Since cash flows can vary greatly over time, this definition does not facilitate the punctual estimation of value added in a given year. Periodic valuations can be done more easily using *economic value added* (EVA). These two notions (RCF and EVA) lead to equivalent definitions of value, which can be understood intuitively by examining them more closely. Let:

T Company's lifespan, in years
R_t Operating revenue of the company in year t
C_t Operating costs of the company for year t
ΔI_t Company's net investment in assets during year t (net of asset sales)

Dp_t Depreciation of company's assets for year t
Tx_t Tax paid by company for year t

The value of the company V^{E} expressed in terms of RCFs (RCF_t) for years $t = 1,\ldots,T$, is given by the following expression:

$$V^{\mathrm{E}} = \sum_{t=1}^{T} \alpha^t RCF_t, RCF_t = (R_t - C_t - Tx_t) - \Delta I_t, Tx_t = \tau(R_t - C_t - Dp_t) \quad (2.4)$$

The company's investment ΔI_t in year t is linked to its set M_t of long-term assets and to variations in current assets. The long-term resources $i \in M_t$ in use come from past investment projects. As shown above, each of these investments can be converted into a series of annuities covering the useful life of the resource. For a year l of the useful life of asset $i \in M_t$, the annuity calculated can be separated into two parts: the asset depreciation Dp_{il} during the year, and the opportunity cost of capital $r\overline{Dp}_{il-1}$, $\overline{Dp}_{il-1}$ being the remaining (non-depreciated) part of the initial investment. Sorting these amounts by years and cumulating them produces the depreciation $Dp_t = \sum_{i \in M_t} Dp_{it}$ and the non-depreciated value $A_{t-1} = \sum_{i \in M_t} \overline{Dp}_{il-1}$ of assets at the beginning of year $t = 1,\ldots,T$ (Shrieves and Wachowicz 2001). Finally, by substituting into (2.4), the following expression is obtained:

$$\begin{aligned} V^{\mathrm{E}} = \sum_{t=1}^{T} \alpha^t EVA_t, \quad EVA_t &= [(R_t - C_t - Dp_t) - Tx_t] - rA_{t-1} \\ &= (R_t - C_t)(1-\tau) - [(Dp_t + rA_{t-1}) - \tau Dp_t] \end{aligned} \quad (2.5)$$

The relationships between the economic value added EVA_t defined in (2.5) and the revenues, expenditures, and assets of a supply chain system are represented in Fig. 2.8. Operating costs and revenues are generated by primary activities affecting products in time (storage), space (transport and handling) and form (production), by transactions with business partners (purchases, sales), and by support activities like maintenance and quality control. The cost of assets are calculated by multiplying the value of the company's assets A_{t-1} (land, facilities, equipment, inventory, and other current assets) by the cost of capital r. Figure 2.9 shows how these costs and assets associate with material flows and resources in a simplified supply chain system. This section will study the supply chain revenue and cost functions required to calculate EVA_t, particularly those associated with cost C_t, annuity $(Dp_t + rA_{t-1})$, and tax savings τDp_t.

Before studying the costs/revenues associated with the material flows, inventories, and resources identified in Fig. 2.9, it is important to distinguish between those required for the short-term operational decision-making (quantities to be ordered, manufactured, delivered, and stored in the immediate future), as opposed to the long-term strategic decision-making. Towards this end, let us examine factory-specific production decisions for a set of products, considered as an

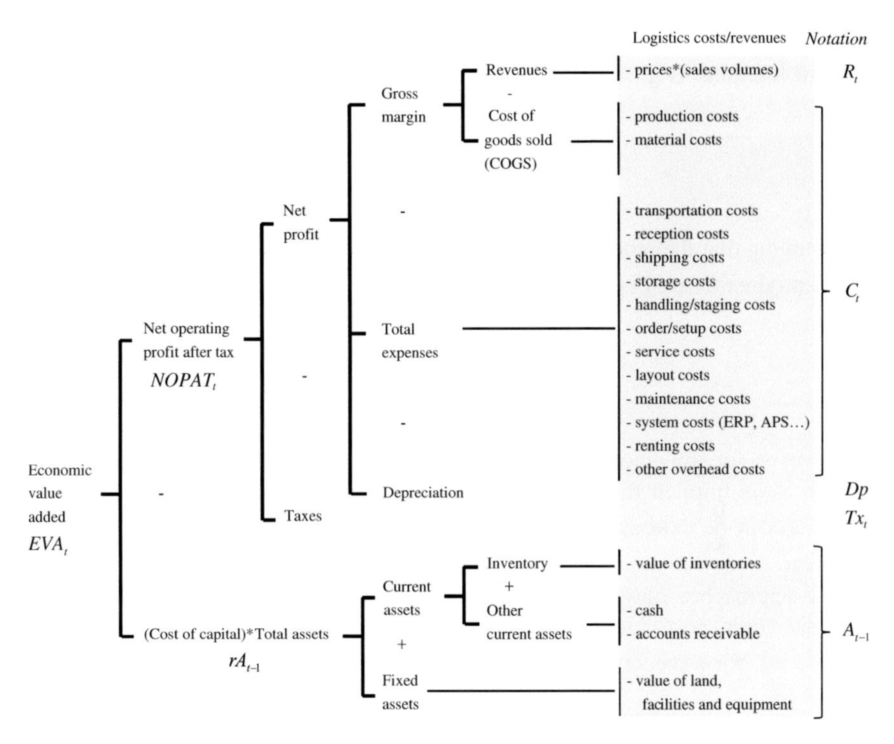

Fig. 2.8 Impact of revenues, costs, and supply chain assets on value creation

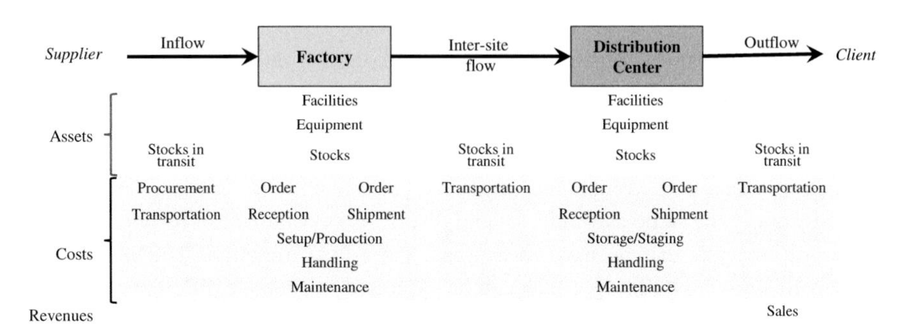

Fig. 2.9 Main assets, costs, and revenues in a supply chain system

aggregate product family for strategic decision-making. Assume that m_p lot-sizing decisions are made for product p during a year, and define the following activity variables:

Q_{pj} Lot size for product p resulting from production decision j ($j = 1, \ldots, m_p$)

X Aggregate quantity of products manufactured during the year for the product family

When these operational and strategic decision variables are all expressed using a standard load unit (e.g., a pallet), we have that

$$X = \sum_{p} \sum_{j=1}^{m_p} Q_{pj} \tag{2.6}$$

Assuming that the production cost function for product p is $c_p^{\mathrm{P}}(Q)$, the aggregate annual production cost is given by

$$C^{\mathrm{P}}(X) = \sum_{p} \sum_{j=1}^{m_p} c_p^{\mathrm{P}}(Q_{pj}) \tag{2.7}$$

However, for strategic decision-making, $C^{\mathrm{P}}(X)$ must be expressed in terms of X, and not as an implicit function of operational lot-sizing variables Q_{pj}. This transformation can be extremely complex, and even impossible, especially if $c_p^{\mathrm{P}}(Q)$ is not linear.

Two approaches may overcome this obstacle. One consists of calculating an aggregate unit cost for product p using historic production decisions Q_{pj}^o, $j = 1, \ldots m_p^o$, for a recent year. This leads to

$$C^{\mathrm{P}}(X) = \bar{c}X, \quad \bar{c} = \sum_{p} \sum_{j=1}^{m_p^o} c_p^{\mathrm{P}}(Q_{pj}^o) \Big/ \sum_{p} \sum_{j=1}^{m_p^o} Q_{pj}^o \tag{2.8}$$

The advantage of this function is that it is linear, but it does suffer from all of the aforementioned unit cost flaws, and it hides economies of scale and scope. An alternative consists of cumulating relevant fund outflows over a typical year, and establishing a relationship between these cumulative expenses and the aggregate production variable X. When responsibility centers with the same mission in a company apply the same methods, several distinct observations can be obtained and regression analysis can be used to estimate the $C^{\mathrm{P}}(X)$ function. The paragraphs below will show how this approach can be exploited to evaluate different aggregate supply chain cost functions.

2.2.3.1 Cost of Adding, Transforming, Using, or Withdrawing Long-Term Resources

When a building is bought to house a factory or distribution center, it is expected that the sum invested will be recovered later from the revenues generated by the building's activities. This reasoning applies not only to the acquisition of a building, but also to the renovation of a facility or to the construction of a new production–distribution center. It also applies to the acquisition/overhauling of production, storage, transportation, or handling equipment. Section 2.2 on discounted cash flows (and the examples it includes on choices between different types of equipment) offers a facility/equipment selection methodology. A few more precisions may be useful, however.

First note that the resources of a supply chain system are often leased or rented and not purchased. Some leasing contracts also enable the acquisition of the facility/equipment after a certain time. Leasing offers certain advantages as follows:

- Companies often consider leasing costs as operational expenditures, which may offer tax advantages and improve certain financial ratios;
- Initial outlays are generally relatively low compared to the value of the good;
- The risk of obsolescence may be reduced;
- Contracts can be adapted to particular situations (like seasonal variations);
- Equipment can be tried out before the purchase, and the project can be abandoned;
- The financing of the investment is the responsibility of the owner.

Among the shortcomings, the full and final cost of a lease can be high. Also, it often leaves the lessee with less room to maneuver in terms of how the resource might be used.

Modifying a supply chain system's long-term resources amounts to a change of state, and the costs incurred depend on the state of the system before the investment or lease takes place. Several situations are possible, and Table 2.2 shows the annual cash flows associated with each. Closing costs can be high when a production or distribution center's operations are stopped. In addition to moving costs, penalty costs are typically incurred including severance packages for departing personnel, the reimbursement of subsidies received, crisis management expenses, and even revenues lost due to dissatisfaction with the decision. On the other hand, where the asset has substantial commercial value, there is a possibility that it can be recovered. If a facility/equipment in place continues to be used, an opportunity cost equal to its commercial value multiplied by the company's cost of capital is incurred. Conversely, if a new factory is being built or an existing one is being purchased and renovated, as discussed above the annuity needed to recover the sums invested during the factory useful life must be calculated. An intermediary situation involves enlarging or renovating an existing building. This kind of project can also involve significant reorganization costs (moving equipment or inventory, installation of IT systems, staff training, etc.) that add nothing to the assets' commercial value. As such, they are operating expenditures that will be disbursed only once when the asset's state changes. Other fixed operating costs associated with ownership of assets —such as heating, insurance, maintenance, and security—are incurred annually.

Overall, the cost function $C^{R}(\,)$ associated with the addition, transformation, usage, or withdrawal of a long-term resource includes a combination of fixed operating expenses and asset ownership costs that do not all have the same fiscal impact. The definition of EVA given in expression (2.5) provides the key to determine how they should be accounted for.

Table 2.2 Examples of the costs of changing the state of long-term resources

		Strategic decision		
Initial state		*Not using the resource*	*Continue using the current resource*	*Using a new resource*
Current resource	*Fixed asset*	– Closing cost – Recovery of commercial value	– Opportunity cost – Fixed operating cost	– Annuity – Opportunity cost – Layout cost – Fixed operating cost
	Lease	– Closing cost – Breach of contract penalty	– Lease – Fixed operating cost	– Lease – Layout cost – Fixed operating cost
Potential resource	*Build*	Nil		– Annuity – Layout cost – Fixed operating cost
	Buy and renovate	Nil		– Annuity – Layout cost – Fixed operating cost
	To lease	Nil		– Lease – Layout cost – Fixed operating cost

2.2.3.2 Inventory Costs

Inventories are inevitable in a supply chain. Whether products are moved within one site (handling) or between sites (transportation), *stocks in transit* are immobilized during the duration of the journey. If products are ordered or manufactured in batches and these batches are not consumed instantly when received/produced, *cyclical stocks* exist during the period of consumption. When demand cannot be forecasted with certainty, *safety stocks* must be kept to avoid shortages. When demand is seasonal but it is not possible to modulate production capacity over the course of the year (due to collective agreements for instance or because the processes are too complex), production must be smoothed thus generating *seasonal stocks*. When raw material prices are volatile, it can be advantageous to invest in *speculative stocks*. If quantity discounts are available, or if substantial savings can be achieved by hauling full loads, it can be advantageous to consolidate purchases/shipments, leading to *scale stocks*. When the supply chain system is vulnerable to natural catastrophes, *insurance stocks* may be worth keeping.

Inventory *holding costs* combine all the costs of keeping inventory over time, thus all relevant costs that vary depending on stock levels. The costs incurred for keeping an item in stock over the course of a year is generally expressed as a percentage of the product value. The unit holding cost is calculated by multiplying this rate by the item's value. For instance, if the rate used is 18 % per annum and if the item is valued at $200, the holding cost is 0.18(200) = $36 per item per annum. Charges incorporated into the calculation of the holding cost can be grouped into four categories: cost of capital, cost of service, cost of risks, and cost of using space. The cost of capital was discussed above, and most companies have their own policy in terms of which rates to use. Stocking items literally immobilize funds that could be invested elsewhere. This is clearly an opportunity cost.

Service costs include expenditures that are incurred to protect stocks, for instance, by insuring them. Other expenses, like cyclical inventory counting or taxes on stocks levied in some states, also come under this category. Companies in Québec (Canada), for instance, must pay taxes on assets, hence on inventories. Risk-related costs are associated with undesirable events that can happen when a company keeps inventories. These can include damage, theft, and obsolescence. Fashion items and perishables are particularly vulnerable to obsolescence. All of these costs can be represented as percentages if they are added up for a given year and divided by the average value of the stocks being held. Lastly, the cost for using warehousing space reflects warehouses' operating expenses. Not all of these costs are relevant. For instance, heating, electricity, and security costs are rarely relevant because they do not vary depending on stock levels. Reception, handling, and shipping costs are also independent of stock levels because they depend on the volume of orders received and issued. If a public warehouse is being used, the amount billed for space rises with stock levels and is therefore relevant. If the space occupied in a private warehouse can be used for other purposes, an opportunity cost is incurred. When calculating the final rate to apply, one must be careful to avoid double counting.

Once the inventory holding cost rate has been established, the annual inventory holding cost $C^{\mathrm{I}}(\bar{I})$ of a product in a supply chain network node or arc can be calculated as follows:

$$C^{\mathrm{I}}(\bar{I}) = (r^{\mathrm{I}}v)\bar{I} \tag{2.9}$$

r^{I} Inventory holding cost rate (in $/$/year)

v Value of the product kept in stock (including value added upstream in the network)

$\bar{I}$ Average inventory level during the year

In inventory management, the average inventory $\bar{I}$ incorporates cyclical stocks associated with the procurement lot size Q as well as safety stocks based on the variance of the demand σ_{LT}^2 during a delivery lead time (LT). For traditional (min, max) inventory control systems,[2] the average level of cyclical stocks is $\bar{I}_C = Q/2$

[2]The *min* is an order point and the *max* a replenishment level.

and the safety stock is $\bar{I}_{SS} = \kappa\sigma_{LT}$, where κ is a safety factor reflecting the required service level (Silver et al. 1998). The value of Q depends on the company's inventory management methods and demand levels. Q is often calculated using the economic order quantity formula $Q = (2q\bar{x})^{0.5}$, where $\bar{x}$ is the average annual demand and q the ratio of fixed order cost over holding cost. It has been shown that the standard deviation of demand during a delivery lead time can be evaluated approximately using the formula $\sigma_{LT} = a_{LT}\bar{x}^{b_{LT}}$, where a_{LT} and b_{LT} are empirically estimated regression coefficients. As a result, when the company's inventories are well managed, an empirical relationship of the form $\bar{I} = a\bar{x}^b$ can be established between the level of cyclical and safety stocks, on the one hand and demand, on the other hand.

Strategic decisions relating to supply chain network design must assess inventory holding costs for all the storage points in the networks considered. The annual demand for a storage point, in this context, corresponds to its product's throughput X, this throughput being one of the variables to be optimized. From the discussion above, it follows that, for a given family of products held in a storage point, an empirical relationship $\bar{I}(X) = aX^b$ can generally be established between the average inventory level and the annual throughput X. Figure 2.10, for instance, illustrates the concave inventory–throughput function estimated for the electric poles held in the storage points of an electricity distributor. This function was estimated using regression analysis, from a set of observations corresponding to historic average inventory and annual throughputs for the company's storage points. For a given historic throughput X^o, the ratio $\varphi = X^o/\bar{I}(X^o)$ corresponds to the inventory turnover ratio. For the poles example, a storage point with an annual throughput of 3,500 poles has an inventory turnover ratio $\varphi = 3500/(7.98(3500)^{0.59}) = 3.8$.

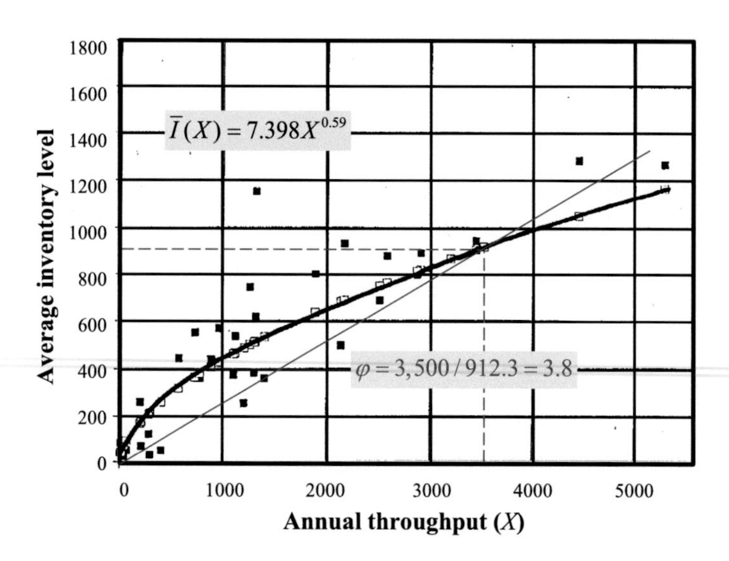

Fig. 2.10 Inventory–throughput function for poles stored by an electricity distributor

Substituting the estimated function $\bar{I}(X)$ in (2.9) produces the following holding cost function:

$$C^{\mathrm{I}}(X) = (r^{\mathrm{I}}\bar{v})aX^{b} \qquad (2.10)$$

where $\bar{v}$ is the average value of the items in the product family being considered. This storage cost curve reflects the economies of scales that can be achieved by using large distribution centers in a supply chain network. Note that, as illustrated in Fig. 2.10, the historic inventory turnover ratio φ can be used to linearize this function, giving $C^{\mathrm{I}}(X) \approx (r^{\mathrm{I}}\bar{v}/\varphi)X$. This linear function suffers, however, from all of the defects associated with the use of unit costs, without, of course, capturing the economies of scale.

To conclude this section, it is worth noting that stock-outs in a supply chain network can also generate opportunity or recourse costs. Stock-outs can lead to either lost sales or backorders (late deliveries). The cost of a lost sale not only includes the margin forgone on the sale but also the NPV of all future contributions to profits lost because of eroded goodwill. This cost is generally very difficult to estimate. Similar costs also arise with backorders. In some cases, emergency measures can be adopted to avoid lost sales or backorders. These recourses can assume different forms: prioritizing an order in the scheduling system, using overtime to increase output, paying for priority shipping services, etc.

2.2.3.3 Transportation Costs

Transportation costs are what a company spends to move materials between different sites in its supply chain network. The mode of transportation and type of transporter used affects these costs. Companies can operate their own vehicle fleet but this involves investment outlays (vehicles and garages), fixed operating costs (maintenance, insurance, permits), and variable costs (drivers, fuel, tires, repairs). Table 2.3 provides a breakdown of the annual cost of trucking in the United States.

Companies can also choose to work with third-party transportation services providers, and transportation costs then depend on the tariffs charged. For road transport, small loads are typically shipped on a less-than-truckload (LTL) basis whereas larger loads are shipped in full truckloads (TL). It is generally possible to achieve significant economies of scale (depending on the load size) for a given origin–destination lane and merchandise type. The typical structure of LTL transportation tariffs is illustrated in Fig. 2.11. Cargo size is generally expressed in kilograms (kg) or hundredweight[3] (cwt). The shipment weight obviously depends on the quantity Q_p, of product p being transported, and on the weight w_p of a unit load. The transportation cost function $c^{\mathrm{T}}(W)$ depends on the total weight being

[3]In North America a hundredweight is equal to 100 pounds (*short* hundredweight) and in Britain to 112 pounds (*long* hundredweight).

Table 2.3 The real cost of a commercial truck in the United States in 2011 *Source* The Truckers Report (2015)

Cost elements	Total cost/year	Cost/mile	Proportion (%)
Truck cab and trailer (amortized over 5 years)	$30,600	$0.24	17
Fuel (20,500 gallons a year)	$70,200	$0.54	39
Driver salary	$46,800	$0.36	26
Maintenance and repairs	$15,000	$0.12	10
Insurance	$6500	$0.05	4
Tires	$4000	$0.03	3
Permits, licenses and tolls	$3600	$0.02	2
Total	$180,000	$1.38	100

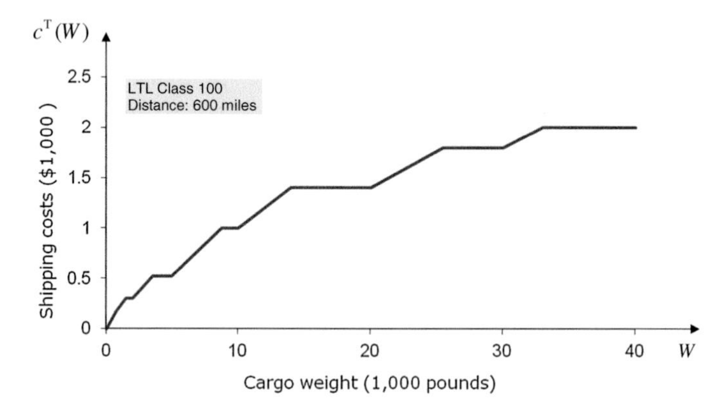

Fig. 2.11 Transportation costs as a function of load

shipped $W = \Sigma_p w_p Q_p$. The function is piecewise linear with a gradually decreasing slope to reflects the fact that as the load increases, it becomes possible to use more efficient handling/transportation means. Tariff structures are different for air, rail, road, and maritime transportation, but economies of scale are usually present. Chapter 5 studies the characteristics of these different transportation modes.

Design decisions usually require the estimation of supply chain network's transportation costs. In this context, the focus is on the costs generated on network arcs by the annual flow F of a product family. The situation is similar to the one described above for Eqs. (2.6) and (2.7). If the size of the transportation loads is predetermined (e.g., a full truckload), the cost for a load is fixed and transportation costs $C^{\mathrm{T}}(F)$ can be reasonably evaluated using a unit cost $\bar{c}$ based on historic data, as done for (2.8), or directly reflecting transportation company tariffs. Conversely, when shipping frequencies are fixed (fixed production cycles, fixed delivery schedules, etc.), the size of the loads being shipped depends on F. More specifically, assume that E shipments are made every year. This means that $W = (F/E)\bar{w}$, where $\bar{w}$ is the average weight of a product. The resulting transportation cost function is $C^{\mathrm{T}}(F) = Ec^{\mathrm{T}}(\bar{w}F/E)$ and it is not linear. As part of a long-term transportation contract negotiation, volume discounts could also be offered on the annual flows F on an arc or on a subset of the network's arcs.

2.2.3.4 Throughput Costs in Facilities

Another important aspect of an SCN design is the costing of production–distribution centers' activities. As illustrated in Fig. 2.9, the following are the main activities performed in these centers:

- Issuing orders to internal or external upstream suppliers, leading to reception, inspection, handling (between storage, production or shipping points) and storage activities;
- Receiving orders from internal or external downstream customers, leading to picking, handling, staging, shipping and service activities;
- Production/assembly of components/products, which usually involves upstream and downstream handling activities.

For discrete products, these activities lead to periodic lot-sizing decisions, which generate setup costs and variable execution costs. There are also a number of indirect costs, relating, for instance, to supervision and maintenance.

At an operational level, when a batch Q of products is ordered or manufactured, the variable costs generally incorporate material purchase, direct labor and energy consumption charges. In a production context, fixed costs include tooling and setup expenses. In a procurement context, they include order preparation, transmission, tracking, reception, inspection, and payment expenses that are independent of Q. If the company manages several facilities, regression analysis can be used to estimate this fixed ordering cost, with the facilities annual purchasing expenditures being the dependent variable and the number of orders issued the independent variable. The slope of the regression line obtained provides an estimate of the ordering cost. Note that the operational activity cost function $c_p(Q)$ for a product p is not necessarily linear. If, as discussed above, different equipment can be used to make a given product, the unit production cost for each piece of equipment will differ. Consequently, when rules are elaborated to specify the equipment to use for the production of different quantity Q of products, $c_p(Q)$ will likely be a concave function.

At the strategic level, for a plant, the aggregate annual throughput X of a product family is usually related to production decisions, as shown in (2.4), and what needs to be done to estimate an aggregate production cost function $C^{\mathrm{P}}(X)$ was already discussed. For distribution centers, the throughput X reflects either aggregate inbound receptions or outbound shipments, which, on a yearly basis, should be roughly equal. This requires the estimation of an aggregate throughput costs function $C^{\mathrm{D}}(X)$ encompassing all relevant reception, handling, and shipment

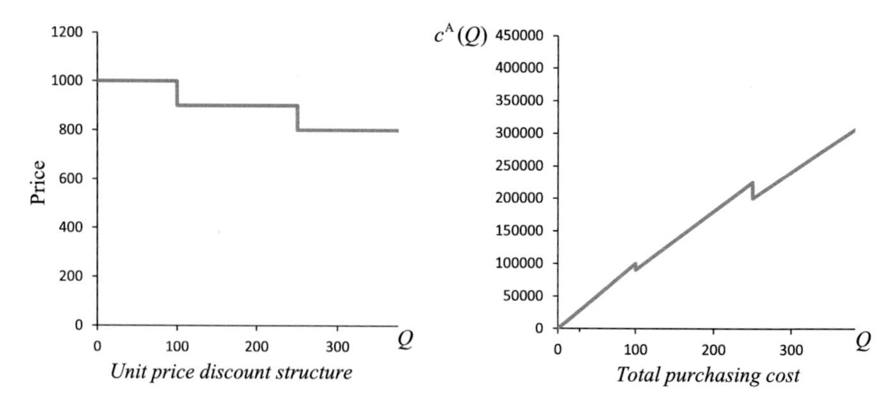

Fig. 2.12 Quantity discounts example

expenses. This function should reflect fixed lot-sizing costs as well as variable operational costs. Lastly note that as annual flows increase, it becomes possible to use better performing production/storage/handling technologies, occasioning possible economies of scale.

2.2.3.5 Purchasing and Sales Prices

Purchasing price structures are often complex because of quantity discounts, temporary low prices, or payment conditions. A typical quantity discounts example is illustrated in Fig. 2.12. The acquisition cost function $c^A(Q)$ represented is piecewise linear. If the price charged by the supplier for a product does not vary over time or as a function of the lot size Q being ordered, it is not relevant to the procurement decision. However, it is relevant if one has to choose between different supply sources. There are also situations where discounts are based on the annual purchase volume X, especially in contexts where supplies are provided through long-term contracts. Negotiating and managing this type of contract, however, does generate fixed costs. This being the case, the acquisition cost function $C^A(X)$ to use at the strategic level is usually concave and often piecewise linear.

This discussion was presented from a procurement prices perspective, but it also applies to the prices that companies charge customers for products sold. Sales price structures can be described by unit price functions $\pi(Q)$ and revenue functions $R(X)$ similar to the acquisition costs functions in Fig. 2.12. It is clear that these functions may also depend on other order winners like response times or service levels. An appropriate pricing policy can be a source of considerable additional revenues due to its impact on unit margins or market share.

2.3 Value Creation

We studied costs and revenues relevant to the optimization of supply chain networks. The total cost of a company's supply chain system depends on the nature of its primary activities, its planning and control processes, and the structure of its supply chain network. Chapter 1 discussed the processes and structures of supply chain systems in detail. We have also seen that the market share of a company, and ensuing revenues, generally depends on the order-winning attributes (price, quality, response times, flexibility, coverage, availability, etc.) the supply chain system can offer to potential customers. Mastering these concepts is important to value creation. The present section examines *value chains* and it shows how value creation results directly from the minimization of total supply chain costs and the elaboration of order winners that maximize corporate revenues. Figure 2.13 illustrates the impact of a supply chain network structure on the cost of value-creating activities and on response time, an order winner favored by world-class customers. Much of the following discussion relates to this figure.

2.3.1 Corporate Value Chain

The first chart in Fig. 2.13 describes how costs accumulate at each stage of a simple production–distribution network, thus forming a *value chain* (Porter 1985). The symbols used to describe the stages are the same as in Fig. 1.8. The chart shows how raw materials, production, warehousing, transportation, and storage costs relate to primary activities. The relation between the cost functions previously discussed and the cost added at each stage of the supply chain is illustrated using the notation introduced above. Support activity costs, long-term resource costs, and taxes are also added to obtain the total system cost.

The chart also illustrates the definition of EVA displayed in Eq. (2.5). It is worth emphasizing that the value chain is a snapshot of the situation of a company at a particular point in time. If the company alters supply contracts, if it adopts new production–distribution technologies, if it changes transportation means or if it modifies planning and control processes, the total system cost curve will change. Even if none of these elements is modified during a given time period, the cost curve can change due to learning phenomena.

2.3.1.1 Response Times and Inventory

Fig 2.13's second chart illustrates the relationship between a company's response times, the types of inventory it keeps (finished products, semi-processed goods, raw materials), and where it is positioned (in a factory, national/regional DC or store) in relation to the marketplace. As Chap. 1 discussed, the first inventory location in a

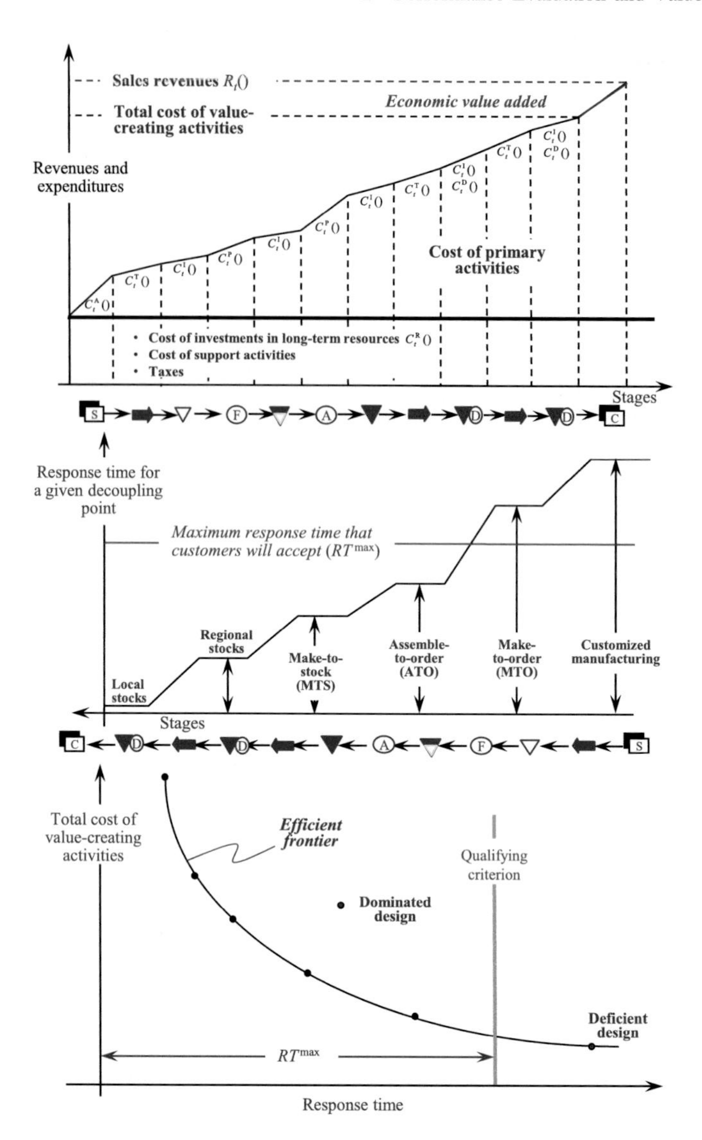

Fig. 2.13 Value chain and efficient frontier

supply chain is a *decoupling point* that determines response times. The closer a company keeps its stocks to customers, the shorter its response times. When a stock of finished products is kept locally, delivery times can be very short. On the other hand, if no inventory is kept (customized manufacturing) the delivery times will necessarily be longer, even if efficient and flexible production technologies are used. For a given company, the best supply chain strategy (local/regional stocks vs make-to-stock vs make-to-order, etc.) depends on several factors discussed below. Clearly, customer expectations are paramount. If the total customer order-to-delivery cycle time under a given strategy (say, make-to-order) is longer than the maximum response time RT^{max} that customers require, this deficient strategy cannot be used. For consumer goods like grocery, inventories must be kept in-store to ensure satisfactory service.

This raises questions about the role inventories play within a supply chain network. In just a few years, industry has gone from using inventories as a protection against all system defects to considering them as a calamity to be avoided at any price. It is true that inventories should not be used as a remedy for all system defects because, as claimed by partisans of the *zero-stock* paradigm, this hides operational problems and inhibits their solution. It is just as clear, however, that in certain contexts, inventories can create value. The zero-stock target can only be approached if there is no uncertainty about future demand and procurement/production lead times, and if demand is relatively stable. This is difficult to imagine, for instance, in the retail trade. Uncertainty and variability can be reduced by working on the environmental factors that create them (through strategic alliances with customers, suppliers, subcontractors, and third-party logistic providers, for example), but they can rarely be eliminated.

2.3.1.2 Efficient Supply Chain Networks

The first two charts in Fig. 2.13 reveal the impact of the structure of a supply chain network on costs and response times. Any potential SCN design is characterized by value and response time curves of this type. Each of the points on the third chart of Fig. 2.13 specifies the total cost and response time associated with a possible design. The line joining the points (designs) not dominated simultaneously on the cost and response time axes form an *efficient frontier*. This implies that several potential SCN designs (possibly including the status quo) are inefficient. For instance, a MTO supply chain using air freight to deliver massive low value products is unlikely to be efficient. The form of the efficient frontier depends on several factors, not all of which are controlled by the company. Before examining the impact of these factors, two important observations must be made, however. On the one hand, not all efficient designs are necessarily valid. For instance, the design to the right of the efficient frontier is defective because it does not enable the RT^{max} response time that customers require. On the other hand, all the designs on the efficient frontier respecting qualifying criteria provide a value-creating posture. Other elements, such as the curve's shape and the response times' impact on revenues, must be examined to reach a final decision.

The shape of the efficient frontier is influenced first by the nature of the products being manufactured/distributed and by the company's industrial environment. The value/weight ratio of the products sold has a crucial effect. A high value product that is light and not very voluminous (jewelry, electronic instruments, etc.) is expensive to stock but can be transported quickly for little cost. Conversely, a low value product that is heavy and voluminous (paper, cement, etc.) is expensive to transport but it can be held in inventory without incurring excessive costs.

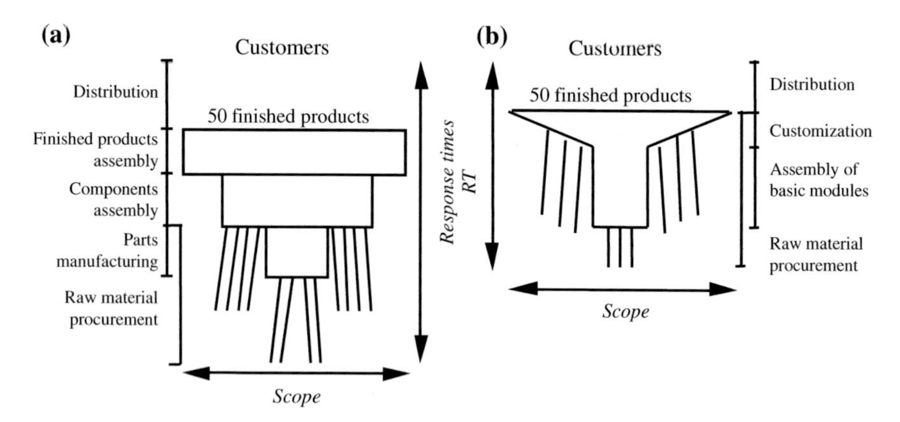

Fig. 2.14 Bill-of-material (BOM) of an assortment of products

Risks associated with the business environment and particularly with the variability of demand also have a significant effect. Where demand is stable and foreseeable, it is possible to provide a good service with very low inventory levels.

Most of the elements in a company's supply chain strategy described in Fig. 1.14, also affect the shape of the efficient frontier. Improvements in sourcing, production/distribution technologies, transportation means, or planning and control system push the efficient frontier to the bottom left. The scope of the assortment of products sold and the form of their bill-of-material (BOM) are also crucial. Figure 2.14 illustrates two extreme cases. The BOM on the left reflects the assembly of finished products from different components incorporating several distinctive parts. This results in very long response times and a need for highly diversified manufacturing, assembly, and sourcing processes. On the other hand, the BOM on the right looks like a mushroom. Here, the finished products are assembled at the very end of the process, using only a limited number of modules. Reducing components' variety makes it possible to manufacture modules in focused factories located where production factors are most advantageous, and it mitigates the negative impacts of uncertainty. This type of BOM leads to flexible supply chain networks with much shorter response times.

The repartition of the costs and response times associated with primary activities along the supply chain also affects the efficient frontier. Since inventory holding costs increase at each stage of the supply chain network (due to the value added during the preceding stages), it is desirable to keep inventory (if needed) before stages that add substantial costs, especially if this does not increase response times significantly. In addition, the value added at a particular stage depends on the economies of scale and scope associated with the technologies being used. Some technologies require significant initial investments (e.g., assembly lines) but they significantly lower marginal production costs. This provides a good reason for concentrating (focusing) production of all products/components requiring similar technological systems in a single facility, which also facilitates learning. When the

BOM is mushroom-shaped (Fig. 2.14b), locating final assembly operations close to the final market enables excellent response times. Economies of scale are also enhanced when components production is concentrated in a few focused factories.

2.3.2 Choosing the Design Maximizing Value Creation

Fig 2.13 shows that the structure of an SCN has a significant impact on two primordial order winners: low prices (via low costs) and quick responses. These are however not the only value criteria to consider (see Table 1.1) and the impact of all order winners on revenue generation is crucial to the final decision. Most order winners depend on the structure of the supply chain network, and the notion of an efficient frontier can be extended to analyze their impact on value creation. As illustrated in Fig. 2.15, the order winners offered by a company affect its revenues and costs, and consequently the value added by its supply chain network. When engaged in a SCN reengineering project, a company should therefore select the design maximizing long-term value creation, that is

$$V^{\text{SCN}} = \max V^{\text{SCN}}(RT) = \sum_{t=1}^{T} \alpha^t EVA_t^{\text{SCN}}(RT), \tag{2.11}$$

$$EVA_t^{\text{SCN}}(RT) = R_t^{\text{SCN}}(\tau, RT) - C_t^{\text{SCN}}(\tau, RT) \tag{2.12}$$

$$R_t^{\text{SCN}}(\tau, RT) = (1-\tau)R_t^{\text{SCN}}() \tag{2.13}$$

$$C_t^{\text{SCN}}(\tau, RT) = (1-\tau)\sum_{PA} C_t^{PA}() + \sum_{LTR} C_t^{LTR}(\tau,) \tag{2.14}$$

In these expressions, the label SCN (Supply Chain Network) is used to indicate that in a reengineering project, the only cash flows to consider are relevant revenues

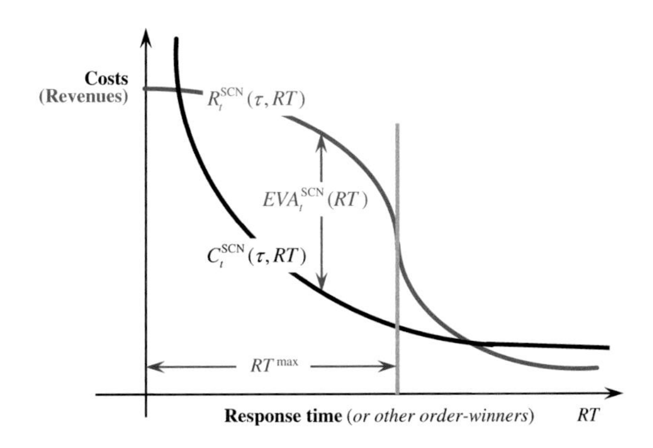

Fig. 2.15 Economic value added for a given period

and costs, and not all the company's inflows and outflows as in (2.5). The value is calculated over the planning horizon of T years considered in the design project. In the definition of $EVA_t^{\mathrm{SCN}}(RT)$, the revenue and cost functions, $R_t^{\mathrm{SCN}}(\tau, RT)$ and $C_t^{\mathrm{SCN}}(\tau, RT)$, depend on τ since tax effects must be considered. In expression (2.14), the index PA denotes the network's *primary activities*, and the costs $C_t^{PA}()$ correspond to the generic functions defined above for purchases, stocks, transportation, production, and throughputs (see the top chart in Fig. 2.13). The LTR index represents the network's *long-term resources* with $C_t^{LTR}(\tau,)$ referring to the generic cost function $C^{\mathrm{R}}()$ defined above. The LTR cost is a function of the tax rate τ since it must account for the tax reductions obtained for the ownership of resources.

The total sales revenue curve $R_t^{\mathrm{SCN}}(\tau, RT)$ provides the value that customers attach to the response times (and to other network-dependent order winners price, flexibility and service) given by the efficient SCN designs. Note that the shape of this curve reflects the fact that beyond a certain response time $RT^{\max}$, there is a sharp decline in customer interest. When delays are too long, if the company does not cut prices, demand and revenues decline. If it cuts prices, market shares may be preserved, but revenues still fall. If the revenue and cost curves are relatively flat, several designs are capable of producing more or less the same value-added, meaning that the company still has much leeway and that other factors must be considered to select a design.[4] One important factor to consider is the fact that the revenue curve is not only affected by the company's posture but also by the position of its competitors. A company wanting to penetrate a market that is already partially occupied by other companies has a revenue curve defined by residual demand and, to differentiate itself, it may be inclined to offer prices or response times that are different from its rivals. It might even find itself in a situation where all the most profitable positions are already taken.

The impact that the position of supply chain partners and competitors may have on a company's strategic posture[5] is illustrated in Fig. 2.16 for the forest products industry. Each actor covers a subset of the supply chain and it has its own total cost and response time curves. If a company's suppliers and customers have neither cost nor time advantages, or if they retain too high margins, they undermine the company's competitive position. Conversely, it is important to have an idea of competitors' costs and response times, even if they do not cover the same supply chain subset (competitors A and C in Fig. 2.16). This helps to identify the SCN segments that the company should improve. This could bring, for example, the company to subcontract activities that it does not do very well or, to the contrary, to develop new competencies that were not considered particularly important in the past.

---

[4]See Rosenfield et al. (1985) for a good discussion of the strategic implications of different forms of cost and revenue curves.

[5]Shank and Govindarajan (1993) show how to construct and use value chains covering the whole of an industry.

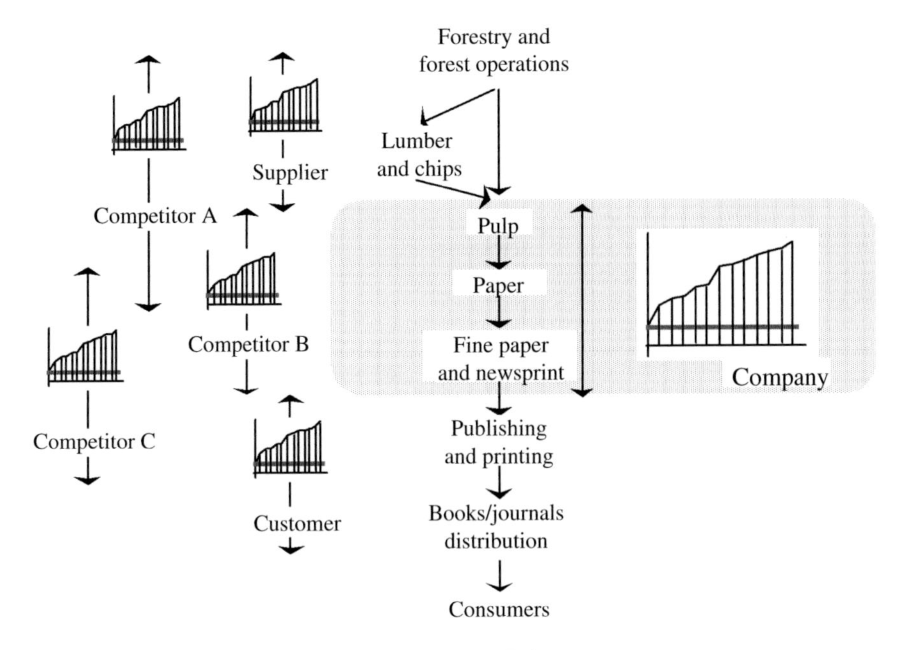

Fig. 2.16 Partial value chains for the forest products industry

2.4 Risk and Social Responsibility

The preceding discussion ignores the impact of risk. Since we want to design robust supply chain networks that will perform well for several years irrespective of the plausible future that will occur, it is clear that we have a decision problem under uncertainty. This needs to be taken into account when evaluating potential SCN designs. The cash flows to consider are at best random variables, and it is not sufficient to use forecasted cash flows (or the average of random flows) in the evaluation process. The probability distribution of relevant cash flows and the decision-makers' attitude toward risk should also be considered. Most decision-makers are risk averse and ready to pay insurance to guard against uncertainty. When they do take risks, it is because they are hoping to increase the gains they get in return.

In finance, investment project evaluations often get around this problem by using expected cash flows (forecasts) to calculate a project's NPV, while increasing the opportunity cost of capital to account for risk. Thus, when estimating the WACC, a risk premium is added to the return r_S that shareholders want to receive. This premium is not easy to evaluate, and it is hard to generalize this approach to choose between projects with different risk levels. Another approach derived from portfolio theory (Markowitz 1959) is more suitable for our needs. It suggests evaluating a potential decision using a performance indicator based on a compromise between its expected value and its risk, the latter being measured by the volatility of the random outcomes associated with the decision. In an SCN design context, this

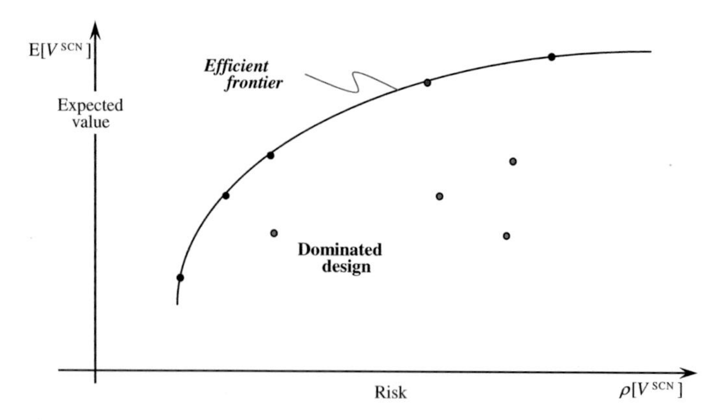

Fig. 2.17 Value–risk efficient frontier

means that when demand, price, or any other problem parameters are random variables, the value of the network V^{SCN} defined by relationship (2.11) is also a random variable. Let $E[V^{SCN}]$ be the expected value (average) of V^{SCN} and $\rho[V^{SCN}]$ a measure of risk for V^{SCN}. When considering alternative SCN designs, their expected value and risk can be plotted on a Cartesian coordinate plane as shown in Fig. 2.17. By taking the upper envelope of these points, an *efficient frontier* similar to the one discussed previously (Fig. 2.13) is obtained.

To identify the points of this efficient frontier, it is necessary to find the designs that maximize expected value for several risk level ρ^o in an interval acceptable to the company. For a set N of potential designs and a given value of ρ^o, this is tantamount to solving the mathematical program

$$\max_{n \in N} \; E[V_n^{SCN}] \text{ subject to } \rho[V_n^{SCN}] \leq \rho^o, \tag{2.15}$$

Equivalently, the efficient frontier can be elaborated by maximizing a linear combination of value and risk

$$\max_{n \in N} \; E[V_n^{SCN}] - \lambda \rho[V_n^{SCN}], \quad \lambda \geq 0, \tag{2.16}$$

for different weights λ. However, in order to be able to solve program (2.15) or (2.16), a suitable risk measure $\rho[V^{SCN}]$ must first be selected.

In his work, Markowitz used the *variance* as a risk measure but he was the first to recognize that this is not the best approach. Since the variance is defined by expression $E[(V^{SCN} - E[V^{SCN}])^2]$, it gives equal weighting to below and above average results, even though no one is against doing better than average. This explains the preference for downside risk measures, the most common of which are the *semivariance* $E[(\min\{0, V^{SCN} - E[V^{SCN}]\})^2]$ and the *absolute semideviation* $E[\max\{0, E[V^{SCN}] - V^{SCN}\}]$. Two other popular risk measures are *worst-case-risk* and *value-at-risk*. The former corresponds to the lowest value of V^{SCN} and it often leads to highly conservative decisions. The second is a quantile of the probability distribution of V^{SCN}. Regardless of the risk measure chosen, to use it in (2.15) or (2.16), one must be able to anticipate certain characteristics of the probability distribution of a design's value V^{SCN}. This generally requires the use of Monte Carlo simulation methods, a topic studied in detail in Chap. 10.

That being said, it should be clear that in order to design a robust and resilient SCN capable of creating value irrespective of the plausible future that eventually occurs, it is not sufficient to maximize expected value as suggested by (2.11). Instead, an efficient frontier first needs to be elaborated using program (2.15) or (2.16). Then, an efficient design can be selected based on the decision-makers' attitude to risk. The value–risk compromise found using this approach should provide an excellent SCN design for all business stakeholders. This design methodology may however be perceived as egotistical since it focuses on the interests of the company, without any explicit regard for its social responsibilities. Clearly, inasmuch as the order winners that customers value can include elements such as sustainability (see Table 1.1), the value creation objective causes the company to prefer all outcomes that customers consider important, which could include low environmental footprint, preserving natural resources, full employment, regional development, fair trade, ethical finance, and so on. Nonetheless, some companies wanting to act as leaders in this area may be willing to sacrifice some of their profits to contribute to the environmental, social, and economic development of the planet. According to a recent report of a corporate social responsibility (CSR) observatory in France (ORSE, www.orse.org), 64 % of world-class companies' rank sustainable procurement as a priority in their CSR policy. This report also indicates that 51 % of these companies set quantitative sustainable procurement objectives.

For companies wanting to display CSR leadership, the aforementioned sustainable value creation objective may no longer suffice. The additional performance indicators to consider depend, in principle, on the company's *ecosystemic* posture, although it may be quite vague. It often takes the form of a statement emphasizing certain socio-environmental values. Within an SCN design context, a company's socio-environmental performance can be improved in three different ways. The first is to artificially inflate certain costs used in decision processes to induce sustainable behavior. This involves, for instance, subjectively increasing shipping costs because of the negative effects of transportation on the environment. Another example is increasing factory shutdown costs to account for the significant social impact of closing a production center. The second avenue involves imposing explicit or implicit socio-environmental constraints on decision-making. It is possible, for instance, to include a constraint on greenhouse gases emission in SCN design models, or even to eliminate any option that is not eco-efficient from the very outset. The advantage of adding an explicit environmental constraint in a design model is that it can then be used to elaborate an efficient (value, eco-efficiency) frontier by proceeding along the lines outlined above in relation to risk. A third option is to consider the reengineering of SCNs as a multi-criteria decision-making problem. This approach will be examined in Chap. 12.

Review Questions

2.1. What is the conceptual relationship between *value drivers* and *performance indicators*?
2.2. What is a relevant cost? Suggest a few examples and explain them.
2.3. Explain the difference between fixed and variable costs using an example.
2.4. What causes economies of scale and scope?
2.5. How might diseconomies of scale be observed in a company?
2.6. What is ABC?
2.7. What is NPV?
2.8. The cost functions needed for operational decision-making and strategic decision-making are not the same. Can you explain why?
2.9. Estimate the annual inventory holding cost rate for your company or for a company of your choice.
2.10. Based on the preceding answer, estimate the annual inventory holding cost of all the products sold by the company you have chosen.
2.11. Why it is often difficult, in practice, to get the data needed to calculate supply chain costs?
2.12. Why is EVA important for the design of supply chain networks?
2.13. What is a value chain?
2.14. What impact does a (costs, response times) efficient frontier has on companies' supply chain strategy?
2.15. What impact does a (value, risk) efficient frontier has on companies' supply chain strategy?
2.16. How might efficient frontiers be elaborated in practice?
2.17. How could CSR be considered in an SCN reengineering project?

Exercises

Exercise 2.1 A company manufactures a product in batches. Every time the production of a batch is started, a setup cost of $350 is incurred. The controller estimates the variable production cost at $12 per unit. If the company produces a batch of 500 units, what is the total cost incurred? If 500-unit batches are always produced, what is the unit cost of this product?

Exercise 2.2 Considering that the variable cost for the first unit produced of an item is $10 and that economies of scale of 95 % apply (that is, economies of scale follow the power function $10x^{0.95}$), what is the total cost incurred for the production of 1000 units? What is the marginal production cost for the 1000th unit manufactured? What is the marginal production cost for the 2000th unit?

Exercise 2.3 Plans exist for a $10 million investment project. The company requires a return of 15 % on its project, and its corporate tax rate is 20 %. If the project involves annual spending of $250,000 for each of the first 5 years and revenues of $4 million over the following 10 years, what is its NPV? Is it profitable to invest in this project? Explain your answer.

Table 2.4 Transportation data

Origin	Destination	Distance	Weight	Cost
AL	UT	1530	852	$676.57
AL	UT	1530	1682	$1335.68
AL	UT	1530	1683	$1124.08
AL	UT	1530	4345	$2902.03
AL	UT	1530	4346	$2522.85
AL	UT	1530	8203	$4 761.84
AL	VA	613	783	$381.32
AL	VA	613	1639	$798.19
AL	VA	613	1640	$654.52
AL	VA	613	3935	$1570.46
AL	VA	613	3936	$1236.30
AL	VA	613	8338	$2618.97
AR	NH	1257	835	$667.92
AR	NH	1257	1773	$1418.22
AR	NH	1257	1774	$1259.19
AR	NH	1257	3965	$2814.36
AR	NH	1257	3968	$2235.17
AR	NH	1257	7963	$4485.56
AR	OR	1760	851	$735.01
AR	OR	1760	1680	$1451.02
AR	OR	1760	1681	$1220.74
AR	OR	1760	4346	$3156.07
AR	OR	1760	4347	$2744.70
AR	OR	1760	8199	$5176.85
CA	MA	2626	852	$889.66
CA	MA	2626	1681	$1755.30
CA	MA	2626	1682	$1476.80
CA	MA	2626	4346	$3815.79
CA	MA	2626	4347	$3318.07
CA	MA	2626	8199	$6258.30

Exercise 2.4 You need to buy or lease a forklift for a distribution center. The lease that you are considering involves payments of $2,000 at the beginning of each month for 36 months. The forklift can be returned at the end of 3 years without any extra cost. The purchasing price for the same forklift is $65,000, tax included. You estimate that the resale value in 3 years will be about $20,000. The company's average cost of capital is 10 %. If the two solutions are equivalent in terms of tax, insurance and maintenance,

(a) Which solution would you choose?
(b) Given that the only uncertain data is the forklift's resale value, what must it be for the two options to be equivalent?

Exercise 2.5 In the road transportation industry, it is well known that less-than-truckload (LTL) shipping costs reflect economies of scale. Using an American LTL-rate benchmarking tool, the costs, distances, and weights associated with a sample of shipments on origin–destination lanes were obtained. The data collected is available in Table 2.4.

Your assignment is to:

(a) Estimate a power function by regression (using Excel) giving the total cost of a shipment as a function of the weight of the load shipped and the distance between the origin and destination;
(b) Show the existence of economies of scale using this function.

Bibliography

Bender P (1985) Logistics system design. In: Robeson J, House R (eds) The distribution handbook. The Free Press, New York, pp 143–224

Bhagwat R, Sharma M (2007) Performance measurement of supply chain management: a balanced scorecard approach. Comput Ind Eng 53:43–62

Brimson J (1991) Activity accounting: an activity-based costing approach. Wiley, New York

Christopher M (2010) Logistics and supply chain management, 4th edn. FT Press, Upper Saddle River

Crundwell F (2008) Finance for engineers: Evaluation and funding of capital projects. Springer, Berlin

Gascon A, Martel A (1995) Le déploiement logistique des opérations. In: Martel A, Oral M (eds) Les défis de la compétitivité: Vision et stratégies. Publi-Relais, Montréal

Giard V (2003) Gestion de la production et des flux, 3rd edn. Economica

Horngren C, Datar S, Rajar M (2011) Cost accounting: A managerial emphasis, 14th edn. Prentice Hall, Upper Saddle River

Kaplan R, Norton D (1992) The balanced scorecard: measures that drive performance. Harvard Bus Rev

Kaplan R, Norton D (1996) Using the balanced scorecard as a strategic management system. Harvard Bus Rev

Krokhmal P, Zabarankin M, Uryasev S (2011) Modeling and optimization of risk. Surv Oper Res Manag Sci 16:49–66

Lambert D (1985) Distribution cost, productivity, and performance analysis. In: Robeson J, House R (eds) The distribution handbook. The Free Press, New York, pp 275–318

Lambert D, Burduroglu R (2000) Measuring and selling the value of logistics. Int. J Logistics Manag 11(1):1–17

Magee J, Copacino W, Rosenfield D (1985) Modern logistics management. Wiley, New York

Markowitz H (1959) Portfolio selection. Wiley, New York

Maskell B (1991) Performance measurement for world class manufacturing. Productivity Press, UK

Melnyk S, Stewart D, Swink M (2004) Metrics and performance measurement in operations management: dealing with the metric maze. J Oper Manag 22:209–217

Miller J, DeMeyer A, Nakane J (1992) Benchmarking global manufacturing. Irwin, Ontario

Peterson D (1969) A quantitative framework for financial management. Irwin, Ontario

Pohlen T (2005) Supply chain metrics. CSCMP Explores, 2

Pohlen T, Klammer T, Cokins G (2009) The handbook of supply chain costing. CSCMP

Porter M (1985) Competitive advantage. Free Press, New York

Rosenfield D, Shapiro R, Bohn R (1985) Implications of cost-service trade-offs on industry logistics structures. Interfaces 15(6):48–59

Ross S, Westerfield R, Jaffe J (2013) Corporate finance, 10th edn. Business and Economics

Schulze M, Seuring S, Ewering C (2012) Applying activity-based costing in a supply chain environment. Int J Prod Econ 135:716–725

Shank J, Govindarajan V (1993) Strategic cost management: the new tool for competitive advantage. Free Press, New York

Shrieves R, Wachowicz J (2001) Retained cash flow (FCF), economic value added (EVA^{TM}), and net present value (NPV): a reconciliation of variations of discounted-cash_flow (DCF) valuation. Eng Econ 46(1):33–52

Silver E, Pyke D, Peterson R (1998) Inventory management and production planning and scheduling, 3rd edn. Wiley, New York

The Truckers Report (2015) www.thetruckersreport.com/infographics/cost-of-trucking. Accessed 22 Mar 2015

Trigeorgis L (1996) Real options. MIT Press, Cambridge

Wu Z, Pagell M (2011) Balancing priorities: decision-making in sustainable supply chain management. J Oper Manag 29:577–590

Yucesan E (2007) Competitive supply chains: a value-based management perspective. Palgrave Macmilan, Basingstoke

Chapter 3
Supply Chain Planning and Execution

The effective management of supply, production, distribution, and sales activities is essential to develop a sustainable competitive advantage. As we discussed in Chap. 1, beyond the strategic development of SC network resources, it is through the planning and control (P&C) of material, information, and cash flows in its SCN that a company is able to offer the products and services wanted by consumers and to create value. This chapter studies material flow management concepts and systems. Because material flows are driven by activities involving costly inputs, human resources, equipment, and outputs, they necessarily give rise to cash flows. Also, the management of activities is necessarily accompanied by transactional, forecasting, and planning information. Consequently, although we focus on material flow management, this cannot be done without examining associated information and cash flows. Also, our concern in this chapter is not the study of detailed flow P&C methods or models for specific sub-problems (e.g., demand forecasting, production planning, and inventory control) but rather the development of world-class SCN planning and control strategies and systems. We also examine data sharing technologies, and software industry solutions. Because SCs may involve several manufacturing, distribution, and retailing companies, the standards, approaches, and systems available to facilitate collaboration between SC partners are given particular attention.

3.1 Planning Strategies and Processes

A mechanical analog of an SC (the system of pipes and stacks shown in Fig. 1.10) was introduced in Sect. 1.2.2 to explain fundamental P&C concepts, namely, capacity, volume, throughputs, and bottlenecks. It was also used to explain that, to respond efficiently to customer demands (i.e., to ensure that outflows match demand), the transportation, handling, fabrication, assembly, storage, and sales activities occurring along the SC must be coordinated. This is possible only if there

A. Martel and W. Klibi, *Designing Value-Creating Supply Chain Networks*,
DOI 10.1007/978-3-319-28146-9_3

is an exchange of information between controllable activities and external entities. Moreover, because of the variability and randomness of customer demands, procurement lead times, and so on, in practice it is usually not possible to get good results by simply reacting to events as they occur. To obtain superior performances, customer needs must be anticipated (forecasted) and system throughputs and volumes must be planned.

This analogy shows that material flows management in an SCN requires a set of P&C processes. It shows that capacity, volume, and throughputs are fundamental variables of any P&C system and that information is its raw material. In practice, several P&C *strategies* can be considered, depending on the following:

- The P&C *process* used, that is, how information is transmitted in the SC and what planning and control decisions are made
- The P&C *methods* (procedures, rules, algorithms) used, that is, how decisions on capacity, volumes, and throughputs are made and executed.

In a given industrial context, SC costs, the value offered to customers (see Table 1.1) and, in particular, SC lead times, order cycle times, and inventory levels depend directly on the P&C strategy deployed. As illustrated in Fig. 3.1, the SC of a finished product can be seen as a series of functional cycles among selling, distribution, production, and supply centers. Each of these functional cycles gives rise to a sequence of activities initiated by a triggering event (e.g., customer or order arrival, reaching the reorder level for a stocking point) and typically include setup (e.g., order entry, production scheduling), execution (e.g., picking, manufacturing, shipping), and closing (e.g., reception, payment) activities. Obviously, all these activities take time and the cycle time depends on the P&C strategy adopted. One wants the customer order cycle time to be shorter than the response time expected by customers and, ideally, in order to avoid unnecessary inventory, the total SC lead time (the sum of all the cycles time) to be as close as possible to the customer order cycle time. This requires an efficient transmission of information in the SC, which is now possible, thanks to modern communication technologies and standards.

The P&C strategy adopted by a company depends on the nature and practices of its industry as well as on its SC strategy (see Sect. 1.3.1). Some P&C strategies perform very well in an environment where demand is stable and predictable but poorly when demand is highly variable; some strategies are well adapted to process industries (chemical, food, etc.) but not to discrete manufacturing industries (aeronautic, automotive, electronic, construction, etc.). Some companies tend to use

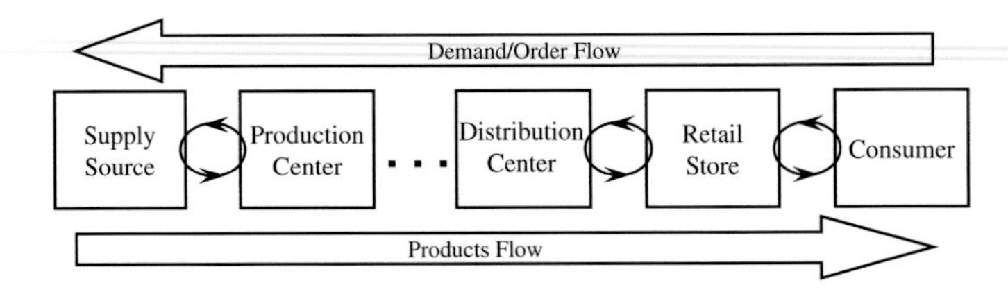

Fig. 3.1 Interconnected functional cycles in a supply chain

the dominant approach in their industry without questioning it (followers). Other companies, such as Dell (www.dell.com), Procter & Gamble (www.pg.com), and Walmart (www.walmart.com), tend rather to question common practices and to try to develop more efficient new P&C strategies (leaders). Professional or company associations such as the APICS supply chain council (www.apics.org)and GS1 (www.gs1.org) are continually working to improve industry P&C standards and practices. One thing is certain; in a given context some P&C processes and methods are more efficient than others.

3.1.1 Behaviors Induced by P&C Strategies

Two decades ago, Proctor & Gamble (P&G) examined the order behavior in its SC for baby diapers (Clark and McKenney 1995) and it found that, although retailer sales were relatively stable, orders received by DCs and production centers showed significant variations. This demand variability amplification phenomenon is illustrated in Fig. 3.2. When P&G examined its raw material orders to suppliers such as

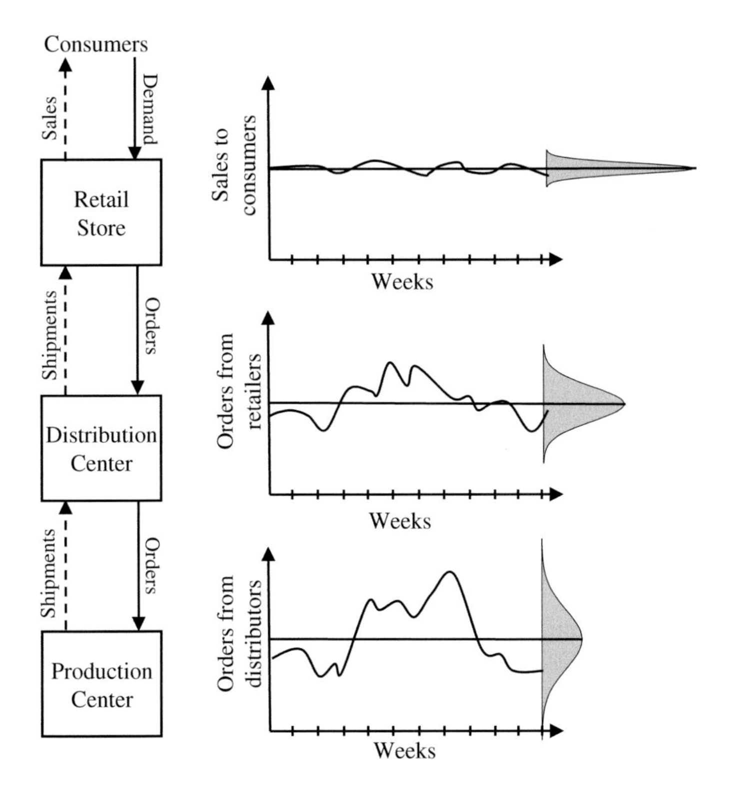

Fig. 3.2 The bullwhip effect

3M, it found that they fluctuated even more. They were observing what is now known as the *bullwhip effect*. This typically occurs when an SC uses a P&C strategy of the type illustrated in Fig. 3.1, that is, when the P&C decisions of each functional cycle are made independently of the others. When an activity center in an SC has no information on end-product sales, it must rely on the orders received from its downstream neighbor to forecast demand, plan capacity, control inventories, and/or schedule production. The typical consequences of this myopic P&C approach are excessive inventories, inadequate capacity, poor customer service, and frequent short-term adjustments to plans. None of this is desirable.

One of the first to observe this phenomenon was Forrester (1961) in his industrial dynamics studies. Based on Forrester's work the System Dynamics Group at MIT developed the *Beer Distribution Game* (Sterman 1989), which has since been used throughout the world to introduce students, managers, and executives to SC management concepts and in particular to the consequences of the bullwhip effect. Extensive research has been conducted (e.g., Chen and Lee 2012; Lee et al. 1997) on the bullwhip effect. The main causes and effects identified are the following:

- *Myopic forecasts*. In practice, SC managers often rely on orders received from their direct customers (downstream neighbor) to forecast demand. If these are not end consumers, they typically order less often than their customers and in larger quantities. When the series of order received is observed, it consequently gives the impression that consumer demand is much more sporadic than it is in reality. If raw material buyers proceed in the same way, they also contribute to this demand distortion phenomenon when they order from suppliers.
- *Long lead times*. When demand fluctuates, intermediate centers must keep safety stocks to be able to provide a good service. The safety stocks required depend on the variance of the demand during a procurement lead time (see Sect. 2.2.3). When lead times are long, more safety stocks must be kept. Moreover, if managers keep ordering while they wait for the reception of their orders, they eventually have too much stock and they tend to correct this by reducing their orders. When lead times are long, these overreactions increase and they generate oscillations in the demand of upstream centers.
- *Lot sizing*. If, whenever a consumer buys some items, the quantity sold was immediately reordered to the next center in the SC, all centers would have an exact vision (shifted by lead times) of consumer demand. Obviously, in the majority of real SCNs, such an approach is neither practical nor economic. In practice, to reduce order costs and to take advantage of quantity or truckload discounts, goods are ordered in lots of a non-negligible size. Unfortunately, for the aforementioned reasons, the lot sizing of orders contributes to the bullwhip effect.
- *Temporary low prices*. Promotional sales of all kinds are a widespread practice in several sectors (e.g., food industry). Planned promotions encourage members of the SC to buy in advance (forward buying) to take advantage of low prices. This increases logistics costs significantly and has been called the "dumbest marketing ploy ever" (Sellers 1992). When prices are reduced, consumers

(and other players in the SC) buy in quantities higher than necessary. However, when prices return to normal no purchases are made until accumulated stocks are depleted. These oscillations amplify the bullwhip effect.
- *Shortage gaming*. When demand for a product exceeds supply (new product more popular than expected, difficulties in raw material supply, SC breakdown, etc.), manufacturers tend to ration sales, and they allocate available products proportionately to orders received. Knowing this, clients not wanting to be out of stock tend to inflate their orders, hoping that the supplier will allocate an amount that corresponds to their actual needs. Later, when supply increases, large orders suddenly cease and several cancellations are made.

Past research also provides insights to try to eliminate the adverse consequences of the bullwhip effect. The main lesson learned is that one should avoid P&C processes that distort the information transmitted in the SC, thus giving wrong signals to upstream activity centers. Several guidelines were proposed to improve P&C strategies and systems:

- Make information on demand and inventories at consumption points available to all members of the SC (via an Intranet or the Internet, for example) so that they can base demand forecasts and plans on real consumer needs.
- Use P&C methods that enable the upstream actors in the SC to ensure that the needs of downstream members will be satisfied.
- Leverage information and communication technologies to reduce information transmission delays and delivery times. When the identity of end customers is attached to shipping units early in the SC, using barcodes or RFID chips, effective handling technologies capable of reading product and customer codes can be used to sort and route products in the SCN much faster, and often without the need to keep intermediate inventories.
- Use just-in-time methods to decrease upstream order sizes. To be able to continue to take advantage of full-load (truckload, carload, container, etc.) transportation savings, consolidate the shipments of several products (possibly from several suppliers).
- Avoid promotions as much as possible and move gradually toward everyday low prices (EDLP). Even if rebates are periodically offered to consumers, an unavoidable practice in some sectors, several upstream SC actors are now negotiating agreements with partners based on constant prices or annual volume discounts.
- When an SC actor has a shortage, allocate products available based on past sales rather than on orders received. This eliminates the incentive to distort information to try to obtain more products. Transmitting quality information on anticipated needs to upstream partners leads to the elaboration of better plans and thus to the reduction of shortages.

These guidelines emphasize the strategic importance of the choice of adequate P&C processes and methods. The approaches presented in the rest of the chapter build on the ideas introduced in this section.

3.1.2 Generic P&C Processes

Several types of P&C processes were devised and tested over the years. Figure 3.3 illustrates the essence of the most popular ones. The symbols used to represent SC entities in the figure are the same as those defined in Fig. 1.8. Sales and inventory information flows are displayed with dotted lines. As explained in Chap. 1, to satisfy customer demands, two basic P&C instruments are used: demand orders and execution orders. To avoid any confusion, in the figure, the former are called *orders* and the latter *commands*. Note that commands are not necessarily associated directly with orders. Some strategies are based on forecasting and planning processes that anticipate orders. A command for the production of a product lot can, for example, be given to build up inventory in anticipation of subsequent demand orders. The P&C processes schematized in the figure are explained briefly in the next paragraphs.

Independent control. This approach was examined in the previous section. It is used very often in practice, probably because it is simple to implement and it is the first idea that comes naturally to those who are not experts in SC management. Each activity is managed independently to respond to the orders received from its immediate predecessor. We explained in the previous section that this short-sighted approach should be avoided because it amplifies the bullwhip effect.

JIT (*Kanban*). This is the *pull* approach usually associated with the just-in-time paradigm. Whenever a product or a small product batch flows out of an activity in response to a customer order or to a command from a downstream activity, a production/transport command is issued to the next upstream activity. These upstream commands thus pull products through the network. This approach has two enormous advantages: it is very simple to implement (manually with cards called kanbans by the Japanese, for example) and it reduces the volume of inventory in the system to a minimum. However, for JIT to work well, the demand from external customers must be relatively stable and the SC very reliable. This reliability requirement is considered by many to be another significant advantage. In some contexts, however, market demands are not sufficiently stable and predictable to use this approach.

Coordinated P&C. This third approach seeks to follow the guidelines provided in the previous section by ensuring that information on sales and downstream inventories is forwarded to upstream activities. Each activity can then prepare forecasts and plans and issue orders and commands with a clear visibility of downstream SC echelons. A variant of this approach, known as the *Base Stock System*, was proposed by Sir George Kimball in the middle of the twentieth century (Magee and Boodman 1967). Its main problem at the time was that the technologies required to transmit information efficiently in the SC were not available so that it had been pretty much forgotten. It was later reinvented to support modern SCM paradigms, which we will look at more closely in the next section.

MRP/DRP. This approach, based on the Material Requirements Planning (MRP) (Orlicky 1975) and Distribution Resource Planning (DRP) (Martin 1983)

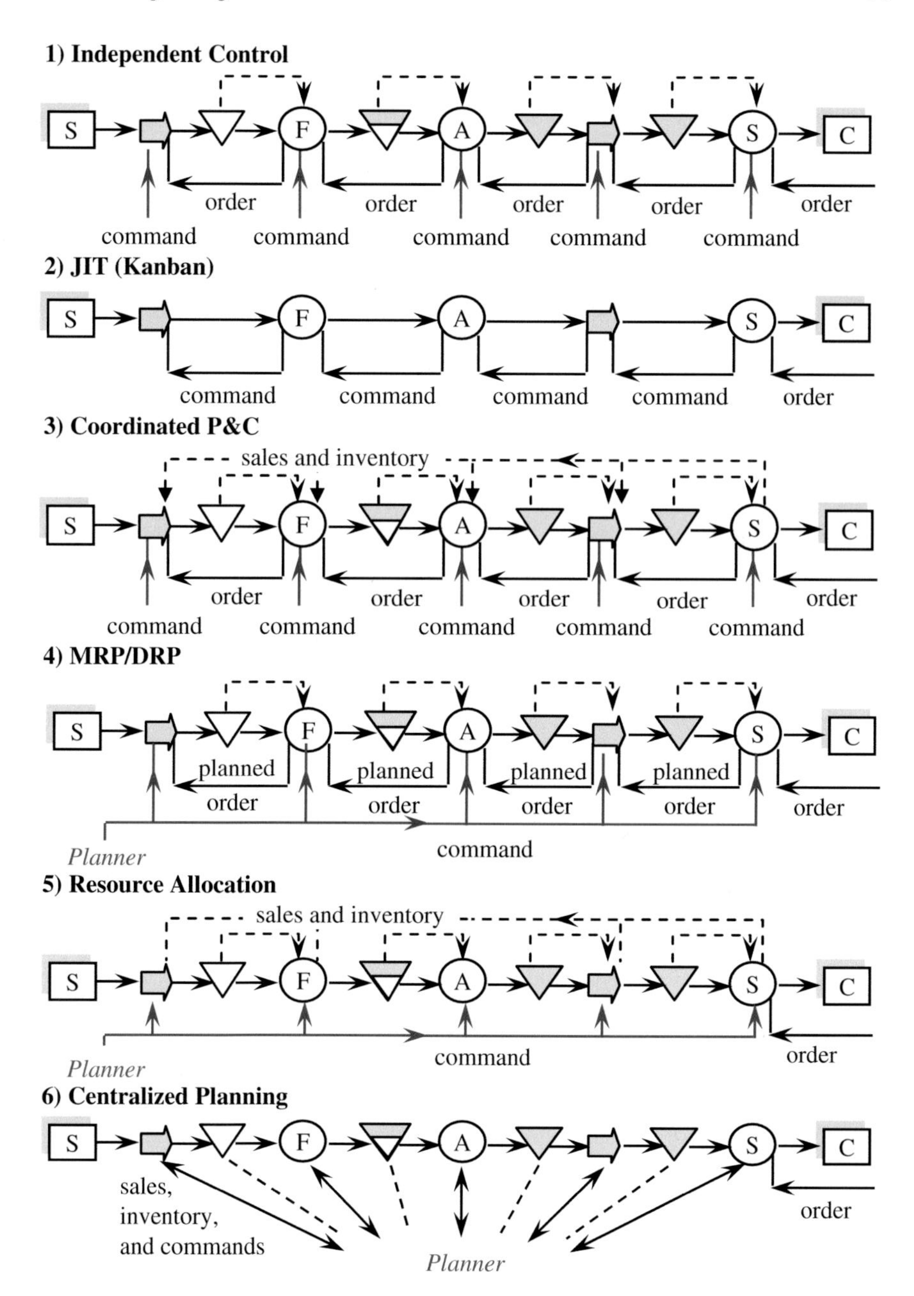

Fig. 3.3 Generic flow P&C processes

paradigm, also sends information upstream, but step by step through *planned orders* for a given planning horizon. These planned orders are based on a calculation of gross requirements that depend explicitly on orders (or planned orders) from downstream activities. When planned orders are available for all SC stages, a system planner verifies that they are doable with available SC resources. If they are, execution commands are given to downstream activities. If they are not, the planner modifies planned orders by trial and error until they are feasible and then issues downstream commands. Because commands are issued by a system planner, MRP/DRP is considered a *push* process. This approach involves the manipulation of a large amount of data, and it requires the implementation of a computerized system. It gives good results when customer demand varies from period to period but is known in advance (dynamic demand) and when customers can wait for a significant time before their orders are fulfilled, which is common for shopping goods and industrial goods.

Resource allocation. One of the problems of the preceding P&C approaches is that they do not consider capacity or volume constraints explicitly. In fact these processes were designed to plan and control the flow of each product separately, and they do not take into account the resources that are shared by several products. Some push approaches, however, incorporate shared resource allocation decisions. As for coordinated P&C, these approaches typically base their decisions on customer demand and downstream inventory information, but they also use resource allocation algorithms, step by step, to optimize execution commands. OPT (Optimized Production Technique), which is based on the theory of constraints (TOC) developed by Eli Goldratt (Goldratt and Cox 1984), is such a finite capacity planning approach. It is based on the allocation of critical bottleneck resources. The *fair share* distribution strategy proposed by Brown (1977) is also a finite resource allocation approach. Modern SCM solutions are often based on variants of this approach.

Centralized planning. This pushes the previous approach to the limit by taking all products, activities, and resources into account simultaneously. It involves the use of large-scale mathematical programming models and, for this reason, applies mainly to companies whose SCN incorporates only a moderate number of stages. It is found mainly in the process industry, and it usually requires the use of hierarchical planning methods (Hax and Candea 1984).

The generic P&C processes introduced in Fig. 3.3 are not mutually exclusive. They can be classified into two broad categories, push and pull, depending on how they fulfill customer demand. With pull processes, upstream execution commands are sequentially given in response to customer orders. Upstream activities are therefore performed under certainty. With push processes, commands are given in expectation of customer orders. Upstream activities anticipate demand and issue commands based on forecasts and plans, and the products purchased or produced are stocked pending demand. Most SCs incorporate a mix of push and pull processes. In many cases, the sales cycle only is in pull mode and all the rest of the chain pushes products. This is typical for consumer goods sold in retail stores or by e-tailers. Today, several companies also operate their distribution and assembly cycle in pull mode. In this case, the distribution cycle sometimes becomes unnecessary and it can be discarded.

Dell (www.dell.com) is a good example to illustrate these possibilities. The company started out by selling its computers directly to consumers on the Internet, and it is often cited as one of the initiators of the direct-only business model. There were merely three cycles in Dell's SC: sales, assembly, and procurement. The arrival of a customer order raised a custom computer assembly command. Components and subassembly procured from suppliers were stocked close to the assembly plants. This was the decoupling point (see Sect. 1.3.1) in the SC. Customer orders penetrated up to the assembly point, which was managed in a pull mode. Part inventories, however, were managed using a push approach. A decade ago, Dell revised its SC strategy to accommodate multiple distribution channels designed to offer specific order winners (speed, assortment, or price) to distinct market segments. One of these channels aims to provide configurable products to customers with specific needs, and the others provide pre-configured products either online via sales agents or through retail chains such as Best Buy. This new SC strategy is summarized in Fig. 3.4. Each retail channel now has a distinct decoupling point, and the global P&C system of the company had to be reengineered to accommodate all of them. Companies are now increasingly looking for P&C strategies that provide seamless supply for *omni-channel* retailing (Carroll and Guzman 2013), that is, for shoppers using all available channels (store, catalog, call center, web, mobile, etc.) simultaneously.

The described P&C processes are often accompanied by an SC management doctrine (just-in-time for pull processes, theory of constraints for OPT, etc.) and are perceived as distinct paradigms. Except for the first process (independent control), which is rarely the most appropriate, each of the other approaches is excellent in

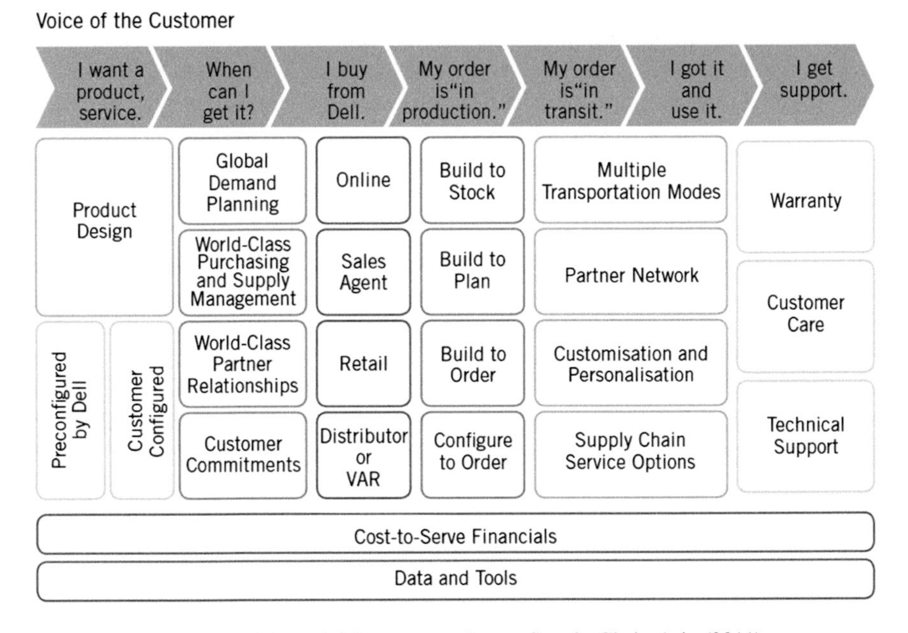

Fig. 3.4 Dell's new multichannel SC strategy—*Source* Supply Chain Asia (2011)

some contexts and inadequate in others. They all have their success stories and their horror stories. In a given context, the results obtained depend on the quality of the decision methods used with the selected P&C process. The elaboration of a winning P&C strategy involves not only the design of effective information-sharing processes providing end-to-end SC visibility but also the choice of efficient decision methods. Bad experiences often come from the fact that the P&C strategy adopted is not supported by an adequate information and decision system and not congruent with the enterprise organizational structure and culture.

3.1.3 SCM Best Practices

Today's dominant SCM paradigm has developed in the last three decades through a number of industry initiatives, such as *Quick Response* (QR) in the apparel industry, and *Efficient Consumer Response* (ECR) in the grocery industry, and through the implementation of improved planning and execution processes in world-class companies. Its basic principles were gradually introduced in previous sections and chapters. They include sharing information in the SC to improve the system visibility of all actors, reducing order/production lot sizes and total SC lead times to improve responsiveness and minimize inventories, pushing the order penetration point as far as possible in the SC to eliminate unnecessary intermediaries, and so on. SCs usually involve several manufacturing, logistics, and retailing companies, and applying SCM principles requires a close collaboration among all of them. Also, to implement best SCM practices, company organizational structures must be adapted and facilitating material handling, information, and communication technologies must be leveraged. This section examines some important organizational, material handling, and collaboration practices that support the implementation of SCM concepts. Facilitating ICT (Information and Communication Technology) standards and systems are studied in Sect. 3.2.

3.1.3.1 Consumer-Driven Category Management

A critical SCM challenge is to shape market offers and optimize order fulfillment processes to provide qualifiers and order winners for individual product-market segments. This requires visibility and synchronization across the SC and traditional functionally organized companies are ill structured to do this. Efficient SCM requires tearing down barriers between functional silos and managing categories of similar products using multi-functional teams to favor internal cooperation and collaboration with SC partners. *Category management* was introduced initially in the grocery industry to encourage supplier–retailer cooperation in finished product marketing and replenishment, but it has since been adopted in other retailing sectors, and even by purchasing organizations. A category is a group of products which meets similar consumer needs and that consumers perceive as being interrelated or substitutable. Sweet spreads, for example, could form a product category. Category management is a SCM process that involves managing product category as a strategic business unit (SBU), focusing on the creation of value for the consumer. The idea is to eliminate internal and external adversarial relationships and to favor SC collaboration.

The adoption of category management requires major strategic decisions as well as the implementation of appropriate business processes. These can take different forms. For example, a retailer may reorganize its buying and merchandising departments in category teams responsible for the integrated management of procurement, operations, and sales in collaboration with suppliers. Manufacturers may restructure their profit centers to no longer manage brands but rather categories of complementary products supported by policies and practices designed to maximize profitability. Finally, coordinated forecasting, planning, and replenishment processes must be jointly elaborated by retailers, distributors, and manufacturer to manage product categories. Categories thus become major resource (capital, time, space, etc.) users.

At the retailer level, categories are defined by identifying member products, structuring them into subcategories and segments, and specifying merchandising roles (e.g., destination, routine, occasional, seasonal, convenience) and tactics (e.g., turf protecting, traffic building, image enhancing, transaction building, etc.). For example, one could define a "pet care" category including all forms of foods and supplies for pets. This category could then be partitioned into the subcategories dog foods, cat foods, and pet supplies, and dog foods could be further subdivided into dry food, canned food, and treats segments. The retailer may then give a *routine* role to the category, meaning that its stores should be considered as a preferred provider of pet care products delivering consistent and competitive customer value. At the subcategory level, the retailer may want to use dog foods for turf protection and pet supplies for transaction building. Specific marketing and logistics plans are elaborated to comply with the category roles and tactics adopted. This involves fixing sale prices, allocating store space to products, designing shelves layouts (planograms), forecasting consumer demand, managing store and DC inventories, processing supply orders and payments, moving products in the SC, and so on. Suppliers, however, must develop strategies and business processes aligned with the category roles and tactics of retailers. This is not obvious because each supplier is serving several retailers and vice versa. Suppliers tend to align their categories with those of their major accounts.

3.1.3.2 Cross-Docking

When considering retailer–manufacturer links, as in the previous section, it is usually necessary to introduce an intermediate DC between the manufacturers' shipping points and the stores to be able to break the bulk shipments received from manufacturers, sort products, and consolidate store shipments. Depending on the SC strategy and P&C processes adopted, this DC could be owned by the retail chain, an independent distributor, or a 3PL, and it could hold inventory or serve essentially as a mixing and consolidation terminal. Under the SCM paradigm, which seeks to eliminate intermediate inventories, the latter is usually preferred. This necessitates *cross-docking* operations, that is, packing and labeling products at the source so that they can be easily sorted and consolidated into store shipments at the intermediate DC. Unit loads arriving at the DC are carried from the reception door to the outgoing vehicle dock without being inventoried. Reaching this level of operational excellence requires a harmonious blend of information, communication, and handling technologies. The handling system implemented can be highly mechanized, as illustrated in Fig. 3.5. However, it is not necessary to achieve this degree of sophistication to implement the concept. For instance, as discussed in Chap. 1, cross-docking is one of the distinctive competencies of Walmart, which operates large platforms with more than 1 million square feet and 5–12 miles of conveyor belts.

As illustrated in Fig. 3.5, most cross-docking terminals are long narrow rectangles but other shapes may be used. They have multiple doors (docks) where trucks (or trailers) can be loaded or unloaded. Incoming trucks are assigned to an unloading door. Then the unit loads are moved to a *stack* door and loaded on an outbound truck. Usually, there is no special equipment to stage freight. If products have to be stored temporarily, they are typically placed on the floor in front of the door assigned to the departing truck. Unit loads can also be reconfigured before they are loaded on vehicles. The following are the key requirements for cross-docking operations:

- Updating the database of the warehouse management system (WMS) when *Advanced Shipping Notices* are received from suppliers so that incoming product information is available to program consolidations

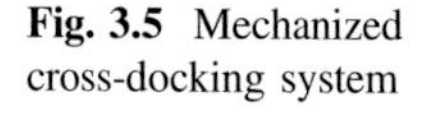
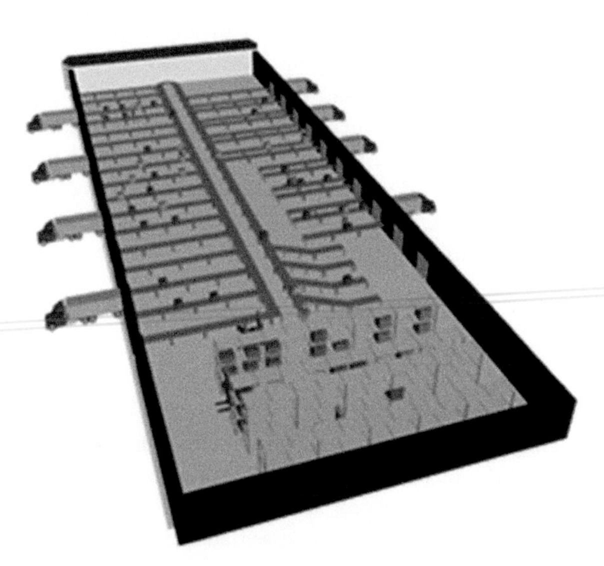

Fig. 3.5 Mechanized cross-docking system

- Optical reading of final destination identifiers on barcodes of unit loads (boxes, pallets, crates, etc.) received from suppliers so that they can be easily sorted
- The application of identification and routing barcodes to packages not already labeled when received to be able to route them to a staging or loading area
- The synchronization of the timing of deliveries to stores with the timing of receptions from suppliers to maximize transshipment and truck loading efficiency.

The product and location identification and data sharing standards required to do this are described in Sect. 3.2.1.

3.1.3.3 Collaborative SCM

Again, efficient exchange of information and collaboration among SC partners are required to apply SCM concepts. For retailer–manufacturer links, three related approaches have been developed to facilitate this: *vendor-managed inventory* (VMI)—or CRP (*continuous replenishment program*) as it is called in the grocery industry—*co-managed inventory*, and *collaborative planning, forecasting, and replenishment* (CPFR).

With VMI, the responsibility for the management of material flows between a supplier and a retailer is given to the former. In some cases, the retailer prepares demand forecasts and transfers them to the supplier, but in other cases forecasts are made by vendors. Forecasts are often based on DC shipment data, even if they should ideally be based on point-of-sale data. The main functions performed by each party and the data exchanged under a VMI approach are summarized in Fig. 3.6.

VMI does not remove the need to manage DC inventories; it simply transfers that responsibility to the vendor. Studies on the subject indicate that the same benefits can be obtained when inventory is managed by the retailer (Cachon and Fisher 1997). Gains do not depend on who makes the decisions but on the extent of the information shared by partners and on the synchronization of their operations.

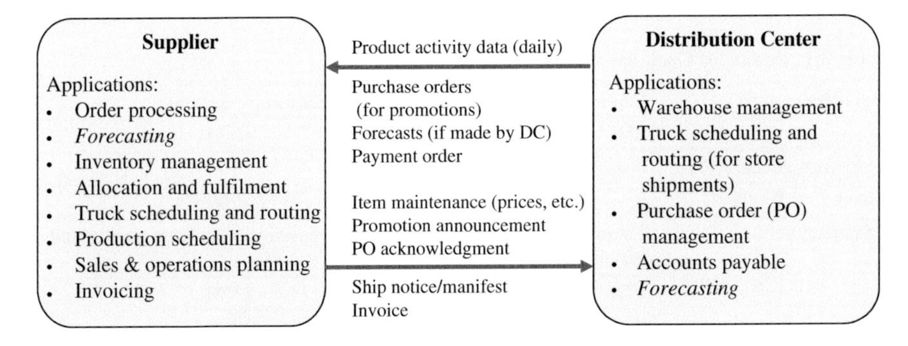

Fig. 3.6 Information interchange between supplier and DC under VMI

Improvements come from the use of decision support tools to generate optimal shipping schedules based on known store requirements, demand forecasts, forecast error statistics, planned promotions, stocks-on-hand at the DC and the supplier shipping point, required service levels, as well as transportation constraints (available vehicles, vehicles capacity, loading-routing restrictions, etc.). Conceptually, the joint management of product flows must consider the state, constraints, and objectives of the supplier–retailer subsystem, including any third-party involved (e.g., transportation resources provided by a 3PL). It may thus be under the control of the supplier or retailer or under the responsibility of both (*co-managed inventory*). In the latter case, required decision support and information-sharing tools may be provided by a service bureau or a 3PL.

The potential gains of VMI are significant. Clark and Hammond (1997) conducted a comparative study of several companies that adopted the approach. Part of the results obtained are summarized in Table 3.1, which shows the impact on inventory turns of the implementation of EDI (*electronic data interchange*), EDLP, and VMI for two manufacturers and two retailers. The results show that, although the use of EDI (or Internet) is necessary to implement VMI, it does not bring any substantial benefits in itself. It is only when EDLP and especially VMI are introduced that important benefits are obtained. Despite these significant paybacks, VMI is only a step toward the ultimate objective of SCM. Typical implementations have several limitations:

- Collaboration between partners is limited and replenishment plans are basic
- Forecasts are typically based on DC shipments and not on point-of-sale data, which may generate the bullwhip effect
- Manufacturers are rarely able to synchronize their production with retailer needs and they continue to produce to stock
- Manufacturers tend to give priority to VMI customers, which may generate shortages for other customers.

The most complete attempt to apply SCM concepts to retailer–manufacturer SCs is found in the work of the CPFR (*Collaborative Planning*, *Forecasting*, *and Replenishment*) workgroup of GS1 US (www.gs1us.org), formerly VICS (*Voluntary Interindustry Commerce Standards*). The generic CPFR framework developed by

Table 3.1 Results of Clark and Hammond (1997) study

Retailers	Estimated inventory turns increase from implementing:		
	EDI	EDLP	VMI
H.E. Butt Grocery Co.	0–10 %	30–50 %	100–150 %
Hannaford Brothers	0 %	50 %	100 %
Manufacturers	Estimated retailer inventory turns increase resulting from retailer and manufacturer implementation of:		
	EDI	EDLP	VMI
Procter & Gamble	0–20 %	40–50 %	100 %+
Campbell Soup Co.	0 %	20–30 %	50–100 %

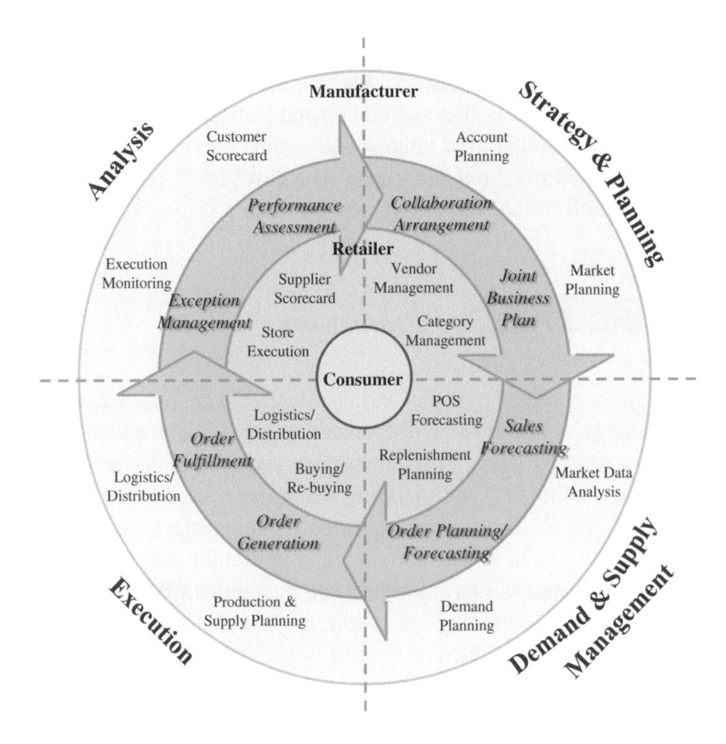

Fig. 3.7 CPFR framework for retailer–manufacturer SC (VICS 2004)

VICS is presented in Fig. 3.7. Its outer circle specifies the activities performed by manufacturers and its inner circle those performed by retailers. The intermediate circle identifies associated pairwise collaboration tasks. These are grouped into four major SCM processes:

- *Strategy and planning*. Establishing the roles of the partners in the collaborative relationship, designing P&C processes, elaborating product categories, and developing aggregate plans for coming months
- *Demand and supply management*. Preparing joint sales forecasts from point-of-sale data and demand shaping plans, and elaborating replenishment and production plans for coming weeks
- *Execution*. Manufacturing products, placing orders, shipping products, receiving and stocking products at DCs and stores, recording transactions, and making payments
- *Analysis*. Monitoring planning and execution activities, calculating scorecard performance metrics, and sharing insights to improve plans.

The customer is at the center of the framework. Partners aim to improve the whole SC and to add customer value. The partners have different capabilities based on their respective strategies and investments. They also have different sources of information and different perceptions of the market. Given that each partner has a partial vision of the SC, and thus a different perception of consumer needs, each can substantially improve demand forecasts and plans by exchanging data and business information without disclosing trade secrets. The shared demand plan elaborated becomes the basis on which all partners develop their internal plans, thus providing an effective way to synchronize activities throughout the chain. The flow P&C process (see Fig. 3.3) jointly selected is used to manage the SC as if it were a single business entity.

3.1.4 Distributed Planning Frameworks

SCNs are extremely complex systems and, as illustrated by Fig. 3.7, independently of the P&C strategy used, several interrelated planning and execution decisions must be made to manage them. Also, managers and executives at different levels have different responsibilities, and planning structures must be designed to fit with organizational structures. This is usually done by adopting a hierarchical or distributed planning approach, the former being a particular case of the latter. Two complexity reduction dimensions are vital when designing a P&C system: planning horizon hierarchies and functional interrelations. The length and granularity of a planning horizon is directly related to the responsibility level of the planner. At the strategic planning level, multiple-year planning horizons are usually considered, but at the operational planning level multiple-day or multiple-week horizons are typically used. The link among planning levels, horizon lengths, and the nature of the planning decisions made is illustrated in Fig. 3.8. Note the downward arrows indicating that the decisions made at one level result in instructions to be followed by the next level. Note also the upward arrows indicating that decisions at a given level depend on data obtained from lower levels and on an anticipation of how lower levels will eventually react to the given instructions.

The SCN design problems addressed at the strategic level were introduced in Sect. 1.2.1 and the following chapters will examine them in detail and propose models and approaches to solve them. The *sales and operations planning* (S&OP) problems addressed at the tactical level are dealing mainly with the allocation of the resources acquired and deployed at the strategic level to product-markets, and with the preparation of aggregate monthly demand, distribution, production, and supply plans. The decisions made at this level require aggregated product, customer, demand, capacity, and time period information. They are typically based on monthly demand forecasts or plans by product families and demand zones, and they involve the preparation of plans at the activity center level. Typical tactical planning deliverables include the following:

- Demand shaping decisions and aggregate sales forecasts and plans
- Allocations of plant capacity to products in order to meet demand plans
- Mid-term capacity decisions related to manpower levels, overtime budgets, and subcontracting
- Target inventory levels for stocking points

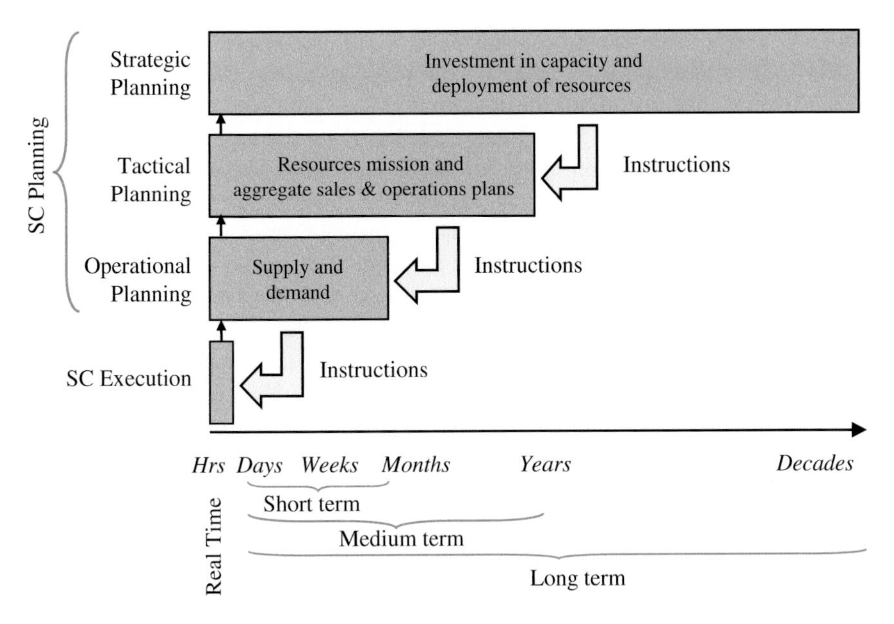

Fig. 3.8 SC planning levels hierarchy

- Assignment of customer ship-to points to distribution centers
- Shipping volumes on SCN arcs, and transmission of planned transportation capacity requirements to third-party logistics (3PL) providers
- Planned raw material and subassembly requirements, and transmission of supply plans to vendors and contract manufacturers

Operational planning and execution decisions are intimately related to the P&C strategy used. For retail chains, the order fulfillment process is usually relatively simple, but for manufacturing firms, it may involve sophisticated order-promising processes (e.g., ATP—available-to-promise, CTP—capable-to-promise). At the production-level detailed production lot-sizing, scheduling, and shop floor control decisions are made. Distribution activities typically involve shipment loading and routing decisions, order-picking, and tracking processes. Inventory control decisions are also made for all stocking points along the SC.

Functional interrelations are directly linked to the functional cycles of the SC (Fig. 3.1) and to the P&C strategy adopted. They relate, more specifically, to how internal responsibility centers cooperate to manage the SC and to how they collaborate with external SC partners. We saw in Sect. 3.1.2 that, between the *independent control* approach in which all activity centers make local decisions at one extreme and the *centralized planning* approach, which tries to optimize all activity plans simultaneously at the other extreme, several intermediate strategies are possible. A given strategy is naturally associated with a set of specific functional decision problems at the tactical and operational levels, and some of them are much easier to solve than others. However, as discussed in Sect. 3.1.1, an oversimplification may lead to poor

P&C and induce undesirable behaviors such as the bullwhip effect. A good P&C system is thus necessarily an adequate compromise among effectiveness, ease of implementation, and organizational fit.

A simplified distributed P&C framework based on the type of bi-dimensional organizational decomposition approach discussed is shown in Fig. 3.9 for a pulp and paper company. The planning modules in P&C frameworks are often supported by descriptive and normative models. The former are used for forecasting, rule propagation, and simulation purposes and the latter to optimize plans. The arrows in the framework represent instructions received from the upper level (e.g., available resources, aggregate plans, inventory targets) and downstream (e.g., demand forecasts, planned requirements) modules, as well as data (e.g., costs, resource usage, delays) and anticipated reactions from the lower level and upstream modules. The performance indicators used for a module depend on the mission and priorities of the associated responsibility centers, as explained in Sect. 2.1. In the decision models, plans are represented by decision variables, instructions become constraints, and performance indicators shape the objective function. Additional decision variables, performance indicators, and constraints must also be defined to anticipate the reaction of lower upstream modules (responsibility centers) to potential plans. This means that the decision models formulated to support planning modules are an amalgam of a planning sub-model and an anticipation sub-model.

To make these abstract concepts more concrete, consider the strategic SCN planning (design) module of a P&C framework, which is the main focus of this book. Once an SCN design has been implemented, the resources (suppliers, facilities, equipment, etc.) deployed are used on a daily basis to perform operations such as sales, warehousing, transportation, production, procurement, and so on. In fact, the revenues and expenses generated by an SCN over its useful life are directly

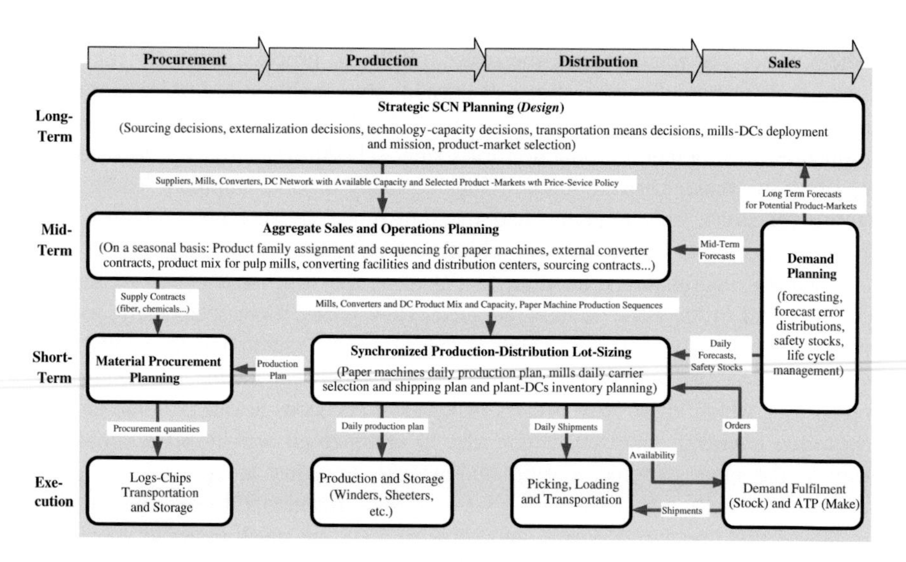

Fig. 3.9 Distributed P&C framework for a pulp and paper company

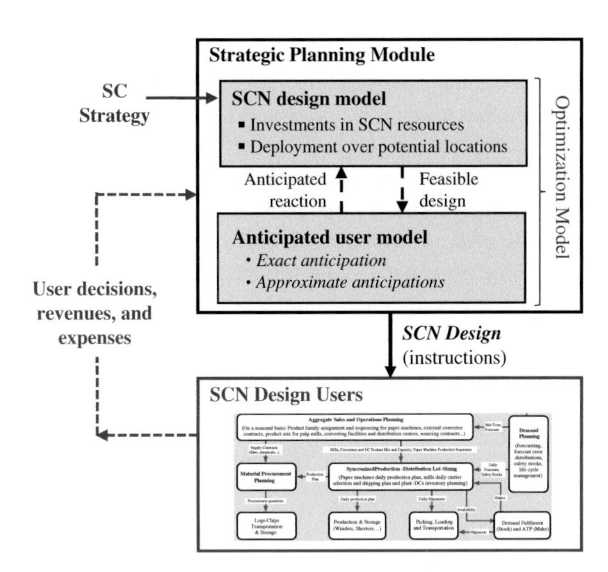

Fig. 3.10 Operations anticipation in distributed planning models

related to these *user* operations. Thus alternative SCN designs cannot be evaluated without anticipating how operations will eventually be performed. More specifically, this anticipation pertains to how operational revenues and costs are captured in the optimization model used to generate and evaluate alternative SCN designs. Figure 3.10 illustrates these concepts and it shows how a module in a planning framework depends on other interrelated modules (user modules).

Given the complexity of user decisions, at least for strategic and tactical plans, it is rarely possible to include an exact anticipation of user reactions in a planning model. Approximate anticipations must typically be used, but the precision of the anticipation has an impact on the quality of the plans obtained. For example, the basic location-allocation models that will be introduced in Chap. 7 to support SCN design decisions are single-period (typically a year), deterministic, mixed-integer programs, and they anticipate operational costs using SCN flow variables aggregated over yearly periods, product families, and demand zones. This is a gross approximation and we will see, in Chap. 11, how it can be improved using stochastic programming techniques. The notion of anticipation was studied in depth in a distributed decision-making context (Schneeweiss 2003) and in an SCN design context (Klibi et al. 2010).

3.2 Enterprise Planning and Execution Systems

In practice, the application of a P&C strategy generally requires the implementation of planning and execution systems (PES). Strategies are concepts and, to obtain concrete results, companies must convert them into a set of operational instruments (policies, plans, procedures, retrieval and transmission mechanisms, databases, decision support tools, etc.). Also, the demand and execution orders triggering actions in the SC generate a high volume of transactional data that must be manipulated efficiently. Nowadays, enterprise PES are typically computerized and they constitute an important part of an enterprise application software (EAS) portfolio. In this section, we examine P&C from a technological system perspective. More specifically, we look at the basic functionalities of P&C systems and at the type of P&C software currently offered on the market. However, before we do that, some essential facilitating technologies must be examined. We explained that material flows in an SC are managed using information. Information must be attached to products and unit loads, and messages must be exchanged between SC actors. This requires standardized physical entity codes, labeling symbolisms, and message interchange protocols, as well as technologies for the attachment of data to physical objects, for encoding and reading this data, and for the transmission of coded messages. We thus start by examining these standards and technologies.

3.2.1 Facilitating Technologies and Standards

As illustrated in Fig. 3.11, several communication standards and technologies are required to support commercial transactions and flow planning in global supply chains. Four main types of standards and technologies can be distinguished:

- *Entity identifiers*. These are codes for the identification of products, logistics objects (cases, pallets, containers, etc.), and places. They are structured to include check digits for validation purposes.
- *Labeling technologies*. These include encoding symbolisms, such as data bars, used to label products and logistics objects, as well as data carriers, that is, media that can hold symbolic data (barcode, datamatrix, RFID tag).

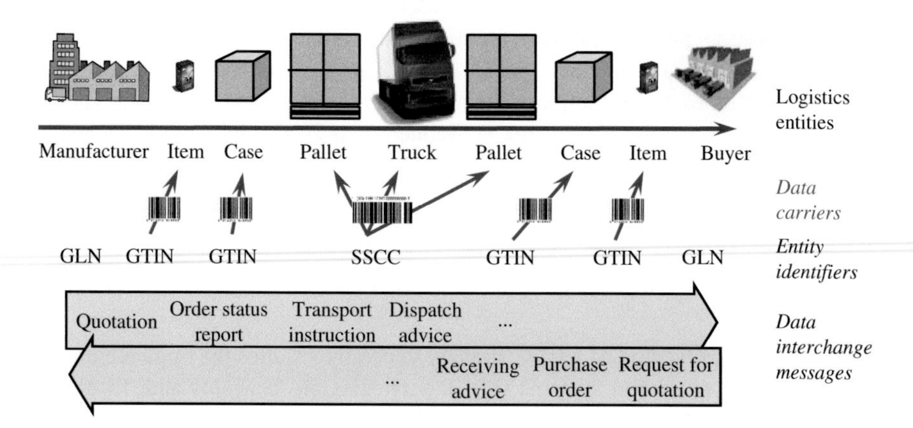

Fig. 3.11 Physical and information flows for a supply lane

- *Data sharing messages*. These are international communication standards for data sharing between SC partners (EANCOM, ASC X12, XML). They are accompanied by data synchronization mechanisms.
- *Information highways*. These are *Intranets* or *Extranets* powered either by the Internet or by private telecommunication networks and used to transmit standard commercial messages between SC actors.

Initially, high volume electronic data interchanges (EDI) were performed mainly between large SC partners using value-added networks (VAN). Nowadays, data sharing can be done between actors of all size using the Internet. Internet is part of our everyday life, and we assume that it does not require any further discussion. The following paragraphs thus concentrate on the three other facilitating technologies listed. To be effective, these technologies must be based on international standards. These standards have evolved dramatically over the past decades. In 2005, *EAN International* and the *Uniform Code Council* (*UCC*) merged to create a single global body for the elaboration of standards: *GS1* (*Global Standard 1*). This not-for-profit organization is now serving 150 countries either from its global office in Belgium (www.gs1.org) or from its 107 national member organizations (e.g., www.gs1.fr for France, www.gs1uk.org for the United Kingdom, www.gs1us.org for the United States).

3.2.1.1 Products and Logistics Objects Identifiers

GS1 operates a system of global keys and application codes that can be used to uniquely identify products, unit loads, assets, and locations in an SC. Encoding is the first step of any collaborative commerce or traceability initiative, and its importance and difficulty must not be underestimated. The main identification keys available to name and distinguish logistics entities are the following:

- *GTIN* (*Global Trade Item Number*). This key is used to uniquely identify any distinct trade item so that it can be recognized at any point in an SC, via a data base query, for example. Each product has its GTIN, but the GTIN is also used to identify the content of unit loads (case, pallet, etc.). A *serialized* GTIN can also be employed to trace specific instances of a product when required. The allocation of a GTIN to a product is the responsibility of its manufacturer. Four GTIN variants are available: GTIN-14, GTIN-13, GTIN-12, and GTIN-8. The GTIN-13, for example, is a 13-character code including this data:
  - A six-digit company prefix; these identifiers are assigned by the GS1 member organization in the country of the manufacturer
  - A manufacturer-controlled six-digit item number; this code must be unique for each distinct consumer product
  - A control digit calculated from the other numbers for validation purposes
- *SSCC* (*Serial Shipping Container Code*). When a unit load (carton, crate, pallet, etc.) is moved or stocked in the SC, it is identified with an SSCC. This code

enables a unit load to be tracked individually through the SC. Thus, two pallets of a given product would have the same GTIN, because their content is the same, but they would receive two distinct SSCC if they are shipped to different locations. The SSCC includes the following data:

  - An application identifier (AI), that is, a two- or three-digit code specifying the structure of the following data in a data carrier (e.g., the AI (00) at the beginning of a bar code indicates that an 18-digit SSCC data structure is following)
  - An extension digit used to increase the length of the serial reference number
  - A company identification code
  - A unique serial reference number allocated by the originator of the unit load
  - A control digit

- *GLN* (*Global Location Number*). The GLN is used to identify locations and legal entities. A location can be a building (e.g., plant, warehouse), but it can also be an area within a building (e.g., storage area, store shelf). A legal entity can be a company, a division, a department, and so on. The GLN includes the following data:

  - An application identifier
  - A company identification code
  - A unique location reference number allocated by the location owner
  - A control digit.

Several other, more specialized, codes are also available:

- GRAI (*Global Returnable Asset Identifier*) to identify and track containers or reusable transport equipment
- GIAI (*Global Individual Asset Identifier*) to identify fixed assets
- GSRN (*Global Service Relation Number*) to identify a type of relationship between business partners
- GSIN (*Global Shipment Identification Number*) to identify a grouping of unit loads (bill of lading) for transportation purposes.

As indicated, application identifiers (AI) can also be used to allow the inclusion of several distinct codes on a same data carrier (bar code, RFID tag). They are simply predefined codes indicating the content and format of the following data field. AI examples are 00 for a SSCC, 01 for a GTIN, 12 for a due date, 21 for a serial number, 400 for a purchase order number, and so on.

The previous identification keys were originally designed to be barcoded and they are not able to fully exploit the capacity of RFID tags. For this reason, GS1 is now developing an *electronic product code* (EPC) fully compatible with previously defined codes but also including additional features to facilitate SC traceability and SCM solutions. The development of this new standard is leaded by EPCglobal (www.gs1.org/epcglobal).

Ensuring that all businesses use a same codification system and that the codes defined are known to all and updated continually is a considerable challenge. To address this, GS1, in collaboration with its partners, has developed the GDSN, a *Global Data Synchronization Network* (www.gs1.org/gdsn). The GDSN includes three basic elements:

- A *Global Product Classification* (GPC) to ensure that products are classified correctly and uniformly and to provide a common language for all SC actors. A four-level product hierarchy is used based on segments, families, classes, and bricks. The foundation of the classification is the brick and every GTIN is assigned to a single brick. For example, the brick *Meal Replacement* is part of the class *Dietary Aid*, which in turn is part of the family *Health Enhancement* in the *Healthcare* segment. Bricks can also be characterized by attributes (e.g., the attribute *base of meal replacement*, with possible values *carbohydrate*, *fiber*, *protein*, or *unclassified*). The latest version of the classification can be explored using the GPC Browser (gpcbrowser.gs1.org).
- Certified *Data Pools*. These are electronic catalogs serving as a source or destination of standardized product data (stable data such as product GTIN, manufacturer GLN, GPC, product name, price, dimensions, picture, etc.). They are operated by GS1 national organizations or by solution providers. For example, two large independent data pools, Transora and ECCnet, merged in 2005 to create 1WorldSync (www.1worldsync).
- The GS1 *Global Registry*, a proprietary network integration and integrity directory that guarantees the uniqueness of registered products and actors. It ensures that all the data pools in the network comply with standards and validation rules.

The functionalities of the GDSN are illustrated in Fig. 3.12.

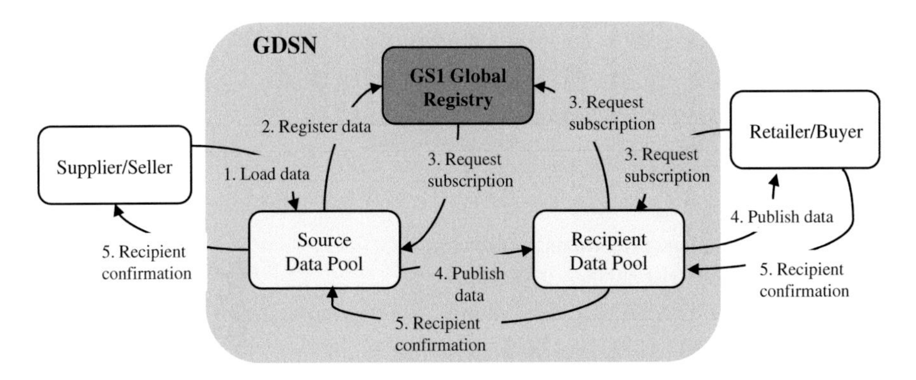

Fig. 3.12 GDSN synchronization mechanism (www.gs1.org/gdsn)

3.2.1.2 Labeling Technologies

Clearly, the previously defined codes are of no use if they cannot be attached to products or logistics objects and follow them in the SC. Nowadays, the main media used to attach data to physical entities are databars, datamatrices, and radio-frequency identification (RFID) tags. Several variants are available for a different use and we examine them briefly thereafter.

Barcodes such as those illustrated in Fig. 3.13 are a succession of light and dark bars of variable thickness encoding a series of numeric or alphanumeric characters based on a standard symbology and intended to be scanned by optical devices. Several types of barcode are available. The three most common are the EAN-13, ITF-14, and GS1-128. EAN-13 barcodes are used primarily to display the GTIN-13 on consumer products. They accommodate only numeric data, but they have omnidirectional scanning capabilities required for retail outlets. ITF barcodes are also used to display GTINs, but they are easier to read in coarse and dirty conditions. The GS1-128 standard is necessary when using application identifiers to include several codes or data fields on the same databar, as illustrated in Fig. 3.13d. Other specialized standards available include smaller stackable barcodes used for marking loose products such as fruits or variable measure items (e.g., meat, cheese) in a grocery store.

Unlike databars, datamatrices such as the one illustrated in Fig. 3.13e are two-dimensional symbologies that can be used to encode a lot of data in a compact space. They can be applied directly on objects using etching or laser-engraving technologies; however, they can only be read by camera-based scanners. Obviously, barcodes and datamatrices would be useless without equipment to read them. We are all familiar with the different types of scanners used in retail points. There are also a wide variety of industrial equipment that exploits this technology to

Fig. 3.13 Databar and datamatrix examples

facilitate the storage and handling of goods in SCNs. For example, the use of barcodes that can be read by handling equipment enables the cross-docking of packages on conveyors in distribution centers.

RFID tags are called to replace barcodes in the future, but their adoption for SC execution and management has been slower than expected. The technology was developed in the 1950s, but it is only in the last two decades that its commercial applications have proliferated. RFID tags contain at least two parts: a microchip for storing and processing information and for modulating or demodulating a radio-frequency signal and an antenna for receiving and transmitting the signal. There are several categories of RFID tags. Some of them are passive; they have no battery and to operate they rely on the radio energy transmitted by the reader. Others are active; they are powered by a battery and they periodically transmit an ID signal. Also, some tags are read-only and they have a factory-assigned ID, and others are read–write and they can memorize object-specific data captured by users. Depending on the type of tag, their price may vary between \$0.10 and \$100, which is still relatively expensive. This is one of the main factors limiting their large-scale adoption. RFID data can be read by passive or active readers (radio transmitter or receiver) connected to an antenna, as shown in Fig. 3.14. Passive readers can read only active tags.

RFID tags have several advantages over barcodes. Among other things, because RF signals can pass through solid material, they can be read at a distance even if they are inside a package or not visible. Also, several tags can be read simultaneously so that an entire container can be scanned in the blink of an eye without having to unpack individual products. Finally, RFID tags can provide the exact position of products, and they are relatively insensitive to the environment (dust, humidity, etc.). In 2003, Walmart was one of the first companies to implement the technology in its SC. In 2009, however, the project was abandoned because it did not provide the expected value added. Walmart wanted to be able to take instantaneous photographs of store and DC inventories. They found that radio frequencies do not pass well through liquids and metal, and they could not get the reliability level originally expected. However, the technology is continually improving and becoming cheaper, and a wide adoption will inevitably come. EPCglobal (www.gs1.org/epcglobal) is leading the development of RFID standards.

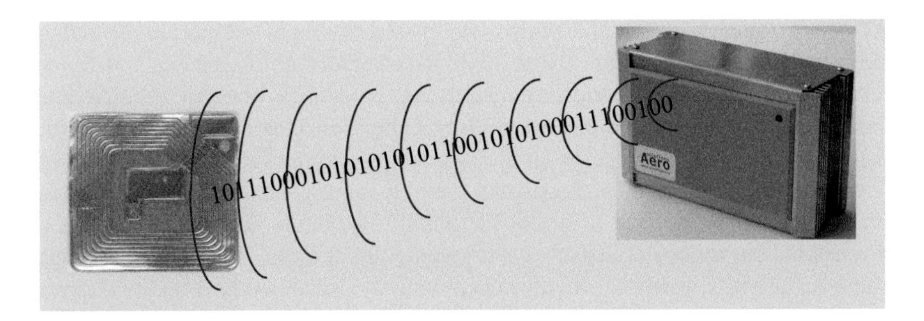

Fig. 3.14 RFID tag and reader

3.2.1.3 Data Sharing Messages

SC partners must continually exchange forecasting, planning, and transactional information. This is usually done through the exchange of structured data between computers, using standard message formats for purchase orders, dispatch advices, status reports, invoices, and so on. Because SC partners do not necessarily use the same application software and the same database entity identifiers, electronic data interchange (EDI) requires a double translation of messages through a standard language. The technology used must also be secure and speedy. When developing an EDI solution, partners must specify the commercial information to be exchanged and the electronic message standard used to codify it. They must also address technical issues such as the choice of an information highway (Internet, VAN, or direct connection) and of message transmission protocols (how data are packetized, addressed, transmitted, routed, and received at destination).

The two most widely used electronic messaging standards are ASC X12 (www.x12.org), developed in the United States, and GS1 EANCOM (www.gs1.org/ecom), a subset of UN/EDIFACT (www.unece.org/cefact) maintained by the United Nations. Several industrial sectors use their own standards, such as UCS for the US grocery industry, ODETTE for the automotive industry in Europe, and RosettaNet for the high-tech and electronics sector. These standards were originally developed to facilitate the exchange of a high volume of messages in value-added networks. With the advent of the Internet, a generic language called XML (*eXtensible Markup Language*) has been independently developed to promote the exchange of messages on the web. Several communities have subsequently emerged to promote the development of standard XML business messages. Nowadays, most commercial standard development organizations, such as GS1 and ASC, propose XML versions of their standard in parallel with their original offering. One of the benefits of XML, when compared to X12 and EANCOM, is that the language is more structured and therefore messages are much easier to write and read.

3.2.2 Technological System Perspective

SC systems are organizational organisms composed of a set of internal *entities* acting and interacting in an environment also involving external entities (see Fig. 1.5). Some of these entities are tangible and they are thus easy to identify. Material resources and human resources are part of this entity class. Others, such as P&C objects (orders, plans, etc.) are more abstract. From an information and decision system's point of view, an entity is a user-defined construct useful to represent and manage a business. Each entity is characterized by a set of attributes that define the *state* of associated physical or abstract object instances at a given point in time. This state can change whenever an entity instance is involved in an activity, which has the effect of altering the value of its attributes. For example, 'mileage' is an attribute of the entity 'vehicle' and, for a specific truck instance, its value changes whenever the vehicle is used to make a delivery. The state of an SC system at any given time is therefore defined by the value of the attributes of all its entities. It is a snapshot of the system at a point in time.

To manage a business, it is not sufficient to know its state; its activity also must be understood and characterized. The activity of a company is linked to how its resources perform *processes*. Processes specify how entities act and interact over time to produce outputs. They are a bit like a movie script. Processes necessarily induce a change of state for at least a subset of the entities involved. As will be seen in Chap. 4, processes involve input, output, and processing entities. It is through its processes that the system acts. Each process performs a specific function. However, several distinct *methods* can generally be used to accomplish a given function. Specific methods are defined using *procedures*, that is, a sequence of ordered rules specifying how the processing entities change inputs into outputs (e.g., administrative procedures, routings, workflows, algorithms, etc.). For example, several distinct time series analysis methods can be devised to implement a demand forecasting process and several workflows can be elaborated to implement an order-picking process.

For a process to be activated, a triggering *event* must occur. This event can take the form of an incoming demand or execution order, a predetermined time (e.g., the beginning of a month), or a predetermined state (e.g., the order point in an inventory system). When a process ends, its outputs often become the input to another process, the state of the system has changed, and time has passed. Any one of these events can trigger one or more other processes and the business continues to operate. Some processes are linked to the execution of operational supply, production, storage, transportation, and sales activities. Others are needed to plan and control operational activities. For this reason, PES must rely on up-to-date information not only on the management processes of the company but also on its operational processes and on the state of all its entities.

As explained in Sect. 3.1, P&C systems support a network of planning modules (see Fig. 3.9). Each module produces *plans* in order to satisfy the *needs* forecasted (based on past-state history and environmental trends) or specified by downstream modules and to attain *performance* targets provided by top modules and responsibility center mission. Plans, however, cannot always be applied exactly because they may be imprecise and because the SC system may unexpectedly change during the implementation lead time. Hence, the *actions* taken do not always correspond to the plans. It is thus actual actions that do change the state of the system and not plans. Control activities are necessary to ensure that actions stick to plans as much as possible and that expected performances are attained. Based on these concepts, Fig. 3.15 illustrates the fundamental functions of a PES. PES are partly formal, but they generally also incorporate a lot of informal processes. Today, thanks to the capabilities of modern ICTs, most companies support computerized PES.

To better understand the functionalities depicted in Fig. 3.15, let us examine them more closely for the different planning levels defined in Fig. 3.5.

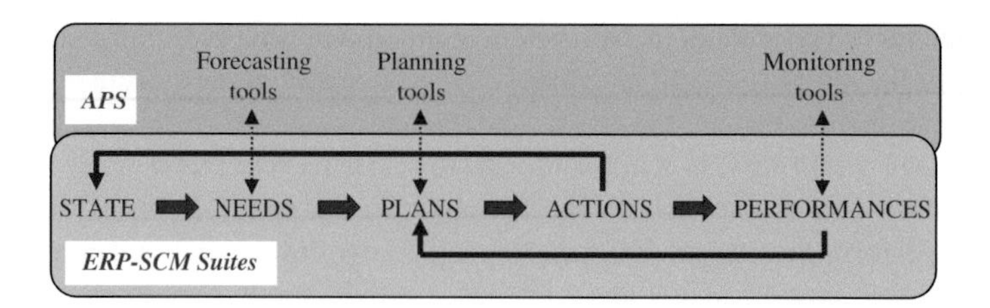

Fig. 3.15 Functionalities of an enterprise planning and execution system (PES)

- *At the strategic level.* Starting from the current *state* of the system (current customers, suppliers, and partners; available facilities, equipment, and personnel; internal competencies, competitive posture, etc.), emerging technological and environmental trends, and strategic directions adopted, the future *needs* of the company (market offer, resources required to support this offer) are specified. Strategic *plans* and policies are then elaborated to meet these needs (sourcing strategy, acquisition, deployment of long-term resources, etc.). The implementation of these plans leads to concrete *actions* related to the arrival or departure of partners and resources and to the development of competencies. Obviously, these actions are changing the state of the system. Resulting *performances* are monitored (see Fig. 2.2) to ensure that corporate objectives are reached, which may lead to revisions of plans and corrective actions.
- *At the tactical and operational level.* Given the constraints and objectives resulting from strategic plans and actions, from the *state* of processed entities (inventory on hand, cash, order portfolio, etc.), and from forecasted short- and medium-term *needs* for capacity, products, and funds, *plans* are developed to be able to meet needs effectively (allocation of resources to product-markets, optimization of production and procurement lot sizes, production schedules, delivery picking and routing schedules, etc.). These plans give rise to purchasing, reception, production, selling, handling, and shipping *actions* associated with the operations of the company, among others, and they change the state of processes entities and resources. *Performance* measures (costs, quality, service, productivity, inventory turnover, etc.) are evaluated in parallel to ensure that SC operations proceed as expected and to take corrective action if necessary.

Although the fundamental functions of PES presented in Fig. 3.15 seem to be pretty straightforward, in practice the developments of value-creating PES proves to be extremely difficult. Having elaborated its SC strategy, a company should ideally develop a tailor-made PES designed to leverage the benefits of this strategy. This seems to be the best approach to obtain a competitive advantage. Unfortunately, although leading companies were able to do this, given the complexity of such projects, several companies attempting to develop proprietary PES have failed, and often with disastrous consequences, so much so that top executives are now largely avoiding PES developments because they perceive them as too risky. This has given rise to a multibillion-dollar PES software solution industry. Software giants such as SAP (www.sap.com), ORACLE (www.oracle.com), and Microsoft (www.microsoft.com/en-us/dynamics) are now offering diversified enterprise software solutions and most companies are selecting such solutions instead of developing them in-house. Horror stories related to the adoption of these packaged solutions also abound, however, and developing a competitive advantage using the same tools as competitors is not obvious.

3.2.3 Software Solutions

The application software industry has evolved considerably over the last decades through mergers and acquisitions, the emergence of new information and communication technologies, and the development of new management paradigms, so much so that the offer of today's industry participants is not easy to characterize. Referring to Fig. 3.15, we can say that most large enterprise software companies are now offering ERP (*enterprise resource planning*) and SCM suites covering basic planning and execution functionalities, and that niche players are offering *advanced planning systems* (APS) for specific sub-problems, based on recent optimization, simulation, statistical, and data-mining technologies. However, as these applications become largely accepted by users, they tend to be integrated to the offer of large industry players. For example, most of the current SCM applications started as APS but, given their success, they have been incorporated into mainstream vendors' software suites.

ERP suites are a legacy of the MRP systems introduced in the 1970s to support manufacturing operations (Orlicky 1975). These systems included functionalities such as BOM management, master production scheduling, MRP, rough-cut capacity planning, shop floor control, purchasing, and inventory control. In the 1980s, additional accounting, financial, asset, and human resource management functionalities were added to these software packages, which became known as *manufacturing resource planning* (MRP II) systems. In the 1990s, with the advent of client–server architectures, MRP II systems were transformed, their functionalities were enlarged to cover most of the resources and processes of modern enterprises, and the acronym ERP was coined. ERP systems are mostly internally focused. They are good to memorize states, needs, plans, actions, and performances and to support execution and basic planning processes. However, their forecasting, planning, and performance monitoring tools are often rudimentary.

As ERP systems were largely adopted by companies in the 1990s, several new players, such as i2 technologies and Manugistic (now both part of JDA—www.jda.com), started offering APSs to improve the decision support capabilities of ERP systems. These original APS evolved to constitute complete SCM suites. SCM suites are more externally focused than classical ERP systems, that is, they are designed to improve collaboration between partners and to provide end-to-end SC visibility. Typical SCM suites include modules for strategic network planning, tactical S&OP, category management, demand planning, distribution planning, production planning, supply planning, order-promising and fulfillment, global trade management, warehouse execution, transportation execution, production scheduling, inventory control, store management, and so on. As these modules gained in popularity, major ERP vendors introduced similar functionalities in their offer to protect their market share. For example, SAP introduced APO (*Advanced Planner and Optimizer*) to complement its ERP offer. Another class of enterprise software sold by most large vendors is CRM (*customer relationship management*) systems. These systems are designed to support a company's interaction with current and future customers. Their functionalities typically cut across the customer-oriented features of ERP–SCM suites.

Most large software vendors also offer vertical solutions for specific industrial sectors. The needs of an industry depend, among other things, on the nature of its products (discrete, process-based, project-based, hybrid) and on dominant SC strategies (e.g., make-to-stock, assemble-to-order, make-to-order, and engineer-to-order). To meet these needs, each industry relies on specific P&C strategies and processes. Implementing PES that fit these characteristics is important and the selection of standard ERP–SCM suites may not be sufficient to do this. To achieve better fits, most suppliers offer software specialized for specific industrial sectors. Oracle, for example, offers specific solutions for the agriculture, apparel, chemical, construction, grocery, oil, and pharmaceutical industries. The solutions of most vendors are also modularized, and their modules can be customized using tools designed to facilitate this process. *Inbound Logistics* publishes a list and classification of top 100 logistics IT providers on an annual basis (O'Reilly 2014). Critical success factors for the implementation of ERP–SCM suites are discussed in Umble et al. (2003), who also propose implementation guidelines.

In the past, ERP–SCM suites were installed in memory on the computer systems of companies. With the advent of Cloud computing, in recent years, this has started to change. Cloud computing is a metaphor used to refer to the practice of delivering application software as a service (SaaS) and providing the hardware and operating system required to run these applications via the web. This is in principle attractive for companies because there are no servers to set up, and no software programs to install and maintain. Also, attractive is the fact that software applications are sold with a pay-as-you-go approach and that computing resources are easily scalable. However, cloud computing raises obvious customization, control, security, and reliability issues. Most ERP–SCM vendors are now providing hybrid solutions involving some cloud services. Mixed solutions are also appropriate for companies: they can run customized applications providing a competitive advantage internally and contract cloud services to support activities that can or should be treated in the same way by all industry players.

3.2.4 SCOR Model

As explained previously, developing an enterprise PES from scratch to obtain a competitive advantage is typically perceived as a risky venture. However, the adoption of an ERP–SCM suite directly from the shelf may prove to be difficult to fit with a company's culture, organization structure, and SC strategy. An intermediate solution is to engage in a P&C system reengineering project using best practice templates to lessen system analysis and design times and costs, improve design quality, and reduce risks. The planning and execution solutions thus adopted may involve a mix of in-house development, package acquisition and customization, and cloud computing services. The best-known framework of this type is the *Supply Chain Operations Reference* (SCOR) model developed over the last two decades by the *Supply Chain Council* (SCC). The SCC has recently merged with APICS (www.apics.org), which is now supporting the development of the framework and offering training and certification programs.

The SCOR model is a tool designed to help evaluate and improve SCM processes and systems. It documents a hierarchical set of standard SCM processes, practices (way to configure a process), performance measures (see Sect. 2.1), and skills that can be adapted to specific company contexts. It spans all customer interactions from order entry to paid invoice, all material movements from supplier's supplier to customer's customer, and all planning levels from strategic design issues to execution. It does not, however, address sales and marketing, research and development, and post-delivery customer support (SCC 2012). As shown in Fig. 3.16, it involves three imbricated industry-neutral process levels.

	Level		Examples	Comments
	No.	Description		
Within scope of SCOR	1	Process Types (Scope)	Plan, Source, Make, Deliver, Return and Enable	Level-1 defines scope and content of a supply chain. At level-1 the basis-of-competition performance targets for a supply chain are set.
	2	Process Categories (Configuration)	Make-to-Stock, Make-to-Order, Engineer-to-Order, Defective Products, MRO products, Excess Products	Level-2 defines the operations strategy. At level-2 the process capabilities for a supply chain are set (Make-to-Stock, Make-to-Order).
	3	Process Elements (Steps)	• Schedule Deliveries • Receive Product • Verify Product • Transfer Product • Authorize Payment	Level-3 defines the configuration of individual processes. At level-3 the ability to execute is set. At level-3 the focus is on the right: • Processes • Inputs and Outputs • Process performance • Practices • Technology capabilities • Skills of staff
Not in scope	4	Activities (Implementation)	Industry-, company-, location- and/or technology specific steps	Level-4 describes the activities performed within the supply chain. Companies implement industry-,company-, and/or location-specific processes and practices to achieve required performance.

Fig. 3.16 SCOR model process hierarchy (SCC 2012)

Companies using the model need to extend it at least for one additional level to design company-specific processes and practices.

The top level is organized around six primary SCM processes:

- *Plan*. Gathering information on the state of available resources, forecasting requirements, balancing requirements and resources to plan capabilities, and identifying actions to correct demand or resource gaps
- *Source*. Ordering and receipt of goods and services from suppliers
- *Make*. Conversion of materials (assembly, processing, maintenance, repair, recycling, refurbishment, etc.) or creation of contents for services
- *Deliver*. Receipt, creation, maintenance, and fulfillment (scheduling, picking, packing, shipping, invoicing) of customer orders
- *Return*. Reverse logistics (scheduling, shipment, and receipt of returned goods)

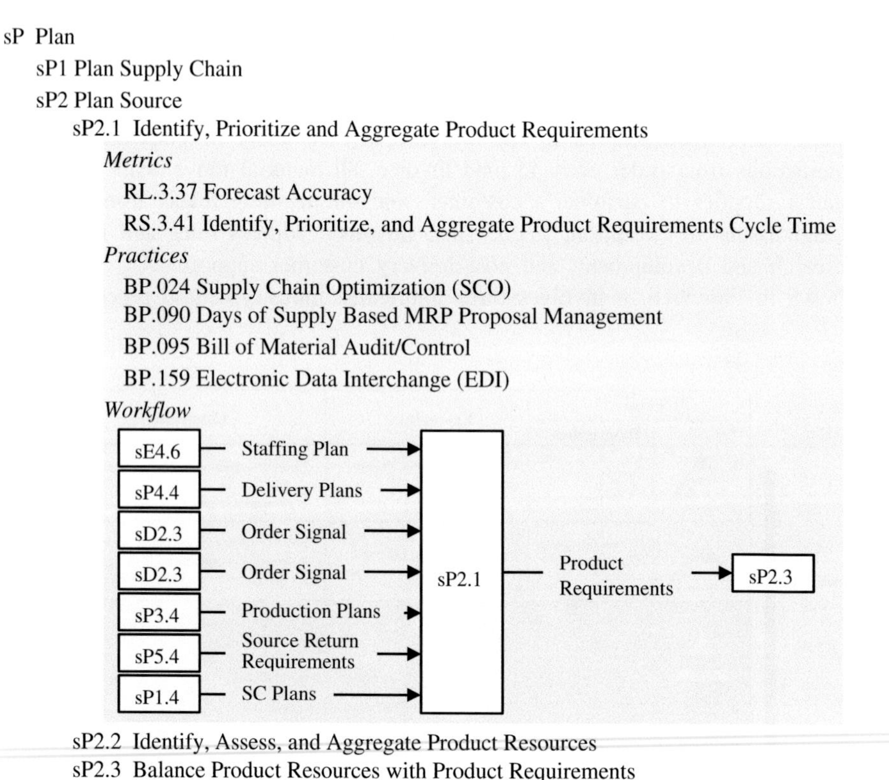

Fig. 3.17 SCOR process hierarchy example

- *Enable*. Management of SC entities, including business rules, performance metrics, data, resources, facilities, contracts, SCN, regulatory compliance, and risks

The SCOR process hierarchy structure is illustrated in Fig. 3.17. The example presented shows how level-1 process Plan decomposes into five level-2 processes (Plan SC, Plan Source, Plan Make, Plan Deliver, Plan Return), and how level-2 process Plan Source decomposes into four level-3 processes. Level-3 processes are documented by specifying related performance metrics, practices, and skills (not shown in the figure). A workflow is also included to show how the process exchanges information with other level-3 processes.

Review Questions

3.1. Using a real company example, how would you describe the various functional cycles of its SCs?
3.2. What is the difference between a P&C process and a P&C method?
3.3. Using a real SC example, how would you explain the behavior induced in the SC when each node of the network is managed independently?
3.4. What are the causes of the bullwhip effect?
3.5. How can the bullwhip effect be avoided?
3.6. After studying the generic P&C process described in Fig. 3.3, how would you describe the P&C process adopted by a few companies that you know?
3.7. Why do several generic P&C processes not explicitly consider capacity constraints?
3.8. What is category management and what does it involve? Based on a food sector example, identify possible product categories and subcategories.
3.9. Under the VMI approach, how are merchandise flows between manufacturers and retailers synchronized?
3.10. To obtain VMI benefits, is it necessary that retailer inventories are managed by suppliers? Explain your answer.
3.11. Why is cross-docking an important element of SCM?
3.12. What are the differences between the VMI and CPFR approaches?
3.13. What are GTIN, GLN, and SSCC?
3.14. What is EDI?
3.15. How are barcodes and datamatrices working?
3.16. What is an RFID tag?
3.17. How can the interaction among product, location, and logistics object identifiers, barcodes, and EDI improve the performance of an SC?
3.18. What are the advantages of the Internet on EDI?
3.19. What is a PES?
3.20. What are the characteristics of the PES of a company you know?
3.21. On the ERP–SCM suite market, what kind of trends do you expect over the next decade?
3.22. How can the SCOR model be used by a company?

Bibliography

Barratt M, Oliveira A (2001) Exploring the experiences of collaborative planning initiatives. Int J Phys Distrib Logistics Manag 31(4):266–289

Bartholdi J, Gue K (2004) The best shape for a crossdock. Transp Sci 38(2):235–244

Bolstorff P, Rosenbaum R (2007) Supply chain excellence: a handbook for dramatic improvement using the SCOR model, 2nd edn. Amacom, USA

Brown R (1977) Materials management systems. Wiley, New York

Cachon G, Fisher M (1997) Campbell soup's continuous replenishment program: evaluation and enhanced inventory decision rules. Prod Oper Manag 6(3):266–276

Carroll D, Guzman I (2013) A new omni-channel approach to serving customers. Accenture

Chen L, Lee H (2012) Bullwhip effect measurement and its implications. Oper Res 60(4):771–784

Chopra S, Meindl P (2012) Supply chain management, 5th edn. Prentice-Hall, Upper Saddle River

Clark T, Hammond J (1997) Reengineering channel reordering processes to improve total supply chain performance. Prod Oper Manag 6(3):248–265

Clark T, McKenney J (1995) Procter & Gamble: improving customer value through process redesign. HBS Case 9-195-126. Harvard Business School, Boston

Croxton K, Lambert D, Garcia-Dastugue S, Rogers D (2002) The demand management process. Int J Logistics Manag 13(2):51–66

De Kok A, Graves S (2003) Supply chain management: design, coordination and operations. In: Handbooks in operations research and management science, vol 11. Elsevier, Philadelphia

De Treville S, Shapiro R, Hameri A (2004) From supply chain to demand chain: the role of lead time reduction in improving demand chain performance. J Oper Manag 21:613–627

Forrester J (1961) Industrial dynamics. MIT Press, Cambridge, MA

Frankel R, Goldsby T, Whipple J (2002) Grocery industry collaboration in the wake of ECR. Int J Logistics Manag 13(1):57–72

Goldratt E, Cox J (1984) The goal. North-River Press, Massachusetts

Hax A, Candea D (1984) Production and inventory management. Prentice-Hall, Englewood

Klibi W, Martel A, Guitouni A (2010) The impact of operations anticipations on the quality of supply chain network design models. Working Paper CIRRELT-2010-45, CIRRELT

Kurt Salmon Associates (1993) Efficient consumer response: enhancing customer value in the grocery industry. Food Marketing Institute

Le Moigne J-L (1994) La théorie du système général. Théorie de la modélisation, 4th edn. PUF

Lee H, Padmanabhan V, Whang S (1997) Information distortion in a supply chain: the bullwhip effect. Manage Sci 43(4):546–558

Magee J, Boodman D (1967) Production planning and inventory control, 2nd edn. McGraw-Hill, New York

Martin A (1983) Distribution resource planning. Prentice-Hall, Upper Saddle River

Miller T (2002) Hierarchical operations and supply chain planning. Springer, Berlin

O'Reilly J (2014) Top 100 logistics IT providers. Inbound Logistics, pp 50–63

Olhager J, Selldin E (2007) Manufacturing planning and control approaches: market alignment and performance. Int J Prod Res 45:1469–1484

Oliva R, Watson N (2011) Cross-functional alignment in supply chain planning: a case study of sales and operations planning. J Oper Manag 29:434–448

Orlicky J (1975) Material requirements planning. Mc-Graw-Hill, New York

Pyke D, Cohen M (1990) Push and pull in manufacturing and distribution systems. J Oper Manag 9(1):24–43

SCC (2012) SCOR: Supply chain operations reference model, revision 11.0. Supply Chain Council

Schneeweiss C (2003). Distributed decision making, 2nd edn. Springer, Berlin

Sellers P (1992) The dumbest marketing ploy. Fortune, pp 88–93

Stadtler H, Kilger C, Meyer H (eds) (2014) Supply chain management and advanced planning: concepts, models, software, and case studies, 5th edn. Springer, Berlin

Sterman J (1989) Modeling managerial behavior: misperceptions of feedback in a dynamic decision making experiment. Manage Sci 35(3):321–339

Supply Chain Asia (2011) Dell's supply chain transformation. Supply Chain Asia, pp 30–32

Umble E, Haft R, Umble M (2003) Enterprise resource planning: implementation procedures and critical success factors. Eur J Oper Res 146:241–257

Van Belle J, Valckenaers P, Cattrysse D (2012) Cross-docking: state of the art. Omega 40(6): 827–846

VICS (2004) Collaborative planning, forecasting, and replenishment (CPFR): an overview. VICS

Wallace T (2004) Sales & operations planning, TF Wallace & Co., Montgomery

Chapter 4
SC Facilities and Activity Centers

In order to be competitive in today's and tomorrow's business world, companies must deploy their resources in activity centers located in industrial and commercial facilities designed for sustainable value creation. These facilities typically stage primary and support activities, and they take several forms: plants, distribution centers, warehouses, points of sale, transportation terminals, transshipment centers, and so on. They are installed in buildings positioned on industrial or commercial sites that must be located with care. They are the nodes of company's internal supply chain networks. In this chapter, after examining the nature of facilities and activity centers, we study design problems related to facility layout and location.

4.1 Company's Facilities and Centers

Figure 4.1 portrays the network of facilities of a company and it shows that these facilities themselves take the form of a network of activity centers that, as we shall see, encapsulate the resources required to perform their work. Industrial or commercial installations typically perform a set of primary activities and a number of support activities. According to their mission in the supply chain network (SCN) of a company, primary activities are concerned with supply, production, distribution, sales, or customer service. To develop and manage its primary activities, a company must also engage in support activities involving the acquisition and management of resources and knowledge, the planning and control of material flows, and communications between activity centers as well as with the environment (suppliers, clients, subcontractors, 3PLs, capital markets, etc.). Support activities are typically positioned in industrial or commercial facilities, but they can also be grouped in dedicated facilities (e.g., a head office). Materials, information, and currencies flow between centers and facilities; however, to lighten the schema, only physical flows are represented in Fig. 4.1.

A. Martel and W. Klibi, *Designing Value-Creating Supply Chain Networks*,
DOI 10.1007/978-3-319-28146-9_4

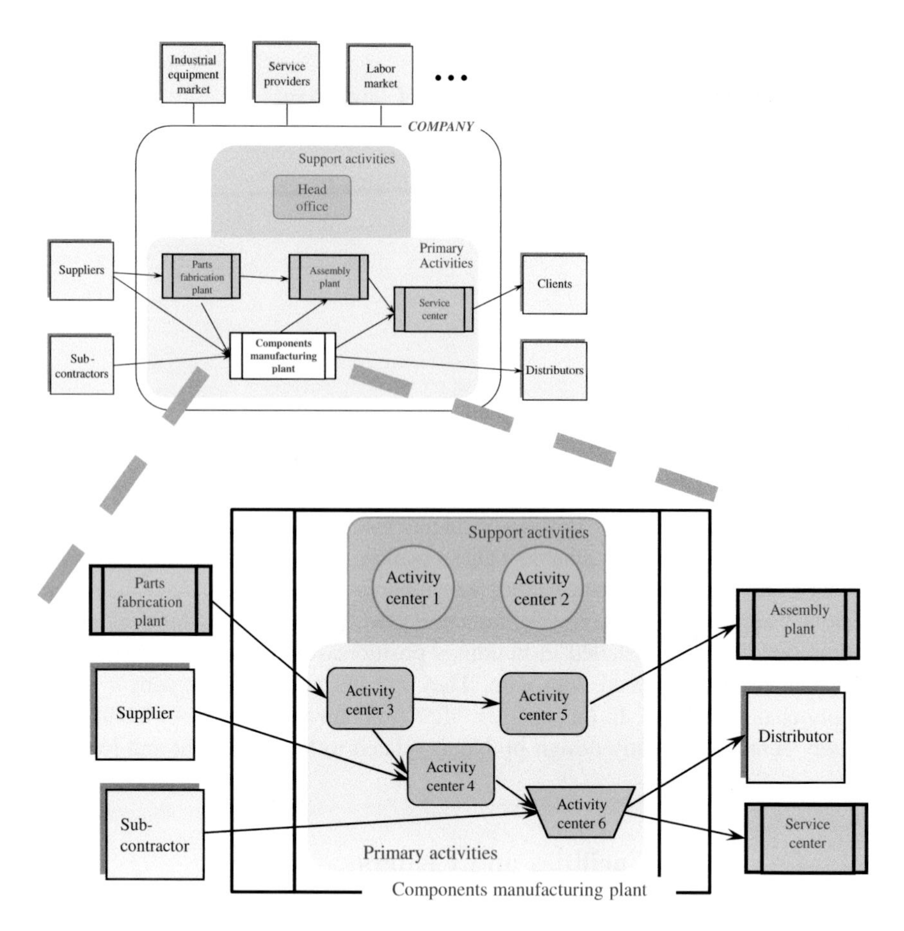

Fig. 4.1 Activity centers and facilities of a company

The mission of a facility relates to the role it plays in the company's SCN (markets and facilities to service, products to manufacture and/or distribute, order-winners to offer internal/external customers, etc.), to the activities it must perform and to required performance targets. The facility must determine which and how to use resources to carry out this mission as effectively as possible. We address this basic issue in this chapter. Two fundamental factors affect the productivity of facilities: the technology in place and the motivation and knowhow of human resources. These two factors are interrelated and we can hardly study one without considering the other. In this chapter, although we consider both, we study facilities and centers mainly from the point of view of technology. We examine the technological issues a production–distribution center must address to fulfill its mission.

The word *technology* refers to a generic set of techniques, methods, tools, machines, materials, and so forth associated with a particular domain. This notion is too vague to guide action in our context. In the following, we discuss a more precise construct: *technological systems*. Unfortunately, a technology is often perceived as equipment or a sophisticated process that can be purchased or developed for a facility. However, the use of new equipment in itself does not guarantee an improvement in productivity or value. To add value, equipment must be inserted efficiently in the structures and processes of the facility, which involves the alignment of resources (work area, equipment, operators, materials, etc.), resources configuration (layout), and methods/procedures with needs. All the elements of the resources–configuration–methods triplet must thus be reengineered to add value, and this is what we call a *technological system*. The term *technology* will be used in the text to refer indifferently to various elements of a technological system.

4.1.1 Activity Graph of a Company's SCN

The network of activity centers of a facility (shown at the bottom of Fig. 4.1) results from the assignment of the activities a company must perform to its SCN facilities. To design these networks, using an activity mapping formalism, one must first elaborate a conceptual model showing how the SC doctrine adopted by the company affects its activities. The mapping formalism used in this book is an *activity graph*, that is, a directed graph of the internal and external activities of the company reflecting strategic supply chain choices, as well as constraints imposed by technology or industry practices on the production and distribution of products.

Figure 4.2 shows an activity graph example for a sawmill in the lumber industry. The graphic symbols used to represent the different types of *activities* (nodes) or *movements* (arcs) of an activity graph are defined on the right-hand side of the figure. The mapping always starts with a generic external *supply* activity, and it always ends with a generic external *demand* activity. In the example, the graph describes the sequence of major sawmill production activities and it indicates where the company wants to keep stocks in the process. The arcs define product movements between activities. The type of arc used specifies whether the end activities can, must, or must not be in the same facility, which will condition the assignment of activities to facilities as well as technology choices. Arc (3, 4) in the graph is dotted (handling), which means that bucking and sawing activities must always be in the same facilities. However, arc (7, 9) is a solid line because boards must always be transported from a finished goods stocking point to customers. The bold arrow for arc (5, 6) indicates that drying and planning/grading activities could be performed in the same facility or on different sites. In a reverse logistics context, the activity graph also includes product return movements and revalorization activities.

More formally, an activity graph is defined over a set $A = \{a\}$ of activities necessary to produce and distribute a set of products $P = \{p\}$ in a supply chain. Two external generic activities are always present: supply ($a = 1$) and demand ($a = \bar{a} = |A|$). Three types of internal activities are distinguished: fabrication-assembly ($a \in A^{\mathrm{F}}$), warehousing-storage ($a \in A^{\mathrm{E}}$), and consolidation-transshipment ($a \in A^{\mathrm{G}}$).

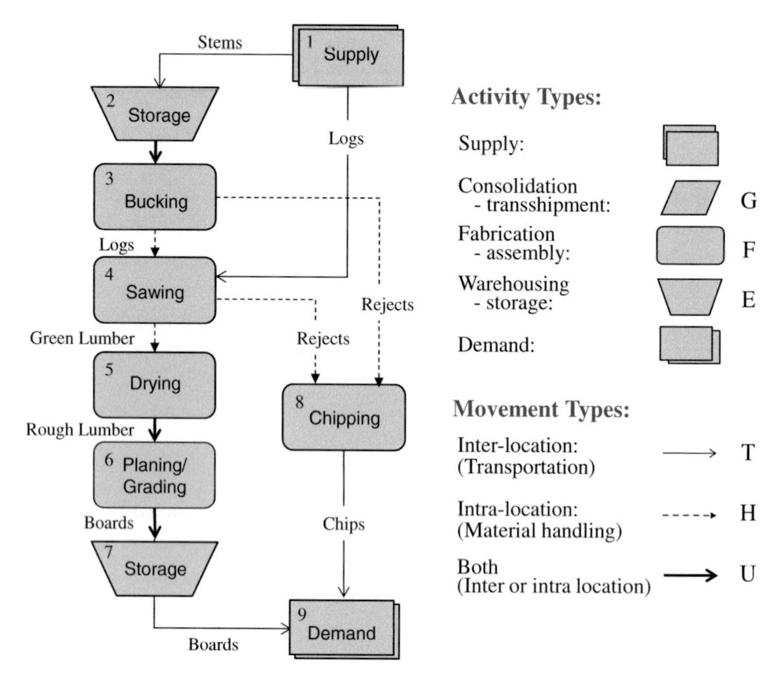

Fig. 4.2 Activity graph example for the Lumber Industry

The company controls these activities but some of them could be outsourced. The activity graph also includes a set of movements $M = \{(a, a')\}$ corresponding to the directed arcs between activities. Movements are associated with a set of product $P_{(a,a')} \subset P$, and they can be restricted a priori to interlocation transportation $M^{\mathrm{T}} \subset M$ or intralocation handling $M^{\mathrm{H}} \subset M$. Some movements (M^{U}) may also be unrestricted. Movements may be subject to infrastructure restrictions and depend on transportation modes availability and compatibility conditions (see Chap. 5). As we shall see in Chap. 7, recipes can also be associated to activities to specify the quantity of input products required to make output products.

Activity graphs map only physical flows and the primary activities related to these flows (production, storage, consolidation, movements[1]), all represented with a high level of aggregation. In other words, this conceptual model provides an integrated representation of the bills-of-material (BOM), decoupling points, and distribution channels of a company. It captures distribution strategies (direct shipments from production sites, distribution through intermediate storage or consolidation centers, deliveries through pick-up points or retailers, etc.), as well as production strategies (dedicated versus flexible facilities, centralized versus decentralized, etc.). When an activity or movement mapped in the graph is assigned

[1]Note that although the mapping formalism portrays movements as arcs to provide a spatial perspective, conceptually, movements are also primary activities.

to long-term resources (floor space, equipment, etc.), an activity center emerges. The assignment of several activities and movements to the resources of a facility gives rise to a network of centers, as shown in Fig. 4.1.

4.1.2 Technological Systems

At first sight, one might think that activity centers and facilities, each at their own level, merely exploit resources to transform inputs into outputs. Clearly, this is not false, but to design and manage facilities providing a competitive advantage, one must be more perceptive and appreciate the growing complexity of the *objects processed* in their centers, of *transformation processes*, and of their *product*. A technological system encompassing these three fundamental concepts is shown in Fig. 4.3. This section begins by examining the elements of modern technological systems and it builds on the insights gained to clarify their mission.

To clarify the nature of the product of a technological system, outputs must be examined from the point of view of the producer and of the client. For the producer, outputs are the result of a transformation process of physical (materials, parts, components, energy, etc.) and symbolic (data, information, and knowledge) inputs. For example, a cardboard box is manufactured by cutting, folding, and gluing cardboard but also by printing what the client wants on the box. Even this elementary example forces one to recognize that processed objects are not all physical. In this case, cardboard, glue, and ink are physical inputs, but the text the customer wants to see on the box is a symbolic input. The spectrum of products incorporating symbolic inputs going beyond primary data and incorporating value added information (e.g., newspapers) or knowledge is continually widening. Symbolic inputs are vital for most companies offering services. They are becoming just as important for companies producing goods. For products such as *Adobe Creative Suite* (www.adobe.com), the flight simulators of CAE Electronics (www.cae.ca), and most innovative products, informational, and cognitive inputs dominate whereas the physical provides only support and packaging.

From the point of view of the producer, one must also understand that inputs can be processed in three basic dimensions: shape, space, and time. Shape transformations include machining parts, assembling components, processing materials or information, or even the encapsulation of knowledge. Space transformations

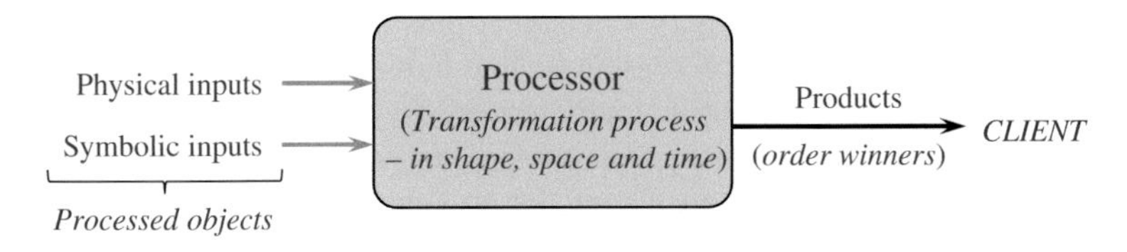

Fig. 4.3 Facilities (activity centers) as technological systems

involve moving inputs from one place to another; transportation, materials handling, and transmission of information are common examples. Time processing involves keeping processed or finished products during some time: the storage of goods or information is an obvious example. The product of a technological system is thus the output of a programmed sequence of such transformations.

When looking at the product of a technological system from the point of view of the customer, another fundamental dimension becomes apparent. When customers purchase a product, they do not limit themselves to a narrow assessment of the physical object, but also consider associated value drivers. As explained in Chap. 1, the purchasing decisions of customers are based on a set of order-winners associated with products. Two main categories of order-winners can be distinguished: those related to the products offered (attractiveness, innovation, etc.), and those related to production–distribution center capabilities (quality, cost, response time, flexibility, service, and greening). When the product of a center is considered as a family of value drivers, it becomes obvious that the performance of the center does not depend only on its physical transformation processes, but also on its service processes and on its operations management processes. Indeed, most order-winners can be provided only through an effective planning and control of activities and flows in production-distribution facilities.

The vision that emerges from this discussion leads to define a facility as a node of the SCN of a company having a dual responsibility:

- To develop the knowhow and technological systems necessary to transform physical and symbolic inputs into finished products;
- To exploit this knowhow and these technological systems to provide the order-winners required to create value.

By studying facilities such as distribution centers that, at first glance, seem relatively simple, one can appreciate all the ramifications of this definition. A distribution center is not only a platform receiving the goods that it stores and picking products for delivery but also a complex of sociotechnical activities and centers that must be operated judiciously by motivated management and staff to fulfill its mission.

The previous discussion focused on the nature of the inputs and outputs of technological systems and on the role of their processors, without examining the structure of the latter. We stressed that a technological system can be productive and create value only if it is designed and managed as a resources–configuration–methods triplet. To appreciate the implications of this statement, it is necessary to understand that each technological system is an amalgam of technological subsystems and that, as shown in Fig. 4.4, a facility thus decomposes in a hierarchy of processors, having elementary processors such as human resources, equipment, and tools at its base. The decomposition into intermediate subsystems (activity centers) can be very different from one context to another, as well as the number of levels of the resulting hierarchical structure. Figure 4.4 also shows that the distinction made between processors, technological systems, and resources is a matter of point of

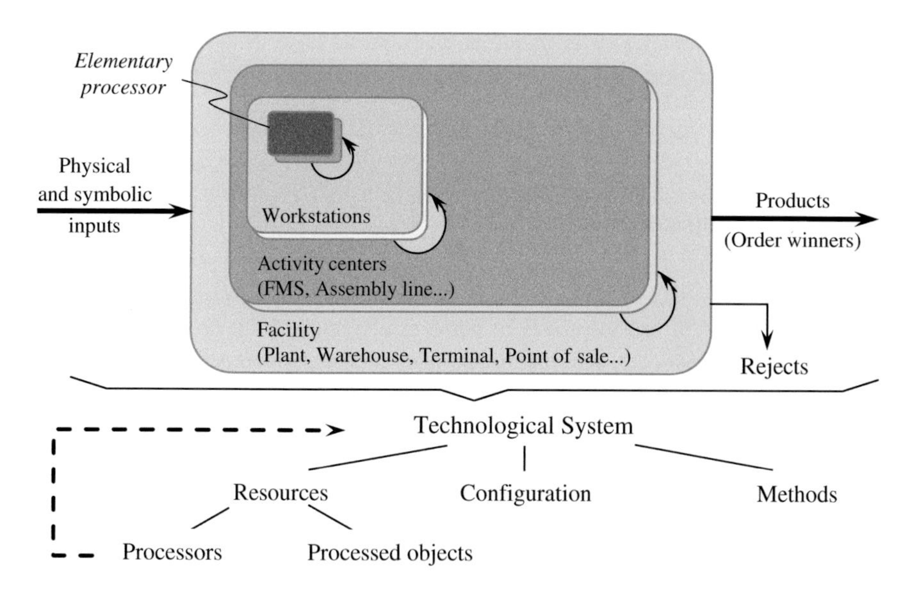

Fig. 4.4 An industrial facility seen as a hierarchical technological system

view, that is, it depends on the level from which the system is observed. The same nuances apply to the mapping of SCNs using activity graphs. The activities depicted can be more or less detailed depending on the objective pursued. A detailed activity graph can also be condensed into more aggregated mappings thus yielding a hierarchy of representations. The representations used for production planning or facilities design would typically be more detailed than the one used for SCN reengineering.

The resources of a technological system may include buildings, production, handling, storage, information processing and communication equipment, tools, software, labor, raw materials, stocks, energy, information, and so forth. Generic enabling technologies such as CNC (computerized numerical control) machines, auto-guided vehicles, and simulators are also part of this resource list. As can be seen, the resources of a technological system can be processed objects, elementary processors (nondecomposable resources), or intricate technological subsystems.

Resources are the first target that presents itself to the industrialist eager to improve its plants or distribution centers. This is why the acquisition of new resources is often the first action considered to improve productivity, a mentality rooted deeply in our industrial culture. This inclination is accentuated by the fact that resources are the only tangible assets of the resources–configuration–methods triplet that investors see in the financial statements of the company. Resources are an important part of a technological system, but, in most cases, they are only one of the pieces of the puzzle and it is dangerous to forget it. When neglecting this, disappointing, even catastrophic, results are obtained. One should not believe either that a more sophisticated technology is necessarily more efficient. All depends on the needs; for example, in many cases manual carts may be much more appropriate than auto-guided vehicles. It is therefore essential to take all the elements of the resources–configuration–methods triplet into account in productivity improvement projects.

4.2 Technological System Methods

The methods of a technological system are procedures governing its processes. They specify how to use resources to perform activities. They do not cover only primary activities related to physical operations but also support activities. They can be classified into five broad categories: design, learning, planning and control, maintenance, and operations methods. *Design* methods focus on the products and services provided by a facility or center and on the processors implemented to produce these outputs. Much of this book deals with technological systems design methods. When new products or processes are introduced, workers and managers must develop the knowhow needed to produce or apply them. This requires *learning* methods. These methods may also include the development of generic knowledge useful to advance SC systems. Flow *planning and control* methods were studied in detail in Chap. 3 and they concern the management of operations. *Maintenance* methods focus on the preventive and curative care of equipment and facilities. They are also concerned with processor replacement, which is related to the capacity planning problems examined in Chap. 8. In this section, we look more specifically at *operations* methods.

4.2.1 Methods for the Production of Discrete Goods

Operations methods relate to shape, space, and time processes. They specify how to perform the operations of an activity graph (Fig. 4.2) using existing resources. Figure 4.5 uses a standard manufacturing process representation formalism to illustrate a generic transformation process. When designing a new product, only value-added shape transformation operations (represented here by circles) are specified, usually using engineering plans and assembly charts. When the product design imposes constraints (the fact that a component must be assembled before another, for example), they are often specified using a precedence diagram similar to Fig. 4.5, but including only operations. Finally, when a manufacturing process is designed, one must specify how to use processors to produce parts, subassemblies, and finished products. In addition to preparation, fabrication, and assembly operations, one must then also specify storage (triangles), movement (arrows), and inspection (squares) activities. This gives rise to a *routing*, that is, a diagram similar to that of Fig. 4.5, but in which unwanted delays (semi-circles), are not represented. This crucial document specifies resources and time required to manufacture a product.

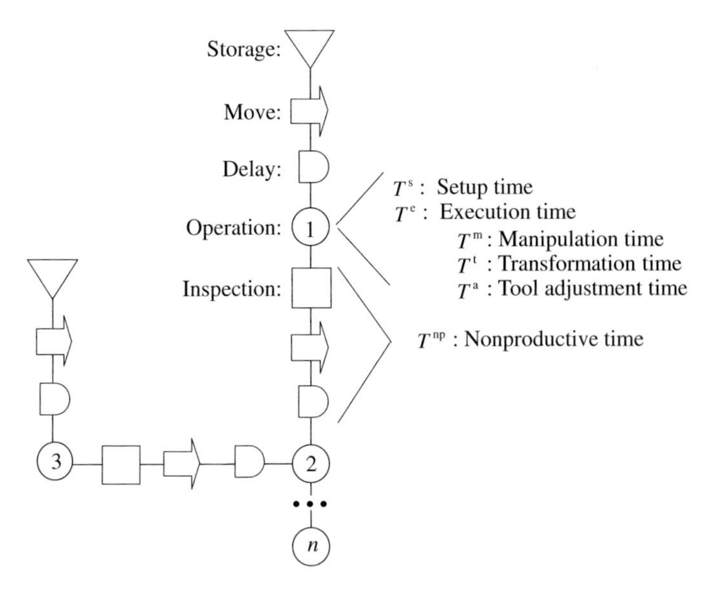

Fig. 4.5 Generic transformation process

A routing implicitly specifies a BOM. Indeed, routings define dependencies among the raw materials, components, and assemblies forming a finished product. This dependency structure is typically represented using a product tree. For example, a routing specifying how to make a chair could give rise to the BOM shown in Fig. 4.6. The numbers in parentheses indicate the quantity of product required to make a unit of its immediate predecessor. In this example, it is necessary to have one (1) *back assembly*, one (1) *seat,* and two (2) *front legs* to manufacture a chair. It is by exploiting the routing and BOM of the products manufactured in a production center that ERP and SCM systems (see Chap. 3) are able to compute resource and material requirements.

Several alternative routings may be feasible to make a given product. They are more or less efficient depending on the resources available in the production center and on its configuration. Given the resources and layout of a center, it is the inherent inefficiencies of a routing that give rise to the delays shown in the flow process chart of Fig. 4.5. As explained in the following box, a major challenge is to develop products and processes that eliminate, as much as possible, nonproductive time associated with inspections, moves, and delays, because they do not add value. Obviously, operations time (setup and execution) cannot be eliminated but, as suggested in the box, several tactics can be considered to reduce them significantly, and thus enhance the productivity of a facility.

Note that the operations management methods used to plan and control material flows in response to demand also have a decisive impact on productivity. These methods are often the source of time lost (delays) on the floor. Making good planning and control decisions is difficult, among other things, because some

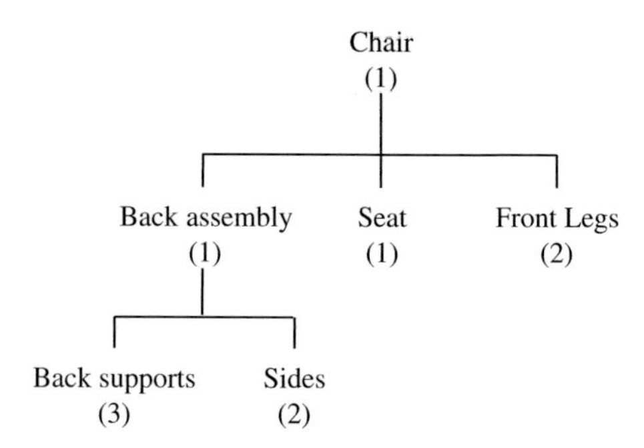

Fig. 4.6 Simplified bill-of-materials (BOM) for a chair

decisions that appear interesting from a point of view may have some unexpected perverse effects. In the following box, for example, we indicate that, under certain conditions, productivity can be improved by increasing lot sizes (Q), or by decreasing them! Strictly speaking, as the cause-and-effect relations examined in the box indicate, increasing batch sizes decreases average manufacturing time. However, it also raises work-in-process (WIP) which, because system flaws are then hidden, tends to increase rejection rates (q) and thus decreases productivity. By contrast, smaller lot sizes reduce WIP and rejection rates, which, especially if setup time (T^{s}) can be reduced, often leads to more pronounced productivity improvements than with the previous approach. This rational is the basis of the just-in-time (JIT) paradigm.

Productivity improvement tactics linked to production methods

When one adds all the relevant times associated to a typical transformation process incorporating n "Move-Delay-Operation-Inspection" sequences, as shown in Fig. 4.5, the time required to produce a batch of Q products is

$$BT = n(T^{\mathrm{s}} + QT^{\mathrm{e}} + T^{\mathrm{np}})$$

where T^{s}, T^{e}, and T^{np} are, respectively, the time required on average for a setup, for execution, and for nonoperational tasks. As a result, the average processing time per product is:

$$T \approx \frac{n}{(1-q)}\left\{T^{\mathrm{e}} + \frac{T^{\mathrm{s}} + T^{\mathrm{np}}}{Q}\right\}$$

where q is the rate of rejection of the process (the proportion of manufactured units not conforming to norms). This simple expression helps understanding how to improve the productivity of a production center. Noting that the arrows pointing down indicate a decrease and those pointing up an increase, the following cause-and-effect relationships are obvious:

$$T \downarrow \quad \text{if} \quad n \downarrow, \quad q \downarrow, \quad T^{s} \downarrow, \quad T^{np} \downarrow, \quad T^{e} \downarrow, \quad \text{or} \quad Q \uparrow$$

However, it should be clear that when the number of storage points in the process increases and when nonproductive times are high, the work-in-process (WIP) tend to be high. Finally, when processing times and WIP are high, direct costs are important and capacity requirements increase.

From these fundamental observations, one is able to understand the merits of several well-known tactics for improving productivity:

- *Engineering improvements*:
  - Combining or integrating operations — Decreases n, T^{m}, T^{np}
  - Simultaneous operations — Decreases n, T^{e}, T^{m}, T^{np}
  - Inspections embedded in operations — Decreases q, T^{np}
  - ...
- *Process improvements*:
  - Reduce handling and storage — Decreases T^{np}, WIP
  - Improve setups (with SMED, for ex.) — Decreases T^{s}, Q, WIP, T
  - ...
- *Operations management improvements*:
  - Larger lot sizes — Increases Q, WIP; decreases T
  - Smaller lot sizes (JIT) — Decreases Q, q, WIP
  - Better planning and control — Decreases T^{np}, WIP; Improves capacity use
  - ...

4.2.2 Methods for Process Industries

Process industries, such as the food, chemical, or pulp and paper industries, have features that distinguish them from discrete-good producers. The food sector for example has, among others, the following distinctive characteristics:

- The raw materials used are often agricultural, so they tend to be seasonal and perishable. Many of the ingredients used are also traded on the stock market, making their price variable from one day to another.
- The manufacturing process often incorporates mixing and cooking activities and its nondiscrete WIP cannot be stored. Systematic loss (e.g., evaporation, saw-dust, falls, tank deposits, remains, etc.) or by-products (e.g., skim milk) can also

be generated. The finished product is often separated in individual portions and packed at the end of the process.

- The finished product is a food, so it is subject to all sorts of constraints imposed by dietary and safety norms. At the same time, some flexibility as to the exact composition of the product often exists.

In these industries, BOMs and routings become formulas, processes, and recipes. The formula corresponds to a BOM, the process to a routing, and the recipe to the juxtaposition of the two (somewhat as in the recipe books used for cooking). The process is usually fixed, that is, it consists of complex monolithic technological systems that require substantial investments. In the pulp and paper industry, for example, these systems take the form of chip mills, pulp mills, paper mills, and so forth. The process generally can be described directly using an activity graph as in Fig. 4.2, the associated BOMs being relatively simple. Many food products, for example, have a three-level BOM:

- Level 0—Finished product
- Level 1—Preparation and packaging components
- Level 2—Ingredients

In several continuous manufacturing processes, however, there are losses, wastes, or by-products. In order to track these elements, an intermediate semi-finished products level may also be required, even if no component is assembled at this stage. Then again, certain losses can sometimes be recycled in the production process and thus become components. These features imply that the BOM of continuous products may partly converge and partly diverge. For example, in an industrial charcuterie, different types of meat are cut first from carcasses (diverging BOM), and then assembled into sausages and pâtés (converging BOM).

Another peculiarity of the process industries comes from the fact that ingredients are often difficult to measure. Many of these ingredients are powders, liquids, pastes, and so on, which can be supplied in various conditioning or different concentrations (a given ingredient may be available in stock in several concentrations). It is therefore necessary to manage stocked items with multiple units of measurement—mass (kg, pounds), volume (liters, m^3), number (box, pallets), density, and so on—and to convert these units as required.

An important feature of several process industries, including the food industry, is the fact that BOMs can be variable, that is, the quantity of each ingredient can change, as illustrated in Fig. 4.7, depending on the conditions prevailing at the time of manufacture. For example, suppose that a company makes a dog food. The preparation of this product includes sago flower, meat, and skim milk enriched with fat. The type of meat used may change depending on availability and prices on the market. Because organ meat (offal, etc.) is commonly used, availability depends on sales of animals for slaughter. Given the price of ingredients for a given week, one wants to use the most economical formula giving the required amounts of protein, fat, carbohydrates, and fiber. In other words, we want to determine the optimal proportion of each ingredient to use in the preparation.

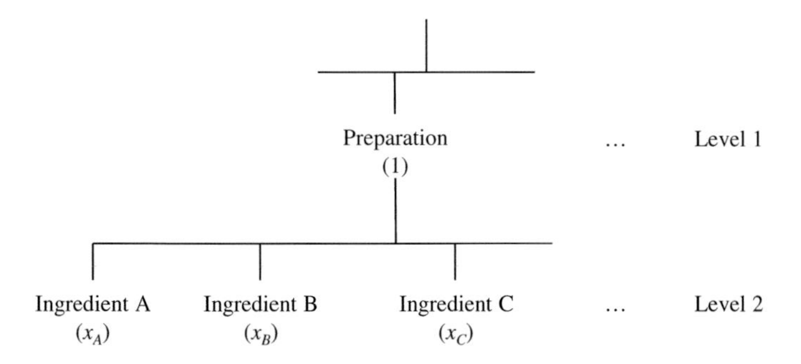

Fig. 4.7 Partial variable BOM

Table 4.1 Data for the food formula optimization example

	Prerequisite	Flower	Meat	Milk
Protein	≥20 %	0	50 %	30 %
Fat	≥10 %	0	10 %	20 %
Carbohydrates	Facultative	100 %	40 %	10 %
Fibers	≤20 %	0	40 %	10 %
Cost ($/100 kilos)		120	150	200

This food-formula optimization problem is a classical linear programming application. To illustrate, suppose that the relevant data for a particular case are those provided in Table 4.1 and define the following decision variables:

x_A Proportion of sago flower in the preparation
x_B Proportion of meat in the preparation
x_C Proportion of enriched milk in the preparation

The linear program (LP) to solve to find the optimal formula is then:

$$\text{Min } 120x_A + 150x_B + 200x_C \quad \text{(Cost)}$$

subject to constraints

$$\begin{aligned} &50x_B + 30x_C \geq 20 && \text{(Protein)} \\ &10x_B + 20x_C \geq 10 && \text{(Fat)} \\ &40x_B + 10x_C \leq 20 && \text{(Fibers)} \\ &x_A + x_B + x_C = 1 && \text{(Proportions)} \\ &x_A \geq 0, \quad x_B \geq 0, \quad x_C \geq 0 \end{aligned}$$

This kind of linear program (LP) can be solved easily with specialized tools such as the Excel spreadsheet Solver. In our case, one can also find the solution graphically: the proportion constraint in the model implies that $x_A = 1 - x_B - x_C$,

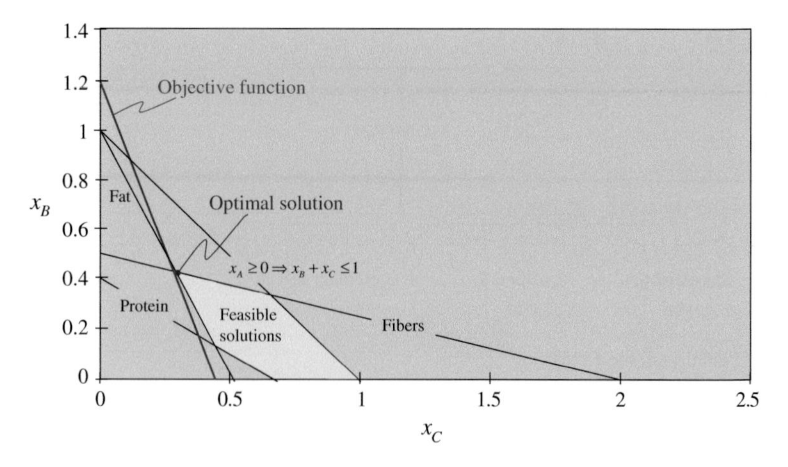

Fig. 4.8 Solution of the linear programming problem

which allows eliminating variable x_A by substituting in the objective function and in constraint $x_A \geq 0$. The new objective function thus obtained is

$$\text{Min } 30x_B + 80x_C$$

and the resulting two-variable LP is graphically stated in Fig. 4.8. The intersection of the regions defined by each constraint defines a set of feasible solutions. By moving the objective function downward, we see that its value is minimized when the point $[x_B, x_C] = [3/7, 2/7]$ is selected. The optimal value of x_A is then obtained by substitution, giving $x_A = 2/7$.

The total cost of the optimal solution thus obtained is \$155.71 per 100 kilos of dog food. This is a very simple example, but there are all sorts of variations of this problem in the food industry. A linear programming model similar to the one presented can also be formulated to find an optimal diet.

4.3 Technological System Configuration

The configuration of a technological system relates to the structural links between its resources (specifically between processors), to its spatial organization and to its redeployment flexibility when needs differ in times. For static resources (workstations, heavy equipment, etc.), the configuration is linked mainly to the physical layout of a facility. For resources that are naturally mobile, such as human resources, the configuration is rather associated to organizational factors such as the authority structure, work relationships, or customer–supplier relations. In what follows, we focus on the layout of SCN facilities.

4.3.1 Generic Layouts

Conceptually, six main types of layouts may be envisioned. Five of them are illustrated schematically in Fig. 4.9. In this figure, geometrical forms are used to represent different types of processors. The main characteristics of each layout type are discussed in the following:

- **Process layout**. The simplest way to organize a facility is to group all processors with similar functionalities in the same activity center. The process layout in Fig. 4.9a incorporates five activity centers. This type of layout is very common in practice. An electronic modules plant could, for example, be

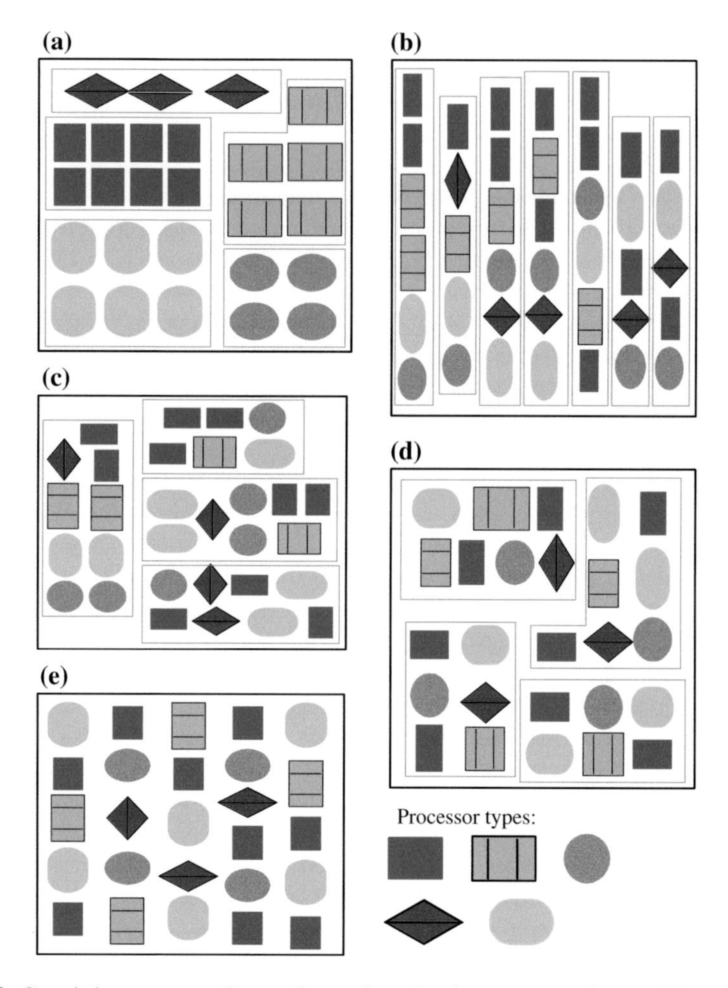

Fig. 4.9 Generic layout types. **a** Process layout, **b** product layout, **c** group layout, **d** fractal layout, **e** holographic layout. *Source* Montreuil et al. (1995)

organized into receiving-shipping, storage, surface components insertion, wave welding, assembly, and inspection centers. Each center processes all products requiring the activity it performs.

- **Product layout**. Another intuitive way to organize a facility is to group all processors necessary for the production of a given product in the same activity center. Figure 4.9b presents a layout of this type for seven distinct products. A doors and windows plant organized for mass production could, for example, include an activity center for each door and window model: one for French doors, one for casement windows, one for sliding windows, and so on. With this type of layout, each activity center is autonomous and covers all operations from the reception of input materials to the shipping of finished products. Although some plants manage to reach this ideal for a few products, this type of design is usually complemented by process centers for reception, storage, and shipping activities, as well as for activities focused on an expensive processor sharing its capacity among several products. The archetype of the product layout is the assembly line or, if the movement of products on the line is mechanized, the automated assembly line.
- **Group layout**. A group layout is a variant of the product layout in which centers are dedicated to families of products rather than specific products; wood handles for all shovel models, for example. The difference between these two types of layout appears when examining Fig. 4.9b, c. Cellular layouts with a U shape are omnipresent in modern factories requiring product and volume flexibility. Their automated form, known as flexible manufacturing systems (FMS), integrate flexible numerical processors with automated handling and storage systems and a computerized control system. This configuration enables the optimization of the performance of the cell for all products in the family under any possible product-mix variation.
- **Fractal layout**. Another way to design a facility is to divide it into multiple nearly identical microfacilities, each able to perform most of the required processes. For example, the layout in Fig. 4.9d includes four almost identical activity centers, each providing roughly a quarter of the capacity of the facility. The implementation of this type of layout involves a repetitive *pasting* of almost identical physical cells, thus its fractal layout designation.
- **Holographic layout**. The holographic layout, as opposed to the fractal one, does not incorporate any explicit physical cells. As shown in Fig. 4.9e the basic cell is the processor. Holographic layouts are appropriate for plants operating in a highly volatile environment, without stable product lines and dominant routings. The objective is to be robust under chaos. To do this, the layout distributes processors of the same type through the factory so that near each processor there is at least a processor of any other type. This enables one to dynamically create a variety of ephemeral virtual cells depending on the demand for various products.
- **Fixed layout**. Some products are too large to be moved. They are then built in a fixed location and it is the pieces of equipment necessary to manufacture them

that are moved. We refer here to the construction of bulky products such as aircrafts, ships, buildings, and bridges. These last examples take the form of a large construction project because a single product is built at its place of use and the necessary equipment is moved to the site. However, when multiple units are built in a factory, equipment is often positioned on a permanent location on the factory floor around the product.

Few actual facilities contain only a single layout type. Modern facilities are often designed in modular form and they are an amalgam of activity centers with different architecture. For example, Bombardier (www.bombardier.com) rail vehicles production plants incorporate a machine shop, a welding shop, a tooling shop, a shipping and receiving area, and several other work areas with a process layout. In addition, they include several assembly lines with a product layout. Industrial facilities also incorporate several support activity centers (quality assurance, engineering, sales, maintenance, etc.) as well as space for meetings, eating, personal needs, air conditioning, parking, and so on. These additional needs must be taken into account in the configuration process. The effective positioning of all these types of primary or support activity centers in a building is not an easy task.

Five generic categories of technological systems are often distinguished in the literature (project-based, disconnected flow, discontinuous flow, connected flow, and continuous flow) and it is now widely accepted that their selection depends mainly on production volumes and on the nature of the work to be done (number of components, complexity of operations, etc.), as shown in Fig. 4.10. Each generic system category is characterized mainly by the nature of its routings and layout. The main idea conveyed by Fig. 4.10 is that each category is appropriate in a particular context. Technological systems located off-diagonal use technology ill-suited to their needs. This characterization enables the identification of undesirable technological choices, but it is not always rich enough to guide decision making. For example, as shown in Fig. 4.10, several layouts can be considered for a discontinuous flow system. Volume plays a role in the final selection, but the nature of the order-winning attributes (assortment, response time, flexibility, etc.) to provide to customers is just as important. Selecting a category of technological system and its layout is a multicriteria decision problem.

As aforementioned, the majority of modern industrial facilities incorporate several types of technological systems. The appropriate deployment depends on the mission of the facility and more specifically on the characteristics of the products to manufacture. It is often advantageous to operate facilities that focus on a small number of technologies, but this is not always possible. As shown in Fig. 4.11, product–quantity (P–Q) diagrams can be used to analyze needs. If the P–Q curve is very pronounced, a hybrid solution like the one shown at the bottom-left of Fig. 4.11 may be fitting. However, if the curve is flat and volumes are not too high, a more flexible configuration is desirable.

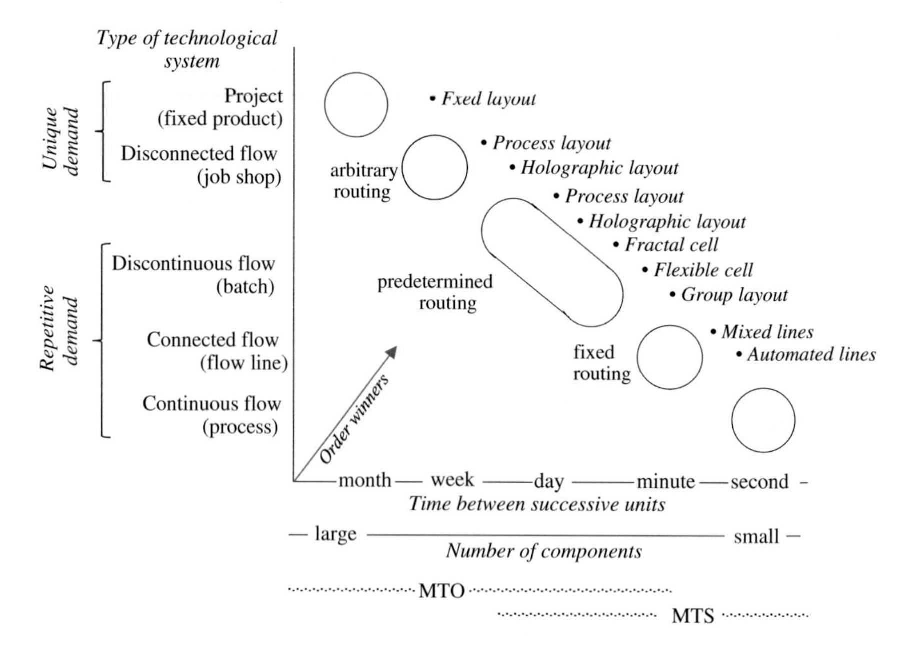

Fig. 4.10 Generic technological systems

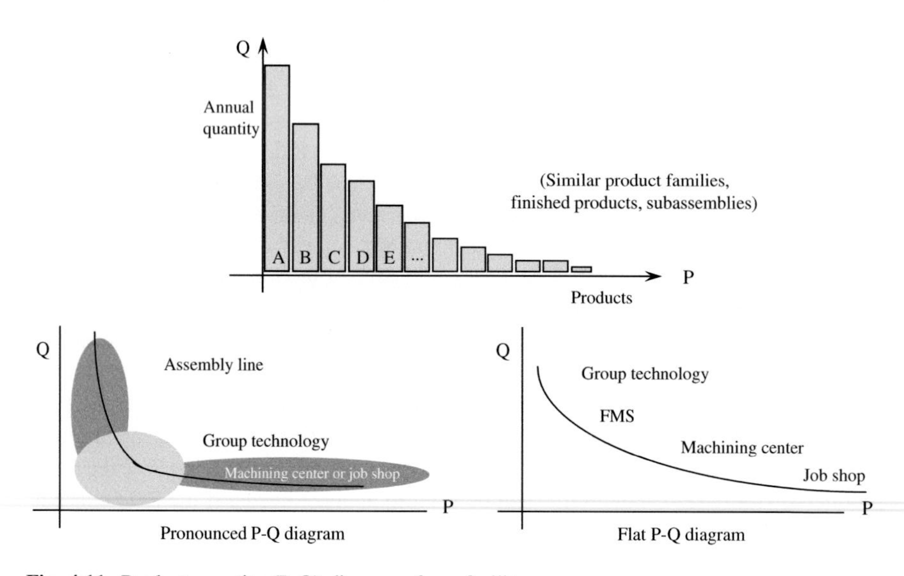

Fig. 4.11 Product-quantity (P-Q) diagrams for a facility

4.3.2 Layout Procedures

A number of approaches have been developed to facilitate the layout of a facility. The majority of them rely on the basic concepts presented with an example in this section. Several of these concepts were introduced by Muther (1961) in his seminal work on the SLP method (Systematic Layout Planning). They are presented here in the context of the layout of activity centers in a facility. Referring to Fig. 4.4, it should be clear that they also apply to the positioning of workstations in an activity center. Several specialized methods were also developed to solve particular layout problems such as products arrangement in order-picking DCs or assembly lines balancing. These particular cases, however, are not studied in this text.

Layout planning problems can be analyzed using space relationship graphs involving nodes associated with activity centers and proximity relations between centers. These relations reflect the intensity of materials flows between centers, and also the qualitative importance of the proximity of centers to facilitate organizational contacts between workers, their movements, and to improve working conditions (security, temperature, noise, humidity, dust, etc.). From this information, and taking into account the space required for each activity, draft layouts in the form of block diagrams are developed. These coarse layouts are then evaluated and refined to get the platform to implement.

The starting point for the development of a relationship diagram is the activity graph of the facility. The activity graph for a simple case is shown in Fig. 4.12. It typically covers the part of the SCN activity graph assigned to the facility during a reengineering project. We will examine the SCN reengineering process in detail in

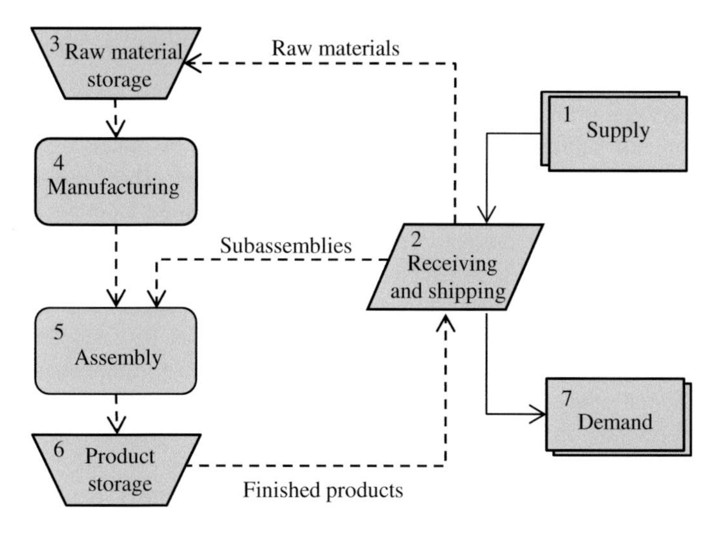

Fig. 4.12 Activity graph of the industrial facility to configure

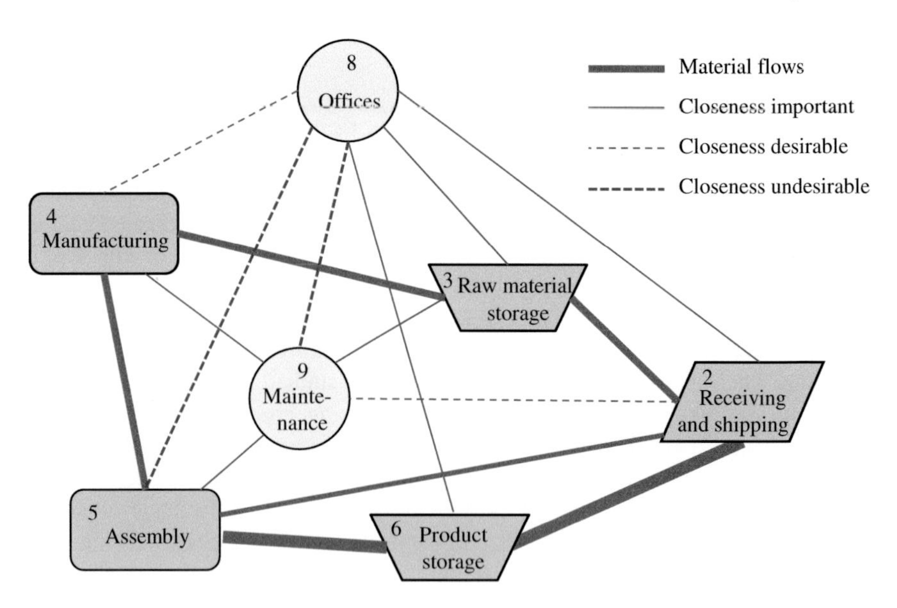

Fig. 4.13 Activity relationship graph

Chaps. 7 and 8. Given this, interactivity flows in terms of a standard load unit (pallet, barrel, kilo, etc.) can be calculated for a predetermined period of time (1 day, 1 month, 1 year, etc.). However, this is not necessarily easy because material flows in the facility can be very disparate and because production–distribution volumes are not known before sales and operation (S&OP) plans are available. One must therefore rely on forecasts. Knowing the quantities to be produced and distributed, as well as product routings/recipes, flows between each pair of centers can be estimated.

Flows between primary activities in a facility are essential to develop a layout but they are not sufficient. Support activities must also be taken into account. Proximity relations between support activity centers and other centers can be subjectively evaluated using a few ordinal closeness scores, for example important, desirable, or undesirable closeness. The *activity relationship graph* so obtained for our example is showed in Fig. 4.13. Note first that support activities 8 (engineering, quality assurance, sales and management offices) and 9 (maintenance and repairs) were inserted in the graph. The width of the black lines between the primary activities is proportional to the intensity of calculated flows. Relationships in red indicate that for safety, noise, or other reasons, adjacent centers should be as far removed as possible from each other.

The space required for the years to come must then be calculated for all the activities of the relationship graph. This calculation obviously depends on the nature of the activity. For storage activities, expected inventory levels, the nature of the products, storage equipment, and handling equipment must be taken into account. The corridors between the storage locations must also be considered. For production activities, one must first specify the configuration of workstations. At this level, ergonomic considerations are important. Generally, a machine tool in a machining station occupies less than half the space allotted so that work can be done properly. Space must also often be provided for WIP. Obviously, corridors for moving between stations must also be planned, as well as space for the personal needs of workers. If the center is an assembly line, its detailed engineering must usually be completed to calculate the necessary space. When a building, or at least a part of the building, is built on several floors (for example, two floors for offices), this of course must also be taken into account. More elaborate layout methods have been proposed for the design of multiple-floor buildings. A detailed discussion on the calculation of space requirements for different types of activities is available in Tompkins and White (1984).

When space calculations are completed, one can construct a *space relationship graph* by replacing the nodes of the activity relationship graph by rectangles with an area proportional to the space required. As shown in Fig. 4.14, the majority of

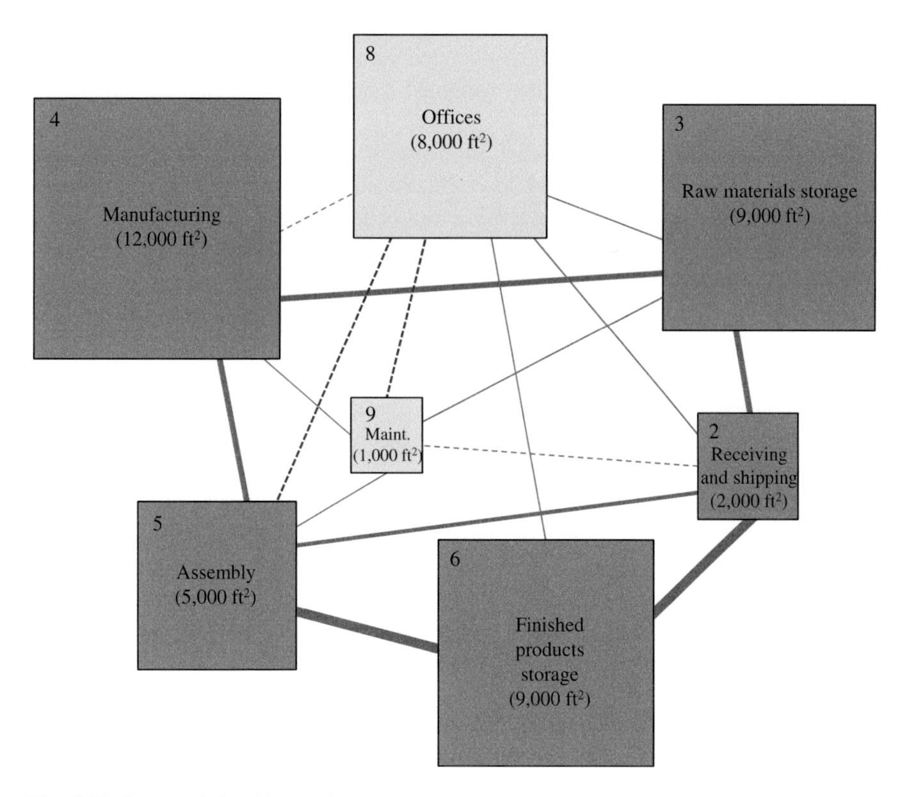

Fig. 4.14 Space relationship graph

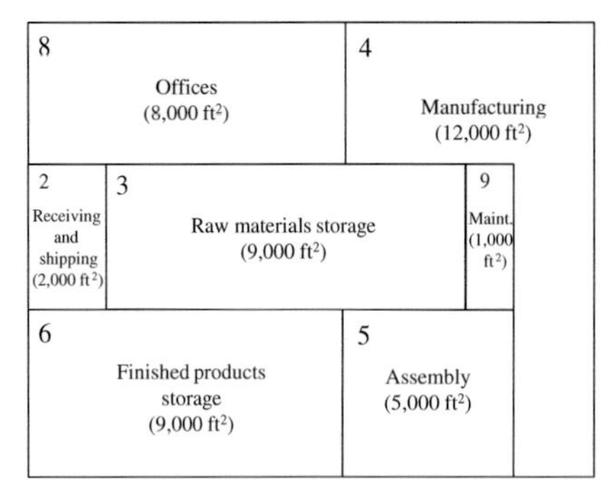

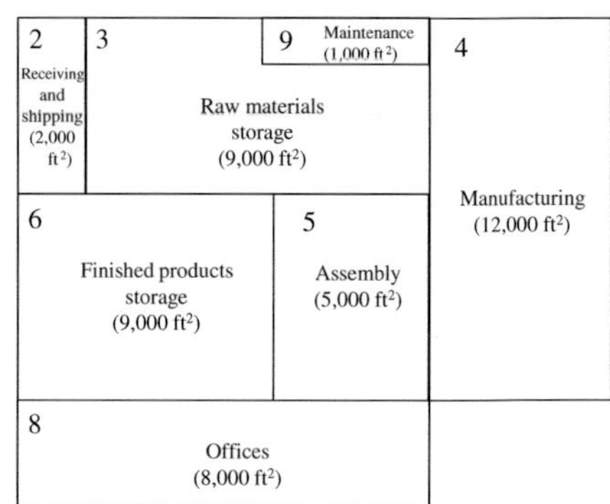

Fig. 4.15 Alternative block layouts

relevant data necessary to develop a layout are summarized in this graph. To develop a layout, one then moves the rectangles on the plane, and changes their shape so that they fit in an existing or planned building, respecting as much as possible the priorities evoked by the relationship lines. By doing so, one can develop several *block layouts* such as those illustrated in Fig. 4.15. Layouts are then evaluated by calculating a total flow-distance metric on the one hand and on the other hand, a score based on the weighted importance-distance of subjective elements.

A large number of heuristics were proposed to automate the development and evaluation of block layouts from a space relationship graph. The majority of them are based on a discretization of space in more or less precise grids. Some of them (construction heuristics) are designed to build layouts from scratch (PLANET, CORELAP, ALDEP, etc.) and others (improvement heuristics) start from a given layout and seek to improve it (CRAFT, COFAD, etc.). Montreuil et al. (1993) also showed that linear programming can be used to develop block layouts. Several empirical studies comparing layouts obtained using heuristics with layouts built manually by professionals have shown that automated methods usually provide better results. A discussion of the state of the art in this field is available in Montreuil (2007).

Additional work remains to be done when a block layout has been selected. To complete the design of an industrial platform, one must also consider building and activity center entry and exit points as well as the corridors required to move between activity centers. One should also position all equipment on the plan. Ultimately, a detailed blueprint such as the one shown in Fig. 4.16 must be produced and the capacity of each primary activity center must be specified in a standard unit suitable for planning purposes. Platform implementation and operation costs must also be estimated. This information is required in the following chapters to optimize the structure of the SCN network of a company.

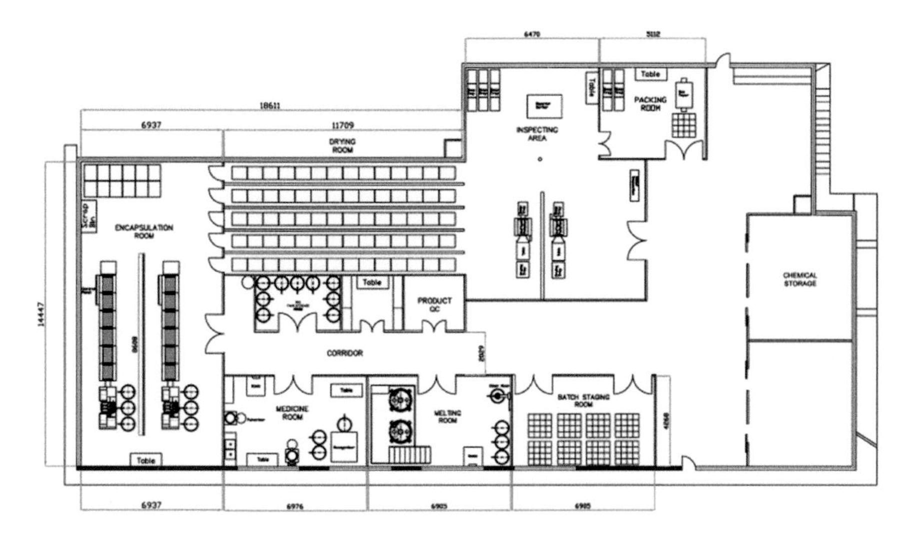

Fig. 4.16 Industrial facility platform. *Source* Cap Plus Technologies (2015)

4.4 Facility Site Selection

In some cases, the layout of an existing facility is revised to take advantage of the acquisition of new equipment or of the expansion of the building in place, among others. On the other hand, companies must also periodically implement new facilities to fulfill the demand of new markets or to take advantage of favorable economic factors. In this event, the choice of a suitable site for the new facility is also an extremely important issue. Site selection decisions are complex, especially for multinational companies, and they require a systematic approach. Site selection projects typically consist of three main tasks:

1. The specification of selection criteria;
2. Geographical exploration to identify potential sites;
3. Facility site selection.

These tasks are discussed in the following paragraphs.

4.4.1 Selection Criteria

Three types of selection criteria may be considered in a site selection project: critical, economic, and qualitative. If a site does not meet a *critical* criterion, it is automatically discarded. *Objective* criteria can be assessed in financial terms for a given processing volume (e.g. transportation costs). *Qualitative* criteria are evaluated using preference indicators. Some of these indicators may be *quantifiable* (e.g., cost of living), but others are purely *subjective* (e.g. quality of life). The nature of the critical criteria depends on the context. The relative importance of the criteria also varies from one case to another and, for a given case, from one level (country, region, specific site) to another.

Several examples of plausible site selection criterion are listed in Table 4.2. In this table, labels are used to indicate whether a criterion is economic (E), quantifiable (Q), or subjective (S). The specification of criteria is a crucial part of the decision-making process because it significantly influences the site selection. A project committee, ideally including representatives of all company stakeholders, typically chooses relevant criteria.

4.4.2 Geographical Exploration to Find Sites

In the current global economy, the search for suitable sites for a facility can be a daunting task. To simplify things, it is generally done top down, gradually refining the geographical areas considered. The focus is first on countries, then on regions or cities, and finally on sites. These days, the initial geographic exploration is usually done using the Internet and informal professional information sources, and it aims to select a limited number of auspicious regions or cities. A few sites in each of these regions or cities are then examined in detail.

In an international context, several factors that do not fall under the control of the company have a huge influence on SC deployment decisions. These factors relate, first, to opportunities offered by some countries relative to company needs and, second, to the inherent benefits of dispersion when a company competes in several markets. One of the main reasons that companies locate in a foreign country is access to low cost or rare resources. For example, the availability of cheap labor has long been a rationale invoked by companies for delocalization decisions. However, labor is not the only resource of importance. Access to raw materials, energy, specialized equipment, or leading-edge technologies is also crucial. For example, given its moderately priced hydroelectric capacity, Canada offers advantages to aluminum manufacturers. At another level, a company requiring strong R&D capabilities may want to locate near a university research center in order to have access to latest technologies and skilled labor.

Moreover, business practices and logistics infrastructures are not the same around the world. Indeed, values, culture, laws, and economic development levels vary widely from one country to the other. Most developed nations, such as G7 countries (Germany, Canada, United States, France, Italy, Japan, and the United Kingdom), are equipped with elaborated road, port, and airport infrastructures. They possess a high level of expertise in information, communication, handling, and transportation technologies. Supply chain managers are well-trained professionals mastering many of the concepts and technologies presented in this book. Wages in these countries are high and consumer markets are sophisticated. They are also more and more concerned with the environment.

Table 4.2 Sample of site selection criteria

Factor	Criterion	Type
Access to markets/distribution centers	• Cost of deliveries	(E)
	• Importance of local markets	(Q)
	• Impact on sales	(Q)
Access to suppliers/resources	• Cost of procurement	(E)
	• Impact on local presence of suppliers	(S)
National/state government	• Political stability	(S)
	• Governmental policies	(S)
	• Taxes and tax credits	(E)
	• Tariff and nontariff barriers	(E,Q)
	• Implementation assistance programs (subsidies, preferential interest rates, protected exchange rates, accelerated depreciation, guaranteed contracts, etc.)	(E,Q)
	• Health system	(S)
	• Related expenses (employment insurance, etc.)	(E)
Community	• Atmosphere and charisma	(S)
	• Crime rate	(S)
	• Cost of living	(Q)
	• Cooperation with local industry	(S)
	• Housing (availability, price)	(Q)
	• Schools, cultural and religious life, sports	(S)
	• Universities and research centers	(S)
	• Municipal taxes	(E)
Competition	• Location of competitors	(Q)
	• Reaction of competitors	(S)
Environment	• Environmental protection laws	(Q)
	• Investment costs required to comply	(E)
	• Deadlines and possible flexibility	(S)
Links with the rest of the company	• Cost of collaboration with existing sites	(E)
	• Need of technical help from the company	(S)
Labor	• Wages	(E)
	• Labor relations	(S)
	• Productivity (turnover, absenteeism, attitude)	(Q)
	• Availability (unemployment, population, local expertise)	(Q)
	• Training opportunities	(S)
Transportation infrastructures	• Quality of road network	(Q)
	• Local carriers (number, rate, quality)	(Q)
	• Access by rail	(Q)
	• Proximity of an airport	(Q)
	• Access to maritime transport	(Q)

(continued)

Table 4.2 (continued)

Factor	Criterion	Type
Telecommunication infrastructure	• Quality of telephone service	(Q)
	• Access to information highways	(Q)
Services	• Quality, price, and availability of energy and water	(E,Q)
	• Wastes disposal	(Q)
	• Quality of roads, police, firefighters, hospital services, etc.	(S)
	• Financial services and financing	(S)
Site specifics	• Available space and layout	(Q)
	• Cost of the site and building	(E)
	• Expansion options	(E)
	• Parking and traffic	(Q)
	• Cost of insurance	(E)

However, developing countries such as BRIC nations (Brazil, Russia, India, and China) are currently experiencing rapid industrialization and improving professional competencies, but the average income of individuals, though increasing, remains moderate. Their transport and communication infrastructures are not always sufficient, but they adopt aggressive industrial plans to improve them. Many of them also have made remarkable advances in some industrial sectors, for example Brazil in the automotive sector and India in software development. They adopt laws governing business transactions and protecting intellectual property (patent, copyright, etc.), but they are not always respected. For all these reasons, it is not always easy to implement advanced SC systems in these countries. This is changing quickly, however. For example, since the admission of China in the WTO, barriers preventing global 3PLs to operate in the country fell and the quality of logistics services available is increasing rapidly, especially in major commercial areas such as Beijing, Shanghai, and Guangzhou.

Third-world countries, such as Haiti, Sudan, and Afghanistan, are characterized by a chronic lack of adequate infrastructures, a shortage of trained personnel, weak currencies, and unstable political regimes, which makes any business ventures very risky. Several ethical issues related to child labor, human rights, bribes, fraud, and environmental breach can also arise. A few comparative statistics are provided in Table 4.3 to show the difference between developed and nondeveloped countries.

The political dimension plays a key role in the decision to delocalize offshore, (or even to another state or province in a given country) and in the specific choice of a site. In some cases, a local presence is the only way to fully satisfy various government requirements (tariff and nontariff barriers, nationalist purchase movements, domestic content rules, etc.). This also avoids the costs and time needed to untangle complex laws regulating foreign exchanges. In Canada, for example, despite the adoption in 1995 of an *Agreement on Internal Trade*, there are still many trade barriers between provinces. However, with the liberalization of

Table 4.3 Comparative country statistics. *Source* CIA World Factbook (2013)

	Population (millions)	Life expectancy	Literacy (%)	GDP per capita ($ US)	Internet users (millions)	Paved roadways (km)	Railways (km)	Paved airports
Canada	34.568	81.5	99	$43,400	26.96	415,600	46,552	522
France	65.961	81.6	99	$36,100	45.26	951,200	29,640	297
United States	316.668	78.6	99	$50,700	245	4,374,784	224,792	5194
Brazil	201.009	73.0	88.6	$12,100	75.982	212,798	28,538	713
China	1349.585	74.9	92.2	$9300	389	3,453,890	86,000	452
India	1220.8	67.5	61	$3900	61.338	3,320,410	63,974	251
Afghanistan	31.108	50.1	28.1	$1100	1	12,350	–	23
Haiti	9.893	62.8	52.9	$1300	1	768	–	4
Sudan	34.847	62.9	61.1	$2600	4.2	4320	5978	15

exchanges, some governments are fighting among themselves to attract investors. A company can then negotiate its presence and benefit from various governmental subsidies: loan at a reduced rate, accessibility to some resources at a low price (e.g., electricity), tax rebates, and so on. In return, the company may have to satisfy certain constraints, such as a local hiring floor or limits on exports. These issues are examined in more depth in Chap. 9.

Several annually published comparative studies provide useful information to compare countries. Reports on the competitiveness of nations by IMD (www.imd.org/wcc/) and the *World Economic Forum* (www.weforum.org), as well as the Globalization Index published by *A.T. Kearney and Foreign Policy* (www.foreignpolicy.com), and the Ease of Doing Business Index (www.doingbusiness.org) of *The World Bank* are examples. Some global consulting firms also offer regional comparative studies. For example, KPMG publishes *Competitive Alternatives* (www.competitivealternatives.com) to help businesses analyze and compare location costs in major cities around the world. The 2012 study investigates costs in 113 cities in 9 countries with mature markets (Australia, Canada, France, Germany, Italy, Japan, Netherlands, the United Kingdom and the United States) and 5 countries with high growth markets (Brazil, China, India, Mexico and Russia). It measures 26 cost elements related to geographic location for 19 industrial sectors over a period of 10 years. The study also compares subjective factors such as the qualified labor pool, economic conditions, the degree of innovation, infrastructure, regulations as well as the cost and quality of life. An overview of the results of the 2012 study is provided in Fig. 4.17. The numbers in the figure give the average annual operational costs for the country, expressed as a percentage of the average cost of large US cities. The results are also available by industrial sector and by cities. When considering the results over several years, it is

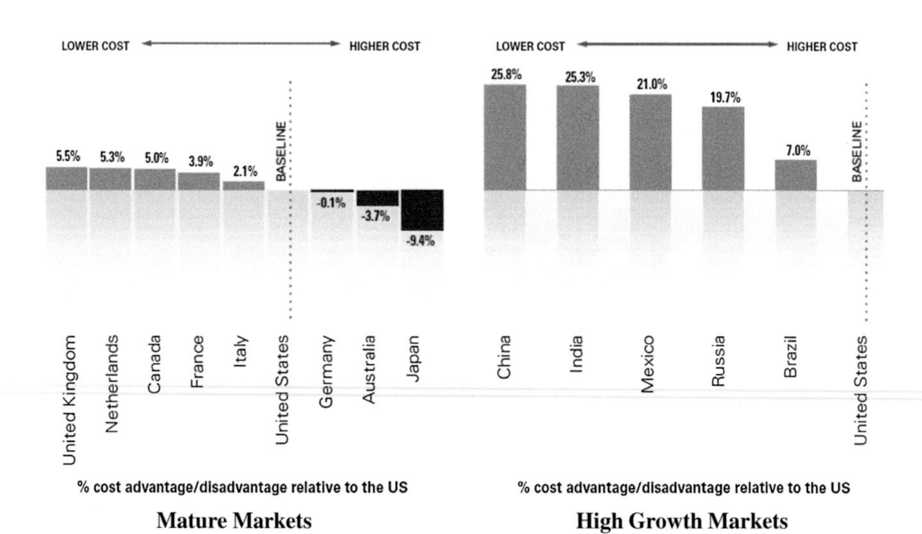

Fig. 4.17 Highlights of results for 2012 KPMG comparative alternatives study. *Source* KPMG (2012)

clear that the comparative advantage of countries and cities can change significantly over time. This is interalia because of the evolution of exchange rates and of countries' internal economic contexts.

Today, the initial geographical exploration can be done effectively using the Internet. In addition to those already mentioned, useful websites for the study of countries include the following:

- United Nations (www.unsystem.org)
- OECD (www.oecd.org)
- The Economist Intelligence Unit (www.eiu.com)
- EUROSTAT (http://ec.europa.eu/eurostat)

To find specific locations, the websites of cities and industrial parks can be useful, as well as some specialized websites such as these:

- Society of Industrial and Office Realtors (www.sior.com)
- Industrial facility search engines (e.g., www.fastfacility.com)
- EFT's North American 3PL Warehousing Map (www.eft.com)

The detailed study of specific sites usually requires visits.

4.4.3 Facility Site Selection

To select a site, the potential locations identified in the geographical exploration phase must be compared using the selection criteria specified. One should start with a rough evaluation of potential sites based on easily accessible information to shorten the list as much as possible. However, to reach a final decision on the site to use, the investigation must be pushed much further. It may involve site visits; detailed information requests to facility owners, municipalities, and governments; meetings with key stakeholders; external expert appraisals; and so on. With the results of these studies in hand, the analysis can be finalized using a multicriteria ranking method.

A simple multicriteria decision-making method proposed by Brown and Gibson (1972) for this kind of study is presented briefly in the following paragraphs. It requires the specification of critical, economic, and qualitative criteria, as discussed previously. The approach is straightforward:

1. Discard any site not complying with a critical criterion.
2. For remaining sites, define a normalized location measure that reflects the relative importance of each criterion and choose the site with the highest score.

The following basic notation is required to describe the approach:

l Site location index ($l \in L^S$, the set of potential site locations)

i Criterion (factor) index

LM_l Location measure computed for site l

OM_l Objective (economic) criteria measure computed for site l

SM_l Subjective criteria measure computed for site l (all qualitative criteria are treated as subjective)

λ Weight given to objective criteria ($0 \leq \lambda \leq 1$), implying that $(1 - \lambda)$ is the weight given to subjective criteria.

The objective criteria measure OM_l must be defined so that $OM_l > OM_{l'}, l \neq l'$, implies that site l is preferred to site l'. Measure SM_l must have the same characteristic. These measures must also be normalized, meaning that they must have the following properties:

$$\sum_{l \in L^S} OM_l = 1, \quad 0 \leq OM_l \leq 1, \quad \text{and} \quad \sum_{l \in L^S} SM_l = 1, \quad 0 \leq SM_l \leq 1$$

An appropriate location measure for site l is obtained by taking a linear combination of the objective and subjective criteria measures, that is, by computing:

$$LM_l = \lambda OM_l + (1 - \lambda) SM_l \tag{4.1}$$

The weight λ does not have to be specified explicitly. A sensitivity analysis for weight values $\lambda \in [0, 1]$ will instead be performed.

Because the objective criteria are expressed in terms of costs, a normalized measure is obtained simply by inverting the total cost, TC_l, for each site l, and then dividing the resulting value $(1/TC_l)$ by the sum of inverted values, that is, by computing:

$$OM_l = 1 \bigg/ TC_l \left(\sum\nolimits_{l' \in L^S} 1/TC_{l'} \right), \quad TC_l = \sum_i C_{il} \tag{4.2}$$

where C_{il} is the cost corresponding to objective criterion i for site l.

The calculations required can be illustrated with an example. Suppose that, following an in-depth study, four potential sites for a facility are considered. Each site was assessed for seven objective criteria. The costs obtained for these criteria are provided in Table 4.4 (in millions of dollars). Evaluating Eq. (4.2) with site 1 data, the following normalized measure is obtained:

$$OM_1 = \frac{1}{2.25(1/2.25 + 1/2.55 + 1/2.7 + 1/4.3)} = 0.309$$

The application of Eq. (4.2) to the other sites gives $OM_2 = 0.272$, $OM_3 = 0.257$ and $OM_4 = 0.162$.

One easily verifies that the sum of these measures is equal to 1 and that they are all between 0 and 1. Site 1 having the highest score ($OM_1 = 0.309$), it is preferred from the point of view of the objective criteria. We would have reached the same conclusion by observing the total costs TC_l, $l \in L^S$, but normalization is required to be able to compute the LM_l's with Eq. (4.1).

Table 4.4 Annual costs in millions of dollars for the objective criteria

	Criterion (i)							
Site (l)	1	2	3	4	5	6	7	$\sum_i C_{il}$
1	1.5	0.6	0.03	0.04	0.02	0.04	0.02	2.25
2	1.6	0.7	0.05	0.06	0.04	0.07	0.03	2.55
3	1.85	0.6	0.06	0.06	0.05	0.05	0.03	2.7
4	3.45	0.5	0.1	0.06	0.08	0.06	0.05	4.3
	Labor	Transport	Municipal taxes	Fuel	Income taxes	Electricity	Water	

A normalized measure for the subjective criteria is obtained by weighing the attractiveness of each site for each subjective criterion using the relative importance of criteria, that is, by calculating:

$$SM_l = \sum_i W_i U_{li} \tag{4.3}$$

where

W_i Weight of criterion i relative to all subjective criteria
U_{li} Preference measure of site l relative to all potential sites for criterion i

To obtain these measures, we use basic concepts from preference theory. The project team must first elaborate a binary preference matrix **P** with elements

$$p_{ij} = \begin{cases} 0 & \text{if criterion } j \text{ is given more importance than criterion } i \\ 1 & \text{if criterion } i \text{ is given more importance than criterion } j \text{ or we are indifferent} \end{cases}$$

The criterion weights, are then given by:

$$W_i = \sum_j p_{ij} \Big/ \left(\sum_i \sum_j p_{ij} \right) \tag{4.4}$$

and they have the following characteristics: $0 \leq W_i \leq 1$, for all i, and $\sum_i W_i = 1$.

For the previous example, assuming that seven subjective criterion were specified, the matrix **P** obtained using pairwise comparisons and the weights calculated are given in Table 4.5. The fact that $p_{21} = 1$, for example, means that the project team considers that *labor competency* (criterion 2) is at least as important as the *availability of labor* (criterion 1). It is easily verified that the weights calculated sum to 1.

The preference measures U_{ik} are obtained similarly by comparing sites two by two for each criterion. If the criterion is quantifiable, then the comparisons are based on its value. For housing, for example, the comparison could be based on regional home price statistics. This process is illustrated in Fig. 4.18. For the example, it yields the normalized measures matrix $\mathbf{U} = [U_{ik}]$ provided in Fig. 4.19. Finally,

Table 4.5 Preference matrix and subjective criteria weights

	Criterion (*j*)							
(*i*)	1	2	3	4	5	6	7	W_i
1		1	1	1	1	1		5/24
2	1		1	1		1	1	5/24
3				1				1/24
4						1		1/24
5	1	1	1	1		1		5/24
6			1					1/24
7	1	1	1	1	1	1		6/24
	Availability of labor	Labor competency	Labor attitude	Quality of life	Housing	Competition	Labor unions	

Criterion 1 (Availability of labor):

Site	1	2	3	4	U_1
1					0.000
2	1		1		0.333
3		1		1	0.333
4	1	1			0.333
					1.000

Criterion 2 (Labor competency):

Site	1	2	3	4	U_2
1					0.000
2	1		1		0.333
3		1		1	0.333
4	1	1			0.333
					1.000

Criterion 3 (Labor attitude):

Site	1	2	3	4	U_3
1		1			0.125
2	1		1	1	0.375
3	1	1		1	0.375
4			1		0.125
					1.000

Criterion 4 (Quality of life):

Site	1	2	3	4	U_4
1					0.000
2	1		1		0.333
3		1		1	0.333
4	1	1			0.333
					1.000

Criterion 5 (Housing):

Site	1	2	3	4	U_5
1		1			0.125
2	1		1	1	0.375
3				1	0.125
4	1	1	1		0.375
					1.000

Criterion 6 (Competition):

Site	1	2	3	4	U_6
1					0.000
2	1		1		0.333
3		1		1	0.333
4	1	1			0.333
					1.000

Criterion 7 (Labor unions):

Site	1	2	3	4	U_7
1		1			0.143
2	1		1	1	0.429
3	1	1		1	0.429
4					0.000
					1.000

Fig. 4.18 Excel site preference matrix for each criterion

Site	Criterion 1	2	3	4	5	6	7
1	0	0	0.125	0	0.125	0	0.143
2	0.333	0.333	0.375	0.333	0.375	0.333	0.429
3	0.333	0.333	0.375	0.333	0.125	0.333	0.429
4	0.333	0.333	0.125	0.333	0.375	0.333	0

Fig. 4.19 Excel matrix (**U**) of preference measures

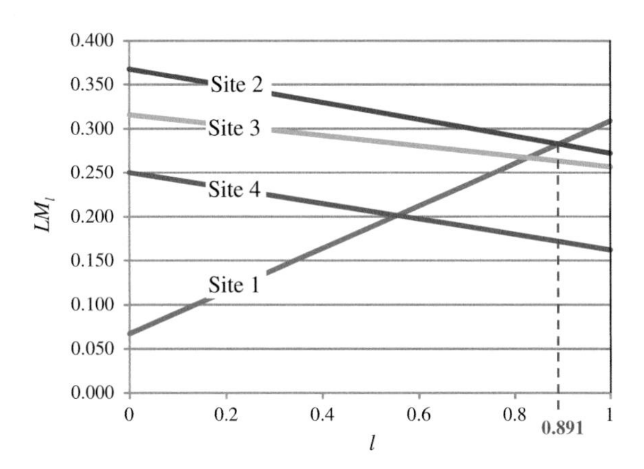

Fig. 4.20 Multi-criteria analysis chart

normalized measures for the subjective criteria are obtained with Eq. (4.3), which boils down to multiplying matrix **U** (Fig. 4.19) by vector $\mathbf{W} = [W_i]$ (right end column in Table 4.5). This yields the following vector of subjective criteria measures for the sites:

$$\mathbf{SM} = [0.067, 0.3676, 0.3155, 0.25]$$

The selection decision is finally made by comparing the site location measures $LM_l, l \in L^S$, computed with Eq. (4.1) for all possible values $\lambda \in [0, 1]$ of the weight given to objective criteria. This comparison is done visually by plotting the lines $LM_l(\lambda) = \lambda OM_l + (1 - \lambda)SM_l$ on a graph for all sites l. Note first that $LM_l(0) = SM_l$ and $LM_l(1) = OM_l$, and that the two extreme points provided by these relations are sufficient to plot the location measure line for each site. The lines obtained for our example are plotted in Fig. 4.20. By examining this figure, it is seen immediately that site 2 dominates sites 3 and 4, which means that sites 3 and 4 can be discarded. It is also clear that unless objective criteria have a weight exceeding 0.89, it would be better to choose site 2.

4.5 Technological Systems and Value Creation

We explained in this chapter that to improve the performance of an industrial facility, it is not sufficient to think in terms of peocessor selection: we need rather to focus on the design of effective resources–configuration–methods systems. In addition, improving the productivity of a facility does not necessarily lead to value creation. It does only if the provided order-winners match the expectations of customers, and exceeds those offered by competitors for at least one of the attributes valued by the customers. To conclude this chapter, we show the interrelationship among all these concepts in a more formal way.

The productivity of an activity center for a period of time, say year t, is the ratio of a measure of its outputs during this period, on a measure of the resources used during the same period (i.e., $\text{Productivity}_t = \text{Output}_t/\text{Resources}_t$). In practice, it is never easy to find an adequate measure for resources used and outputs produced, and it is for this reason that commonly used productivity ratios, such as the productivity of human resources ($ income/$ salaries), provide only a partial view. Nevertheless, the implications are clear: to improve productivity, one must design activity centers that produce more outputs with fewer resources. However, in the previous sections, we indicated that

$$\text{Outputs}_t = \text{function}_t(\text{Resources}_t, \text{Configuration}_t, \text{Methods}_t)$$

where, for a given technological system, *function*$_t$() evolves with time because, as the system gains experience, it is usually able to produce more outputs with the same resources (i.e., the function depends on the system learning curve). It is therefore by designing resources–configuration–methods systems adapted to needs, and more specifically to demand and products complexity (which determine work contents), that a facility can improve its productivity. Often, it is not necessary to upgrade resources, it is sufficient to align layouts and methods with needs. However, when resources are modified to reap expected gains, one must not forget to adapt the facility configuration and methods.

This discussion focusing on the intrinsic efficiency of a technological system misses an extremely important part of the problem. Activity centers and facilities are part of a company and their mission is not only to produce goods but also to provide all the order winners (price, quality, response time, flexibility, service, etc.) valued by clients. Many of these order winners not only depend on the productivity of a facility but also on its location relative to sources of supply, subcontractors, and markets. However, customers purchasing decisions are also affected by the order winners offered by competitors. As explained in Chap. 1, to create value, an activity center must develop the capabilities necessary to meet all qualification criteria required to compete on the market, and it must also dominate competitors for at least one order-winning criterion. Improving the performance of a facility thus requires an understanding of its impact on all value drivers, that is, an investigation of the following relationship:

$$\text{Order-Winners}_t = \text{function}_t(\text{Resources}_t, \text{Configuration}_t, \text{Methods}_t, \text{Localisation}_t)$$

In addition, while creating value for customers, the company must ensure that it remains profitable, that is, it also creates value for its owners. Productivity improvement thus remains an important issue. Consequently, the critical factors driving the contribution of a facility to a company's value creation can be summarized as follows:

$$\text{Value Added}_t = \text{function}_t(\text{Productivity}_t, \text{Order-Winners}_t, \text{Customer-Needs}_t, \text{Competition}_t)$$

The study of present and future impacts on these competitiveness factors should prevail in any facility development project.

Review Questions

4.1. Give a definition of an activity center. Present a concrete example to illustrate your definition.
4.2. Compare the definition of technology provided in this chapter to the definition found in a dictionary.
4.3. What are the roles of processed objects and processors in a technological system?
4.4. Three basic processor types are identified in this chapter. What are they? Give specific examples for each from your company or a company you know.
4.5. Why do we say that a facility can be seen as a hierarchical technological system?
4.6. What do we mean by a resources-configuration-methods triplet?
4.7. Different layout types were presented in this chapter. What type of layout is currently used in your company? Is this a desirable configuration for your business? Explain.
4.8. What is a method?
4.9. How can we improve productivity by revising methods? Try to see how your answer may apply to your business.
4.10. Suppose you must select a site for a new facility. How would you proceed?
4.11. Is it necessary to apply Brown and Gibson's method to choose between several potential sites? Could we use another multicriteria decision-making method?

Exercises

Exercise 4.1 A farming company raises sheep to harvest wool. For sheep to grow properly, they must receive a minimum quantity of each of four foods. Three types

Table 4.6 Food production data

Food	Type of grain 1	2	3	Minimum daily quantities
A	20	30	70	110
B	10	10	0	18
C	50	30	0	90
D	6	2.5	10	14
Cost ($/kg)	1.10	0.90	2.25	

of grains are considered to feed the sheep flock. Table 4.6 shows minimum daily requirements by food and food contents by type of grain. It also shows the cost of each type of grain in $/kg. As long as the sheeps eat the amount of grain required to meet the minimum daily food needs, they will be healthy and will produce a normal amount of wool. What daily quantity of each grain should each sheep receive in order to minimize costs while respecting minimal daily food needs?

Exercise 4.2 A production center involves six machining workstations and three service areas listed in Table 4.7. The center mainly produces the five components described in Table 4.8. The scores given in Table 4.9 reveal the importance of closeness between service areas and workstations.

(a) Calculate the monthly traffic (in pallets) between each pair of workstations.
(b) Develop a space relationship graph for the production center.
(c) Based on your space relationship graph, develop two feasible block layouts for the center.
(d) Which block layout is the best? Why?

Exercise 4.3 A technological SME in the Quebec City suburbs saw its sales increase rapidly during the past 2 years and it must significantly increase its production capacity. This growth results from the gradual geographical expansion of the company sales market, which now has customers in the Quebec City, Montreal, Toronto, and Boston regions. As a consequence, the company must now build a

Table 4.7 Production center work areas

Workstations/areas	Required space (ft^2)
A—Band saw	50
B—Folder	50
C—Press	100
D—Turret lathe	100
E—Drill	40
F—Polisher	60
G—Production control	60
H—Supervisor	30
I—Quality assurance	80

Table 4.8 Component routings and production data

Component	Routing	Production per month	Components/pallet
1	A → B → D → E → F	1000	100
2	B → D → E	500	250
3	C → E → F	300	50
4	A → B → C	2000	100
5	B → C → E	600	50

Table 4.9 Proximity scores

	A	B	C	D	E	F	G	H	I
G	I/2		X/3					I/1,2	
H	D/1	D/1	D/1						
I			X/3			I/2			

Scores (I) Closeness important, (D) closeness desirable, (X) closeness undesirable
Reasons 1—ease of supervision, 2—document flow, 3—noise and vibrations

new plant and the management is wondering if it should be located in Montreal, Toronto, or Boston instead of Quebec City. You are asked to find an appropriate site in each region and to assess the costs of construction, supply, production, distribution, and services for each site. You are also asked to identify the subjective criteria to take into account in the study and to propose an adequate approach for choosing a site.

Having studied the multicriteria ranking method of Brown and Gibson at the university, you decide to suggest it to your boss, who accepts your proposal. After compiling all relevant costs, evaluating the relative importance of specified subjective criteria, and the attractiveness of each site against these criteria (using pairwise comparisons), you have all the data required to make a decision. These data are summarized in Fig. 4.21. Knowing that owners give as much importance to subjective criteria as to economic criteria, which site would you recommend?

Exercise 4.4 Plasbec manufactures plastic bottles and its sales increased so much during the last year that it is now forecasted that its Montreal plant will not have enough capacity to meet future demand. The company has customers in Quebec and Ontario, as well as in the northeast of the United States. By consulting the Comparative Alternatives Study published by KPMG (2012) for the plastic products industry, Plasbec sees that the cities with the lowest location costs in its sector are Quebec City (QC), Montreal (QC), Niagara (ON), and Charleston (WV). The company therefore wonders if it should build a new plant in Quebec City, Niagara, or Charleston or expand its Montreal factory. To reach a decision, the CEO asks you to find sites that could be suitable in each region, to identify the criteria to be taken into account in the study, and to propose an adequate site selection approach. To keep the costs and delays of the study to a minimum, it is agreed that instead of doing a detailed cost analysis for each potential site, the following relative cost

Criteria	Description	Weight
1	Availability of labor	0.22
2	Labor productivity	0.21
3	Labor attitude	0.06
4	Quality of life	0.05
5	Housing	0.21
6	Labor unions	0.25

	Preference measures						Costs
Site	1	2	3	4	5	6	M$
Quebec	0	0	0.12	0	0.13	0.14	2.1
Montreal	0.34	0.32	0.38	0.34	0.37	0.44	2.4
Toronto	0.4	0.34	0.37	0.33	0.14	0.42	2.9
Boston	0.26	0.34	0.13	0.33	0.36	0	4.1

Fig. 4.21 Excel tables for subjective criteria

index found in the KPMG study can be used as a basis for relevant total costs comparisons:

Quebec City : 91 Montreal : 93.4 Niagara : 93.6 Charleston : 94.1

After discussion, the subjective criteria and pairwise criterion comparison results at the top of Fig. 4.22 are obtained. Proceeding by pairwise comparison with Excel, the preference measures at the bottom of Fig. 4.22 for each site in relation to each subjective criteria are also obtained. Which site should be selected?

Exercise 4.5 *EcoloSoap* is a company specialized in selling ecological soaps and cleaning products for commercial and industrial use. It sells bath soaps, bath gels, powder detergents, and liquid laundry soaps. A secondary market targeted by the company is fatty oils and processed glycerin for industrial use. Glycerin, is a semi-finished product obtained during the transformation of soap and it is used by other companies to produce cosmetics. Processed fats are also semi-finished products that the company can sell to other cleaning products manufacturers.

Five raw materials are required to produce soap: fatty oils, two laundry ingredients (sodium and potassium), and two additives (detergent and perfume). The production is generally divided into four key processes: purification, saponification, crystallization, and finishing (cutting and curing). The last three steps require different equipment for the transformation of liquid soaps and solid soaps, and are grouped under a hot production process and a cold production process, respectively. The fatty oils are first transformed into fats by the purification process. Processed fats are considered as semi-finished products since they are produced by the company through the purification activity. However, if need be, they can also be procured from external suppliers. During the cold process, an optional separation activity can be used to separate glycerin from other oils. This activity allows the production of glycerin to sell on the market. Fats turn into soaps when coupled with

Pairwise Comparison Results

Criterion	Description	1	2	3	4	5	6
1	Availability of labor		1	1	1		1
2	Labor competency	1		1	1		
3	Municipal regulations				1		
4	Quality of city servicies			1			
5	Labor union	1	1	1	1		1
6	Quality of life			1			

Preference measures for subjective criteria

Site	1	2	3	4	5	6
Quebec City	0.15	0.05	0.12	0	0.15	0.18
Montreal	0.33	0.29	0.38	0.3	0.37	0.4
Niagara	0.27	0.34	0.37	0.34	0.12	0.42
Charleston	0.25	0.32	0.13	0.36	0.36	0

Fig. 4.22 Excel tables for pairwise comparison and preference measures

sodium or potassium, through the hot or cold saponification process, respectively. The hot process gives liquid soaps and the cold process solid soaps.

After production, all finished products are stored before being shipped to customers. Remaining products are sold under a make-to-order strategy and are transshipped from the plants through 3PL platforms. Finally, since most of the suppliers for fatty oils are located in Asia, this raw material is the only one stored at plants. Based on the previous description, and using the mapping formalism illustrated in Fig. 4.2, prepare an activity graph for Ecolosoap.

Bibliography

Benjaafar S, Heragu S, Irani S (2002) Next generation factory layouts: research challenges and recent progress. Interfaces 32(6):58–76

Brown P, Gibson D (1972) A quantified model for facility site selection. AIIE Trans 4:1–10

Bruce J (1985) Developing a framework for location analysis. Ind Eng 60–65

Canbolat Y, Chelst K, Garg N (2007) Combining decision tree and MAUT for selecting a country for a global manufacturing facility. Omega 35:312–325

Cap Plus Technologies (2015) www.capplustech.com/softgel/plant_layout.htm. Accessed 29 Mar 2015

Chapman K, Walker D (1991) Industrial location: principles and policies. Blackwell, Oxford

CIA World Factbook (2013) https://www.cia.gov/library/publications/resources/the-world-factbook/. Accessed 19 Feb 2013

Francis R, McGinnis L, White J (1992) Facility layout and location, 2nd edn. Prentice-Hall, Englewood Cliffs

Harrington J, Warf B (1995) Industrial location: principles and practice. Routledge, London

Hayes R, Wheelwright S (1984) Restoring our competitive edge: Competing through manufacturing. Wiley, New York

Kathawala Y, Gholamnezhad H (1987) New approach to facility locations decisions. Int J Syst Sci 18–2:389–402
KPMG (2012) Comparative alternatives: KPMG's guide to international location costs. KPMG LLP
Montreuil B (2007) Layout and location of facilities. In: Taylor G (ed) Handbook on logistics engineering. CRC Press, Boca Raton
Montreuil B, Boctor F, Martel A (1995) La maîtrise des technologies de production. In: Oral M (ed) Martel A. Vision et stratégies, Publi-Relais, Les défis de la compétitivité
Montreuil B, Venkatadri U, Ratliff D (1993) Generating a layout from a design skeleton. IIE Trans 25(1):3–15
Muther R (1961) Systematic layout planning. Industrial Education Institute
Partovi F (2006) An analytic model for locating facilities strategically. Omega 34:41–55
Porter M (1990) The competitive advantage of nations. Free Press, New York
Roper K, Ha Kim J, Lee S-H (2009) Strategic facility planning: a white paper. International Facility Management Association
Salvendy G (ed) (2001) Handbook of industrial engineering, 3rd edn. Wiley, New York
Schmenner R (1982) Making business location decisions. Prentice-Hall, Englewood Cliffs
Sosef D, Nassiri A (2013) Europes's most desirable logistics locations. Prologis
Tompkins J, White J (1984) Facilities planning. Wiley, New York
Treillon R, Lecomte C (1996) Gestion industrielle des entreprises alimentaires. Tec & Doc
Wood D, Barone A, Murphy P, Wardlow D (1995) International logistics. Chapman & Hall, London

Chapter 5
Transportation in the Supply Chain

A supply chain network can be seen as a geographically deployed set of internal or external resource nodes interconnected by transportation arcs. Internal nodes correspond to the production–distribution facilities and centers studied in Chap. 4. External nodes include customer and supplier locations and, as we shall see in Chap. 6, any other SC partner premises performing some supply chain activity. This chapter focuses on SCN arcs and it studies transportation means available to move materials in a SC. Transportation is an essential economic activity, and it generates a large proportion of logistics costs in most SCs. According to the CSCMP's State of Logistics Report (Wilson 2013), in 2012 freight transportation costs accounted for about 5 % of the United States' GDP, 7 % of Europe's GDP, and 10 % of Asia's GDP.

Our focus in this chapter is on strategic transportation decisions made by production–distribution centers, and not on operational shipping decisions, or on planning-execution decisions made by transportation service providers. However, to make judicious transportation choices production–distribution companies must have an extensive knowledge of the services offered by the transportation industry. The chapter thus starts by presenting the elements of modern transportation systems, it highlights the characteristics of alternative freight transportation modes (road, rail, water, air, intermodal), and it gives an overview of processes, regulations, and services associated with shipping freight. Transportation pricing mechanisms are examined, and various cost modeling approaches are proposed to facilitate transportation means comparisons. Finally, an approach for the evaluation and selection of transportation means is presented.

A. Martel and W. Klibi, *Designing Value-Creating Supply Chain Networks*,
DOI 10.1007/978-3-319-28146-9_5

5.1 Freight Transportation in Perspective

5.1.1 Modern Transport Systems

Modern transport systems incorporate five basic elements, as shown in Fig. 5.1: (1) SC shipment demand sources, (2) freight transportation providers, (3) transport infrastructures, (4) transportation management systems, and (5) telecommunication infrastructures. Each of these subsystems is examined briefly in the following paragraphs.

The activities of production–distribution companies (transportation service users) require the movement of raw materials, semi-finished products, and finished products between the nodes (suppliers, plants, DCs, sales points, customers) of their SC network, that is, they spawn a material flow demand on the arcs of the SCN. The material to be moved may take several forms: it can be solid or liquid, it may require refrigeration, it may be dangerous (hazardous material), and so on. It may also be unpacked or packed into unit loads (boxes, bags, crates, drums, barrels, pallets, etc.) to facilitate its manipulation. The pickup and delivery points for a SCN arc may be in the same landlocked area, thus permitting ground transportation, but it may also require overseas transportation. All this generates a demand for transportation services involving different transportation modes (road, rail, water, air, and intermodal)[1] and equipment (dry van, flatbed, open-top and tank truck/trailer; rail boxcar, open wagon, flatcar and tank-car; barge, tanker and other cargo ships; containers, etc.). Moreover, depending on the volumes involved, SC shipments may require a full-vehicle load or less than a truck/car/container load. The specific shape of the demand depends on the flow planning and control processes used by companies, as discussed in Chap. 3. Our objective in this chapter is to show how a company should proceed to meet its transportation needs.

Materials are moved on SCN arcs by carriers using vehicles and transportation equipment. Manufacturing and distribution firms may have a private fleet and/or they may use the services of common carriers. As we shall see, all sorts of asset-based and non-asset-based carriers and third-parties provide transportation services. Collectively, they form the transportation industry. The vehicles used to apply steering and pull loads (tractors, locomotives, towboats, etc.) are often physically detached from the equipment containing freight (semi-trailers, wagons, barges, etc.). The activity of freight carriers gives rise to flows of vehicles and equipment between the nodes of a transportation network. These nodes can be customer ship-to or -from points, transport terminals (used for cross-docking, sorting, consolidation and unbundling of loads), intermodal transfer facilities (cargo hub), ports, airports, and so on. Note that vehicles or equipment may move on network arcs even in the absence of material flows (e.g., to return empty containers or trailers, or to relocate a tractor).

[1]Liquid products such as oil may also be transported using pipelines. This particular case, however, is not examined in this book.

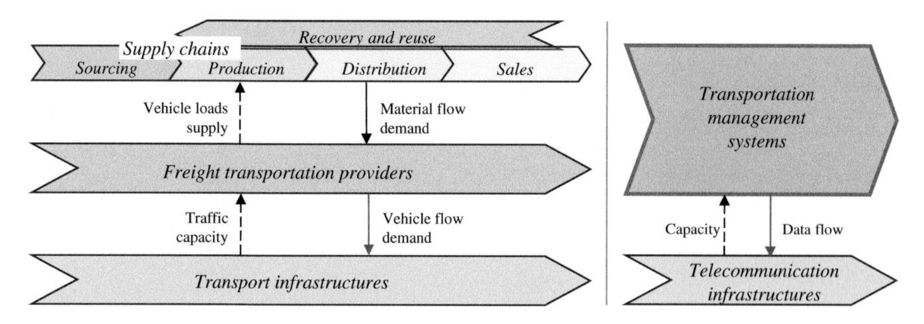

Fig. 5.1 Modern transport. Adapted from OECD (1992)

Transportation networks (TNs) can take several forms, as shown in Fig. 5.2. The simplest case is point-to-point shipments (TN a) from an origin (consignor) to a destination (consignee). Long-haul full-vehicle load shipments often take this form. Multiple pickup and/or delivery routes (TN b) are commonly used for local distribution and in the parcel carrier industry. When both pickups and deliveries are required, a transshipment platform (TN c) is often set up to reorganize the flow of goods in order to improve the use of loading units and vehicles. To cover larger areas, local consolidation terminals are usually implemented near large cities and

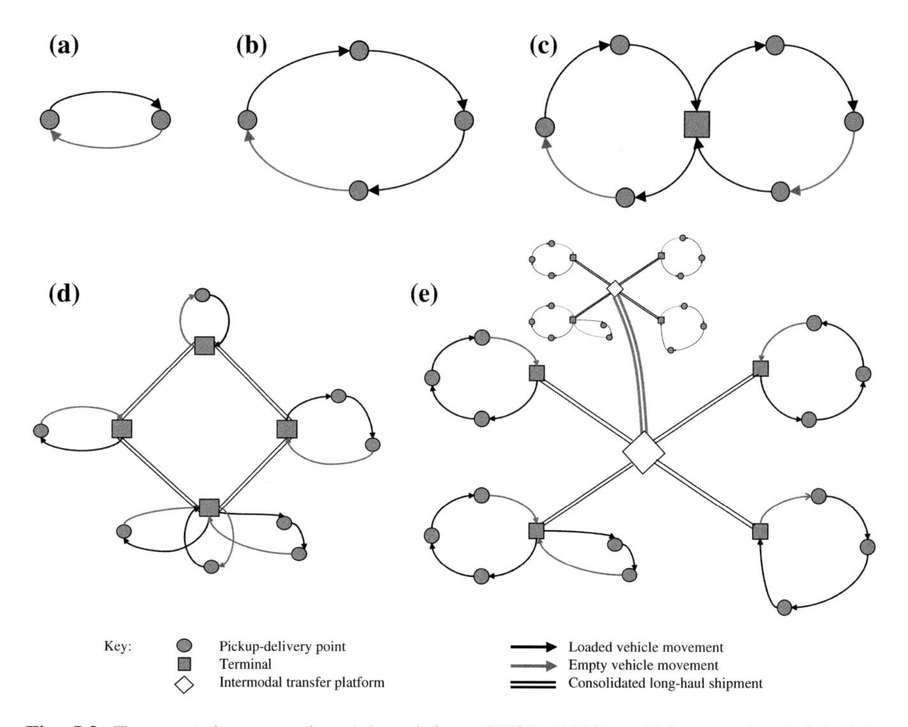

Fig. 5.2 Transportation examples. Adapted from OECD (1992). **a** Point to point. **b** Multiple pickup-delivery stops. **c** Pickup and delivery with a transshipment. **d** Network with long-haul inter-terminal lanes. **e** Hub and spoke network with intermodal transfer platforms

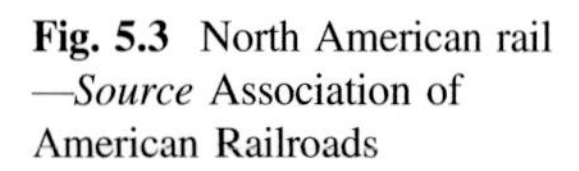

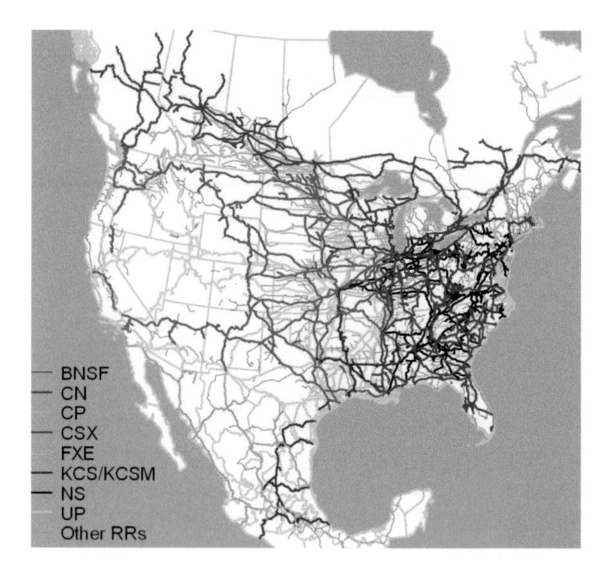

Fig. 5.3 North American rail —*Source* Association of American Railroads

full-load shipments are made between terminals (TN d). This may involve intermodal transport and one must then determine the mode to use on each lane. Hub-and-spoke networks (TN e) are frequently used for national and international shipping. In addition to terminals, they involve high-speed and high-volume multimodal transshipment facilities. Vehicles and equipment movements on TNs are constrained by the traffic capacity provided by transport infrastructures.

Transport infrastructures include roads, bridges, interchanges, and ports, as well as regulations concerning their usage (soft infrastructures). Roads encompass highways, railways, sea lanes, and air corridors; ports comprise airports, seaports, and rail yards. Each transportation mode has its road network. We are all familiar with highway networks. The North American rail network is illustrated in Fig. 5.3. Examples of regulations include vehicle dimensions, speed and load standards, one-way street restrictions, traffic priorities, road and bridge tolls, right of way on water, hazmat (hazardous material) route restrictions, and so on. Transport infrastructures offer the physical structures required for the movement of vehicles, and the magnitude and state of these structures provide a spatiotemporal traffic capacity. This capacity is shared, however, by vehicles for the transportation of persons as well as goods. The quality and capacity of transport infrastructures results mostly from public investments and government regulations.

The management of the former three subsystems—supply chain networks, transportation networks, and transport infrastructures—requires information as well as planning and control decisions. (TMSs) are used to collect primary transportation data generated at each level to transform it into managerial information and to provide decision support. Several of these systems operate in parallel at the shipper, carrier, and government levels and they must exchange information to make planning and execution decisions. Some of them are developed and operated by shippers and carriers but they may also be outsourced. The exchange of information between companies generate data flows that require a telecommunication infrastructure. As explained in Chap. 3, most of this exchange of data and information is done nowadays using the Internet or private EDI systems.

5.1.2 The State of the Transportation Industry

Efficient transportation is required to design value-creating supply chain networks. Raw material sources located in non-accessible regions are at a disadvantage. Similarly, regions with inefficient transportation systems are not likely candidates for the location of production–distribution facilities. This is true at both the national and international levels. The importance of national traffic in different parts of the world can be appreciated by looking at Table 5.1, which provides freight shipment volumes by inland surface transportation modes for developed and developing economies. The volumes involved are huge. International transport is a prerequisite for international trade. According to the WTO (World Trade Organization), global merchandise exports reached US$17,930 billion in 2012. The proportions of inter and intraregional merchandise flows associated with these global exports are shown in Fig. 5.4. A significant part of intercontinental flows involve seaborne or airborne shipments. According to UNCTAD (United Nation Conference on Trade and Development), 46,160 billion ton-miles of sea cargo was shipped in 2012. Of these, 31 % involved flows of major bulk commodities (iron ore, coal, grain, bauxite/alumina, and phosphate rock) and 27 % tanker oil and gas shipments. They state that container shipments accounts for about 16 % of global sea trade by volume, but more than 50 % by value. In the same year (2012), according to IATA (International Air Transport Association), air freight transport neared 170 billion tonne-kilometers.[2]

Despite these large numbers, demand for transportation services is expected to continue to expand, driven by higher GDP, global trade, and larger populations. Figure 5.5 illustrates current trends for inland (road, rail, and waterways) freight transport (in billion tonne-km) in the United States and China. Figure 5.6 show world seaborne shipments (in billion tons) by cargo type. Both figures display significant upward trends. Following the economic shock in 2008–2009, some regions have recovered and some are still struggling. However, there is a marked difference in transport growth between developing and developed economies, with the former faring better than the latter. For the long term, the OECD (ITF Transport Outlook, 2012) is forecasting that surface freight shipments (road and rail) will be about six times as large in 2050 as in 2010 in non-OECD regions, and more than two times as large in OEDC countries. Also, according to OECD projections, air freight could triple and container traffic in ports could quadruple within the next 15 years (ITF 2013).

[2]A "tonne" is a metric measure equal to 1000 kg. In the United States it is referred to as a "metric ton."

Table 5.1 Freight shipments by country and mode for 2011

Billion tonne-kilometers	Developed economies		Developing countries		
	European Union	United States	China	India	Russia
Rail	2607	2519	2947	669	2128
Road	1672	3934	5137	1212	223
Waterways	140	465	2607	4	59
Total	4419	6918	10,691	1885	2410

Data Source ITF (2014)

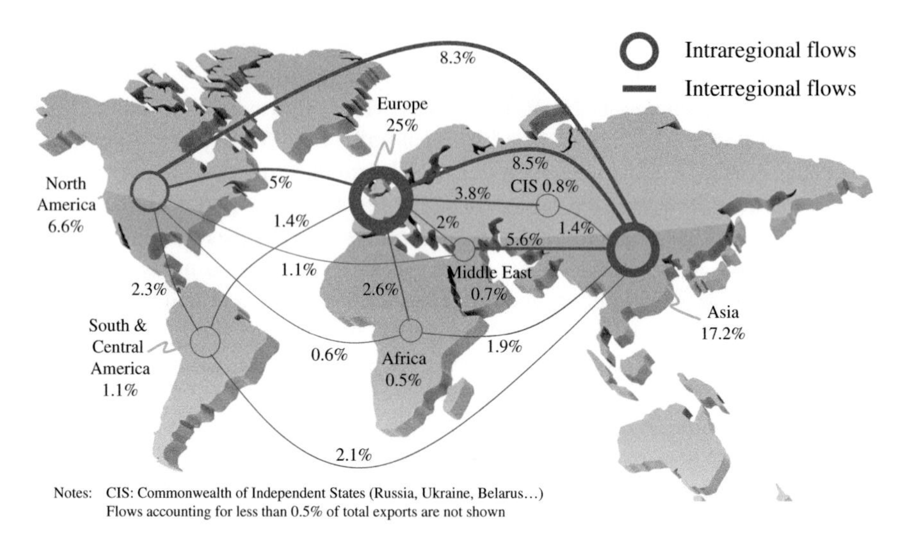

Fig. 5.4 Merchandise flows generated by global 2012 exports—data compiled from WTO's world and regional export profiles 2012

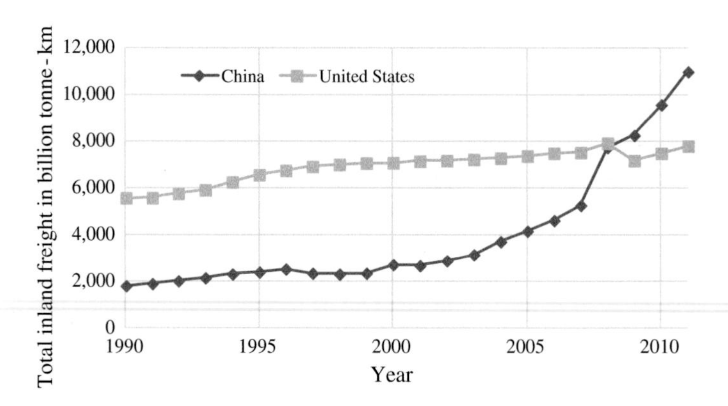

Fig. 5.5 National freight shipments trends—data *Source* International Transport Forum Data Base (2011 data)

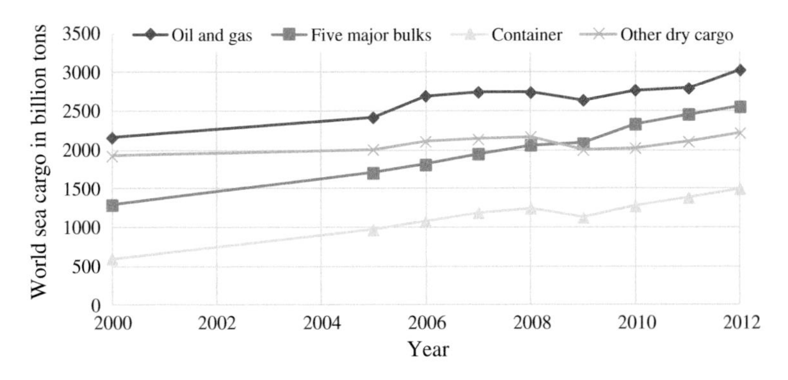

Fig. 5.6 World seaborne freight shipments trends—data *Source* UNCTAD (2013a)

This forecasted growth in demand is likely to require adjustments to transport infrastructures, services, equipment, operations, and underlying legal and regulatory frameworks. Current infrastructures are far from being able to cope with such an increase in demand. The road and rail networks of advanced countries were developed in the early (rail) and mid (road) twentieth century and their structures are not optimized for today's economy. This has become a serious problem because, until recent years, few new infrastructure projects were undertaken, and existing networks are subject to degradation because of the postponement of necessary maintenance. In developing countries, the situation is different. To sustain fast economic growth, investment in transportation infrastructures has been a priority. In China, for example, during the second half of the twentieth century, the road network length has multiplied by 15, the number of seaport terminals by 7.6, and cargo transportation volumes by 155 (Wang 2006). Currently, eight out of the ten busiest container ports in the world are in Asia and, among those, six are in China (HHLA 2013). The 2013 container throughput in Shanghai was 33.6 million (m) TEU.[3] In comparison, the largest port area in the United States (Los Angeles with Long Beach) reached 14.6 m TEU, and the busiest European port (Rotterdam) 11.6 m TEU.

A number of major infrastructure projects currently under way will have a significant impact on freight flow structures. The Panama Canal expansion, scheduled for completion in 2016, is possibly the most impactful contemporary transport infrastructure project. Currently limited to "Panamax" ships with capacities of up to 4400 TEU, the canal will be able to accommodate so-called post-Panamax vessels capable of transporting up to 12,600 TEU. This "is likely to change transportation flow patterns through North and South America, as well as port loads and transportation flows inland in the Americas" (Rivera and Sheffi 2013). Several eastern-USA and Gulf of Mexico ports (e.g., Miami, Norfolk, and New York) are presently improving their facilities to be able to attract the largest

[3]Container traffic is measured in 20-foot equivalent units (TEUs).

vessels. Increased road congestion and pollution has also sparked several projects for the development of intermodal freight transport corridors relying more heavily on means other than trucking. The expansion of the Crescent Corridor that runs through 13 states and connects New Orleans to New Jersey, is an example in the Unite States. The EU is working on the implementation of nine trans-European corridors involving at least three transportation modes (European Commission 2013). More efficient multimodal freight hubs are also under development all over the world.

Despite all these enhancements, the transportation industry is facing several problems. The expected growth of transport activities cannot but accentuate the depletion of fossil fuels, as well as pollution, congestion, noise, and security problems. The depletion of oil reserves will push fuel prices up. Further traffic will increase road and city network congestion, as well as accidents and noise. Safety and security will also become a major concern both at the national and international levels. This will create additional transport delays and may lead governments to impose more restrictive regulations and additional taxes. The transportation sector accounts for about 15 % of overall greenhouse gas (GHG) emissions and roughly one-third of those are attributed to freight transport (OECD/ITF 2010). There is a growing need for environmentally clean transportation which creates pressure to use less polluting modes (rail and sea). This pressure may also lead to additional taxes. Finally, in developed countries, it is becoming more difficult to find truck drivers and their wages are increasing. Solving all these problems will require significant improvements in vehicle, infrastructure, and traffic management technologies. These changes will in turn encourage the reengineering of supply chain networks.

5.1.3 Strategic Transportation Decisions

Given the diversity of transportation means available and the state of the transport industry, production–distribution companies have a wide range of strategic options on how to move products in their supply chain networks. Decisions to be made include the following:

- **The selection of a transportation mode**. Should a company use road, air, rail, sea, and/or intermodal transport to move its products on the different internal and external arcs of its SC network? More specifically, what transport equipment should be used? Transporting products on an arc of a SCN is usually divided in three segments: (1) *pre-haul,* that is, the first miles between pickup location and terminal/port; (2) *long-haul*, that is, inter-terminal or port move; and (3) *end-haul,* that is, the last miles between the port/terminal and the ship-to-point. The pre-haul and end-haul segments usually require road transportation but the long-haul segment may involve any mode. Mode selection decisions therefore usually require the evaluation of multimodal chains. The main advantages and disadvantages of transportation modes are examined in the next section.

- **The selection of unit loads**. Should the products manipulated be packed in boxes, crates, drums, barrels, and so on; should they be stacked on pallets, slip sheets, and so forth or shipped in bulk? This decision may seem at first glance without consequence, but it has a significant impact on how easily unit loads fit in storage racks, intermodal containers, trucks and boxcars, on how easily they can be moved with pallet jacks, forklift trucks, or conveyer belts, as well as on how easily they can be broken apart at distribution or sales points. This has a direct impact on distribution costs.
- **Investing in a private transport fleet or using carriers**. This issue has all sorts of ramifications. In an era when businesses seek to focus on their core competencies and delegate logistics to third-parties, the decision to operate their own fleet should not be taken lightly. Some reasons to do so will be discussed in the next section. Also, each alternative can take different forms. Should we use a parcel carrier (UPS, FedEx, postal services, etc.) or a common carrier? If we decide to operate a private fleet, should we buy or lease equipment? Should we adopt an intermediate solution, such as a long-term agreement with a 3PL that owns and operates the fleet on our behalf?
- **The selection or development of an adequate TMS**. This is directly related to the strategic selection of flow planning and control processes and systems discussed in Chap. 3. In order to ship material from origins to destinations in a SC, *products* are packed into *unit loads* and moved in transportation *equipment*. The link among these three encapsulation levels and the spatiotemporal movements of products, unit loads, and transportation equipment must be managed by a TMS. This requires the implementation of specialized planning and execution tools to do the following:
  - Provide information on tariffs, schedules, routes, shipping requirements, delivery orders, and so on
  - Track shipments and equipment, using on-board computers and geographical positioning systems (GPSs), for example, and provide delivery status information to clients
  - Optimize the consolidation of the unit loads to be transported, in space (loading, routing, and transshipment problems) and in time (lot-sizing and scheduling problems)
  - Minimize the flow of empty equipment and vehicles (backhauling)

Obviously, strategic decisions on modes, unit loads, and insourcing-outsourcing are interrelated. In practice, one will have to choose between alternative transportation *means*, which provide a complete solution specifying the type of equipment (thus the mode) and unit load to use, as well as implementation modalities. The main purpose of this chapter is to present approaches and modeling tools to help make these types of decisions.

5.2 Transportation Modes

Each transportation mode has its advantages, disadvantages, and operational requirements. In this section, we examine the main features of transportation modes and compare them.

5.2.1 Road

Road transport is the most common form of transportation. It has several advantages: its speed (about 50 miles per hour on average for long hauls), its availability, the ability to reach remote locations, door-to-door pickups and deliveries, a single bill of lading (B/L) and carrier, consolidation opportunities, and so on. A driver can cover about 500 miles a day; however, some carriers are using two-driver teams for long hauls. The average length of truck hauls in the United States used to be close to 1000 miles but, since, it has fallen to about 550 miles because of the increasing number of DCs and the reemergence of rail as a viable alternative for cross-country shipments (Perry 2010). Road transport is not without drawbacks. Among other things, weight and size restrictions limit the type of goods that can be transported, the costs incurred may be excessive for long distances, and GHG emissions are significant.

The equipment available are either straight trucks (for city deliveries) or tractor-semitrailer combinations (for long hauls), as shown in Fig. 5.7. Additional trailers may be hooked to each configuration. Several types of bodies are used depending on the nature of the commodities transported. The most common are dry van, flatbed, bulk (dump, open-top van, hoppers), reefer (insulated refrigerated), and tanks (for dry bulk, liquid, or gases), but all sorts of specialized body types also exist (for logging, livestock, automobiles, etc.). A 53-foot semitrailer can contain two rows of 13 pallets, that is, 52 pallets if the products can be double-stacked. In the United States, the weight limit for these trailers is about 45,000 pounds.

Road transportation used to be highly regulated but much of these barriers were removed in the late twentieth century so that the industry now operates mostly on a free-market basis. Following this deregulation, several companies dismantled their private fleets because they can now buy transportation services at competitive prices. The trucking market can be segmented based on load volume (parcel, LTL—less-than-truckload, and TL—truckload), coverage (regional, national, and international), and equipment type. TL carriers, the largest segment, target shipments in the 15,000–50,000 pound range and they provide door-to-door service. LTL carriers specialize in smaller loads in the 150–20,000 pound range and they share trailer capacity among several shippers. To do this, they must operate terminals to collect and sort freight as well as to station drivers and equipment.

The previous paragraph has focused on business-to-business (B2B) shipments. With the advent of Internet retailing and mass customization, the volume of

Fig. 5.7 Typical road transportation equipment—*Source* commons.wikimedia.org. **a** Dry van. **b** Delivery truck. **c** Flatbed truck with trailer. **d** Tanker trailer

business-to-consumer (B2C) shipments has increased tremendously. This usually involves the shipment of parcels (items weighing less than 70 pounds), which used to be the domain of postal services. With the JIT paradigm (see Chap. 1) adopted in several industries, B2B shipment loads have decreased considerably and a lot of them now involve small packages. Parcel carriers are typically offering same-day, next-day, or two-day express services. This market has been the fastest-growing segment in the past two decades and, in the United States, it is now larger than the airline and railroad freight shipping industry (Dennis 2011). It has given birth to global giants such as DHL, UPS, and FedEx, which are now offering services in multiple market segments. Although the visible end of the parcel carrier industry involves customer delivery trucks (Fig. 5.7b), its back office involves complex multimodal networks of the type illustrated in Fig. 5.2e). In several European and Asian cities, parcel carriers are now testing cargo bikes for urban distribution (e.g., DHL's cubicycle with its one cubic meter container).

5.2.2 *Rail*

The main advantages of shipping freight by train are its low cost, the possibility of shipping heavy and/or bulk products (paper, wood, grain, chemicals, etc.) over long

Fig. 5.8 Typical railroad transportation equipment—**a** Box car. **b** Hopper car. **c** Flat car. **d** Tank car—*Source* commons.wikimedia.org

distances, the existence of railcars adapted to several specialized needs, and its fuel consumption, and thus its GHG emission efficiency for long-haul high-volume movements. Its main drawbacks are its slowness, its poor reliability in terms of delivery times, and the significant risk of damage because of frequent car coupling, decoupling, and triage operations subjecting freight to shocks. This comes from the fact that railroad companies combine many cars to form trains that can cover long distances efficiently. Freight trains can include more than 100 cars. It can take several days to form a train with cars from different shippers. A given car may also switch between several trains before reaching its destination, which creates delays. An exception is when a company hires a rail carrier for dedicated service between an origin and a destination (unit-trains). Also, railroads are not accessible from multiple locations because of the limited coverage of rail networks.

Several types of equipment are offered by railroad carriers, as seen in Fig. 5.8. Box cars are typically employed for packaged freight. A box car can carry roughly three times more material than a dry van. However, railroads are used mainly for bulk shipments. These require hopper, flat, or tank cars. A large number of specialty cars are also available (center beam, refrigerated, mineral, auto rack, coil, etc.). Unfortunately, different track gauges (spacing between the rails) are used around the world. Six major gauges are found, the Standard Gauge (1435 mm) being the most common. Shippers are responsible for picking, loading, and anchoring their goods in the cars. Some bulk shippers have their own tracks, but in most cases the freight must be carried to and from rail yards. It is estimated that the average pickup and delivery cost of a typical rail move is about $700 (Perry 2010), which explains why rail is neither economic nor practical for short distances. Also, because of the way the rail industry operates, filling backhauls is very difficult, with the consequence that a large portion of the cars on some rail-lanes return empty.

The freight railroad industry also used to be highly regulated. Following deregulation, several railroad operators abandoned parts of their network considered unprofitable. In some countries, however, they are forced to operate unprofitable rail segments for public interest. After several years of instability, North American railroad companies have become profitable because mainly of regulatory freedom, improved labor flexibility, and better management. In Europe, the industry suffers from the fact that it has to share rail infrastructures with passenger traffic, that different track gauges are used in the Iberian Peninsula and the former Soviet Republics, and because it is not sufficiently customer oriented. The development of freight rail corridors launched by the European Commission addresses these problems, among other things, by giving adequate priority to freight traffic.

5.2.3 Water

Water transportation occurs from port to port on oceans, but also inland on lakes, rivers, and canals. Ocean shipping is essential for international trade and it is the cheapest and most reliable way to ship freight overseas. Compared to other modes, it is also environmentally friendly. Its main disadvantages are its low speed (in the 10–26 knots per hour range) and the fact that additional transportation is required between ports and end-user locations. A current tendency is to reduce speed to improve fuel efficiency. Weeks may be required to reach a destination, which is prohibitive for some industrial sectors. Although rare, problems with sinking or crashes can occur, and piracy is an issue in some parts of the world (e.g., off the coast of Indonesia, Somalia, and Bangladesh).

Several types of ocean vessels are in operation, as illustrated in Fig. 5.9. They are classified in categories reflecting their size and purpose. Single or multiple hold (cavity) *bulk carriers* (42 % of 2013 world fleet capacity in DWT[4] according to UNCTAD 2013b) are used to ship unpackaged cargo such as grain, coal, bauxite, iron ore, cement, and so on. Their size ranges from small 10,000 DWT *handysize* carriers to giant ships with a capacity of 400,000 DWT. *Container ships* (13 % of world fleet capacity in DWT), as the name suggest, carry containers. Modern container ships, such as Maersk (www.maersk.com) Triple E class, are fuel efficient and have a capacity of 18,000 TEU. *Tankers* (34 % of world fleet capacity in DWT) are used for transporting crude oil, petroleum products, liquefied natural gas, chemicals, vegetable oils, and other liquids in bulk. They also come in varied sizes

[4]Dead Weight Tonnage—Weight in tonnes a ship can carry without riding dangerously low in the water.

Fig. 5.9 Typical water transportation equipment—**a** Bulk carrier. **b** Container ship. **c** Tanker. **d** River barge—*Source* commons.wikimedia.org

ranging from handy size to ultra-large with a capacity of up to 555,000 DWT. All sort of specialized vessels are also in service.

Most countries possess and operate merchant ships. However, they are often registered in a nation other than the ship's owners (e.g. Panama, Liberia) to reduce operating costs and avoid regulations. "Most of the top 35 ship-owning countries have more than half their tonnage under a foreign flag" (UNCTAD 2013b). The world fleet has more than doubled between 2001 and 2012 reaching 1.63 billion DWT. This is the largest ship building cycle in recorded history (UNCTAD 2013b). In 2013, however, new ship building orders were at an historical low. Currently, ship owners from five countries (Greece, Japan, China, Germany, and the Republic of Korea, in order of decreasing tonnage) account for 53 % of the world tonnage. The size of ships has increased during that period, but the number of shipping companies is decreasing. Three European companies (Maersk in Denmark, MSC in Switzerland, and CMA CGM in France) provide one-third of the global container-carrying capacity. Shipping can be arranged either through a regular liner or tramp services. The former has fixed routes and schedules, a bit like a bus line, and it provides quality services. The later carries mainly bulk cargo on a per-load on-demand basis, like a taxi service. Ships can also be chartered. The www.marinetraffic.com platform provides the position and destination of all cargo vessels and depicts congested waters.

Inland water transportation involves mainly freighters on lakes and motorized or non-motorized barges (flat-bottom vessels) on rivers (Fig. 5.9d). The later requires the use of a towboat. For larger rivers, several barges can be linked in a single tow. Inland waterways are used mainly to transport bulk commodities over a long distance. Their chief disadvantages are their low speed and limited access. In the Americas, the main waterways are the Great Lakes, the Saint Lawrence Seaway, the Mississippi River, the Amazon, and their tributaries. In Europe, the Rhine, the Danube, and the Thames are the focal waterways. The Garonne River in France is used for example to transport the Airbus 380 aisles assembled in Toulouse. The most used inland waterway in the world for the transport of goods is the Yangtze in China (850 million tons per year). Rivers linked by canals are developed across the world to increase the reach of waterway networks. Many canals require locks to enable ships to travel from low to high water areas or vice versa. Ice-breaking operations are required to keep northern waterways active during the winter.

5.2.4 *Air*

The main advantages of air transport are its speed and reliability which, as we shall see, reduces storage costs. It is often used for the export of perishable products (e.g., fish, flowers). Shipments consolidation is possible through international freight forwarders. As can be expected, its main disadvantage is its high cost. In addition, pickup and delivery fees are incurred to move products to and from airports. Air cargo is quite sensitive to backhaul imbalances, since freight flowing from manufacturers to markets tend to be much greater than opposite flows.

Air cargo is carried in three different types of plane: wide-body jets, narrow-body jets, and narrow-body turboprop aircrafts (feeder aircrafts). Wide-body jets such as the Antonov 225 (the largest plane in the world, with a maximum payload of 250 tons), the Airbus A380-800F (payload of 150 tons), and the Boeing B747-8F (payload of 135 tons) are assigned mainly to overseas or transcontinental routes. They are often employed for the deployment of military or humanitarian relief missions. Narrow-body jets (payloads in the 10–45 ton range) and feeder aircrafts (payloads in the 1–5 ton range) are used mainly on domestic routes and for small niche markets. Air cargo can also travel in the baggage compartment or lower deck of passenger aircrafts. A Boeing 777 can carry 9.5 tons of cargo in its "belly". FedEx uses this means to move Apple's high-value products from Asia to the US. An important attribute of wide-body and narrow-body aircrafts is their capacity to house containers and pallets. They have large doors and deck-rollers to facilitate the loading and unloading of freight.

Air cargo services are provided by three basic types of carriers. *All-cargo carriers* (e.g., Cargolux) generally operate scheduled wide-body jets primarily on long-haul routes between major international airports. Nowadays, they are often divisions of passenger airlines (e.g., Emirates SkyCargo, Air France-KLM-Martinair Cargo). *Combination carriers* are passenger airlines offering cargo services (e.g., Korean

Air, United). *Ad hoc cargo charter carriers* are unscheduled operators. Their market share is small and decreasing. Leading parcel carriers such as DHL, FedEx, and UPS also operate large aircraft fleets to provide fast long-haul transportation in their extensive hub-and-spoke networks. In fact, according to IATA, in terms of freight ton-kilometers (FTKs) shipped, FedEx and UPS now operate the two largest cargo airlines in the world.

5.2.5 Multimodal and Intermodal

Multimodal transport involves the movement of goods with at least two different modes, in order to benefit from their respective forces, but under a single contract with a carrier (known as an MTO—Multimodal Transport Operator) legally liable for the entire move. To provide this service, the MTO signs contracts with transporters, cargo consolidators, ports, airports, and so on; coordinates custom procedures; and manages the flow of goods from end to end. For the consignor, this simplifies the shipping process considerably. The burden of documentation and formalities is significantly reduced, time is saved, and costs as well as risks are reduced. *Intermodal* freight transportation occurs when the freight is loaded in containers or trailers that can be moved by all the transportation modes without any handling of the freight itself when changing mode. Its main advantage is that it reduces cargo handling, thus improving security, reducing damage and loss, and reducing transit times, costs, and carbon footprint.

Several types of multimodal and intermodal transportation are encountered:

- **Road-rail transportation**. That is, the serial coupling of rail and road using trailers that can be loaded on flat cars (TOFC or piggyback—Fig. 5.10a), or containers that can ride both on flatbed trucks and on flat railcars (COFC). For the latter, specialized flat cars with a container-sized depression can be used for a double stack arrangement (Fig. 5.10b). This opportunity has led to the development of multimodal hub-and-spoke networks. In North America, for example, CN (www.cn.ca) currently operates 23 hubs in major cities between the Canadian Atlantic and Pacific Coasts, and between the Great Lakes and the Gulf of Mexico.
- **Ocean-ground transportation**. That is, the combination of maritime transport with rail and/or road. Two types of vessels are currently used to facilitate load transfers: "LoLo" (load-on, load-off) containerships and "RoRo" (roll-on, roll-off) ships equipped to receive cargo on wheels. This type of service has led to the implementation of ocean-to-ocean *landbridges*. For example, to deliver containers from Europe to Asia, instead of going through the Panama Canal, shippers can use the North American Landbridge, involving first a trans-Atlantic passage by ship, then a move across the continent by train, and finally a trans-Pacific shipment, all with a single B/L.

Fig. 5.10 Intermodal—*Sources* **a**, **b**, and **c** commons.wikimedia.org, **d** authors. **a** Tanker trailer piggyback on flatcar. **b** Double-stack container train. **c** Ship-to-rail transfer. **d** Truck-to-ship transfer

- **Ocean-inland waterway transportation**. Barges navigating on inland waterways can be loaded onto lighter aboard ships (LASH) to cross the ocean and, at the destination, unloaded to continue their way on rivers or canals.
- **Air-surface transportation**. Practically all air cargo is multimodal in the sense that it requires pickup and delivery by truck. However, as previously mentioned, leading parcel carriers have set up air-surface hub-and-spoke networks enabling them to ship parcels and small packages quickly across the globe.

The rapid growth of intermodal transportation over the past decades was made possible by the use of containers. Intermodal container sizes and specifications are defined by ISO standards. They can be stacked, loaded, unloaded, and transferred efficiently from mode to mode. ISO steel containers are 8 ft. wide by 8 ft. 6 in. high. *High cube* (9 ft. 6 in. high) containers can also be used. The most common lengths are 20 ft. (one TEU), 40, 45, 48, and 53 ft., but other lengths exist. The maximum payload for a 20 ft. container is about 22,000 kg. Various specialized containers also exist, such as refrigerated containers (reefers) for perishable products, open-topped containers, and tank-inside containers. The majority of containers are owned by maritime shipping companies (about two-thirds) and container-leasing companies. As for other transportation equipment, containers do not always travel with cargo. Empty containers must be repositioned from areas of low demand, such as the US West Coast, to areas with high demand, such as Chinese ports. A large proportion of available containers is always sitting idle in yards and depots around the world, which is a serious problem. Recent foldable containers could improve the situation (e.g., www.hcinnovations.nl).

Specialized handling equipment is used to facilitate the transfer of containers from one mode to the other. These include gantry cranes (Fig. 5.10d) for trans-loading containers from seagoing vessels onto either trucks or railcars, straddle carriers (Fig. 5.10c), reach stackers, sidelifters, and so on. Specialized automated guided vehicles (AGVs) are also used. The loading and unloading of containers from trucks or trains to containerships is difficult to synchronize because it depends on ship schedules and on the availability of ground transportation. For this reason most container ports incorporate large areas where containers can be stored while waiting for intermodal trans-loading operations. Recent initiatives such as *synchromodality* (www.dinalog.nl/en/themes/synchromodal_transport) may help resolve these issues.

5.2.6 Mode Comparison

As can be seen, the advantages and disadvantages of specific transportation modes are not the same so that, in a particular context, some modes are generally preferable to others. Modal options available are summarized in Fig. 5.11. Several criteria come into play in mode selection: costs, delays, reliability, accessibility, load damage, safety, carbon emissions, noise, administrative burden, and so on. Basic performance statistics for each mode are provided in Table 5.2. Speed influences transit times, however, and some equipment (e.g., railcars) must stop often, which has a significant impact. Note also that average time is not the only important statistic for making a good decision. The variability of transit times must also be taken into account. The reliability of delivery times, often measured by its *coefficient of variation* (CV = variance/mean), is usually considered more important than speed. A comparison of transit times for selected transportation means is provided in Fig. 5.12.

The cost per ton-mile given in the table corresponds to freight tariffs. When making comparisons, however, one must also consider the cost of packing, pre-haul, end-haul, loading, unloading, storage, handling, insurance, and all other relevant costs. The total costs analysis approach will be studied in detail in Sect. 5.5. Note that, as could be expected, there is a strong negative correlation between ton-miles per gallon and CO_2 emissions. There is also a strong correlation between costs and these two variables. For this reason, when selecting a transportation means based on costs, the solution obtained tends to be the best in terms of CO_2 emissions. Some modes also perform much better than others in terms of losses and damage. Rail transport is particularly bad from that point of view. Based on the previous discussion, Table 5.3 provides a mode ranking for critical performance factors.

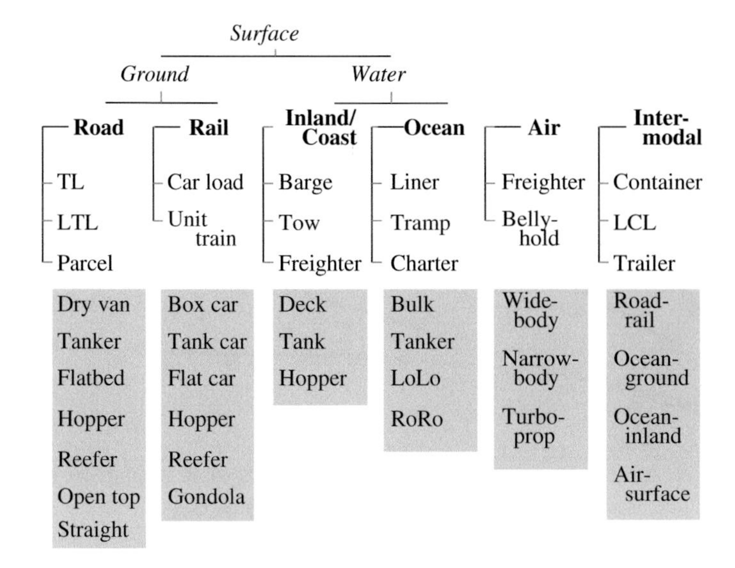

Fig. 5.11 Summary of modal options

Table 5.2 Basic mode performance statistics (wrought mean estimates over all equipment, loads, and routes obtained from various sources)

Mode	Capacity (tons)	Speed (mph)	Ton-miles/gal	$/ Ton-mile[a]	CO_2 (g/tkm)[b]
Road	25/trailer	56	150	17	75
Rail	110/car	22	450	3	20
Inland	1500/barge	7	575	2	60
Ocean	–	22	750	–	15
Air	135/B747	500	15	89	1000

[a]2007 average carrier revenue per mode, US Bureau of transportation statistics (2004 for inland)
[b]2011 CO_2 emissions, TERM27, European environmental agency (excluding air)

In some contexts, a company can obtain better performances in terms of some of the factors discussed (costs, lead times, deadlines, carbon emissions, and damages), by investing in a private fleet instead of using carriers. This typically involves specific needs that cannot be adequately satisfied by carriers. In a highly competitive environment, for example, a company may want to differentiate itself by offering flexibility, speed, and/or delivery reliability levels that exceed industry standards. It may also be easier to avoid empty backhauls with a private fleet. In some cases, for security reasons or because of the nature of the product, its handling and transportation may require special equipment that is not available otherwise. Some shippers use private fleets simply because it facilitates control and guarantees capacity. A private fleet also provides a marketing instrument permitting advertisements on trucks. It is estimated (CSCMP et al. 2014) that more than 30,000

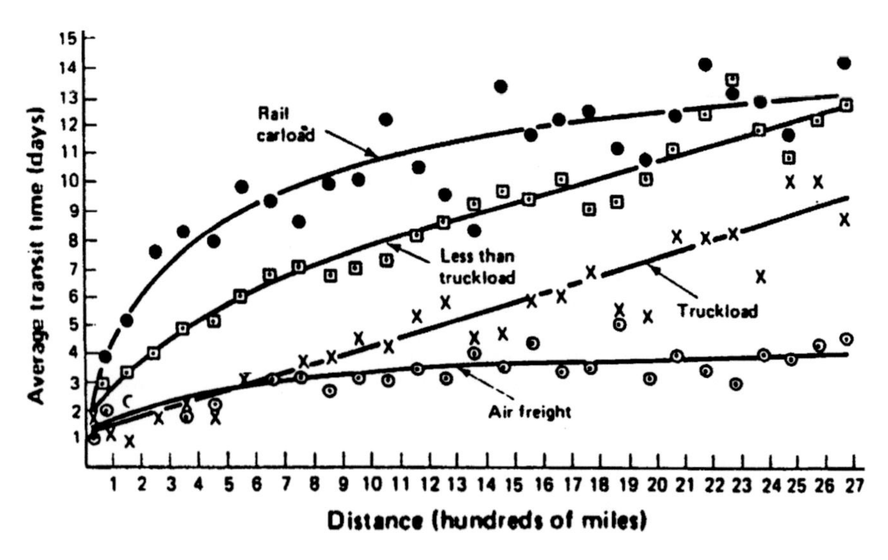

Fig. 5.12 Transit times by transportation means—*Source* Ballou (1992)

Table 5.3 Relative rankings of transportation modes

Mode of transportation	Performance factors (1 = best, 4 = worst)				
	Cost	Average delivery Time	CV of delivery-time[a]	Carbon emissions	Loss and Damage[a]
Rail	2	3	2	2	4
Road	3	2	1	3	3
Water	1	4	3	1	1
Air	4	1	4	4	2

[a]Ranking based on Ballou (1992)

companies have private fleets (ten vehicles or more) in the United States (e.g., PepsiCo, Coca-Cola, and Walmart). Note, however, that some for-hire carriers now offer dedicated fleet services that provide the same advantages as a private fleet.

5.3 Freight Shipping

The transportation industry is complex and shipping freight is rarely simple. This section reviews the main elements of sales and shipping processes. Particular attention is given to transportation tariffs. Also, the intermediaries that can be used to simplify the process are presented.

5.3.1 *Terms of Sale*

Shipping necessarily involves two parties, namely an origin (consignor) and a destination (consignee). This raises four important questions: (1) Who pays for transportation? (2) Who controls the shipment in transit? (3) Who bears the risk for loss, damage, or delays? (4) Who has the burden of filing claims when something goes wrong? If the origin and the destination belong to the same company, this may not be critical, although, as we shall see in Chap. 9, it has significant impacts in an international context. If one of the parties is a customer, a supplier, or a subcontractor, however, these issues are contractual matters that must be settled in court in case of dispute. The basic issue here is to determine when the ownership of the goods is transferred from the seller to the buyer, that is, who assumes *title* during the shipping process.

The most common terms used for domestic shipments are *FOB origin* or *FOB destination*.[5] With the former, the title is transferred to the buyer at the seller's shipping point when the goods are tendered to the carrier, which act as the buyer's agent. The buyer assumes risks and responsibilities during transportation and, unless otherwise specified, pays the carrier fees. Under FOB destination terms, the title is transferred when the goods arrive at the ship-to-point, and the seller is responsible for transportation. In addition, the parties may use the terms "prepaid" or "collect" to specify who pays the carrier for transportation.

In an international context, many more options are available and *Incoterms* (International commerce terms) are used to specify them. Incoterms are managed by the International Chamber of Commerce (www.iccwbo.org) and a major revision of associated rules was made in 2010. They clarify the responsibility for tasks, costs, and risks involved in the delivery of goods from a seller to a buyer. They belong to four families (E-Ex Works, F-Free carriage, C-Carriage paid, D-Delivered) designed to facilitate multimodal or maritime shipping. Responsibility rules for *any mode or modes* Incoterms are specified in Fig. 5.13, and those for *sea transport* in Fig. 5.14. CIF and CIP are the only terms imposing an obligation for the seller to provide cargo insurance; however, it is usually a good idea to purchase insurance to protect against risk.

Terms of sale affect carrier selection, operations management, consolidation opportunities, cash flows, and claims administration. When a company is sufficiently large to profit from economies of scale, it is often advantageous to keep control over transportation. When materials are purchased FOB destination, the transportation costs are hidden in product prices, and it may be difficult to determine if they are minimized. At the other end, by keeping control over outbound shipments, the company may profit from consolidation opportunities. However, if products are sold FOB origin, the buyer assumes title at the beginning of the process and the products are payable immediately. Consequently there is no capital

---

[5]FOB stands for Free-On-Board and it originally applied to international maritime shipments. Although not really appropriate for domestic shipments, it is used profusely in practice.

<table>
<tr><th></th><th></th><th>Packaging by seller</th><th>Loading on vehicle</th><th>Carriage to terminal</th><th>Customs</th><th>Terminal operations</th><th>Loading on vehicle</th><th>Long-haul carriage</th><th>Unloading from vehicle</th><th>Terminal operations</th><th>Carriage to buyer premises</th><th>Customs</th><th>Unloading by buyer</th></tr>
<tr><td rowspan="2">EXW</td><td>Cost</td><td>Seller</td><td colspan="11">Buyer</td></tr>
<tr><td>Risk</td><td>Seller</td><td colspan="11">Buyer</td></tr>
<tr><td rowspan="2">FCA</td><td>Cost</td><td colspan="4">Seller</td><td colspan="8">Buyer</td></tr>
<tr><td>Risk</td><td colspan="4">Seller</td><td colspan="8">Buyer</td></tr>
<tr><td rowspan="2">CPT</td><td>Cost</td><td colspan="10">Carriage paid by seller</td><td colspan="2">Buyer</td></tr>
<tr><td>Risk</td><td colspan="5">Seller</td><td colspan="7">Buyer</td></tr>
<tr><td rowspan="2">CIP</td><td>Cost</td><td colspan="10">Carriage and insurance paid by seller</td><td colspan="2">Buyer</td></tr>
<tr><td>Risk</td><td colspan="5">Seller</td><td colspan="7">Buyer</td></tr>
<tr><td rowspan="2">DAT</td><td>Cost</td><td colspan="9">Delivered at terminal by seller</td><td colspan="3">Buyer</td></tr>
<tr><td>Risk</td><td colspan="9">Seller</td><td colspan="3">Buyer</td></tr>
<tr><td rowspan="2">DAP</td><td>Cost</td><td colspan="10">Delivered at place by seller</td><td colspan="2">Buyer</td></tr>
<tr><td>Risk</td><td colspan="10">Seller</td><td colspan="2">Buyer</td></tr>
<tr><td rowspan="2">DDP</td><td>Cost</td><td colspan="11">Delivered duty paid by seller</td><td>Buyer</td></tr>
<tr><td>Risk</td><td colspan="11">Seller</td><td>Buyer</td></tr>
</table>

Fig. 5.13 Incoterms 2010 rules for any mode or modes of transport

<table>
<tr><th></th><th></th><th>Packaging by seller</th><th>Loading on vehicle</th><th>Carriage to port</th><th>Customs</th><th>Port operations</th><th>Loading on vessel</th><th>Long-haul carriage</th><th>Unloading from vessel</th><th>Port operations</th><th>Carriage to buyer premises</th><th>Customs</th><th>Unloading by buyer</th></tr>
<tr><td rowspan="2">FAS</td><td>Cost</td><td colspan="5">Seller</td><td colspan="7">Buyer</td></tr>
<tr><td>Risk</td><td colspan="5">Seller</td><td colspan="7">Buyer</td></tr>
<tr><td rowspan="2">FOB</td><td>Cost</td><td colspan="6">Seller</td><td colspan="6">Buyer</td></tr>
<tr><td>Risk</td><td colspan="6">Seller</td><td colspan="6">Buyer</td></tr>
<tr><td rowspan="2">CFR</td><td>Cost</td><td colspan="7">Cost and freight to destination port paid by seller</td><td colspan="5">Buyer</td></tr>
<tr><td>Risk</td><td colspan="6">Seller</td><td colspan="6">Buyer</td></tr>
<tr><td rowspan="2">CIF</td><td>Cost</td><td colspan="7">Cost, insurance and freight to destination port paid by seller</td><td colspan="5">Buyer</td></tr>
<tr><td>Risk</td><td colspan="6">Seller</td><td colspan="6">Buyer</td></tr>
</table>

Fig. 5.14 Incoterms 2010 rules for sea transport

tied up in the inventory in transit. All these elements must be weighed carefully before choosing terms of sale.

5.3.2 Shipping Process

Depending on the circumstances, ad hoc transportation services may be required or regular shipments between SC origins and destinations may be necessary. When a company does not own its own fleet, it usually negotiates long-term agreements (contracts) with transportation service providers. Such contracts typically specify the period covered, the scope of required services (type of commodities, volume, equipment, geographic region, expected performance levels, penalties for nonperformance, etc.), rate schedules, surcharges, payment terms (including discounts or penalties for early or late payments), documentation requirements, liability for loss, damage or delay, insurance levels, dispute resolution mechanisms, and so on. Specific terms of the contract depend on the transportation mode and on whether shipments are local or international. Individual shipments are then made according to the terms of the agreement.

Making a shipment involves several tasks. The products to ship must first be packaged and identified adequately. This may require the use of an SSCC (Serial Shipping Container Code) bar code as explained in Chap. 3. Consolidation opportunities must then be examined, mainly if shipments are made in full-vehicle loads. The transportation rates must also be retrieved. As we shall see in the next section, transportation tariffs depend on a lot of factors, and several freight rating systems and websites are available to support the process. Documentation can also become a heavy burden when making a shipment. The documents required depend on the nature of the cargo, shipment, and trading partners. Some documents are required for all shipments:

- *Packing list*: Prepared by shippers, it lists the products in the shipment as well as their quantity, weight, and size. A copy is often attached to the shipment in a waterproof envelope and another one is sent to the consignee.
- *Declaration of dangerous goods*: A letter prepared by the shipper, when applicable, and providing information on hazardous material.
- *Bill of lading (B/L)*: Issued by carriers, it acknowledges that the goods identified have been received onboard from a shipper as cargo to be delivered to a consignee at a named place. It also serves as a contract of carriage. *Negotiable* B/Ls transfer ownership rights to anyone who has possession of the document.
- *Delivery receipt*: Prepared by carriers, it is signed by the consignee to confirm the reception of the shipment.
- *Freight bill*: Prepared by the carrier, this invoice is sent to the shipper/consignee to request payment for services rendered.

When freight crosses borders, additional documents are required. These may include a shipper's export declaration, export and import licenses, a commercial invoice, a certificate of origin, a consular invoice, and so forth. All documents must be accurate and timely to guarantee problem free transportation. Several of these documents can now be transmitted electronically to reduce paperwork.

5.3.3 Transportation Tariffs

The rating mechanisms used by most carriers are complex and they can vary depending on the particular situation of the parties. In some countries, transportation rates are regulated by government. However, following the 1980 deregulation in the United States, several countries now let market forces determine rates. Rates also depend on the context in which carriers are operating, and more specifically on how they can balance their lane traffic, that is, on how easily they can find backhaul loads. For this reason, certain carriers can offer significantly better rates than others for the same service in the same lanes. Also, published rates usually apply to the spot market, that is, to ad hoc shippers who have no established agreements with the carrier. When a contract ties a carrier and a shipper (or broker) very significant discounts (as high as 80 %!) can apply depending on volumes. One should therefore

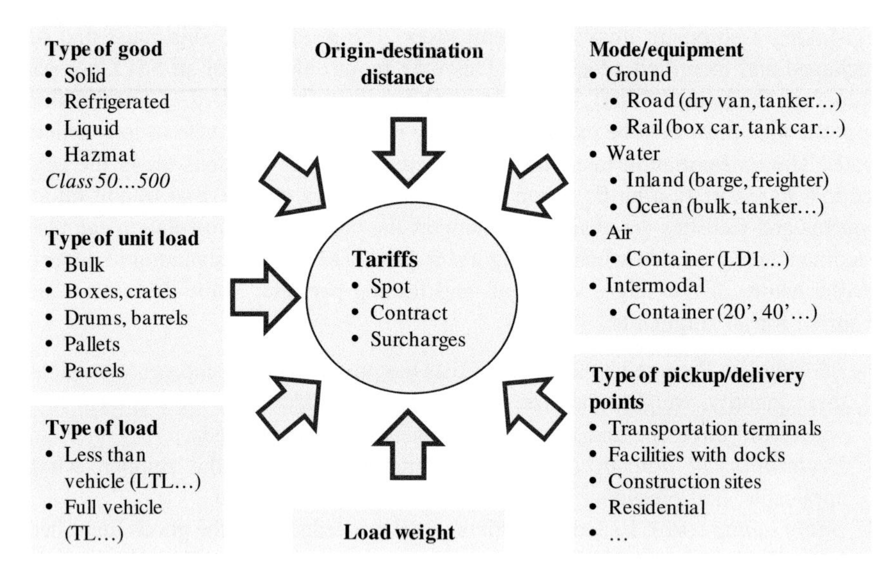

Fig. 5.15 Main variables affecting transportation rates

always examine several options before a carrier is selected. The main factors considered to establish tariffs are presented in Fig. 5.15.

The approach taken to rate shipments first depends on whether the shipper requires full equipment (truck, car, container, etc.) or only a portion of the capacity provided by a piece of equipment (LTL, LCL, etc.). When full equipment is required, the shipper is responsible for loading freight onboard, and the tariff depends essentially on the number of equipment pieces required for a one-way origin–destination trip. If multiple stops are made along the way, this is also factored into the calculation. Otherwise, the carrier is responsible for loading freight onboard, and costs then depend on how easily the freight shipped can be handled and stacked in the equipment, given applicable weight and volume limits. The nature of the freight shipped, that is, the type of goods and unit loads, then becomes an important element of the rating method. Also, there are a virtually infinite number of possible origin–destination pairs and it is not possible to publish rates for all of them. Instead, the territory covered by a carrier is typically divided into geographical zones (e.g., three-digit zip codes), and rates are provided for zone-to-zone pairs. The rate for a given shipment is found by identifying the zones of the origin and destination.

The nature of freight shipped is extremely diversified: it can have a low (Ping-Pong balls) or high (metal) density, a regular (palletized) or irregular (bulk) shape, and a low (bricks) or high (gold) value; it can be fragile or not, perishable or not, dangerous or not, and so on. To take this variability into account, carriers typically base their *less-than-vehicle* rates on a combination of weight and commodity *class*. In North America, classes for LTL transportation are managed by the National Motor Freight Traffic Association (www.nmfta.org). Products are grouped

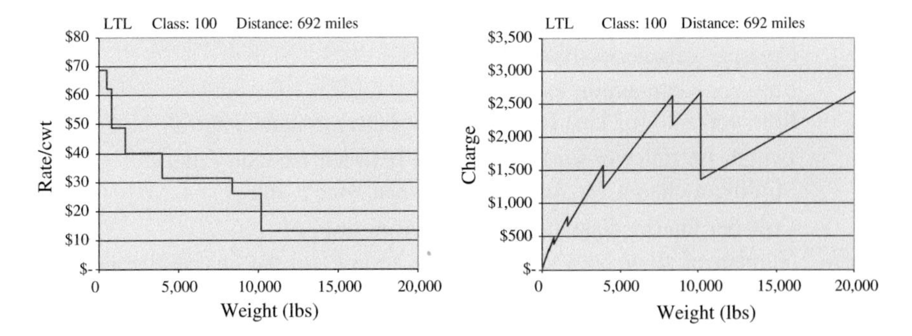

Fig. 5.16 Impact of weight on LTL spot rates for a given class and origin–destination

into 18 different classes ranging from 50 to 500, with Class 50 having the lowest rate per hundredweight (cwt). For a given class and origin–destination, long-haul LTL rates take the form of a piecewise-linear function. As illustrated in Fig. 5.16, the rate per cwt decreases in steps of variable length defined by predetermined weight break values. The charge for a shipment is obtained by multiplying the applicable rate by the cargo weight. Note that when a shipment includes several items of different classes, the charge for each item is calculated separately, but the rate applied depends on the weight of all the items in the shipment. For same-city LTL shipments, a local cartage rate independent of distance usually applies. Some contracts may also be negotiated with *freight-all-kind* (FAK) rates that apply to all commodity classes.

Another approach to take the density and shape of unit loads into account, applied mainly by air and parcel carriers, is the use of *dimensional weights*, that is, the weight of a package based on its volume multiplied by a predetermined density (expressed for example in lb/ft^3 or kg/m^3). The volume is given by (length × width × height), but if the package is irregular, the longest measure is taken for each dimension. The density is predetermined by the carrier, for example, 10.4 lb/ft^3 for most IATA shipments. The density used for domestic and international shipments usually differs. Finally, if the weight thus obtained is higher than the real weight, it is used to determine the transportation rate. For example, a package of (1′ × 1′ × 1.5′) = 1.5 ft^3 could have a dimensional weight of 1.5 (10.4) = 15.6 pounds. If its real weight is 10 lbs, it is rated as if it weighed 15.6 lbs. However, if the real weight is 20 lbs, it is rated with a weight of 20 lbs. This rating method therefore penalizes shippers of low-density freight.

Full-vehicle shipments (truck load, car load, and container load) are subject to maximum weight regulations, but their rates rarely depend on the cargo weight. One-way multiple-stop TL shipments between an origin location l and a destination location l' are typically rated based on the following formula:

$$c_{ll'}^{\mathrm{T}} = \max\left(c_{z(l),z(l')}^{\mathrm{m}} D_{ll'};\, c_{z(l),z(l')}^{\mathrm{min}}\right) + c_l^{\mathrm{s}} Stops_{ll'}$$

where

$c^{\mathrm{T}}_{ll'}$	Cost per vehicle for route (l, l')
$z(l)$	Rate zone containing location l
$c^{\mathrm{m}}_{z,z'}$	Rate per mile (or km) for a shipment between zone z and z'
$D_{ll'}$	Length (in miles or km) of the route between location l and l'
$c^{\min}_{z,z'}$	Minimum charge for shipments between zone z and z'
c^{S}_{l}	Cost per intermediate stop en route from origin l
$Stops_{ll'}$	Number of stops between shipper location l and the final destination l'

The rate per mile, $c^{\min}_{z,z'}$, tends to decrease slightly as the distance increases.

Parcel carriers use zone- and weight-based pricing formulas. However, their rates also depend on the delivery speed required. Expedited shipments (e.g., same-day or next-day deliveries) are more expensive than regular shipments (two or three days for local service and five days for international service). Air carriers, break bulk ocean shippers, LCL rail, and intermodal carriers use an approach similar to LTL carriers to calculate rates, except that it is often based on dimensional weight. They also use different systems to classify commodities. In the United States, for example, rail rates are based on Standard Transportation Commodity Codes (STCC) maintained and published by the Association of America Railroads. Full-container or railcar prices are calculated with an approach similar to TL shipments, but the rate depends on the type of container or railcar used.

All sorts of surcharges (often called *ancillary* or *accessorial* charges) can be added to the basic rates calculated for a transportation mode. These may include pickup and delivery charges, oversize charges, COD (cash on delivery) charges, insurance, detention charges, port charges, and so forth. For example, demurrage charges are added by railroad companies if the time taken to load or unload goods exceeds a standard (48 h). Also, each port has its own rules regarding the inclusion of terminal handling charges (THCs) for freight or container inspection, moving, loading, unloading, and storage. Finally, because of the significant variations of fuel prices observed in recent years, fuel surcharge programs are now applied by most carriers. These surcharges are based on a fuel cost index (e.g., weekly national average fuel cost per gallon or liter) and they reflect differences between the fuel cost used to calculate the rates and the real cost of fuel for a shipment. In the United States and Canada, the increase or decrease observed is usually expressed in dollars per mile and, consequently, it depends on the delivery distance.

5.3.4 Intermediaries

Given the complexity and structure of the transportation industry, several intermediaries offer services to facilitate the efficient and economical movement of goods, through process simplification, shipment consolidation, and so on.

Intermediaries play an important role in multimodal transportation, and cross-border shipping by land, sea, and air. Traditionally, they have been relatively small non-asset-based facilitators such as freight forwarders (providing small shipment consolidation services); ground, ocean, or air freight brokers (agents paid on a commission base); and customs brokers. However, with the restructuration of the transportation industry over the past decades, their number has shrink and the trend is toward fewer and larger players providing a much broader palette of logistics services. As a consequence we now see former intermediaries selling asset-based services and large third-party logistics providers (3PLs) offering all the services required for seamless international multimodal transportation. 3PLs are studied in detail in Chap. 6. Another trend in the past decades has been the entry of so-called *infomediaries*, that is, middlemen who operate in electronic markets to facilitate exchanges. They are particularly useful for carriers wanting to find backhaul shipments.

5.4 Modeling Transportation Costs

In the rest of this chapter, we turn our attention to the selection of transportation means. These decisions must be based on serious cost-benefit analyses. Costs clearly depend on transportation tariffs, however, as we just saw, tariffs for a given transportation equipment and commodity type are complex discontinuous functions of cargo weight and distance shipped. The fixed or variable surcharges applied must also be taken into account. Tariffs can be used directly to make ad hoc transportation cost comparisons for a given load and origin–destination lane. However, when selecting transportation means for several shipments involving variable weights and distances over a planning horizon, or when designing a supply chain network, simplified continuous cost functions must be used. Consequently, the first task required to carry out such studies is the modeling of transportation costs, initially as functions of load and distance and, when designing a SCN, as functions of aggregate annual product flows and distance. As mentioned previously, GHG emissions should also be considered when selecting transportation modes, and to do this they must likewise be related to relevant logistics variables (load weight, distance, and equipment type). An approach to do this will be presented in Chap. 12.

5.4.1 Load Cost Functions Estimation

Shipping costs are not linear: they reflect the economies of scale that carriers can achieve when transporting larger loads over longer distances. Shippers can profit from these potential savings, and economies of scale should therefore be taken into account in any serious economic analysis. In Chap. 2, we showed that economies of scale can be effectively modeled by power functions. For a given type of commodity and transportation mode, an excellent approximation of transportation charges is thus usually given by the following function of weight and distance:

$$c^{\mathrm{T}}(D,W) = c^{\mathrm{o}} D^{\gamma} W^{\upsilon} \tag{5.1}$$

where

D	Round-trip (or one-way, depending on the context) distance between the origin and destination
W	Weight of load shipped
c^{o}, γ , υ	Multiple regression parameters

For a given origin and destination, the distance D is fixed, and we have $c^{\mathrm{T}}(W) = \chi W^{\upsilon}$ with $\chi = c^{\mathrm{o}} D^{\gamma}$. The intuitive interpretation of (5.1), however, is not always obvious and its use can lead to difficult nonlinear models.

A simpler adequate model of transportation charges for a given range of cargo weights (say LTL shipments between 5000 and 20,000 pounds) is provided by the following linear approximation[6]:

$$c^{\mathrm{T}}(D,W) = c^{\mathrm{s}} + c^{\mathrm{d}} D + c^{\mathrm{w}} W \tag{5.2}$$

where c^{s}, c^{d}, and c^{w} are multiple linear regression parameters. For a given distance, D, we have $c^{\mathrm{T}}(W) = c^{\mathrm{o}} + c^{\mathrm{w}} W$ with $c^{\mathrm{o}} = c^{\mathrm{s}} + c^{\mathrm{d}} D$.

Besides being simpler, the practical interpretation of the parameters of this model can be given the following practical interpretation:

c^{s} Cost of *stopping* (independent of distance and load) incurred because of the assignment of a vehicle to a shipment and, in particular, to the time lost while the vehicle is waiting at the departure and arrival.

c^{d} Cost of *distance* (independent of the content of the vehicle), that is, driver, fuel, and wear costs per mile/kilometer.

c^{w} Cost of *load* (independent of the distance traveled) incurred for each weight unit shipped in the vehicle (for loading and unloading unit loads and handling in the vehicle).

This model can also be used to characterize the cost of point-to-point deliveries with a private fleet.

Models (5.1) and (5.2) can be estimated from published tariffs or from historical transportation data. When a new option is considered, carrier tariffs must be used.

[6]Note that a fourth term capturing combined weight-distance effects ($c^{\mathrm{dw}} DW$) could be added to (5.2) to improve the approximation. However, this term would introduce nonlinearities and, because it is usually relatively small, it will be ignored.

However, when evaluating a transportation means already in use, it may be more appropriate to base the estimation on historical shipment data. Examples illustrating these two cases are given in the following.

Example 5.1
Suppose that a company having facilities in Alabama, Arkansas, and California in the United States, must regularly make LTL shipments between 1000 and 8000 pounds to customers. Carrier tariffs, and relevant weight breaks, for five popular lanes are illustrated in Fig. 5.17. The corresponding distances, weights and charges were given in Exercise 2.5. Models (5.1) and (5.2) can be easily fitted to this data using the regression tool provided in the Data Analysis Add-in of Excel or any other statistical package. As can be seen in Fig. 5.17, the linear regression lines $c^{\mathrm{T}} = c^{\mathrm{o}} + c^{\mathrm{w}}W$ for individual lanes are very good approximations ($R^2 > 0.97$). The multiple linear regression model (5.2) for this data is $c^{\mathrm{T}} = 0.4D + 0.48W$. The value of the cost of stopping c^{s} originally obtained by regression was negative, which does not make sense. To avoid this, the regression constant was forced to take a zero value to yield the previous equation. Although the multiple R^2 value for the equation fitted is 0.96, the mean absolute deviation (MAD) between the estimated cost and the real cost is 24.7 %, which is not very good. This is because the estimation was based on five lanes only to reduce the size of the example. In practice, one would use all the shipping lanes of the company.

In Exercise 2.5, you were asked to fit a power function to the transportation data provided. This corresponds to model (5.1). The best model fitted to the data using Excel is $c^{\mathrm{T}} = 0.04D^{0.56}W^{0.84}$. For this model, $R^2 = 0.99$ and MAD = 7.8 %, which is much better than for the linear case.

Surcharges must also be taken into account in the estimation. If these surcharges depend on the origin, destination, or any other shipment variable, they must be added to the basic tariffs before the model is estimated. For linear models, if they apply to all shipments, they can be added afterward to the relevant model parameter. Suppose, for example, that a carrier charges a security fee of $0.04/pound for all shipments. For our example, this surcharge would be added to the estimated cost of load (c^{w}), giving $c^{\mathrm{T}} = 0.4D + (0.48 + 0.04)W$. Also, shippers must request and control the shipments made by carriers. Suppose that the administrative costs thus incurred are estimated at $50 per shipment. This fixed cost should then also be added to the model to get $c^{\mathrm{T}} = 50 + 0.4D + 0.52W$.

Example 5.2
The cost and distance of a sample of long-haul TL shipments made from a US plant to DCs and major accounts across the country are plotted on a graph in Fig. 5.18, and Excel was used to fit a regression line to the data. The model

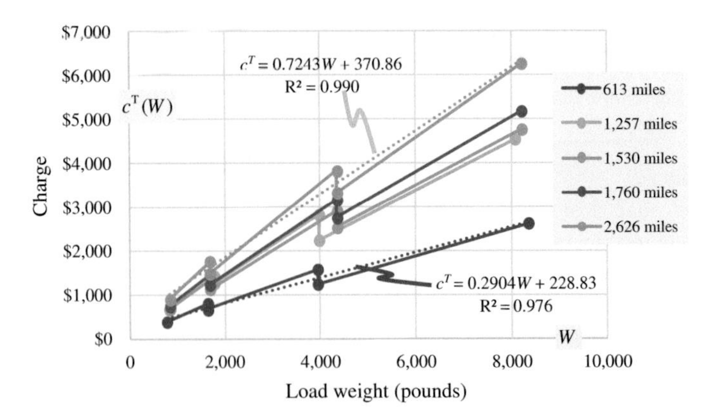

Fig. 5.17 Linear regression models for LTL tariffs

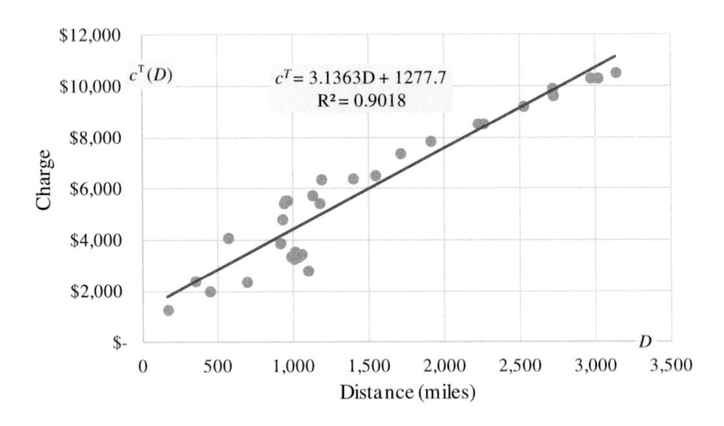

Fig. 5.18 Linear regression model for TL shipments costs

fitted corresponded to (5.2). Note, however, that the last term ($c^w W$) is absent because TL shipment costs do not depend on the weight of the load. A cost per pound can be calculated by dividing the amount paid by the weight of the load shipped. However, the rate obtained then depends on how the shipper loads the trucks. For a given origin–destination lane it will change from shipment to shipment depending on the loading efficiency of the shipper.

When several deliveries are made on a multi-stop route, the charge incurred can be modeled using the following variant of (5.2):

$$c^T(D, W) = c^s(m+1) + c^d D + c^w W \tag{5.3}$$

where m is the number of stops on the route. In this context, c^{s} is a cost per stop, W is the weight of the products loaded at the origin, and D is the length of the delivery route. This distance depends on the route chosen to make the delivery. Dagenzo (1984) derived an expression to estimate the length of the routes used to deliver products to clients in a demand zone from a shipping point when each vehicle makes a maximum of m stops:

$$D \cong \zeta(2\rho/m + 0.57\delta^{-0.5}) \tag{5.4}$$

where

- ζ Number of clients in the demand zone.
- ρ Average distance between the shipping point and the clients in the demand zone.
- δ Expected client density in the demand zone, that is, number of client per unit area (e.g. per square-km).

5.4.2 *Flow Cost Functions Estimation*

When designing a supply chain network, or selecting a transportation means for several lanes during a planning horizon, say a year, the annual flow F of the family of products shipped on a lane, expressed in a given load unit (e.g., pallets, cwt), is not necessarily predetermined and it may be a decision variable. Moreover, several thousand potential lanes may be considered in these problems, and it is not possible to estimate a transportation cost function for all these lanes individually. Instead, as in the previous section, the length D of a lane is considered as an explanatory variable in the estimation process. To address these problems, one therefore needs to estimate aggregate lane flow cost functions $C^{\mathrm{T}}(D, F)$. A cost function for a specific lane is then obtained simply by calculating its length D and substituting in $C^{\mathrm{T}}(D, F)$. The structure of $C^{\mathrm{T}}(D, F)$ depends essentially on how individual shipments are made during the year. Three typical cases are examined thereafter.

Suppose that E shipments are made on a lane during a year. If this lane involves ad hoc shipments of varying weights, in response to customer orders for example, these are probably made on-demand in less-than-vehicle loads (or as part of a TL shipment with multiple stops). The annual flow is then given by $F = \Sigma_{j=1}^{E} Q_j$, where Q_j is the size of the jth shipment in load units, and the annual cost by $C^{\mathrm{T}}(F) = \Sigma_{j=1}^{E} c^{\mathrm{T}}(\bar{w}Q_j)$, where $\bar{w}$ is the average weight of a unit load. However, when the transportation decisions are made, the lot sizes $Q_j, j = 1, \ldots, E$, are not known and a function $C^{\mathrm{T}}(D, F)$ must be estimated either from tariffs or from historical data. In this context, it is reasonable to assume that the function takes the form:

$$C^{\mathrm{T}}(D,F) = \bar{c}(D)F \tag{5.5}$$

where $\bar{c}(D)$ provides a cost per unit load as a function of distance. Equation (5.5) is a linear function of annual flows, and the modeling problem then boils down to estimating $\bar{c}(D)$ from tariff-based or historical shipment rates.

The rates are usually modeled as linear functions $\bar{c}(D) = \bar{c}^{\mathrm{o}} + \bar{c}^{\mathrm{d}}D$ or power functions $\bar{c}(D) = \bar{c}^{\mathrm{o}}D^{a}$ of distance, where $\bar{c}^{\mathrm{o}}$, $\bar{c}^{\mathrm{d}}$, and a are regression parameters. Empirical studies (Ballou 1991) have shown that better approximations are obtained when distinct rate functions are estimated for each SCN origin (e.g., suppliers, plants, and DCs). Also, better fits are obtained when the average shipment weights on the lanes considered are all in one or two tariff weight intervals. When this is not the case it is best to estimate distinct functions for predefined weight intervals, say for small, medium, and large shipment loads. Another possibility is to consider the probability distribution of the random regression errors explicitly in the design model. This leads to the use of more complex stochastic design models such as those studied in Chaps. 10 and 11.

Example 5.3

A leading North American fine paper manufacturer is shipping finished products with trucks or railcars on internal lanes between mills and DCs, and with trucks on external lanes between mills or DCs and customer ship-to-points. The company has contracts with several carriers, and DC-to-customer deliveries are made mainly with multiple-stop TL shipments. Finished products are shipped on pallets, on skids, or in rolls, but the industry practice is to measure product flows in tons. For an existing lane and product type, transportation rates, in dollars per ton, are computed from historical data by dividing the total annual transportation costs for that lane by the total number of tons shipped during a typical year. The data thus obtained is plotted in Fig. 5.19 for 355 lanes between a DC in the Montreal area and its ship-to points with average order sizes between 10 and 25 tons. The regression line fitted has an $R^2 = 0.85$, which is acceptable, but, as can be seen in the random errors histogram on the right-hand side, the rate variation is relatively high, because of the disparity in load weights and the use of different equipment (semi-trailers of diverse length provided by various carriers).

If all the shipments made on the lanes considered involve full-vehicle loads of size Q, the annual flow on a lane is given by $F = EQ$, and the annual cost is $C^{\mathrm{T}}(F) = Ec^{\mathrm{T}}(D) = [c^{\mathrm{T}}(D)/Q]F$, where D is the length of the lane. Recall, that

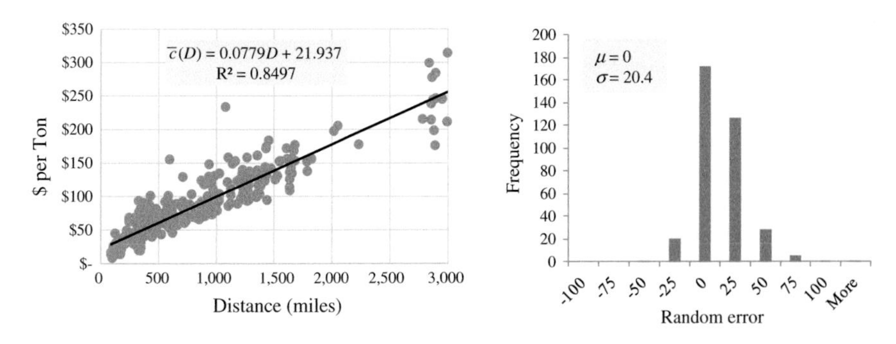

Fig. 5.19 Transportation rates as a function of distance

TL-shipment costs $c^{\mathrm{T}}(D)$ do not depend on the weight of the load. Because Q is predetermined, $C^{\mathrm{T}}(D,F)$ is a linear function of F, and model (5.5) still applies. However, in this case, $\bar{c}(D) = c^{\mathrm{T}}(D)/Q$, and it can be derived directly from the load cost functions (5.1) or (5.2).

Example 5.4
In Example 5.2, the cost function $c^{\mathrm{T}}(D) = 1277.7 + 3.14D$ was estimated for TL shipments. Now suppose that products are shipped in pallets and that a truck load contains 40 pallets, that is, that $Q = 40$. Then the required cost per pallet function is given simply by $\bar{c}(D) = c^{\mathrm{T}}(D)/40 = 31.9 + 0.078D$.

A third possibility is that shipments are made on a fixed schedule, say every Monday. In this case, the number of shipments per year E is predetermined but the shipment size depends on the annual flow, that is, $Q = F/E$. The cost of a shipment is then $c^{\mathrm{T}}(D, \bar{w}Q) = c^{\mathrm{T}}(D, \bar{w}F/E)$, and the total cost of transportation on a lane of length D for the year $C^{\mathrm{T}}(D,F) = Ec^{\mathrm{T}}(D, \bar{w}F/E)$ is a nonlinear function of F.

Example 5.5
In Example 5.1, the best fit for the load shipment costs was the power function $c^{\mathrm{T}}(D,W) = 0.04D^{0.56}W^{0.84}$. Suppose that weekly shipments are scheduled, that is, that $E = 52$, and that the unit loads used are pallets, each pallet weighing about $\bar{w} = 1000$ pounds. The resulting annual flow cost function is then:

$$
\begin{aligned}
C^{\mathrm{T}}(D,F) &= E\left[0.04D^{0.56}(\bar{w}F/E)^{0.84}\right] = 52(1000/52)^{0.84}0.04D^{0.56}F^{0.84} \\
&= 24.9D^{0.56}F^{0.84}
\end{aligned}
$$

5.5 Transportation Means Selection

Several methods were proposed to select a transportation means (mode selection, external carrier versus owned or leased private fleet, direct deliveries versus shipping through consolidation terminals, etc.) among a set of possible options. The two most common approaches are *total cost analysis* and *multi-criteria analysis*. In both cases, the following costs must be evaluated for all the options considered:

- *Shipping* costs (incurred when moving products in *space* between origins and destinations): These costs depend on who is responsible for the various steps involved (vehicle loading, delivery to the port of embarkation, boarding, carriage, debarkation, delivery to buyer premises, unloading, insurances, custom clearance, demurrage, etc.) as specified by selected Incoterms or transportation contract terms.
- *Operating* costs (incurred when making or moving products in facilities—plants, transshipment terminals, DCs, etc.): Only operating costs differing for the potential solutions considered are relevant.
- *Holding* costs (incurred when products are immobilized during a period of *time*): This includes the costs of inventory in transit as well as cycle and safety stocks in the shipper and destination facilities affected by the means selection decision (if they are under the control of the party making the analysis).
- The *investment* costs necessary to set up the transportation means, if any (or the annuity required to recover the initial outlay, with interest): These costs are evaluated based on the discounted cash flow principles studied in Chap. 2.

When a company is selecting a transportation means, the complexity of the analysis depends on several factors:

- The part of the transportation network controlled by the company, that is, the steps of the transportation process that fall under its responsibility
- The type of production and inventory P&C system used (especially whether production and transportation planning are *independent* or *synchronized*)
- The nature of the information available (*deterministic* versus *stochastic* costs, demand, and transit times)
- The complexity of the transportation network considered
- Whether the means selected will be used to ship a single or multiple products

In the current highly competitive business context, it is often more appropriate to minimize all supply chain costs, even if some of them are not under the responsibility of the company. This approach amounts to maximizing the revenues generated, and it gives better results in the long term. The solution selected when the costs minimized are limited to those under the responsibility of the company may not be valued by clients who may then decide to accept an offer from a competitor. By including all costs in the analysis, better offers are made to customers, which eventually improve market shares and profits.

The details of the total cost analysis approach to use to analyze transportation options can vary from one context to another. However, the previous principles should always be followed. In the next section, we illustrate the approach with a few particular cases.

5.5.1 Carrier Selection

Suppose that we want to choose a carrier to deliver on a regular basis products manufactured in a factory to a distant DC. This could be the case of a manufacturer that has a plant in Chicago and needs to ship products regularly to a DC in Orlando, from which it serves the Florida market. More specifically, the context is the following:

- The plant and the DC are part of the same company.
- The DC demand and the delivery times are known and constant.
- The factory makes products just in time at a constant rate; production and transportation are synchronized, that is, products are shipped as soon as the production of a shipping lot is complete; production setup costs are negligible.
- The products manufactured being similar, they are all shipped in the same unit load (e.g., pallet), and each load has the same weight.
- The carriers considered can be used without any initial investment and they involve the same operating costs.

In order to formalize the approach proposed to select a carrier, the following notation is required:

d	Annual DC demand in unit loads
b	Annual production capacity of the plant in unit loads
Q	Number of unit loads in a shipment (depending on the type of transportation means used—for example, the number of pallets in a full truck load)
r	Cost of capital in \$/\$/year, as defined in Chap. 2
r_u	Storage cost for the plant in \$/\$/year
r_w	Storage cost for the DC in \$/\$/year
v_u	Value of a unit load when shipped from the plant
c_u^{I}	Annual inventory holding cost at the plant in \$/unit load: $c_u^{\mathrm{I}} = (r + r_u)v_u$
$c^{\mathrm{T}}(Q)$	Load-shipping cost based on (5.1) or (5.2) (because the plant to DC distance and the weight of a unit load are known, it can be expressed as a function of Q)
v_w	Value of a unit load at the DC: $v_w = v_u + c^{\mathrm{T}}(Q)/Q$
c_w^{I}	Annual inventory holding cost at the DC in \$/unit: $c_w^{\mathrm{I}} = (r + r_w)v_w$
τ	Plant to DC delivery time (expressed in years)
$C^{\mathrm{T}}(Q)$	Total shipping cost per year
$C^{\mathrm{I}}(Q)$	Total inventory holding cost per year

$C^{\mathrm{M}}(Q)$	Total cost of the means considered per year: $C^{\mathrm{M}}(Q) = C^{\mathrm{T}}(Q) + C^{\mathrm{I}}(Q)$
E	Number of shipments per year: $E = d/Q$
G	Duration (in years) of a replenishment cycle: $G = 1/E = Q/d$

Our goal is to calculate the total annual cost, $C^{\mathrm{M}}(Q)$, for each option considered in order to compare them and determine which one is the cheapest. In our context, this cost has two fundamental components: the total annual shipping cost, $C^{\mathrm{T}}(Q)$, and the total annual inventory holding cost, $C^{\mathrm{I}}(Q)$. We must therefore calculate each of these costs. In practice, in this kind of analysis, inventory holding costs are often overlooked. This error can be very expensive.

Shipping costs calculations are straightforward. We simply multiply the number of shipments to make during the year by the cost of a shipment, giving:

$$C^{\mathrm{T}}(Q) = Ec^{\mathrm{T}}(Q) = c^{\mathrm{T}}(Q)d/Q \tag{5.6}$$

The calculation of the total inventory holding cost is a little more complicated. It has three distinct components: (1) the cost for the inventory sitting at the factory while waiting to be shipped, (2) the cost of the inventory in transit, and (3) the cost of the stock kept at the DC while waiting to be sold. These three costs are calculated separately but using the same rationale: the holding cost for the year is equal to the average inventory level during the year multiplied by the inventory holding cost of a product at the place where the stock is kept. This method is applied in turn to the plant, the stock in vehicles during transportation, and the DC. The behavior of inventories in the transportation system is illustrated in Fig. 5.20.

The triangles in the top schema of Fig. 5.20 show how stocks accumulate at the plant at a rate of b units per year, once production has begun, while waiting for the shipping quantity Q to be completed. A new production lot is started every G years and it takes (Q/b) years to produce a lot. The average inventory level during a year is therefore equal to the area of a triangle multiplied by the number of triangle in a year, that is, $(Q/b)(Q/2)E = dQ/2b$. Multiplying this by the unit holding cost at the plant $c_u^{\mathrm{I}} = (r + r_u)v_u$ gives $c_u^{\mathrm{I}}dQ/2b$, which is the total inventory holding cost at the plant for the year.

With regard to the stock in transit, it is clear that d products must be shipped from the factory to the DC during the year. Each product shipped is immobilized in transit during τ years. The average level of inventory in transit during a year is therefore $d\tau$. Because the value of a product in transit is v_u and the cost of capital is r, the unit inventory holding cost in transit is rv_u. The total holding cost for the inventory in transit during the year is therefore $(rv_u)d\tau$.

The situation for the DC is illustrated in the bottom schema of Fig. 5.20. Shipments must be synchronized so that a lot arrives at the DC as soon as it runs out of stock. Each replenishment cycle, the inventory level in the DC thus varies continuously between Q and 0, which implies that the average stock level during a year is $Q/2$. Multiplying this by the unit DC holding cost, $c_w^{\mathrm{I}} = (r + r_w)v_w$, with

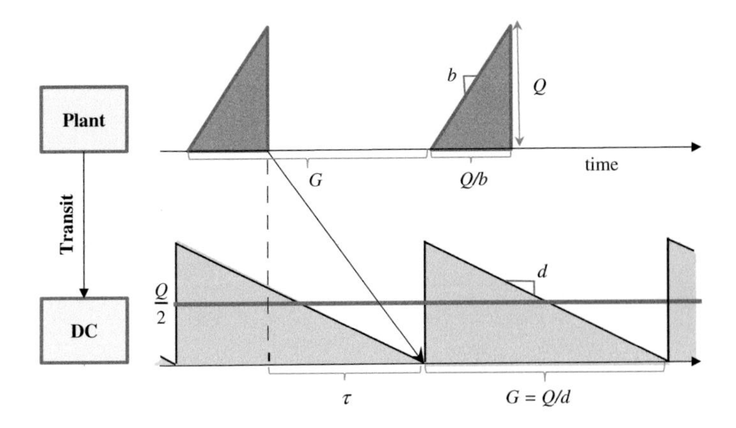

Fig. 5.20 Inventories in the transportation network

$v_w = v_u + c^{\mathrm{T}}(Q)/Q$, gives the total inventory holding cost at the DC for the year: $(r + r_w)[v_u Q + c^{\mathrm{T}}(Q)]/2$.

Taking the sum of these three terms, we see that the total inventory holding cost per year for the transportation system is:

$$C^{\mathrm{I}}(Q) = \{c_u^{\mathrm{I}} dQ/2b\} + \{(rv_u)d\tau\} + \{(r + r_w)[v_u Q + c^{\mathrm{T}}(Q)]/2\} \tag{5.7}$$

The total annual cost, $C^{\mathrm{M}}(Q) = C^{\mathrm{T}}(Q) + C^{\mathrm{I}}(Q)$, for each transportation means considered can now be calculated using (5.6) and (5.7) in order to compare them and determine the best one.

Example 5.6
A manufacturer must send 200,000 kg of products per year from its Montreal plant to its Miami DC. The production capacity of the factory is 2000 kg per working day (240 days/year). The value of the product at the plant is \$20/kg and the cost of capital used by the company is 10 %. The storage cost at the plant and DC is also 10 %. Three transportation options are considered: (1) once a week by plane (τ = 2 days) in batches of 4000 kg; (2) once per month per truck (τ = 3 days) in batches of 20,000 kg; and (3) eight times a year by rail (τ = 10 days) in batches of 30,000 kg. Shipping costs (packaging, pickup, loading, long-haul carriage, delivery, damage, and administration fees) were assessed on an annual basis from the bids obtained from different carriers. The result of the shipping costs analysis are given in Table 5.4, as well as the parameters of the resulting load cost function $c^{\mathrm{T}}(Q) = c^{\mathrm{o}} + c^{\mathrm{w}}Q$ for the transportation modes considered.

To select the best option, the total annual cost, $C^{\mathrm{M}}(Q)$, must be evaluated for each of them. This is easily done using an Excel spreadsheet such as illustrated in Fig. 5.21. Note that it is important that all the problem data be

Table 5.4 Shipping costs for the options considered

Shipping cost ($)	Plane	Truck	Rail
Carriage—fixed cost per shipment	1600	1800	4350
Carriage—variable cost per kg	0.35	0.12	0.03
Pick-up and delivery	160		300
Packaging, loading, and damage (per kg)	0.0115	0.0375	0.06
Administration cost (per shipment)	40	150	150
Total fixed cost per shipment (c^0)	1800	1950	4800
Total variable cost per kilo (c^w)	0.3615	0.1575	0.09

	Plane	Truck	Rail
Basic data			
Year demand (d)	200,000	200,000	200,000
Production rate (b)	480,000	480,000	480,000
Product value at plant (v_u)	$20.00	$20.00	$20.00
Shipment lot size (Q)	4,000	20,000	30,000
Transit time in years (τ = days/365)	0.00548	0.00822	0.02740
Fixed load carriage cost (c^o)	$1,800.00	$1,950.00	$4,800.00
Variable load carriage cost (c^w)	$0.36	$0.16	$0.09
Cost of capital (r):	10%	10%	10%
Plant storage cost (r_u)	10%	10%	10%
DC storage cost (r_w)	10%	10%	10%
Plant inventory holding cost ($c^I_u = (r+r_u)v_u$)	$4.00	$4.00	$4.00
Shipping costs			
Load shipping cost ($c^T = c^o + c^w Q$)	$3,246.00	$5,100.00	$7,500.00
Number of shipments ($E = d/Q$)	50.0	10.0	6.7
Total shipping cost for year ($C^T = c^T E$)	$162,300.00	$51,000.00	$50,000.00
Inventory holding costs			
Plant holding cost ($c^I_u dQ/2b$)	$3,333.33	$16,666.67	$25,000.00
Inventory in transit cost ($rv_u d\tau$)	$2,191.78	$3,287.67	$10,958.90
DC holding cost ($(r+r_w)(v_u Q+c^T)/2$)	$8,324.60	$40,510.00	$60,750.00
Total inventory holding cost for year (C^I)	$13,849.71	$60,464.34	$96,708.90
Total mode cost for year ($C^M = C^T + C^I$)	**$176,150**	**$111,464**	**$146,709**

Fig. 5.21 Excel spreadsheet for the evaluation of transportation means

expressed in years because it is our basis for comparison. Given that a year includes 240 working days, the annual production capacity is b = 2000 (240) = 480,000 kg/year. Transit times are also converted to years.

The analysis indicates that the best choice is to use trucks. Note that if inventory holding costs were not taken into account, rail would be selected.

This erroneous conclusion would lead to a loss of approximately $35,000 by year. Note also that, by taking into account the inventory holding costs, the impact of transit times is implicitly considered in the analysis.

5.5.2 Load Size Optimization

The evaluation approach presented in the previous section assumed that size of shipping lots, Q, is predetermined for each transportation means. In some context, the P and C strategy or carrier used may impose a load size, but in other contexts, the load size Q may be a decision variable to optimize. To find the optimal load size for a given transportation means, when the load cost function used is linear, that is, when $c^{\mathrm{T}}(Q) = c^{\mathrm{o}} + c^{\mathrm{w}}Q$, one has to find the minimum of the convex total cost function $C^{\mathrm{M}}(Q)$. Adding (5.6) and (5.7) this function reduces to:

$$C^{\mathrm{M}}(Q) = c^{\mathrm{o}}d/Q + \left[c_u^{\mathrm{I}}d + (c^{\mathrm{w}} + v_u)(r + r_w)b\right]Q/2b + \mathrm{cst} \tag{5.8}$$

where the last term is a constant ($\mathrm{cst} = (c^{\mathrm{w}} + rv_u\tau)d + (r + r_w)c^{\mathrm{o}}/2$) independent of Q.

The minimum of this kind of univariate function lies at the point where its slope is zero. This point is found by taking the derivative of $C^{\mathrm{M}}(Q)$ with respect to Q and by finding the value of Q for which it is nil, that is, by solving:

$$\frac{dC^{\mathrm{M}}(Q)}{dQ} = \frac{c_u^{\mathrm{I}}d + (c^{\mathrm{w}} + v_u)(r + r_w)b}{2b} - \frac{c^{\mathrm{o}}d}{Q^2} = 0$$

while taking into account the values that Q can take, given the unit load and the transportation equipment used. The solution of this equation is:

$$Q^* = \sqrt{2bc^{\mathrm{o}}d/c_u^{\mathrm{I}}d + (c^{\mathrm{w}} + v_u)(r + r_w)b} \tag{5.9}$$

Example 5.7 To illustrate the impact of (5.9), we can calculate Q^* for the trucking option considered in Example 5.6. The optimum lot size obtained is $Q^* = 11,699.83$. Now suppose that, given the nature of the trucking services offered, a new option involving truck shipping lots of 12,000 kg, instead of 20,000 kg, is evaluated. When Q = 12,000 is used in the "Truck" column of the evaluation spreadsheet (Fig. 5.21), the total annual cost obtained is $101,671, which shows that this option could save about $10,000 per year.

5.5.3 Random Demand and Transit Times

In the previous sections, we assumed that the annual demand is known with certainty and that there is no variability in transit times, which is not very realistic. The approach is, however, easily modified to take demand and transit time randomness into account. Suppose that, in addition to the characteristics already detailed, the transportation means selection problem considered has the following features:

- The DC annual demand is a stationary stochastic process with a known average μ^d and standard deviation σ^d.
- The transit time is a random variable with a known average μ^τ and standard deviation σ^τ.
- The DC controls its inventory with a (min, max) continuous review system, in which min is an order point and max an order-up-to level. The min is calculated so as to ensure that the probability of no stockout occurring during a replenishment cycle is at least α.

Under these assumptions, transportation means can be compared by calculating their total annual *expected* costs $\mathrm{E}[C^{\mathrm{M}}(Q)] = \mathrm{E}[C^{\mathrm{T}}(Q)] + \mathrm{E}[C^{\mathrm{I}}(Q)] + \mathrm{E}[C^{\mathrm{SS}}]$ for each option, where E[.] denotes the expected value, and $\mathrm{E}[C^{\mathrm{SS}}]$ is the expected cost of the safety stocks kept at the DC as a protection against risk during delivery lead times. The expected shipment costs $\mathrm{E}[C^{\mathrm{T}}(Q)]$ and the expected cycle stock costs $\mathrm{E}[C^{\mathrm{I}}(Q)]$ are calculated simply by replacing the demand d in (5.6) and (5.7) by its expected value μ^d.

The expected safety stock cost at the DC is given by $\mathrm{E}[C^{\mathrm{SS}}] = c_w^{\mathrm{I}} SS$ where SS is the safety stock required to protect from risk. Assuming that the demand during a transit time is normally distributed, it has been shown (Silver et al. 1998) that, for the service measure α used, the required safety stock is given by $SS = \Phi^{-1}(\alpha)\sigma_{\mathrm{LT}}$, where the safety factor $\Phi^{-1}(\alpha)$ is the α-level inverse of the standardized normal distribution (mean 0, standard deviation 1), and σ_{LT} is the standard deviation of the demand during a delivery lead time. It can also be shown that, under certain independence conditions, the variance of the lead-time demand is given by $\sigma_{\mathrm{LT}}^2 = \mu_\tau \sigma_d^2 + \mu_d^2 \sigma_\tau^2$.

It should be noted that, in addition to taking expected transit times into account, as done previously, this transportation means selection approach also considers transit time variability (measured by its variance).

5.5.4 Direct Delivery Versus Consolidation

Consider the following situation. Products manufactured in distinct focused factories must be shipped to a set of distribution centers. Would it be better to ship all products directly from their factory to each DC or to ship all products to a mixing and consolidation center in order to sort them in mix loads for each DC? These two

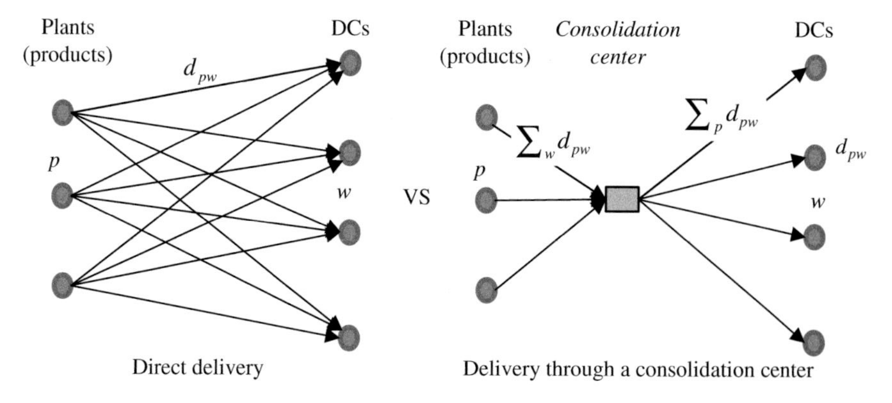

Fig. 5.22 Direct delivery versus mixing and consolidation

options are represented in Fig. 5.22. One sees immediately that direct deliveries require much transport. However, consolidation requires the operation of a cross-docking center, which can be very costly. In a given context, which of these two options should we choose?

Each lane of the direct delivery network involves a plant to DC shipment of the type studied in the previous paragraphs. We assume that the sum of the quantities shipped to DCs by a plant p matches its total production capacity, that is, that $b_p = \Sigma_p d_{pw}$, where b_p is the annual production capacity of the plant and d_{pw} the annual demand for product p at DC w. For direct deliveries, the subnetwork associated with each plant can be optimized separately and, to simplify the following formulas, the subscript p is dropped from the capacity and demand notation.

Each procurement cycle, the products required by all DCs must be produced by the plants. Therefore, the size of DC shipments is $Q_w = Gd_w$, for all w, and the length G of replenishment cycles may be fixed by management or optimized. For a given G, based on relationship (5.8), and replacing Q by Gd_w, for all w, the total annual cost for a plant subnetwork is given by:

$$\begin{aligned} C^{\mathrm{M}}(G) &= \sum_w C^{\mathrm{M}}(Gd_w) \\ &= \sum_w \left\{ \frac{c_w^{\mathrm{o}}}{G} + \left[\frac{[c_p^{\mathrm{I}} d_w + (c_w^{\mathrm{w}} + v_p)(r + r_w)b]d_w}{2b} \right] G + (c_w^{\mathrm{w}} + rv_p\tau)d_w + \frac{(r + r_w)c_w^{\mathrm{o}}}{2} \right\} \end{aligned} \tag{5.10}$$

and the minimization of this univariate convex function gives

$$G^* = \sqrt{2b \sum_w c_w^{\mathrm{o}} \Big/ \sum_w [c_p^{\mathrm{I}} d_w + (c_w^{\mathrm{w}} + v_p)(r + r_w)b]d_w} \tag{5.11}$$

The total cost of the direct delivery option for a given G is finally obtained by calculating the cost of each plant subnetwork using (5.10) and by summing them, that is, by computing $\Sigma_p C_p^{\mathrm{M}}(G)$. Blumenfeld et al. (1985) derive an expression similar to (5.10) to evaluate the total cost of the consolidation network on the right-hand side of Fig. 5.22, for when production and transport planning is not synchronized. Several other strategies for the transportation of products between a number of origins and destinations can also be considered:

- A mixed strategy can be profitable, that is, a strategy that combines the two options described in Fig. 5.22 in an optimal way. Hall (1987) presents an optimization model for this type of strategy.
- A strategy with pickups from a few neighboring origins in the same shipment and/or with deliveries to a number of nearby destinations can also be cost-effective. Burns et al. (1985) show that, with this type of strategy, one should always ship full-vehicle loads.
- Strategies that exploit more than a mixing and consolidation terminal can be envisioned. Klincewicz (1990) suggests an approach for the development of this type of strategy.

For any transportation strategy envisioned, one can choose between them by computing and comparing their total annual relevant costs. Dagenzo (2005) proposes models for the evaluation and optimization of a large assortment of freight transportation strategies.

5.5.5 *Multi-criteria Evaluation*

We stressed in Sect. 5.2.6 that selecting a transportation means is essentially a multi-criteria evaluation problem, and that relevant criteria include shipping costs, transit time durations and variability, loss and damage, safety, GHG emissions, noise, and administrative burden. The total cost analysis approach presented in the previous sections can take most of these criteria into account directly or indirectly. In Ex. 5.6, we saw that loss, damage, and administrative costs can be embedded in the shipping cost functions estimated. In Sect. 5.5.3, we saw that the impact of transit time durations and variability can be assessed by considering the expected holding costs of cycle and safety stocks. However, other criteria, such as the environmental footprint of the solution, the associated risks, and the capability, flexibility, and financial stability of the carriers involved, may be important.

When this is the case, the Brown and Gibson approach presented for facility site selection in Sect. 4.4.3 can be adapted to make transportation means selection decisions. Any other multi-criteria decision-making methods capable of blending objective and subjective criteria could also be used. Moreover, in some contexts, the solutions examined may take the form of a long-term partnership between a shipper and a carrier or 3PL. The selection of supply chain partners is examined in depth in the next chapter.

Review Questions

5.1. What are the elements of a modern transportation system? Give a practical example for each element.
5.2. What strategic decisions must a company make in order to move products in its SC network?
5.3. What transportation modes can be used to move products? Discuss the advantages and disadvantages of each of these modes.
5.4. What is the difference between multimodal and intermodal transportation? Give examples to illustrate your answer.
5.5. What type of transportation intermediaries are offering services to shippers? Do you think that these intermediaries will play a more important role in the future? Explain.
5.6. What are the main criteria to use for the selection of a carrier?
5.7. Why would a company operate a private fleet rather than use a common carrier to transport products in its SC?
5.8. Explain the tariffs structure for road transportation.
5.9. What cost elements should be included in a shipping cost function?
5.10. How can transportation cost functions be estimated?
5.11. What are the costs to be taken into account when a company wants to select a transportation means? Explain the nature of these costs and why it is important to take them into account.
5.12. When conducting a total cost analysis of alternative transportation means, why is it important to express all dates in the same time units?
5.13. How can we proceed to select a transportation means when the demand is not known with certainty?
5.14. What other criteria could be consider to solve Example 5.6? Suggest a multi-criteria method to tackle the problem.

Exercises

Exercise 5.1 A New York food manufacturer currently buys a chemical required in its production process from a local supplier. The chemical is purchased in lots of 300 bags. Each bag weighs 10 kg and it costs $150, shipping included. When an order is passed, a $75 fixed cost is incurred. The company needs 12,000 bags per year, it estimates its annual storage cost at about 10 % of the value of the product, and it uses a cost of capital of 8 %.

During a recent business trip to Mexico, the VP Supply-Chain discovered that an equivalent product is manufactured by a multinational in the Juarez maquiladora. After negotiation, taking exchange rates and import duties into account, he arrives at the conclusion that he could get the chemical from this source for $140 per bag, as long as he takes care of transportation. Two alternatives are available to ship the products: (1) by air in 240 bags lots, once a week (50 weeks per year), or (2) by truck, once a month, in batches of 1000 bags. Air transportation takes two days, and it has a fixed shipment cost (pickup, delivery, reception, and administration) of $350 per trip and a variable cost (freight, packaging, insurance) of $8.58 per bag. For the trucking option, the transit time is five days, the fixed cost per trip $125, and the variable cost $3.08 per bag. Should the company buy the chemical in Mexico and, if so, what transportation means should it use?

Exercise 5.2 WindSports in Montreal is the distributor of F2 windsurfing boards for east Canada. The windsurfing boards are imported from Germany and they can be transported by air or by containership. The average price paid by WindSports for a board is $800. During the windsurfing season, which lasts six months, the distributor sells about 50 boards per month (30 days). It can borrow money at an interest rate of 15 % and it stores its boards in a public warehouse for $10 per board per month.

The owners consider two transportation options. The first one involves making shipments of 50 boards each month by air, with a lead time of two days. The air carrier charges $35 per board and, each time, it costs $500 to clear customs and pickup the shipment at the airport. Alternatively, a 150-board container can be shipped by sea twice during the season. The transit time is then 30 days, freight costs are $2000 per container and the costs for clearing customs, and picking up the shipment in the port of Montreal is $1000. Packaging and administration costs are the same for the two transportation modes. How should WindSports import its boards?

Exercise 5.3[7] Multimedic manufactures steel surgical instruments in Sheffield, UK. The company operates a subsidiary in Istanbul, Turkey, that it must replenish on a regular basis. It ships its instruments in boxes weighing roughly 10 kg and having an average value of €500. For the next year of operation, it forecasts that 480 boxes of instruments will be shipped to Turkey. The company can obtain a 10 % return on its investments and its storage costs in England as in Turkey are 10 % of the inventory value. The company is considering three transportation options:

- It can ship all the required boxes in a 40-foot container once per year by sea, which costs €1935 and entails a delivery lead time of 27 days. The amount of insurance payable in this case is €1131.05.
- It can make LTL truck shipments every 6 months, which costs €1710 per shipment and takes 12 days. The amount of insurance required in this case is €669.41 per shipment.
- It can make monthly shipments by air which costs €846 for cargo and €97.46 for insurance. The transit time in then 4.5 days.

In all cases, the production of the instruments at the plant begins one month before the shipment date. Packaging and administration costs are the same for the three options. How should Multimedic ship its instruments?

[7]This exercise is based on the Polymedic Ltd case found in Taylor (1997).

Exercise 5.4 Go back to Ex. 5.6 which showed how to select a transportation means (plane, truck or rail) based on economic criteria. Consider now that variability of delivery time, carbon emissions, and loss and damage should also be part of the analysis. Based on Table 5.3 (or your own ranking) propose a multi-criteria approach (such as the Brown and Gibson method) to select the best transportation mode. Compare your results with those of the economic approach used in Ex. 5.6.

Exercise 5.5 Going back to Sect. 5.5.4, derive an expression similar to (5.10) for the calculation of the total annual relevant cost when making deliveries through a mixing and consolidation center (right-hand side of Fig. 5.22). Assume that the receptions and shipments in the crossdocking center can be synchronized so that all the products required by a DC are shipped during a procurement cycle.

Bibliography

Ballou R (1991) The accuracy in estimating truck class rates for logistical planning. Transp Res-A 25A(6):327–337

Ballou R (1992) Business logistics management, 3rd edn. Prentice-Hall

Blumenfeld D, Burns L, Diltz D, Daganzo C (1985) Analyzing trade-offs between transportation, inventory and production costs on freight networks. Transp Res-B 19B(5):361–380

Blumenfeld D, Burns L, Daganzo C, Frick M, Hall R (1987) Reducing logistics costs at general motors. Interfaces 17(1):26–47

Burns L, Hall R, Blumenfeld D, Dagenzo C (1985) Distribution strategies that minimize transportation and inventory costs. Ops Res 33(3):469–490

CSCMP, Goldsby TJ, Iyengar D, Rao S (2014) The definitive guide to transportation. Pearson Education

Dagenzo C (1984) The distance needed to visit N points with a maximum of C stops per vehicle: an analytic model and an application. Trans Sci 18–4:331–350

Dagenzo C (2005) Logistics systems analysis, 4th edn. Springer, Berlin

Dennis WT (2011) Parcel and small package delivery industry. CreateSpace, North Charleston

European Commission (2013) The core network corridors. Directorate general for mobility and transport

Hall R (1987) Direct versus terminal freight routing on a network with concave costs. Transp Res-B 21B:287–298

HHLA (2013) Annual report 2013. Hamburger Hafen und Logistik aktiengesellschaft

ITF (2013) 2013 annual summit highlights—funding transport: session summaries. OECD

ITF (2014) International transport forum data base. www.internationaltransportforum.org. Accessed 25 Mar 2014

Klincewicz JG (1990) Solving a freight transport problem using facility location techniques. Ops Res 38–1:99–109

Magee JF, Copacino WC, Rosenfield DB (1985) Modern logistics management. Wiley

Miller T (1991) The international modal decision, Distribution 82–92

Muller G (1989) Intermodal freight transportation, 2nd edn. Eno Foundation

OECD (1992) Advanced logistics and road freight transport. Organisation for Economic Co-operation and Development, Paris

OECD/ITF (2010) Reducing transport greenhouse gas emissions

Perry N (2010) The state of truck/rail modal shares: an analysis for transportation customers. US Xpress Enterprises

Rivera L, Sheffi Y (2013) Panama canal update, In: Essig M, Hülsmann, Kern E, Klein-Schmeink S (eds) Supply chain safety management. Springer, Berlin, pp 213–216
Rodrigue J-P, Comtois C, Slack B (2013) The geography of transport systems, 3rd edn. Routledge
Silver E, Pykc D, Peterson R (1998) Inventory management and production planning and scheduling, 3rd edn. Wiley
SteadieSeifi M, Dellaert NP, Nuijten W, Van Woensel T, Raoufi R (2014) Multimodal freight transportation planning: a literature review. EJOR 233:1–15
Taylor D (ed) (1997) Global cases in logistics and supply chain management. Thomson Business Press
Tompkins J, Harmelink D (eds) (1994) The distribution management handbook. McGraw-Hill
Tyworth J (1991) The inventory theoritic approach in transportation selection models: a critical review. Log Trans Rev 27(4):299–318
Tyworth J (1992) Modeling transportation-inventory trade-offs in a stochastic setting. J Bus Log 13(2):97–124
Tyworth J, Cavinato J, Langley J (1987) Traffic management. Addison Wesley
UNCTAD (2013a) Recent developments and trends in international maritime transport affecting trade of developing countries. Note TD/B/C.I/30, United Nations
UNCTAD (2013b) Review of maritime transport. United Nations
Wang CG (2006) China. CSCMP global perspectives. CSCMP
Wilson R (2013) 24th annual state of logistics report. CSCMP

Chapter 6
Supply Chain Partnerships

Supply chains are integrated arrangements of the resources and activities required to transform raw materials into finished industrial or consumer products. As pointed out in Chap. 1, several companies are usually involved in the SC sourcing, transformation, delivery, and sales activities required to bring products to markets. The part of the supply chains in which a given company is involved defines its internal supply chain network (SCN). A company therefore necessarily maintains relations with numerous SC partners. The boundary of its SCN is shaped by the SCs of the industries in which it competes, but also by strategic decisions on what it should insource, outsource, or coproduce. This chapter studies these decisions and the impact they have on SCN structures, processes, and performances. It starts by examining the nature of inter-company relationships, and in particular the topology of dominant SC structures. Taking the point of view of a particular company, it then looks at the value-added services provided by primary SC partners, namely, suppliers, contract manufacturers, and third-party logistics (3PL) providers. Finally, it presents an approach and a model to facilitate partner selection decisions and it discusses important issues related to the management of partnerships.

6.1 Value-Added Networking

6.1.1 Networked Companies

Chapter 1 points out that strategic alliances are part of the means a company can use to develop order-winning offers for its product-markets (see Fig. 1.14). All companies entertain vertical relationships with their suppliers and customers. Most companies also engage in horizontal relationships with subcontractors, 3PLs, service providers (e.g., financial institutions), facilitators (e.g., governmental research centers), and even competing firms (see Fig. 1.4). Collectively, all these vertical and

A. Martel and W. Klibi, *Designing Value-Creating Supply Chain Networks*,
DOI 10.1007/978-3-319-28146-9_6

horizontal partners form the *external* network of a company. The expression *networked company* is a metaphor used to designate firms with non-negligible external networks aiming to achieve sustainable value creation by leveraging the resources of their partners and by continually seeking the best balance between their internal and external networks.

The nature of inter-firm relationships is quite varied. At one extreme, for transactional supplier–customer exchanges, the parties are involved in *arm's length* relationships governed by terms of sale (see Sect. 5.3.1). At the other extreme, the partners may engage in joint ventures involving the creation of a new business entity. In between, all sorts of mid-term or long-term strategic alliances may be negotiated between two or more parties. The associated objectives, roles, boundaries, and cost or profit sharing rules are then usually specified in a contract. When examining the frontier between the internal and external networks of a company, four distinct internalization-externalization options can be distinguished: insourcing (make), buying, outsourcing, and cooperative alliances. The features of each of these options in terms of resource ownership and activity planning, control, and execution are indicated in Fig. 6.1. As can be seen, this goes beyond the classical make or buy dichotomy.

Insourcing involves the planning, control, and execution of an activity internally using company resources. Conversely, *buying* involves the procurement of raw materials, components, products, or services from external suppliers. The company then has no control over the resources and processes used to deliver the products or service procured. This does not mean, however, that it does not expect vendors to provide competitive costs, quality, delivery, reliability, and service levels. *Outsourcing* involves performing an activity that could be done internally using external resources, usually under a long-term agreement. The company then generally keeps some level of control over the planning and/or execution of the activity. Typical examples are contract manufacturing and the use of 3PLs. Another example would be the fabrication of a product in a company plant using leased equipment or agency workers. A *cooperative alliance* involves performing an activity, in collaboration with a partner, using resources from both parties and shared planning, control, and execution processes. At the limit, this could go as far

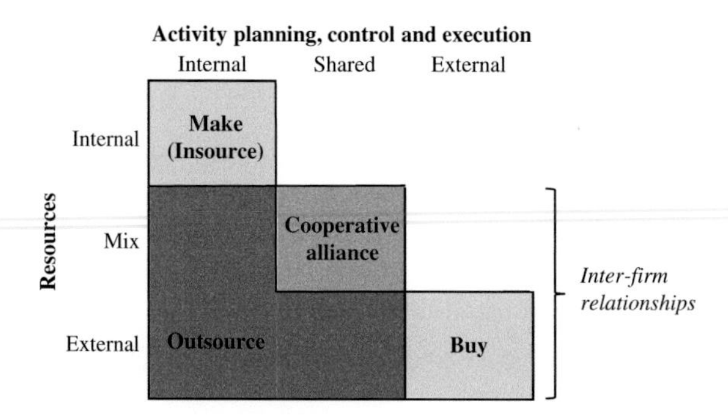

Fig. 6.1 Internalization-externalization options

as a merger or an acquisition. As we shall see, the distinction between these options is not always as clear-cut as shown in Fig. 6.1.

Companies engage in inter-firm relationships mainly to improve their value creation capabilities. Going back to the concepts presented in Chaps. 1 and 2, the reasons invoked by companies to develop partnerships can be linked to five fundamental strategic needs:

- **Access to resources**. This may include the access to remote, rare, or cheaper raw material sources, access to highly qualified and/or low-cost human resources, and access to state-of-the-art technologies. In some contexts, engaging in partnerships may be the only way to acquire critical resources. It may also free important internal resources so that they can contribute to core business endeavors. Using external instead of internal resources reduces the need for capital and liquidities, and it avoids irreversible investments, which usually improves short-term financial results. The ensuing reduction of complexity may also allow the simplification of organizational structures.
- **Process improvement**. This is related to the necessity of reducing costs and current assets and of developing valued capabilities such as short response times and time to market, flexibility, agility, superior product quality, and supply chain resilience. By concentrating on core business activities, companies are able to streamline their processes and to reap the benefits of economies of scale and scope, as well as of so-called economies of *skill*. By externalizing noncore activities (and in particular support activities) to domain specialists having high business volumes and expertise, they can achieve additional economies of scale, scope, and skill that would not be reachable if these activities were insourced. Outsourcing some activities may also enable companies to realize the benefits of process reengineering more quickly.
- **Access to markets**. It is normal for companies to want to grow and enlarge their revenue base. This requires a continuous improvement of their offer to clients (through better order winners) and a widening of product-markets, often by trying to reach new territories where they are not implemented and have no resources. It may also involve offering complementary products to their current clientele. The easiest way to reach new product-markets is to make alliances with companies already well positioned in these product-markets.
- **Knowledge and learning**. In order to improve their competitive position, companies need to develop innovative products and processes and to perfect their competencies. This may require knowledge and know-how not available in the company, and the fastest way to fill this gap may be to collaborate with companies already possessing the required world-class competencies.
- **Risk reduction**. When a company expands, making the large investments required without knowing what the results will be may be very risky. Also, with today's labor laws, it may be very difficult to downsize once resources have been committed. By working with partners, the risk is shared. If the project is not successful, it is much easier to backtrack.

Several of these opportunities can become threats if the results expected from a partnership are not delivered. Partnerships are not without danger. Some of these dangers can be alleviated by a proper management of the relationship but others are more strategic. The disadvantages often cited include the loss of managerial control, hidden costs, threats to security and confidentiality, quality problems, and the dependence on the financial well-being of other companies. Most of these problems can be eliminated by making good partnership choices and negotiating solid contracts. As we shall see in the next section, bad externalization decisions also may lead to losing a strategic advantage, which is a much more fundamental problem.

6.1.2 Internalization–Externalization Decisions

Internalization–externalization decisions are extremely important strategic issues. Welch et al. (1992) examine several North American industrial sectors (consumer electronics (CE), machine tools, semiconductors, and office equipment) negatively affected by outsourcing activities to Asian companies. When they decided to outsource core production activities, these companies were focusing on short-term gain. By doing so, they transferred know-how to subcontractors, who eventually developed into competitors offering better order winners then the original product manufacturer. This led to the loss of entire industrial sectors. Activity externalization decisions cannot be based simply on a comparison of internal standard costs (or transfer prices, for profit centers) with the prices demanded by potential suppliers or subcontractors. There are at least three fundamental problems associated with this approach:

- The calculation of standard costs generally incorporates an allocation of a portion of overhead costs. But there is nothing that guarantees that overhead charges will disappear if the activity is externalized.
- There is no guarantee that the order-winning attributes (quality, lead times, flexibility, etc.) depending on the externalized activity will be preserved. They could be better or worse. In any case they will have to be monitored, which generates costs. The order-winning impact, in terms of added or reduced revenues and expenses, should thus be considered in the decision process.
- The technology associated with the activity can be the source of a major competitive advantage. Externalizing an activity then amounts to sharing the secrets of our success with potential competitors.

A conceptual framework stressing the technological factors to take into account when making activity internalization–externalization decisions is presented in Fig. 6.2. This evaluation grid highlights three critical strategic dimensions to consider: the impact of the technology of the activity on the competitive advantage of the firm, the maturity of this technology across the industry, and its strength when compared to the technology of competitors. Making this type of evaluation clearly requires an in-depth understanding of the position of the company in its

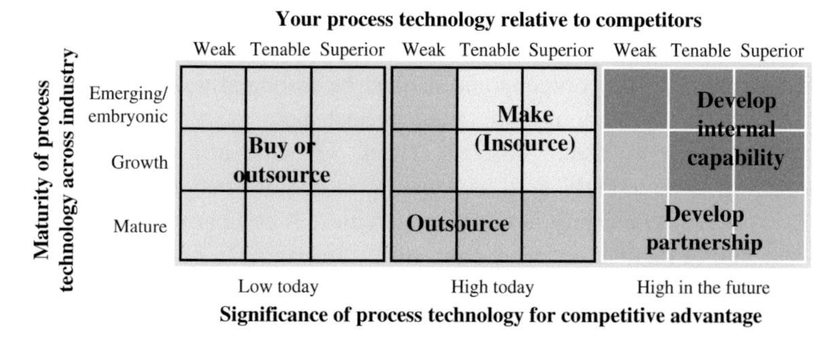

Fig. 6.2 Strategic internalization-externalization model—based on Welch et al. (1992)

industry, as well as of the technology available and under development in competing firms. By examining these technological dimensions, a firm can avoid the pitfalls of decisions based purely on costs. This is not to say, however, that a rigorous value analysis is not required. The two elements are necessary.

Among the numerous possible forms of cooperation between companies, some are more relevant than others to meet specific strategic needs. In a given context, one must therefore determine the most appropriate type of inter-company relationship to develop. Before we examine common forms of buying, outsourcing, and cooperative alliance relationships (see Fig. 6.1), note that insourcing can also take various forms, depending on the context. For example, for support activities (e.g., finance, human resources, information systems), an alternative to outsourcing often considered by large corporations is the setup of one, or more, *shared services* center, that is, the activity is moved out of individual business units and consolidated into a corporate service organization.

Historically, the dominant Western *buying* strategy has been to develop a wide supplier base, which allows negotiation of good prices and protection against disturbances (strikes, bankruptcy, etc.). With the advent of total quality and JIT, this vision has become less common and strategic alliances with suppliers are favored. Four forms of relationships with suppliers can be distinguished: ad hoc, standard, privileged, or specialized. The first applies to transactions made to fulfill ad hoc non-repetitive needs. With the introduction of online catalogs and marketplaces, this approach is now common for the acquisition of consumer goods and business supplies. A standard relationship exists when the link between a company and a vendor (customer) is based solely on a purchase (sale) contract. It involves products and monetary flows between the parties and the exchange of primary information on product characteristics, prices, payment, delivery terms, and so forth. This type of relationship is widespread in practice.

A privileged relationship develops when a company selects suppliers that meet a predetermined set of criteria as part of a strategic sourcing program. Successful suppliers are granted certain advantages in exchange, of course, for equivalent privileges. For a retail store, for example, it could be a reduced price in exchange for a visible position on the store shelves. Finally, a specialized relationship is sought between a company and a supplier when the latter offers a rare or specialized product or service needed by the former. This type of relationship has obvious strategic importance to a company and it must be managed with care. We discuss relationships with suppliers in more detail in Sect. 6.2.

Outsourcing encompasses different types of subcontracting relationships. Subcontracting is commonly used as a means to acquire temporary capacity (see Chap. 8) or to avoid investing in noncore activities. A company specializing in the design and sale of medical imaging equipment, for example, could subcontract the production of electronic components because it is not its core business. In the food industry, retailers subcontract the production of house brand products to food manufacturers. In basic subcontracting relationships, the original company keeps control over the process. When the management of an activity is delegated to a subcontractor, companies engage in a genuine partnership. Trust is the cornerstone of this type of relationship involving a much more intense exchange of information.

Brokerage is another type of outsourcing relationship. It involves asking a third party to engage in a relationship with other parties on our behalf. A transportation broker would, for example, organize the transport of goods from an origin to a destination on our behalf. This may involve the choice of a transportation mode, a carrier, insurers, shipment consolidation, customs clearance, and so on. Activities associated with the following functions are commonly outsourced by companies: information technology, operations (logistics, manufacturing, and procurement), finance, human resources, legal services, real estate, and sales and marketing support. The outsourcing of manufacturing, logistics, and procurement activities is examined more closely in Sect. 6.3.

Cooperative alliances involving two or more companies also can take several forms. These alliances can be informal, formal, or strategic. Informal alliances are flexible and they allow partners to protect their independence. Business-trade associations and sponsorships are typical examples. The former are often formed for lobbying, education, and standardization purposes. Sponsorships usually involve a company helping another to develop the competencies required to obtain adequate services. This is often facilitated by financial assistance from governments.

Formal alliances involve signed agreements defining the obligations of the partners. Typical examples are license agreements and franchises. Companies also sign cooperation agreements to share complementary resources, technologies, and expertise or to fulfill a common need. Cheese or wine producers in a region could organize common advertising campaigns, for example, to improve the visibility and reputation of their products. Co-contracting (or co-sourcing) is another example. It involves the mutual commitment of two or more companies to jointly complete the terms of a contract. This is common for large projects requiring the resources of several companies, such as the construction of a new dam in a developing country.

When the alliance is strategic, it often leads to the creation of a new legal entity such as a consortium or a joint venture. A *consortium* is an alliance between several partners, each bringing specific resources, skills, and technologies, within a shared capital structure. Consortiums are generally time-limited; they are dismantled when they cease to provide benefits. *Joint ventures* refer to long-term value-creating activities beneficial to all the companies involved. They usually take the form of a new company that is not in competition with the partners. The investments required to start the joint venture can take the form of capital, human resources, or equipment. An example of a joint venture is the creation of a purchasing group that allows the participants to get better prices from product and service suppliers because of ensuing economies of scale. Several large retailing chains started this way.

6.1.3 SCN Boundaries

Several of the inter-firm partnerships examined in the previous section are present in any supply chain. As mentioned, these involve vertical and horizontal relationships and they may even concern cooperation with competitors in areas not covered by antitrust laws (e.g., shared distribution resources). They result from negotiations with potential partners as part of a strategic decision process. Internalization-externalization decisions delineate the boundaries of a company's internal supply chain network. Given its industry structure, the position of the company in the industry, its internal competencies and technology, and those of competitors, there are always, at one extreme, activities that cannot or should not be externalized; at the other extreme, there are activities that cannot or should not be internalized; and, in the middle, there is a large gray zone of activities that could be internalized or externalized. An activity should be externalized only if this improves the value creation capability of the company, but there might be several ways of doing this. The structure of the SCN of the company also depends on the industry context and on its strategic posture (see Fig. 1.14). The SCN of a vertically integrated company is necessarily much more elaborate than the internal network of a virtual company. This section shows how activity graphs (see Sect. 4.1.2) can be used to specify the internalization–externalization opportunities of a company.

Figure 6.3 presents a simplified primary activity graph for a fine paper (FP) manufacturer. This graph reflects a number of SC policies and strategies pre-adopted by the company, but it also designates the internalization-externalization decisions to make. Activities represented in non-shaded (white) nodes must be insourced. Pulp and paper making (activities 4 and 6) are core activities for this company, and they use proprietary technologies that are a source of competitive advantage. For this reason the company does not want to outsource them. Activities in shaded nodes can be externalized. In this case, all storage activities could be outsourced. Because chipping and paper-conversion activities involve mature technologies, the company could also outsource them. The fact that a shaded node is used does not mean that the activity must be outsourced. It could be insourced, partially outsourced, or completely outsourced. The decision remains to be made.

Pulping and paper making requires the use of chemicals. The availability of quality chemicals is important for the company but their production is completely outside its field of expertise. Chemicals must thus be purchased outside and their

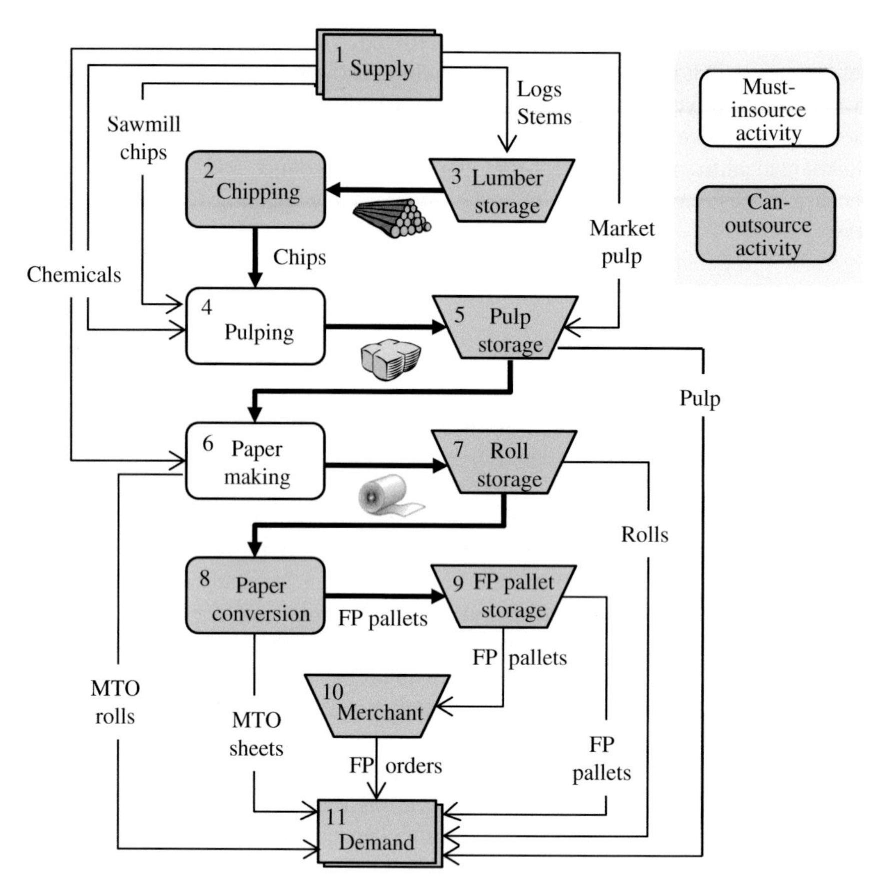

Fig. 6.3 Primary activity graph for a FP manufacturer

production is not included explicitly as an activity in the graph. Chemicals appear on raw material supply arcs to indicate that they must be procured from external chemical suppliers. Logs and stems are obtained from forest operations companies. The fact that no harvesting activity is included in the graph means that the company does not want to get involved in forest operations. In some cases, paper companies are involved in forest operations and sawmilling but these activities are usually managed by independent divisions. Including the activity or not in the activity graph would then depend on whether the SCN design project is done at the division or corporate level. Although chipping is included in the graph, because sawmill residues provide a cheap source of chips, the company also wants to profit from this opportunity. Activities 2 and 3 are thus included as complements necessary to obtain a sufficient quantity of chips. Also, because the company is located in North America, it cannot economically produce some pulp grades made from trees that do not grow in nearby forests. When required to make specialty paper, these pulp grades are imported from abroad.

On the demand side, the company is active in several product-markets. It sells both standard Make-to-stock (MTS) products and specialty Make-to-order (MTO) products. Although the company is essentially a FP producer, it also sells paper rolls and pulp to industrial customers. FP is sold to small institutional clients and stores by local merchants operating a delivery fleet. These customers order boxes or packs and they expect next-day delivery. Larger institutional and industrial customers order FPs in pallets, and they expect two- to three-day delivery times. The company has decided to serve these customers, as well as industrial clients buying MTO or MTS rolls, from its production–distribution centers using multiple-stop TL transportation. Pulp is delivered directly from the mills. These various possibilities drove the company to select the decoupling points specified in Fig. 1.15 (Chap. 1), which in turn explains the nature and position of the semifinished and finished products storage activities in the activity graph. The company can operate its own corporate merchant network, but it can also work with independent merchants under value-added selling or franchising agreements.

Given the opportunities and constraints delineated in Fig. 6.3, the company must decide if it will concentrate on its core pulp and paper-making competencies and rely on external chip mills, converters, merchants, and 3PLs to perform other activities or, on the other end, insource everything. In the latter case, the company would be pretty much vertically integrated. In the former case, it would act as an orchestrator for the whole network and keep its involvement in primary activities minimal. Of course, mix strategies are also possible. The final decision depends on the state of available internal resources and on the offers made by potential partners. These potential partners have their own activity graph and their offers depend on the state of their resources and processes. Simplified activity graphs for typical converters and merchants are shown in Fig. 6.4. Converters being subcontractors for FP companies, their demand for finished products (FP pallets) comes mostly from these companies. This is why in the activity graph converters both receive rolls from external demand entities (movement 4–2) and ship FP pallets back to them

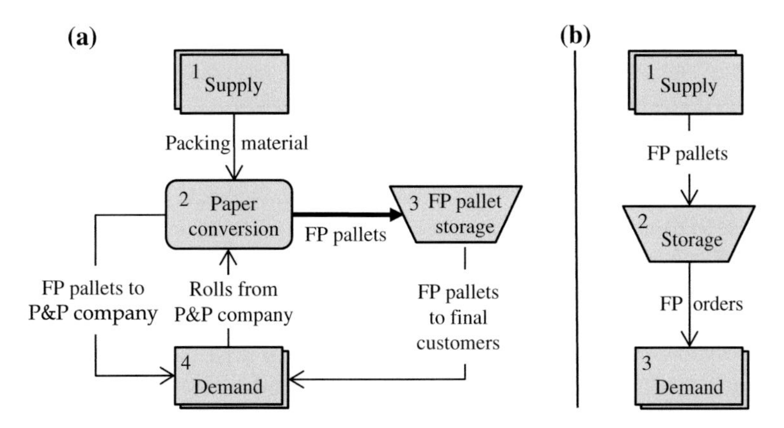

Fig. 6.4 Activity graphs for external converters and merchants **a** Converter **b** Merchant

(movement 2–4). Converters can also, however, offer FP pallet distribution services (movement 3–4). In the next chapters, we will see how to formulate optimization models based on activity graphs, such as those in Figs. 6.3 and 6.4, to make value-creating internalization–externalization decisions and other SCN design decisions.

In some industrial sectors, such as forest products (e.g., International Paper, Domtar) and oil (e.g., Exxon, Shell), several companies are still quite vertically integrated. Over the last decades, however, the tendency has been to externalize more and more activities. Some firms go as far as to externalize most of their activities to essentially become network orchestrators. *Virtual* enterprises are also emerging. They involve pools of specialized companies that dynamically form short-term joint ventures in response to volatile market needs. The coordination of the ventures is made possible through the use of modern information and communication technologies, and specific projects fall under the responsibility of one of the participating firms or of a small HQ. The firms involved share competencies, costs, risks, and profits for the duration of the projects.

The dominant approach is a mix strategy, that is, networked companies insourcing core activities, working jointly with partners on the design and production of products and services, and on their delivery to product-markets (*virtual* integration as opposed to vertical integration). Cisco (www.cisco.com), for example, concentrates on product design and customer relationships and externalizes most of its manufacturing and logistics activities. Networked companies are usually preeminent players in their industry and they act as network orchestrators. They provide web-based platforms across which the network members can collaborate. Cisco uses the Cisco Connection Online (CCC) portal for exchanges with customers, resellers, developers, and distributors, and Engineering & Manufacturing Connection Online (EMCO) enables its employees and suppliers to manage manufacturing and logistics processes from customer order to fulfillment. These companies usually share the technology they want to see become the industry standard, but preserve their core technologies.

Not every company can become a network orchestrator. This requires a valued brand name, a strong position in a market, and a close relationship with final customers. The position of some companies in an industrial sector makes them better suited to a specialist role within a network, and they can be very profitable in this role. Hon Hai/Foxconn (www.foxconn.com), for example, a Taiwanese multinational acting as contract manufacturer for Cisco, Apple, Hewlett Packard, Dell, and several other American, European, and Japanese electronics and IT companies, employs close to a million workers in China alone! A network is successful only if the orchestrator cares for the prosperity of all its partners and customers. Value-sharing mechanisms must be designed to align the members' behavior with the larger interests of the network.

Networked companies, however, are vulnerable to their partners' financial and operational problems. This is becoming more critical as the number of contract manufacturer decreases and their size expands, which leads to higher dependency and increases the possibility of opportunistic behavior. Also, in this context, relationship-specific investments are difficult to justify. An investment in the resources or capabilities of a given company cannot be justified without being related to partner-firm assets and performances. In other words, although the investment is made by a given firm, the expected return and risk depend on the performance of the whole network. The investment decision must be based on forecasts and information provided by partners. But, can these forecasts be trusted? One way to reduce risks is to sign long-term contracts with upstream partners that guarantee a minimal order volume. This means, however, that the upstream company is taking all the risks. Cisco did that during the 1995–2000 boom years but, when the market dropped in 2001, it had to write off $2.2 billion of excess inventory. The only way to mitigate these risks is to negotiate adequate risk-sharing agreements with partners. These may involve joint investment in relationship-specific assets, for example. Despite this, according to Häcki and Lighton (2001), highly networked companies outperform conventional ones even during economic downturns.

The externalization phenomenon described previously gave rise to *multi-tier* supply chains. The typical structure of these supply chains is illustrated in Fig. 6.5. At the center is the network orchestrator, often called the original equipment manufacturer (OEM). In highly networked companies such as Cisco, the final assembly of products is outsourced to first-tier contract manufacturers. In the car industry, the final assembly is usually completed by the network orchestrator. As indicated before, the OEM usually insources core activities. The various supply tiers give rise to networks of supplier–buyer relationships such as the one illustrated in Fig. 1.13. On the outbound side of the orchestrator, products may be delivered to markets using external distributors, 3PLs, value-added resellers (VAR), or even retailers. Figure 6.5 represents supply chains in which the orchestrator is a product designer-manufacturer. Nowadays, with the advent of global retailers such as Walmart, Carrefour, Tesco, The Home Depot, and so on, the network orchestrator for some products may be a retailer.

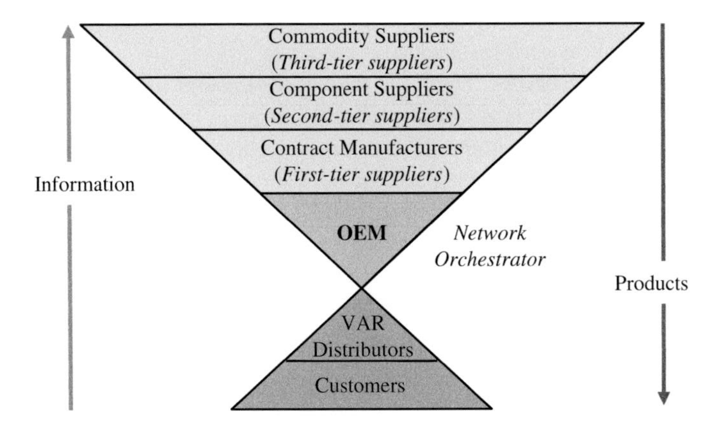

Fig. 6.5 Multi-tier supply chain structure

Outsourcing is often associated to *offshoring*, but outsourcing does not necessarily involve offshoring and vice versa. The notion of offshoring brings a geographical dimension into account. Outsourcing relates to *who* does an activity and offshoring to *where* it is done. An insourced activity is offshored when it is performed by a foreign division of a company. But an outsourced activity is also offshored when it is performed by a foreign partner. All sorts of nuances have been introduced into related business terminology. *Onshoring* describes activities performed in the country of the orchestrator; *near-shoring* designates activities performed in a foreign country close to the country of the company controlling the activity (say activities performed in Mexico for an American company); *farm-shoring* refers to the development of local suppliers; and *reshoring* involves the repatriation of an activity currently offshored.

6.2 Suppliers

6.2.1 Sourcing Strategies

In its broadest sense, a supplier is any external entity providing goods or services to a company, that is, any external party receiving payments from the company. This would include raw material vendors but also accounting firms auditing the company accounts, advertising agencies preparing promotional material, HR enterprises recruiting personal, and IT firms developing software for the company. In our supply chain context, however, we focus on suppliers related to the primary production–distribution activities of a company. When looking at an activity graph such as the one depicting the FP industry SC in Fig. 6.3, five broad categories of suppliers can be distinguished:

- **Facilities and equipment builders or vendors**. These are companies involved in the construction of SCN facilities or in the production–distribution of manufacturing and warehousing equipment. They provide the lasting resources used by the company to perform insourced activities.
- **Contract manufacturers**. These are the subcontractors selected by the firm to perform outsourced manufacturing activities.
- **Logistics service providers**. These are warehousing and/or transportation companies providing storage space for outsourced storage activities and transportation means for the movement of products between activity locations.
- **Material vendors**. These are external raw material, components, or product sources. In a manufacturing context, they provide the material associated with the leaves of BOM-trees. In a distribution context, they provide the products sold to customers. In an activity graph, they provide the material identified on the arcs adjacent to the generic supply activity.

- **MRO supplies vendors**. MROs are maintenance, repair, and operating supplies. They are required to perform activities but they do not become part of the end product or are not central to the firm's output. MRO items include consumables such as cleaning, testing, or office supplies; minor industrial equipment such as measurement instruments and safety equipment; spare parts, lubricants, and repair tools required to maintain facilities and equipment; computers; furniture; and so on. MRO items are usually not shown explicitly on activity graphs.

The nature of the relationships developed with suppliers depends on whether the need for its products or services is repetitive or not. The procurement of plant, equipment, and MRO supplies is often non-repetitive. In the former case, it takes the form of *investment projects*. Alternative proposals are usually evaluated using a life-cycle cost approach, that is, the evaluation is based on the total cost of ownership over the life of the asset (see Sect. 2.2). In addition to the financing of the construction-acquisition costs, the expenditures considered include operations, maintenance, overhaul, and replacement or disposal costs. When relevant, the environmental and social impact of the options considered must also be evaluated. As discussed in Chap. 2, it may also be possible to lease equipment instead of buying it. The acquisition of non-repetitive MRO supplies often takes the form of *spot buys*. This involves finding a low-value item or service quickly to fulfill a short-term need. The item may be purchased from a local store, for example, but, nowadays, e-procurement is often used. When the item is more expensive, a request for proposal (RFP) may be issued either by phone or via Internet to be able to evaluate a few options.

Other than that, most buying and outsourcing relationships involve repetitive needs. They may also involve a single product or a family of products. This is important because significant economies of scope may be generated when several items are procured from the same vendor. In this context, some kind of contract must be negotiated with suppliers. These can range from mid-term commodity-specific contracts to long-term partnerships. For example, contracts with material vendors could take the following forms:

- **Fixed price**. The vendor sets a unit price based on expected volumes and the buyer has the flexibility to order any quantity at any time during the contract period. This type of contract is widely used in practice.
- **Fixed commitment**. A periodic (say monthly) fixed delivery quantity is specified by the contract. Discounts are offered based on the level of the fixed quantity commitment.
- **Minimum quantity commitment**. Periodic minimum quantities are imposed by the contract and the buyer has the flexibility to order any quantity above this minimum in each period of the contract (say each month of a yearly contract). Discounts are offered based on the level of the minimum commitment.
- **Order band commitment**. Periodic minimum and maximum quantities are imposed by the vendor and the buyer has the flexibility to order any quantity in the interval specified. Discounts are offered based on the level of the minimum commitment and range of the interval.

Three contextual factors have a significant influence on the nature of the relationship developed with suppliers: the value creation impact of the product or service, the complexity of the supply process, and the risks involved. The value creation impact is related in particular to technology as previously discussed. A commodity has much less value than a high-tech component based on proprietary technology. Supply complexity depends on how easily a product or service can be obtained. A commodity largely available on the market is easy to procure. However, if only a few vendors qualify to supply the product, if capacity is limited, if JIT deliveries are required, if quality is critical, then it is much more difficult to find good suppliers. Risks are related mainly to the possible disruption of the supply line and to price fluctuations. If an item is procured close by in the same country, risks are lower than if it is purchased overseas in a country with highly fluctuating exchange rates and strong inflation.

For low-value and low-supply-complexity products and services, a traditional *competitive bidding* purchasing strategy structured to leverage economies of scale is usually suitable. However, for high-value and high-supply-complexity products and services, *supplier partnerships* are more appropriate. The main differences between these two sourcing strategies are described in Table 6.1. Companies adopting a competitive bidding strategy feel that for the products and services procured the competition among suppliers leads to lower prices. Companies adopting a supplier partnership strategy feel that this will decrease setup costs, production costs, inventory holding costs, and transportation costs; improve response times and quality; and provide greater flexibility because of the enhancement of the technological systems of both partners and the development of collaborative forecasts and plans. In several industries, first-tier suppliers go as far as implementing dedicated component manufacturing facilities on the assembly sites of the network orchestrator, thus enabling efficient JIT deliveries. In the Brazilian car industry, the newer facilities implemented by car manufacturers (GM, VW/Audi, Renault, Mercedes-Benz) are *industrial*

Table 6.1 Competitive bidding versus supplier partnerships—Adapted from Dornier et al. (1998)

Competitive bidding	Supplier partnerships
Primary emphasis on price	Multiple criteria
Short-term contracts	Long-term contracts
Selection by bids	Selection based on long-term value creation potential
Many suppliers	Small number of suppliers
Proprietary information	Shared information
Periodic evaluation	Supplier improvement programs
Remote product-process engineering	Concurrent product-process engineering
Suppliers responsible for problems	Problems jointly solved
Clear delineation of business responsibility	Quasi-vertical integration
Power-driven relationship	Win-win relationship

condominiums. For example, seven first-tier suppliers share a condominium with Mercedes-Benz in Juiz de Fora.

Risk is a more intricate issue that raises the question of *sole* sourcing versus *portfolio* sourcing. Supplier partnerships reduce uncertainty for the buyer through agreements on prices and discounts, guaranteed quality levels, and shorter reliable lead times. It reduces uncertainty for the supplier through assured demand and a better understanding of client needs. Both partners having converging objectives, there is less opportunistic behavior, and they are less influenced by externalities. However, working with a single partner may introduce some vulnerabilities, mainly in a global sourcing context. Suppliers may be the victim of disastrous events resulting in the disruption of supply lines (see Chap. 10). Contingency plans, possibly involving back-up suppliers, must be elaborated to mitigate this type of risk. In a global sourcing context, currency fluctuations are a major problem. The impact of these fluctuations is so large that it has been suggested that suppliers should be selected to help balance currency flows (see Chap. 9). The idea is to hedge against exchange rate variations by selecting a portfolio of suppliers in different countries. Of course, first-tier suppliers are often multinational companies with manufacturing facilities in different parts of the world. They may therefore be able to provide the flexibility of changing supply sources when required. These considerations are important, and risk-mitigation and risk-sharing agreements are an essential part of partnership contracts.

6.2.2 Contract Manufacturers

Contract manufacturers (CMs) are known under different names depending on the industry (*electronic manufacturing services* in the CE industry, *auto parts manufacturers* in the car industry, *contract manufacturing organizations* in the pharmaceutical and biotech industry, *contract packers* in the food industry, *converters* in the FP industry, etc.) but the concept is essentially the same. It describes companies performing manufacturing activities for other companies. Contract manufacturers are currently widespread in several industrial sectors, but this is a relatively new phenomenon resulting from the disaggregation of large vertically integrated firms to focus on their core competencies and from the present tendency to network companies. All industrial sectors are not at the same stage and they are not structured in the same way. In some industries, the phenomenon is relatively recent and CMs are fairly small companies. In other cases, it is a well-established practice and CMs have become large multinationals.

Table 6.2 provides basic statistics on leading CMs in the CE and car industries. As can be seen, in both cases, major CMs have become large multinationals. The dominant business model in these two sectors, however, is very different. In the CE industry, short-term contracts prevail and there is a lot of competition. CMs are often selected on a competitive bidding basis. In the early 2000s CM sales were

Table 6.2 Large contract manufacturers in CE and car industries

Company	Industry	2012 Sales (Billion USD)	Base country
Foxconn (foxconn.com)	CE	132	Taiwan
Flextronics (flextronics.com)	CE	29.4	Singapore
Jabil (jabil.com)	CE	17.2	USA
Celestica (celestica.com)	CE	6.5	Canada
Benchmark (bench.com)	CE	2.5	USA
Bosch (bosch.com)	Car	36.7	Germany
Denso (globaldenso.com)	Car	34.2	Japan
Continental (conti-online.com)	Car	32.8	Germany
Magna International (magna.com)	Car	30.4	Canada
Aisin Seiki (aisin.co.jp)	Car	30.1	Japan

Sources Supply Chain Insights LLC (CE) and Automotive News (Car)

growing at an annual rate of 18 %, but in 2012 this had fallen to a mere 2 %. This slowdown has resulted in excess capacity in the industry. A large part of a CM's sales is usually generated by a few major accounts. For example, in 2012, 21 % of Benchmark's sales were coming from IBM. Margins are also very low at about 2 %. The industry is therefore relatively unstable and some CMs will have serious difficulties if the brand owners (Apple, Dell, LG Electronics, Motorola, Samsung, etc.) start to back-source their manufacturing activities.

The car industry, however, is engaged in long-term partnerships with its parts manufacturers. Contracts often extend beyond the lifetime of a car model. The industry move to *global platforms* (shared set of common designs and major components over a number of seemingly distinct car models) is favoring standardization and the concentration of the automaker supply base. In 2012, Ford allocated 65 % of its global purchasing budget to the 104 CMs on its preferred suppliers list, 10 % more than in 2010. At General Motors, the top 400 suppliers account for 90 % of the company's global purchase. CMs are key players in the industry global SCs and they build factories around the world. Most of these factories are within 100 miles of a carmaker assembly plant. Currently, Johnson Controls ($22.5 billion sales in 2012) is producing door panels and consoles for Daimler AG in factories in the United States, Germany, Hungary, the Czech Republic, Mexico, China, and South Africa (Sedgwick 2013). Car industry CMs are also heavily involved in R&D, and several new car features are driven by their innovations. In 2012, for example, Bosch allotted 9 % of its revenues to R&D.

In some sectors, such as the pharmaceutical and biotechnology industries, CMs are often also finished product manufacturers. Contract manufacturing is then seen as an opportunity to leverage excess capacity. Although this industry is newer to contract manufacturing, it now commonly outsources analytical and testing services, clinical trials, formulation development, solid-dosage-form manufacturing, and so on. In some contexts, such as the food industry, CMs are involved more at the final production stage, and in particular in packaging operations. The brand owners can then ship their products in bulk to distant markets and use contract packers for local unit packaging. This type of service is now often offered by 3PLs, which is examined in the next section.

6.2.3 Third-Party Logistics Providers

The use of external carriers, public warehouses, customs brokers, and international freight forwarders by production–distribution companies is nothing new. In the past, these external services were typically purchased on a transactional basis from several local providers. Today, however, all these services are offered by *third-party logistics* (3PL) providers. Shippers now often outsource their logistics activities and form strategic alliances with globally deployed 3PLs. 3PLs offer a wide range of operational, management, and IT services, including the following:

- Repetitive operational activities such as
  - Domestic inbound and outbound transportation (TL, LTL)
  - International multimodal transportation
  - Home parcel deliveries
  - Dedicated contract carriage
  - Warehousing
  - Freight forwarding
  - Customs brokerage
- Value-added services
  - Order processing and fulfillment
  - Procurement
  - Cross-docking
  - Product packaging, labeling, assembly, and kitting
  - Freight bill auditing and payment
  - Inventory auditing (cycle counting)
  - Customer service
  - JIT deliveries
  - Vendor-managed inventory (VMI)
  - Reverse logistics (returns management)
  - Safety norms compliance and insurance
  - Service parts logistics
  - Foreign trade zones
  - Collaborative distribution
- Management services
  - Traffic management
  - Inventory management
  - Procurement and distribution planning

  - Fleet management
  - Global trade management
  - Logistics network design
  - Contingency planning
  - Green supply chain services
  - SCM consulting (integrated solutions)
  - Logistics process reengineering

- IT services
  - Bar coding and RFID solutions
  - EDI and web communications
  - Transportation and warehousing execution systems (TMS, WMS)
  - SC visibility (track and trace) solutions
  - Customer ERP and APS interface capabilities
  - Cloud-based solutions
  - Analytics and optimization

Major 3PLs have their own warehouses and transportation fleet. Their assets are often deployed worldwide but they also subcontract shipping and warehousing activities to local suppliers. A distinction is often made between asset-based and non-asset-based providers. The former owns most of the assets (trucks, DCs, etc.) required to run a client's SCN. The latter do not own assets; instead, they offer expertise in developing contracts with asset owners in order to manage their clients' SCN at the lowest possible cost. In other words, they concentrate on the value-added, management, and IT services listed previously. In a recent 3PL survey of *Inbound Logistics* (O'Reilly 2013), 47 % of the participants identified themselves as non-asset-based providers. Non-asset 3PLs concentrating on management and IT services are often called 4PLs. More recently, the label *control tower* has also been used. Companies providing all listed services are now known as lead logistics providers (LLPs). These distinctive labels are somewhat artificial, however, because large 3PLs are continuously widening their service offerings. Their distinctive competencies often come from proprietary advanced logistics execution and planning systems. Among 318 3PLs surveyed by Armstrong and Associates (2014), 185 use in-house TMS and 130 use proprietary WMS. Some 3PLs also distinguish themselves by providing industry-specific services.

The majority of leading 3PLs were initially carriers (e.g., Menlo, Ryder, Schneider), postal services (e.g., Deutsche Post), storage companies (e.g., Exel), or freight forwarders (e.g., C.H. Robinson), and they have diversified through mergers and acquisitions. For example, Deutsche Post acquired US-based DHL in 2002 in order to expand in the United States, and it combined its activities under the DHL brand name. In 2003 DHL purchased the US parcel delivery company Airborne (for 1.4 billion USD), and in 2005 the UK logistics company Exel (for 6.6 billion USD). Currently, Deutsche Post DHL operates four global divisions: Post and Parcel, Express, Global Forwarding, and Supply Chain. In 2013, the group generated 55 billion euros in revenues with about 480,000 employees in 220 countries or

Table 6.3 Largest global 3PLs in 2013

Third-Party logistics provider (3PL)	2013 Sales (m USD)	Third-Party logistics provider (3PL)	2013 Sales (mil. USD)
DHL supply chain	31,432	DACHSER	6627
Kuehne+Nagel	22,587	Toll holdings	6266
DB Schenker logistics	19,732	Expeditors international	6080
Nippon express	17,317	Geodis	5828
C.H. Robinson worldwide	12,752	UPS supply chain solutions	5492
CEVA logistics	8517	GEFCO	5300
DSV	8140	J.B. hunt (JBI, DCS & ICS)	5224
Sinotrans	7738	UTi worldwide	4441
Panalpina	7293	Agility	4415
SDV (Bolloré Group)	7263	Yusen logistics	4042

Source Armstrong and Associates (2014)

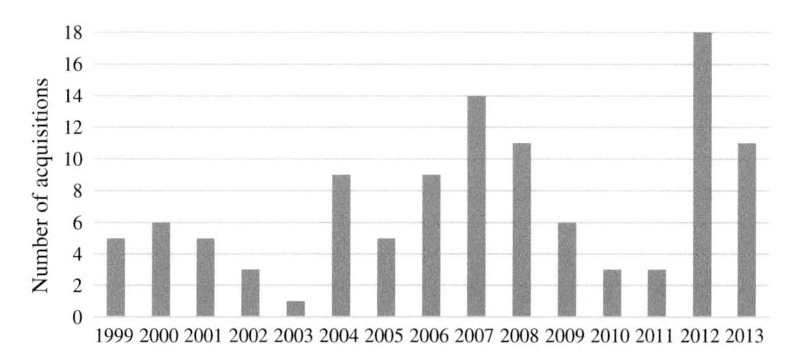

Fig. 6.6 3PL acquisitions since 1999—Data, *source* Armstrong and Associates (2014)

territories (www.dpdhl.com). Table 6.3 provides the 2013 revenues of the current largest global 3PLs. Details on 3PL companies are found in the *Inbound Logistics* Web_Cite City data base (www.inboundlogistics.com). The industry landscape is still under transformation, however, as can be seen in Fig. 6.6.

Manufacturing and retailing companies outsource their logistics function in record numbers. According to Armstrong and Associates (2014), 82 % of US Fortune 500 companies were using 3PLs in 2010 (up from 56 % in 2004), and global 3PL revenues in 2012 were 685 billion USD (171 in North America, 158 in Europe, 243 in Asia Pacific, and 44 in South America). This is a direct result of the current propensity to network companies as explained previously. Companies outsource their logistics for the reasons discussed in Sect. 6.1, and in particular to accomplish the following:

- Cut costs and overhead and improve order winners
- Facilitate B2C e-commerce, *omni-channel* retailing (providing a unified experience for customers in a multi-channel retailing context), and, more specifically, direct-to-consumer fulfillment
- Enable the postponement of packaging and labeling activities for remote markets

In a recent Capgemini 3PL market study (Langley 2014), shippers working with 3PLs report reductions of 11 % of logistics costs, 6 % of inventory costs, and 23 % of fixed logistics assets. They also claim improvements in fill rates and order accuracy, and 55 % of them point out that their use of 3PLs led to year-over-year incremental benefits. When they outsource their logistics activities, shippers have specific expectations. The most often noted are the following:

- Service offer aligned with shipper SC strategy and deep industry knowledge
- Compliance with performance levels specified in contracts
- Competitive pricing with no hidden costs
- Transparent integration with insourced activities
- Trust, openness, and information sharing
- Superior communication and IT capabilities
- Ability to provide continuous improvement
- Access to global markets

It is apparent that relationships between 3PLs and their customers are growing more collaborative and that the shipper's satisfaction level is increasing. According to Eyefortransport (2012), 50 % of shippers are using two to three 3PLs. For instance Airbus has strategic partnerships with DHL and Kuehne+Nagel. As shippers centralize their procurement function, however, there is a tendency to rely on a smaller number of 3PLs. Shipper 3PL contracts are defined mostly over one to three years, and they are usually renewed unless some of the 3PL expectations previously cited are not met. The bid process for the selection of a 3PL lasts on average three to six months. The negotiation of fair and clear contracts is a key success factor. Shippers not using 3PLs usually consider logistics to be a core competency. Some go as far as developing their own logistics services division. In 1986, for example, the motorized equipment manufacturer Caterpillar launched CAT Logistics (now Neovia; www.neovialogistics.com), a division providing spare parts logistics services to several auto manufacturers.

6.3 Industrial Clusters and Resource Pooling

The previous section concentrates on vertical collaborations in the SC, that is, on supplier–buyer relationships. Horizontal collaborations between companies playing more or less the same role in an SC are also taking place, mainly with the objective of developing joint-infrastructures to improve their supply base, share costly resources, and tap industry knowledge, thus gaining economies of scale, scope, and skill. This naturally leads to the development of industrial clusters and in particular of logistics clusters. The emergence of 3PL-operated consolidation centers for mass distribution and of purchasing groups are other forms of horizontal collaboration.

Porter (1998) defines industrial clusters as geographic concentrations of interconnected companies and institutions in a particular sector. Clusters have formed all over the globe for years and they are still proliferating despite the advent of the digital economy characterized by the elimination of the hurdles of distance. Well-known cluster examples include Silicon Valley and Route 128 outside Boston for the IT sector, Taipei for computer products, Detroit and Curitiba (Brazil) for car manufacturers, Northern Italy for the fashion industry, Hollywood for film studios, and so on. Even at a micro-level, we often see competing businesses locating in the same areas in our cities. Co-location of similar companies is a source of reciprocal reinforcement that goes far beyond geographical proximity. People and organizations with similar backgrounds tend to understand and trust each other more easily. Their proximity leads to mutually beneficial tacit knowledge sharing and spillovers. This is often reinforced by university-cluster collaborations favoring innovation and the development of skilled labor pools. For example, the presence of Stanford University in Silicon Valley and of MIT in Boston has been instrumental in the development of numerous IT innovations and the creation of abundant spinoff companies.

Industrial cluster companies are typically involved in joint activities. These include lobbying with local and state governments to intensify infrastructure investments as well as advantageous regulations and incentives. They engage in cluster-specific development and marketing activities to improve their brand image and market potential. They foster focused procurement strategies to obtain better quality inputs, lower prices, improved response times, and to decrease supply risks. Clusters attract suppliers who, because of the volumes involved, see advantages in locating nearby. Cluster companies may even share expensive or scarce resources, such as state-of-the-art testing equipment. Because they may have the same customers, they may also pool distribution resources. In sum, as Porter (1998) puts it, "a cluster allows each member to benefit *as if* it had greater scale or *as if* it had joined with others without sacrificing its flexibility."

Logistics clusters, that is, groupings of logistic service providers and logistics-intensive companies, are particularly important to support and improve the operations of supply chains. These clusters are known under different names, such as *logistics platforms, logistics villages, distribution parks,* and *freight villages,* and they take different forms. They are usually developed and managed by real estate developers, local governments, or public authorities. Some of them, such as the Singapore Port area, Rotterdam in Holland, the Los Angeles–Long Beach area, the Panama Canal Zone, Dubai Maritime City, and Tanger Med in Morocco, are established around major port facilities. Others such as Memphis Airport, Hong Kong International Airport, and Frankfurt Main Airport are air logistics parks. Finally, clusters such as Zaragoza in Spain and Beijing Liangxiang LP in China are major multimodal hubs. Some clusters have international vocations but others are regional or urban distribution parks. Some are free trade zones, others specialize in particular commodities (automotive, agricultural, chemical, pharmaceutical, etc.), and some provide special services (temperature-controlled storage, hazmat, etc.).

Sheffi (2013) stresses that logistics clusters provide substantial transportation and resource-sharing advantages adding "to the reciprocal reinforcing feedback mechanism which makes the cluster more attractive as it grows, leading to further growth." Because of the high volumes involved, cluster-to-cluster transportation costs are lower. The concentration of freight shipments provide better consolidation opportunities, it allows the use of larger transportation equipment, and results in better load profiles. Also, the high shipment frequency leads to superior response times. By sharing resources, clusters can also serve customers better and adjust more easily to fluctuations in business volume. When the daily flight of a 3PL out of an air logistics park is full, customer loads can be shipped immediately on a competitor's plane. Similarly, competitor's warehouses can be used for short periods of time when a 3PL is running temporarily out of space. Conversely, excess capacity can be temporarily leased to competitors. As far as workforce is concerned, several parks contain temporary staffing agencies offering trained workers when additional capacity is required. All this provides considerable flexibility at a reasonable cost.

As pointed out in Chap. 5, flow imbalances generated in freight networks by demand patterns are a source of serious inefficiencies. Doherty and Hoyle (2009) claim that 24 % of truck miles in the EU are empty and that the average load use of nonempty vehicles is only 57 %. Because carriers move what shippers demand, horizontal collaboration between shippers and between carriers is almost the only way to improve the situation. Collaboration between shippers is usually facilitated by one or more third parties (3PLs, 4PLs, and carriers). In France, several consolidation and collaboration centers (CCCs) have been set up to enable the approach. For example, in the hygiene–beauty products sector, five manufacturers (Colgate, Henkel, GSK, Sara Lee, and Eugène Perma) collaborate in the "CHANGES pool" to supply eleven DCs of four mass retailers (Auchon, Carrefour, Match, and Monoprix) serving more than 7000 stores. The shipment consolidation is performed in a CCC operated by the 3PL FM Logistics. Transportation schedules are optimized by a 4PL (bp2r) and the shipments are made by six carriers. The pool members claim that they have tripled their DC delivery frequency with half the transportation resources previously used (*Logistics Magazine* 2011). Carrefour, Europe's leading mass retailer, is now asking its suppliers to adopt this approach. Currently, eight CCCs have been set up for different product categories (ASLOG 2011). Similar initiatives were started by Kimberly-Clark and Unilever in the Benelux countries.

Another type of horizontal collaboration that has long been popular is purchasing groups. This is also known as cooperative purchasing, pooled purchasing, and alliance purchasing. It involves several businesses pooling their buying volumes and sharing resources to improve purchasing activities. By doing so, they obtain economies of scale (mainly through price discounts and transportation consolidation) and efficient purchasing processes. In most cases, this outweighs the costs of cooperating and potential difficulties related to antitrust legal issues and disclosure of sensitive information. Successful purchasing groups can give birth to formal joint ventures. For example, RONA (www.rona.ca), a major Canadian retailer and distributor of hardware, building materials, and home renovation products, started as a purchasing group. In 1939, a group of hardware retailers joined forces to bypass large competitors threatening their access to suppliers. The group incorporated in 1960, and it subsequently grew through acquisitions, by building a major distribution center near Montreal, and by developing a network of big-box stores. In the 1990s, the group enhanced its buying power through associations with Home Hardware Stores (1000 merchants in Canada) and Hardware Wholesalers (3000 retailers in the United States). In 2002 it entered the stock market (TSX: RON). It now has 24,000 employees and generates annual sales of $4.2 billion.

6.4 Partner Selection

Having decided to partially or completely externalize an activity or to procure materials or services from an external supply source, the suppliers or partners to work with must be selected. In fact, these two decisions are interrelated and ideally they should be made jointly. This issue is addressed in the following chapters. In this section, we assume that the decisions are made sequentially and we examine the supplier selection problem. The proposed approach is easily adaptable to other types of partnerships. Several aspects of the problem must be clarified before it can be studied adequately. There is a large amount of professional and scientific literature on the supplier selection problem. However, much of it implicitly assumes point-to-point relationships, that is, that a need occurring in a single buyer location must be satisfied from a single supply location. This greatly simplifies the economic evaluation of potential partnerships.

In real life, partnerships often involve business-to-business relationships and each business has its own supply chain network. A company operating several plants may procure raw materials and components from one or more suppliers, each serving their customers through a distribution network. The evaluation of procurement costs (product prices, transportation costs, and inventory holding costs) and response times for a potential supplier then depends on the location of the supply and manufacturing facilities of the partners and on site-to-site product flows. Besides, the capacity of some potential suppliers may not be sufficient to cover all the company needs. Making choices in this context requires network optimization models of the type studied in the next chapters. Moreover, the sites from which the supplier may ship products are not necessarily predetermined. As mentioned, a potential supplier may be willing to implement new production–distribution facilities close to the buyer facilities to improve service levels. If this is the case, then one has to solve a joint supplier-buyer SCN optimization problem.

Table 6.4 Structured list of selection criteria cited in the literature

Order-winners	e-Commerce capabilities
Product attributes	Strategic purchasing
(appearance, ease-of-use)	Repair service
Products maintainability and warranties	Packing abilities
Portfolio of products-services offered	Productivity
Product-service quality	SC management
Price	Value creation
(structure, cost reduction programs)	*Innovation and learning*
Delivery	Technology level
Flexibility	Research and development
Service	Service innovation
(after-sale, technical support, training)	*Risk sources*
Sustainability	Political and economic stability
Available resources	Financial position
Network locations	Labor relations
Capacity	*Relational abilities*
Equipment characteristics	Past business history with company
Planning and control systems	Reputation and image
Communication & information systems	Legal disputes history
Willingness to expand SCN	Strategic orientations compatibility
Capabilities	Potential for growth
JIT capabilities	

SCN optimization models of the type studied in the next chapters concentrate on the economic and response time evaluation of potential solutions. However, when selecting suppliers, several other not necessarily location dependent performance criteria are usually considered. These are typically related to the order-winning criteria discussed in Chap. 1 and to the value drivers identified in Chap. 2. Table 6.4 provides a structured list of the main criteria cited in the literature. These should be considered as examples, and the criteria to use must be selected by management based on the particularities of their business context. Also, selected criteria must be expressed using a specific performance measure. For example, the delivery criteria could be stated in terms of average delivery times, percent of late deliveries, and so forth. As explained in Sect. 4.4, when studying site selection problems, some measures can be expressed in monetary terms, others are quantifiable, but others are purely subjective. Also, if the supplier considered has worked with the company in the past, its performance may be much easier to evaluate than if it is a new candidate.

As can be seen, the problem is not easy. The network optimization models studied in the next chapters cannot encompass purely subjective criteria. However, multiple-criteria decision-making methods developed to cope with subjective criteria are not capable of making the economic tradeoffs necessary to optimize a supply chain network. The solution to this dilemma is provided by the approach usually taken to address these problems in practice. Companies can rarely find a single supplier capable of satisfying all their needs. They must instead elaborate a *supply base* composed of qualified potential suppliers. This suggests a two-step approach to supplier selection:

(1) *Multi-criteria filtering*. This is analogous to the supplier qualification process used by companies as part of their total quality programs. It involves the multi-criteria evaluation and ranking of potential suppliers to select a qualified subset capable of satisfying the company needs. Surrogate measures (e.g., product prices, distances, percent of late deliveries, etc.) are used to grossly evaluate economic and response factors but network intricacies are neglected. If one needs a single supplier, then the best ranking candidate is selected.
(2) *SCN optimization*. The qualified suppliers are used as potential suppliers in an SCN optimization model capable of evaluating complex economic and response tradeoffs. These models are studied in detail in the next chapters.

Several decision-making approaches were proposed in the literature for supplier evaluation and selection. Detailed reviews of these approaches are found in de Boer et al. (2001) and Ho et al. (2010). They can be classified in four broad categories: scoring methods, multi-criteria decision-making (MCDM) methods, optimization methods, and combined approaches. *Scoring methods* include the rough criteria weighting procedures regularly used in practice. These procedures often violate mathematical scaling and preference theory laws and they may lead to poor decisions. The main *MCDM* approaches proposed are based on variants of the analytic hierarchical process (AHP), data envelopment analysis (DEA), multi-objective programming (MOP), or multi-attribute rating techniques (MART). AHP methods are similar to the pairwise comparison technique presented in Sect. 4.4.3 to derive weights for subjective site location criteria. The main difference is that the judgments made when comparing two factors are more sophisticated. One must specify whether the preference is weak, strong, very strong, or absolute. When several suppliers and criteria must be evaluated, this process becomes very arduous, which explains why it is rarely used in practice (de Boer and van der Wegen 2003).

MOP and MART require that decision-makers exogenously specify weights for individual criterion. Determining weights a priori, however, is very difficult and somewhat arbitrary, which limits the applicability of these approaches. DEA, however, does not require the decision maker to predetermine weights. DEA models set weights endogenously because they optimize the performance scores of suppliers. This approach suffers from the opposite problem because management does not have any control over the importance assigned to criteria. Also, DEA models are input (costs, capabilities)–output (benefits) models and the parting of relevant criteria into inputs and outputs is not always straightforward. The *optimization methods* published for supplier selection can be seen as particular cases of the more general SCN optimization approach developed in the next chapters.

Ng (2008) proposes an interesting compromise in terms of applicability and adequacy, which can be used for the multi-criteria filtering step. It assumes that decision-makers can list criteria in order of importance, but it does not require the specification of weights. Also, the model proposed can be solved easily with a spreadsheet. It does not require the use of optimization software. The following basic notation is required to describe the approach:

l Supplier index ($l \in L^V$, the set of potential vendors)
i Index of supplier selection criterion (the indexes $i = 1, 2, \ldots, I$ must be ordered so that $i < i'$ implies that criterion i is more important than i')
SM_l Performance score calculated for supplier l
PM_{li} Evaluation of supplier l under criterion i (expressed in terms of the economic, quantifiable, or subjective performance measure used to characterize the criterion)
U_{li} Normalized preference measure of supplier l relative to all potential suppliers for criterion i (must be defined so that under criterion i, $U_{li} > U_{l'i}$, $l \neq l'$, implies that supplier l is preferred to supplier l')
W_i Best weight respecting the criteria order of importance that can be assigned to criterion i when a supplier is evaluated (decision variable)

Suppliers are evaluated by combining their performance measures PM_{li} under each criterion into a single score SM_l. The measures used are assumed to be positively related to the performance of a supplier. If the measure for a criterion is negatively related to performance, it must be transformed by inverting its scale or by taking its complement (e.g., converting *% rejected* into *% without defect*). It is also essential to use a common scale for all measures to avoid giving undue importance to a particular criterion. To do this, the performance levels PM_{li} given by decision-makers are transformed into normalized measures $U_{li} \geq 0$ such that $\Sigma_{l \in L^V} U_{li} = 1$, for all i, as in Sect. 4.4.3. For economic or quantifiable criterion i, adequate normalized measures are obtained as follows:

$$U_{li} = PM_{li} / \sum\nolimits_{l \in L^V} PM_{li} \tag{6.1}$$

Subjective normalized measures can be elaborated using the pairwise comparison method presented in Sect. 4.4.3.

Example 6.1
To illustrate the approach, consider the case of a company manufacturing motorized construction equipment. The company wants to select a limited number of powertrain component suppliers among ten candidates. The evaluation criteria designated, in order of importance, are the following:

(1) *Variety*. Product variety is measured by the number of components a supplier can provide, and it reflects the breadth of its products portfolio. Because the company would like to reduce its supply base, this is considered to be crucial.

(2) *Quality*. Quality is measured in terms of the percentage of components delivered without defect. This is based on former contract data and on statistics provided by the industry.
(3) *Distance*. It is assumed that future transportation costs will be proportional to distance. The measure used is a weighted average of the distance between the origins and destinations of deliveries.
(4) *Delivery*. This is the percentage of on-time deliveries based on historical data and statistics provided by the industry.
(5) *Price*. The price performance of a supplier is measured by an index based on its estimated price level relative to average market price.

Relevant measures for each supplier are provided in Fig. 6.7. Note that the measures used for *Distance* and *Price* are inversely related to supplier performance because higher distances (prices) are less desirable. To obtain measures positively related to supplier performance, the inverse of the distance (price index) is taken (columns 5 and 8, instead of 4 and 7, in Fig. 6.7). From now on, these measures are used for the distance and price criteria. Normalization is then performed using relation (6.1). The normalized measures obtained are given in Fig. 6.8.

Supplier	Variety	Quality	Distance	*1/Distance*	Delivery	Price	*1/Price*
1	2	100.00%	249	*0.0040*	90%	100%	*0.0100*
2	24	99.83%	238	*0.0042*	92%	90%	*0.0111*
3	28	96.59%	241	*0.0041*	90%	100%	*0.0100*
4	53	97.54%	588	*0.0017*	100%	95%	*0.0105*
5	10	99.95%	241	*0.0041*	95%	90%	*0.0111*
6	7	99.85%	567	*0.0018*	98%	102%	*0.0098*
7	12	91.89%	967	*0.0010*	90%	100%	*0.0100*
8	33	99.99%	635	*0.0016*	95%	85%	*0.0118*
9	2	100.00%	795	*0.0013*	96%	100%	*0.0100*
10	34	99.99%	689	*0.0015*	95%	80%	*0.0125*
Sum	205	9.8563	5210	0.0253	9.41	9.42	0.107

Fig. 6.7 Spreadsheet with supplier performance measures data

Fig. 6.8 Spreadsheet with calculated normalized supplier performance measures

Supplier	Variety	Quality	Distance	Delivery	Price
1	0.010	0.101	0.159	0.096	0.094
2	0.117	0.101	0.166	0.098	0.104
3	0.137	0.098	0.164	0.096	0.094
4	0.259	0.099	0.067	0.106	0.099
5	0.049	0.101	0.164	0.101	0.104
6	0.034	0.101	0.070	0.104	0.092
7	0.059	0.093	0.041	0.096	0.094
8	0.161	0.101	0.062	0.101	0.110
9	0.010	0.101	0.050	0.102	0.094
10	0.166	0.101	0.057	0.101	0.117
Sum	1.0	1.0	1.0	1.0	1.0

Each supplier is evaluated independently. The score of supplier l is given by the weighted sum $SM_l = \Sigma_i W_i U_{li}$, as in Chap. 4. As indicated previously, the weights used for a supplier do not have to be provided by the decision maker, but they must be calculated so that they respect the relative importance of criteria, that is, so that $W_1 \geq W_2 \geq \ldots \geq W_I \geq 0$. Also, we want these weights to be normalized, that is, to be such that $\Sigma_i W_i = 1$. Finally, the weights used should be fair for the suppliers. This is achieved by calculating weights so that they give the largest possible score to each supplier while respecting previous conditions. For a supplier l, weights having these characteristics can be obtained by solving the following linear program (LP):

$$\max \quad S_l = \sum_{i=1}^{I} W_i U_{li} \tag{6.2}$$

subject to

- Criteria importance ranking constraints

$$W_i - W_{i+1} \geq 0, \quad i = 1, 2, \ldots, (I-1) \tag{6.3}$$

- Weights normalization constraints

$$\sum_{i=1}^{I} W_i = 1 \tag{6.4}$$

- Weights non-negativity constraints

$$W_i \geq 0, \quad i = 1, 2, \ldots, I \tag{6.5}$$

LP (6.2)–(6.5) can be solved with any optimization tool, but Ng (2008) showed that, given its simple structure, its solution can be found without a solver. More specifically, he showed that optimal supplier scores can be calculated with the following formula:

$$S_l = \max_{i=1,2,\ldots,I} \left(\frac{1}{j} \sum_{j=1}^{i} U_{lj}\right), \quad l \in L^{\mathrm{V}} \tag{6.6}$$

Having calculated these scores, it is sufficient to sort them in descending order to determine the suppliers to select. Ng also showed that it is very easy to do some sensitivity analysis using this approach and, in particular, to examine the impact of a modification of criteria ranking. For suppliers that are below the demarcation line between those selected and those excluded, it is also easy to determine what they need to improve to be able to join the qualified supply base.

Supplier	Variety	Quality	Distance	Delivery	Price	Score
4	0.259	0.179	0.142	0.133	0.126	**0.259**
10	0.166	0.134	0.108	0.106	0.108	**0.166**
8	0.161	0.131	0.108	0.106	0.107	**0.161**
3	0.137	0.117	0.133	0.124	0.118	**0.137**
2	0.117	0.109	0.128	0.121	0.117	**0.128**
5	0.049	0.075	0.105	0.104	0.104	**0.105**
1	0.010	0.056	0.090	0.091	0.092	**0.092**
6	0.034	0.068	0.068	0.077	0.081	**0.080**
7	0.059	0.076	0.064	0.072	0.076	**0.076**
9	0.010	0.056	0.054	0.066	0.071	**0.071**

Fig. 6.9 Spreadsheet with supplier partial averages and scores

Example 6.1 (continued)
In order to rank the suppliers, the partial averages $\frac{1}{j}\Sigma_{j=1}^{i} U_{lj}$, $i = 1, 2, \ldots, I$, must first be calculated. This is done in the spreadsheet shown in Fig. 6.9. The *Distance* partial average for supplier 4, for example, is calculated from the three first values of supplier-4 row in Fig. 6.8, giving $(0.259 + 0.099 + 0.067)/3 = 0.142$. The final supplier scores S_l obtained by taking largest partial averages are shown in the last column. The suppliers were sorted based on the last column scores.

As can be seen, if the company wants to restrict itself to four suppliers, then candidates 4, 10, 8, and 3 would be selected. Clearly, the company would have to ensure, however, that these four suppliers have sufficient capacity to fulfill all its needs. Also, note that the scores of suppliers 2 and 3 are relatively close. The company may then consider including supplier 2 in its supply base, provided that it engages in an improvement program to alleviate its shortcomings.

Care should be taken when using multi-criteria decision-making methods, however, because different methods or performance measure scales may lead to different candidate partner rankings.

6.5 Managing Partnerships

When an activity is outsourced to a partner, it may not be necessary to manage it anymore; however, the relationship with the partner selected has to be managed. As mentioned, the needs, roles, and accountabilities governing any partner relationship should be specified in a formal contract. These contracts should provide a shared vision of what the relationship will be. Adequate relationship governance procedures also should be elaborated prior to implementation. During the contract duration, the partner must be monitored, evaluated, motivated, and rewarded, which requires constant communications between the parties. Inevitably, frictions will arise, and if they are not anticipated and managed, the relationship may become dysfunctional.

We insisted previously on the fact that supplier contracts should be clear on issues related to product and service characteristics (offer, quality levels, reliability, etc.), pricing (discount schedules, payment terms, etc.) or benefit sharing, delivery commitments (minimal-maximal quantities, response times, etc.), and risk sharing–mitigation (contingency plans, insurance, etc.). Contracts should also cover agreements on specific resources to be used to perform the activity, information-sharing mechanisms, and collaborative planning. The performance criteria to be used for the evaluation of the supplier should be clearly stated as well as expected performance levels and performance-monitoring mechanisms. Finally, mechanisms for conflict resolution and termination conditions should be included.

The negotiation of winning agreements is not an easy thing and several difficulties are encountered in practice, both during the negotiation process and after implementation. The main difficulties reported include the following:

- *Pressures during negotiation*. Partnerships are often initiated to solve an internal problem. If the problem is serious there may be pressure to reach an agreement quickly, which typically leads to deficient contracts. Partners should take time to explore their respective goals, requirements, capabilities, cultures, and to reconcile differences.
- *Poorly stated expectations*. Companies do not always invest the time necessary to establish realistic performance standards and evaluation mechanisms.
- *No continuous improvement*. Pricing and service levels are established when negotiating the contract and often no mechanisms for continuous improvement and knowledge transfer are elaborated.
- *Lack of confidence*. Companies are often reluctant to reveal strategic or tactical financial and supply goals and plans to partners.
- *Conflicting cultures and strategies*. Differences in company and supplier cultures and business strategies cause misunderstandings, mistrust, and inefficiencies.
- *Lack of flexibility*. Contracts are based on the state of the businesses during negotiations. Most contracts cannot anticipate all occurring changes in an evolving environment. Unanticipated changes may cause dissatisfactions if the contract is too rigid.
- *Opportunistic behavior*. Once the contract is in force, there is a tendency for both parties to optimize operations from their point of view, without concern for the good of the partner.
- *Key employee transfer*. If key members of the team that negotiated the contract or key employees knowledgeable of company operations are transferred or lost, then needs may be poorly understood, leading to disappointing performances.
- *Poor relationship management*. Partners often underestimate the time and effort required to manage the relationship, or worse, the relationship management responsibility is given to the supplier.

Monitoring and evaluating performances are probably the two most important elements of a sound partner relationship management approach. The performance measures used for supplier evaluation can be the same as those employed for partner selection. Examples of typical evaluation criteria were given in Table 6.4. Expected performances should be stated clearly in the contract between the parties, and responsibilities as to who will evaluate each performance indicator should also be specified. Operational indicators calculated by the provider, such as fill rates and on-time deliveries for 3PLs, should be transmitted to the company at least once per month, by location and by customer. Some measures such as defect rates and order compliance levels should be based on sampling and satisfaction studies at internal or customer ship-to points. Costs should also be monitored closely. Performance measures must be clear and easily evaluated with data collectible at a reasonable cost. Feedback on perceived achievements should also be provided to the supplier on a regular basis. This can be done, for example, by sharing a balanced scorecard (see Sect. 2.1) with quadrants designed to highlight strategic partner expectations. As a result of this process, suppliers performing poorly are forced to adopt correcting measures and even pay penalties, and outstanding providers are rewarded (by sharing cost savings, for example).

Good governance is also essential to maintain profitable relationships. It sustains relationships by providing a charter and standard procedures to guide decision-making, issue resolution, and adaptation to change, thus avoiding the inefficiencies of ad hoc recourses. Procedures should be sufficiently detailed to provide guidance, but flexible enough to allow adequate responses to unexpected changes. The governance model should ideally be developed by a cross-functional team with company and provider representation. It should provide a transparent integration between the operations of the partners involved (company, providers and possibly customers). This may involve developing interfaces between the information systems of the respective partners. It may also involve developing collaborative planning, forecasting and replenishment procedures (see Chap. 3).

The people involved in the management of the relationship are also obviously a vital part of the process. They must have strong communication capabilities as well as planning and control competencies. Periodical joint company–supplier relationship reviews should be conducted to assess the quality of the relationship (including positive and negative outcomes, communications, and issue resolution), improve defective processes, and align priorities.

Review Questions

6.1. What is a networked company? Give a practical example to illustrate your answer.
6.2. What is the difference between the internal and external networks of a company? Explain.
6.3. What are the strategic internalization–externalization options available to a company? Explain the main features of each of these options.

6.4. What kind of externalization option should be considered in these circumstances?

   - For an activity requiring a tight control by the company
   - If the company does not have the production resources to do an activity
   - If an activity would be carried out more efficiently and cheaper by a third party

6.5. Why is the choice of partners important for a company?
6.6. How can a company proceed to find and select partners? Explain.
6.7. Does the emergence of electronic marketplaces completely eliminate the interest of establishing strategic alliances with a small number of suppliers?
6.8. What criteria should be considered for supplier selection?
6.9. What is a 3PL?
6.10. Should all businesses outsource their logistic activities to 3PLs?
6.11. What are value-sharing mechanisms and how can they be established?
6.12. Explain the benefits of forming alliances and give a practical example supporting your explanation.

Exercises

Exercise 6.1 (Based on Chen 2011) A fabrics manufacturer in the textile industry must select textured yarn suppliers. After a detailed analysis of the industry, management decides to base its evaluation of candidate suppliers on the criteria and performance measures listed in Table 6.5. The order of importance of the criteria for the project team is the following:

$$\text{Quality} \succ \text{Innovation} \succ \text{Operations management} \succ \text{Price} \succ \text{Value creation}$$

Ten potential suppliers are considered by the company. Based on stock exchange and industry data, the performance values in Table 6.6 are obtained.

Which suppliers should be selected? Knowing that the *Innovation* and *Operations management* criteria are in fact perceived as having the same

Table 6.5 Selection criteria and performance measures

Criteria	Measure	Description
Quality	Return rate	Sales return/gross sales (fewer returns means better quality and higher customer acceptance)
Innovation	R&D rate	R&D expenses/sales (more R&D implies better innovation potential)
Operations management	Inventory turnover ratio	COGS/average inventory (larger turnover indicates better operations management capability)
Price	Discounts	Average discounts in % of list price (higher discounts mean better prices)
Value creation	Gross profit rate	(net sales-COGS)/sales (larger profit indicates better value creation capability)

Table 6.6 Supplier performance evaluations

Supplier	Return rate	Profit rate	Discount (%)	R&D rate	Inventory turns
1	0.06	0.01	7	1.11	0.67
2	0.54	9.69	7	1.13	6.02
3	1.11	6.36	5	2.12	5.8
4	0.15	6.42	5	1.57	6.17
5	0.19	9.51	10	1.5	6.76
6	1.28	13.81	7	3.08	7.48
7	0.01	5.41	8	2	7.04
8	0.42	6.82	7	1.04	11.16
9	0.65	7.51	5	1.66	5.17
10	0.25	1.43	8	2.62	5.16

importance, if their order was reversed in the criteria preference relation, would this change the supplier selection decisions?

Exercise 6.2 (Based on Abdel-Malek and Areeratchakul 2004) Consider the case of a company manufacturing a finished product assembled from eight major components. Four of these components are produced in-house, but it has been decided to outsource the production of the other four (labeled 1–4). Seven contract manufacturers (labeled A to G) have been identified as potential partners, but they do not have the same capabilities. The costs, defect rate, and lead times of potential partners for the components they can manufacture are given in Table 6.7. The order of the *Defects*, *Lead time*, and *Cost* columns in the table reflects the importance given to each criterion. The company wants to minimize the number of contract manufacturers used. Which partners should be selected?

Exercise 6.3 A distributor is considering three suppliers to meet a forecasted annual demand of 4500 units for a family of similar products. Table 6.8 provides information on the supplier's annual capacity and on three performance measures listed in order of importance. Which supplier should be selected?

Exercise 6.4 (Based on Jharkharia and Shankar 2007) A company has decided to outsource its logistics activities. Based on reputation, revenues, potential growth,

Table 6.7 Capabilities and attributes of potential contract manufacturers

Component	Contract manufacturer	Defects (%)	Lead time (days)	Cost ($)
1	A	5	2	$111
	B	4	3	$116
	C	2	1	$125
2	C	5	3	$80
	D	4	1	$90
3	D	8	4	$12
	E	7	3	$16
4	F	3	6	$200
	G	1	3	$255

Table 6.8 Data on potential suppliers

Supplier	Capacity	Defect rate (%)	Price ($)	Late deliveries (%)
1	2500	0.12	$65	0.44
2	2500	0.28	$55	0.41
3	2500	0.19	$60	0.55

Table 6.9 Normalized preference measures for candidate 3PLs

3PL	Compatibility	Cost	Reputation	Quality
1	0.2705	0.1964	0.1605	0.1239
2	0.4061	0.5281	0.3674	0.6494
3	0.3234	0.2755	0.4721	0.2267

geographic coverage, and service offered, six potential 3PLs were identified. These six 3PLs were asked to respond to a request for proposal (RFP) developed by the company. Four 3PLs replied to the RFP and among those, three proposals were judged to comply with company needs. To have a better perspective on the capabilities of these prospects, the selection team members visited the DCs that would be used to support the company logistics activities. This provided sufficient information to be able to make pairwise comparisons of candidate 3PLs for each of the selection criteria retained by the team. The criteria considered, in order of importance, are compatibility, cost, reputation, and quality of service. Proceeding by pairwise comparisons, as in Sect. 4.4.3, normalized preference measures were obtained for each criterion. These measures are given in Table 6.9. Which 3PL should be selected?

Bibliography

Abdel-Malek L, Areeratchakul N (2004) An analytical approach for evaluating and selecting vendors with independent performance in a supply chain. Int J Integr Supply Manag 1(1):64–78

Armstrong and Associates (2014) Current state of the 3PL market. www.3PLogistics.com. Accessed 20 Sep 2014

ASLOG (2011) Libre blanc GMA—Gestion mutualisée des approvisionnements

Bruno G, Esposito E, Genovese A, Passaro R (2012) AHP-based approaches for supplier evaluation: Problems and perspectives. J Purchasing Supply Manag 18:159–172

Chen Y-J (2011) Structured methodology for supplier selection and evaluation in a supply chain. Inf Sci 181:1651–1670

de Boer L, van der Wegen L (2003) Practice and promise of formal supplier selection: A study of four empirical cases. J Purchasing & Supply Manag 9:109–118

de Boer L, Labro E, Morlacchi P (2001) A review of methods supporting supplier selection. Eur J Purchasing & Supply Manag 7:75–89

Deloitte (2012) Outsourcing, today and tomorrow, Deloitte development LLC. http://www2.deloitte.com/content/dam/Deloitte/se/Documents/finance/InsightsfromDeloittes2012globaloutsourcingandinsourcingsurvey_SE.pdf

Doherty S, Hoyle S (2009) Supply chain decarbonization. World Economic Forum, Geneva

Dornier PP, Ernst R, Fender M, Kouvelis P (1998) Global operations and logistics. Wiley

Elmaghraby W (2000) Supply contract competition and sourcing policies. Manufact Serv Oper Manag 2(4):350–371

Esposito E, Evangelista P (2014) Investigating virtual enterprise models: Literature review and empirical findings. Int J Prod Econ 148:145–157

Eyefortransport (2012) 3PL selection & contract renewal—North American focus. http://www.supplychain247.com/paper/3pl_selection_contract_renewal_north_american_focus/eyefortransport

Feng Y, Martel A, D'Amours S, Beauregard R (2013) Coordinated contract decisions in a make-to-order manufacturing supply chain. Prod Oper Manag 22(3):642–660

Girst A, Schleyer R (2005) Managing the outsourcing relationship. TODAY—J Work Process Improv 14–17

Häcki R, Lighton J (2001) The future of the networked company. The Mckensey Quarterly 3:26–39

Ho W, Xu X, Dey P (2010) Multi-criteria decision making approaches for supplier evaluation and selection: A literature review. Eur J Oper Res 202:16–24

Jharkharia S, Shankar R (2007) Selection of logistics service provider: An analytic network process (ANP) approach. Omega 35:274–289

Lakhal S, Martel A, Oral M, Montreuil B (1999) Network companies and competitiveness: A framework for analysis. Eur J Oper Res 118–2:278–294

Langley J (2014) 18th annual third-party logistics study. Capgemini

Leenders M, Fraser Johnson P, Flynn A, Fearon H (2005) Purchasing supply management, 13th edn. Mc-Graw-Hill, Irwin

Logistics Magazine (2011) Un « pooling » réussi dans les produits d'hygiène-beauté. Logistics Magazine 263:34–35

Lynch C (2004) Managing the outsourced relationship.In: Council of logistics management presentation

Mazzola E, Perrone G (2013) A strategic need perspective on operations outsourcing and other inter-firm relationships. Int J Prod Econ 114:256–267

Mena C, Humphries A, Choi T (2013) Toward a theory of multi-tier supply chain management. J SCM 49–2:58–77

Miles R, Snow C (1986) Organizations: New concepts for new forms. Calif Manag Rev 28(3):62–73

Neto M, Pires S (2008) Analysis of the relationship between automaker and systemist in an Industrial Condominium of the automotive industry. J Oper Supply Chain Manag 1(2):41–52

Ng WL (2008) An efficient and simple model for multiple criteria supplier selection problem. Eur J Oper Res 186:1059–1067

O'Reilly J (2013) 3PL perspectives. Inbound logistics 1–14

Ojala M, Hallikas J (2006) Investment decision-making in supplier networks: Management of risk. Int J Prod Econ 104:201–213

Porter M (1998) Clusters and the new economics of competition. Harvard Bus Rev 77–90

Poulin D, Montreuil B, D'Amours S (1995) L'organisation virtuelle en réseau. In: Martel A, Oral M (eds) Les défis de la compétitivité: Vision et stratégies. Publi-Relais, Montréal, pp 59–82

Powell W (1987) Hybrid organizational arrangements: New form or transitional development? Calif Manag Rev 30(1):67–87

Pyndt J, Pedersen T (2006) Managing global offshoring strategies. Copenhagen Business School Press, Denmark

Quinn J, Doorley T, Paquette P (1991) The intellectual holding company: Structuring around core activities. In: Mintzberg H, Quinn J (eds) The strategy process: Concepts, contexts, cases, Prentice Hall

Schotanus F, Telgen J, de Boer L (2010) Critical success factors for managing purchasing groups. J Purchasing Supply Manag 16:51–60

Sedgwick D (2013) Global industry craves megasuppliers. Automot News

Sheffi Y (2013) Logistics-intensive clusters: Global competitiveness and regional growth. In: Bookbinder J (ed) Handbook of global logistics, Springer Science+Business Media, New York, p 463–500

Thorelli H (1986) Networks: Between markets and hierarchies. Strateg Manag J 7(1):37–51

Welch JM, Little AD, Nayak PR (1992) Strategic sourcing: A progressive approach to the make-or-buy decision. Acad Manag Executive 6–1:23–31

Williamson OE (1975) Markets and hierarchies, analysis and antitrust implications: A study in the economics of internal organizations. The Free Press, New York

Chapter 7
Supply Chain Networks Optimization

How many production, assembly, transshipment, distribution, or service centers should a company implement to meet its evolving needs, where should they be located, and what should their mission be? These strategic issues are addressed in this chapter. As indicated in Chap. 1, the supply chain network reengineering methodology proposed in this book requires the modeling and optimization of the physical structure of the SCN. This chapter is intended to clarify the nature of SCN design problems and their link with the company's supply chain strategy. It also presents basic optimization models that can be solved to facilitate the reengineering of the network.

The first section addresses the nature and issues of SCN design problems. The second presents common deployment strategies found in practice and the associated activity graphs. Section 7.3 proposes cost-minimization models for the design of basic production–distribution networks and discusses the modeling of several variants. Section 7.4 focuses on the maximization of the value added by a SCN. It proposes extensions of the basic design models considering order winners valued by customers. Section 7.5 provides an activity-based modeling approach for the design of a generic SC network, and it examines the characteristics of manufacturing networks more closely. Finally, Sect. 7.6 shows how economies of scale can be taken into account in SCN design models. These last four sections are accompanied by practical examples modeled and solved using the Excel Solver.

7.1 Supply Chain Network Design Problems

Chapter 1 explained that the supply chain network of a company covers all its supply, production, distribution, and sale facilities, as well as those of its suppliers (material vendors, contract manufacturers, 3PLs, etc.) and clients. It was also

A. Martel and W. Klibi, *Designing Value-Creating Supply Chain Networks*,
DOI 10.1007/978-3-319-28146-9_7

pointed out that a SCN's propensity to fulfill its mission depends largely on the capability of its physical structure to offer the production and distribution capacity required, as well as desired performance levels in terms of cost, service, responsiveness, and so forth. Manufacturing and distribution capacity is provided by dedicated or flexible fabrication, assembly, handling, storage, and transportation technologies. The physical deployment of these resources on the geographical territory covered by a company offers the costs, scale, flexibility, service, and proximity levers required to improve performance.

When addressing these strategic issues collectively, in addition to deciding where to locate production and distribution centers, which suppliers to use, and what technologies to implement in selected facilities, we must anticipate the quantity of products to manufacture, store, and ship in the network. When the company operates in several countries, it must also consider exchange rates, customs duties, non-tariff barriers, available subsidies, transfer prices, and corporate taxes. Strategic SCN design decisions will be gradually addressed in this chapter and the next two chapters. As Fig. 7.1 illustrates, the first thing to do to optimize these decisions is to conceptually develop a potential logistics network that incorporates all possible options. The nodes of this potential network typically can be partitioned in three fundamental subsets: supply sources, production and/or distribution facilities, and customers, the last being typically clustered in demand zones. Given a potential network of this type, it remains to choose from all the available options those that are likely to add as much value as possible for the company.

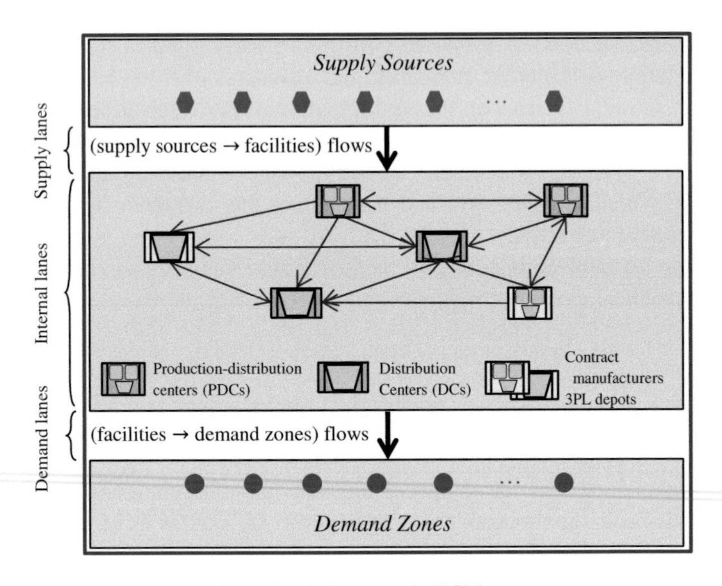

Fig. 7.1 Elements of a potential supply chain network (SCN)

The aim of SCN reengineering projects is to determine how to redeploy an existing SC network and to specify the mission of selected facilities. The potential SC network reflects the company's supply chain strategy by specifying possible redeployment opportunities for existing or prospective internal facilities and partners. As shown schematically in Fig. 7.2, after a potential SCN has been elaborated, decisions must be made on the facilities to close, transform, or build; on the activities to outsource; and on the specific mission of each of the selected sites. This may lead to the addition, closing, or reorganization of some facilities; to the outsourcing of production or distribution activities; and so on. As illustrated in Fig. 7.2, we could, for example, implement a single manufacturing plant among two envisaged worldwide to benefit from economies of scale, subcontract part of the production, and open a new DC to better serve local markets.

SCN design problems are difficult to solve in a comprehensive manner. This is largely because of the complexity of the huge number of options to consider and evaluate. For this reason, it is essential to use an optimization model to develop good designs. Given the magnitude of the problem, it can also be extremely difficult to find the optimal solution of the optimization model. Consequently, most SCN design models address only a subset of identified issues, and they focus on subnetworks that consider only one or two echelons (or stages) of global SCNs. These simplified models typically rely on several approximations (in the representation of relevant revenues and costs, among others), overlook some important decisions, and neglect some aspects of reality (e.g., dynamics, uncertainty, etc.). In order to understand the difficulties encountered, the factors that contribute to the complexity of the models to solve are outlined in the following:

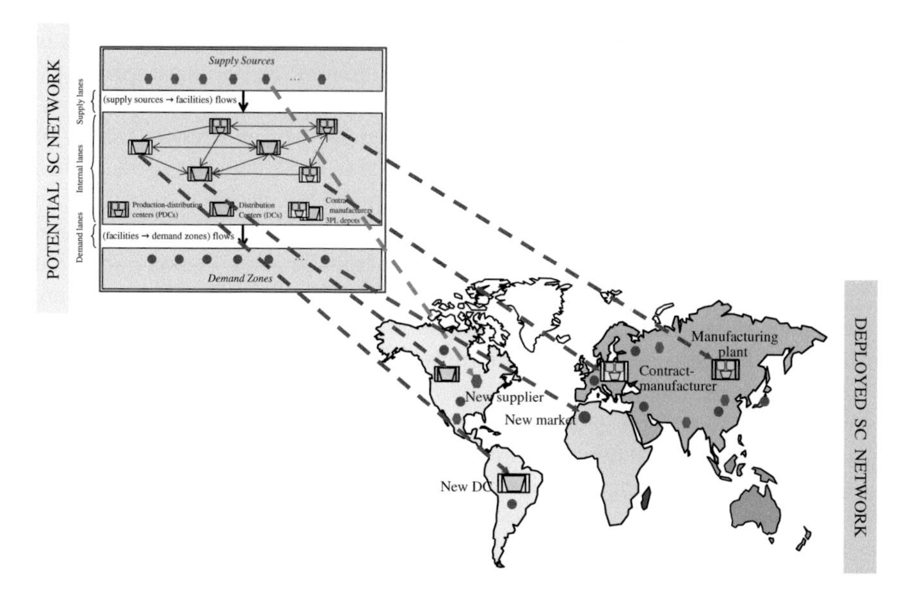

Fig. 7.2 Strategic deployment decisions

- **The scope of the decisions to make**. The complexity of the decision model to solve depends on the number and nature of the decision variables (mainly binary) involved. In some cases, for simplicity, it is assumed that inter-site (depots, plants, and customers) flows are predetermined. One then has a pure *location problem* including only site-selection decisions. In other cases, the locations of the facilities are fixed and it remains only to clarify their mission in terms of products to process and logistics activities to accomplish. This is known as a pure *allocation problem*. In the majority of cases, these two problems are combined. One then has a *location–allocation problem* that is much harder to solve. The problem becomes more complex when sourcing, capacity expansion, technology selection, and product-market offer decisions are also considered. Real-life SCN design problems are thus often far more intricate than classical location–allocation problems.
- **The quality of cost and revenue anticipations**. To assess adequately the performance of strategic deployment decisions, the operational transportation, production, and storage decisions made under feasible designs, as well as related cash flows, must be anticipated as precisely as possible. It is obvious that the explicit inclusion of operational decisions in design models would lead to large mathematical programs that are extremely difficult to solve. Therefore, a judicious compromise between the accuracy of the model and its solvability must be found through astute approximations of relevant operational costs and revenues. In-depth anticipations lead to accurate models that are hard to solve, and vice versa. For example, for the location–allocation problem, many models with different anticipations of transportation costs were proposed (e.g., estimation of origin–destination flow costs by regression analysis from historical data, explicit inclusion of a subset of efficient delivery routes, etc.), including integrated location-routing models that explicitly consider the choice of delivery routes in location models. One should always use the simplest model that captures the essence of the problem.
- **The range of potential resources considered**. If there is a single facility to locate, one supplier to select, or a product-market to choose, the problem then reduces to a classical multi-criteria decision-making problem as shown in Chaps. 4, 5, and 6. When there are several facilities to locate, multiple potential suppliers, different capacity options, or numerous product market expansion opportunities, the combinatorics of the problem gives rise to large, much more complex, mixed-integer programs (MIPs).
- **The number and disparity of products**. When the products manufactured or distributed are relatively uniform, they can be clustered into a family and considered as a single product in the decision models. However, when their manufacture or distribution requires distinct production, storage, or transportation technologies, much more elaborate multi-product models must be formulated.
- **The expectations of the product-markets to serve**. When the fact that the order winners provided by the SCN affect value creation (see Chaps. 1 and 2) is explicitly considered, alternative product-market offers must be specified.

A product-market is a set of clients in a geographical area that values products with similar order winners. Offers relate to a set of order winners that the company may associate with some products. When several potential offers are considered, the complexity and size of the model expand quickly. Also, when different supply chain tactics (e.g., decoupling point or response time offered by depot–customer proximity and transportation mode speed) are needed to serve distinct product-markets properly, it is necessary to differentiate them in the design model. For example, supply chains with make-to-stock (MTS), make-to-order (MTO), or vendor-managed inventory (VMI) decoupling points provide very different response times, flexibility, and prices. Similarly, response times, service quality, prices, and environmental footprint are not the same when using rail or road transportation. This must be captured by the model.

- **The complexity of production–distribution processes**. The number and mission of the facilities of a SCN also depend on the nature of its production–distribution processes. It is clear that when the manufacturing process involves numerous stages (multilevel bill of materials) or when distribution involves several echelons, the formulation of a SCN design model is much more complex. If one wants to find the number, location, and mission of regional DCs, and the rest of the network (plants, national DC, etc.) is not under review, the model to solve is much simpler than if the entire network is reengineered. The problem becomes particularly complex when the production and distribution subnetworks of a company must be redesigned simultaneously, that is, when design decisions must be made for all the elements in the central rectangle shown in Fig. 7.1. When in addition to locating potential resources, current facilities can be reorganized or expanded, and the outsourcing of activities is considered, the problem gives rise to much more complex mathematical programs.
- **The objectives to optimize**. As discussed in Chap. 2, a fundamental objective of a company is the maximization of value added. Taking explicit account of this objective leads to very difficult problems, requiring the modeling of the relationship between revenues and order winners in addition to considering aspects related to economies of scale, greening, and risks in the calculation of the return on capital employed. If one restricts itself to the minimization of the relevant costs, as we shall initially do in this chapter, this simplifies the problem considerably. This simplification is adequate when demand is predetermined and when the order winners influencing the company's revenues are independent of the SC network structure, which is rarely the case in practice. Also, if the economies of scale associated with structural choices are not too pronounced, design decisions can be assumed to generate only fixed and variable costs, which reduces the problem to a mixed-integer linear program (MILP). However, the consideration of performance variability under risk leads to nonlinear objective functions and design models harder to solve.
- **The constraints to consider**. Several constraints must be taken into account to get realistic SCN designs. In addition to customer demands and supply offers, technical and commercial constraints related to production and distribution

activities must be specified. Potential platforms and existing facilities have a limited capacity, which must be taken into account. Service constraints are also required. Typically, we want to make sure that each client may be served within a reasonable time, which obviously depends on the distance between clients and service points. Depending on the context, we may require that each client is served by a single service point (single sourcing) or, to improve resilience in the event of accidents or disasters, we may require that each customer can be served by more than one service point. Under risk, we must also adapt the constraints to account for possible recourses.

- **The uncertainty of the business environment**. When strategic deployment decisions are made, we do not know what will be the state of the business environment when the SCN network designed will be in operation. One ideally would like to be able to anticipate the state of the environment during a multiyear planning horizon covering a few reengineering cycles. To do this, the types of events and trends that are likely to affect the SCN network in the future must be clearly identified, and an approach for decision-making under uncertainty must be used. These approaches often involve the elaboration of a set of plausible future scenarios, leading to very difficult optimization models. Totally neglecting uncertainty in practice, however, may jeopardize the quality of the designs obtained.

Taking these elements into account in a reengineering project requires the formulation and resolution of large mathematical programs. In this chapter, after examining different possible deployment strategies, we show how to model the basic elements of the problem. For simplicity, it is assumed that all problem data are known and that the business environment is stable, that is, that it will remain the same during the life of the network. Under these circumstances, the SCN can be optimized for a typical year. These simplifying assumptions will be relaxed in subsequent chapters. The majority of the models presented can be solved with commercial optimization packages such as CPLEX (www.ibm.com), Xpress (www.fico.com), Gurobi (www.gurobi.com), or Frontline (www.solver.com). In some cases, nonlinearities or the size of the problem may require the use of specialized solution methods.

7.2 Deployment Strategies

The foundation of a supply chain strategy is the distinctive offer a company makes to customers in order to create value (recall Fig. 1.14). The company must, however, develop some means to deliver this offer. Customers are necessarily dispersed geographically, as are raw material sources and the long-term resources employed for the production–distribution of products and services. The spatial organization of long-term resources in the activity centers of production–distribution facilities provides an essential means required to deliver a company's offer. The deployment of resources determines flow patterns between sources and customers in a SCN, and it has a major impact on costs incurred and order winners offered. This being said, what is the best deployment strategy for a company, that is, the best SCN structure? When studying the SCNs operated by companies, a wide variety of structures are observed, depending on the industrial sector, targeted order winners, and the SC doctrine advocated.

Two complementary approaches can be adopted to address this difficult question. On the one hand, by observing the SCN of successful companies, the type of structure required to support specific order winners and doctrines can be identified. This bottom-up intelligence approach that starts from reality to infer concepts is assumed in this section. On the other hand, one can use a top-down reengineering approach going from the conceptual level to the physical level. The basic elements of a company's SC strategy are then molded in a conceptual representation of the desired network, which becomes the blueprint used to build or improve a physical network. This approach will be adopted in subsequent sections of this chapter. In either case, one needs a mapping formalism for the conceptual representation of an SCN and of its doctrinal foundations. In this book, *activity graphs* (see Sects. 4.1.2 and 6.1.3) are used for this purpose, that is, directed graphs of the internal and external activities of the company reflecting strategic design choices as well as constraints imposed by technology and industry practices.

The simplest deployment strategy in a MTS context is to keep finished products in inventory at the production–distribution center (PDC) producing them and to deliver the products ordered by customers directly from these plants. This approach can generate economies of scale for storage costs, but it increases transportation costs significantly because shipments are not consolidated. A SCN structure of this type and the corresponding activity graph are illustrated in Fig. 7.3. This

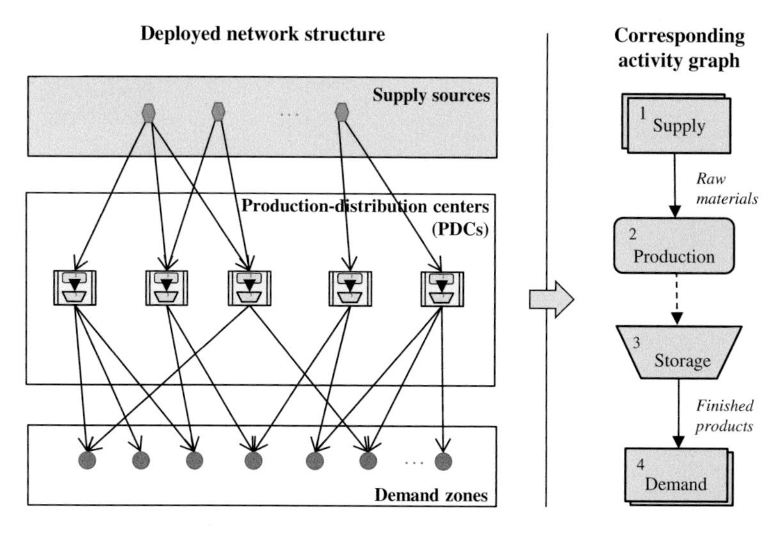

Fig. 7.3 Direct shipments from PDCs

single-echelon deployment strategy would be more appropriate in a MTO or mass customization context. As seen at the end of Chap. 5, a variant of this strategy is to use consolidation centers between factories and customers. This consolidation is typically done in the cross-docking facilities of a 3PL. This would be represented on the activity graph in Fig. 7.3 by adding a consolidation-transfer node after the storage activity with inter-site transportation arcs (movements). The deployed network would then include intermediate transshipment sites. This strategy is often used by companies selling products online.

A second typical deployment strategy is to use separate production and distribution centers, as shown in Fig. 7.4. On the activity graph, production and storage are now linked by an inter-site transportation arc. An additional echelon appears in the deployed network to store the products coming from production centers and shipped to customers. This strategy also applies to pure distribution companies, in which case the production echelon is obviously absent. In this context, transportation and storage are often outsourced. This is what Amazon (www.amazon.com) does, for example, to distribute its products. This strategy positions inventories closer to demand zones, which significantly improves service levels. However, it increases inventory holding costs, mainly when several local depots are opened. When customer demand is high, it may be possible to operate only a few regional DCs and to use truckload (TL) transportation for deliveries. However, when one has several small customers requiring same-day or next-day deliveries, it may be necessary to implement several local depots and to develop delivery routes between ship-to points. This type of network is certainly more efficient in terms of service, but it increases logistics costs: the high number of sites increases fixed costs, the number and length of delivery routes generates significant transportation

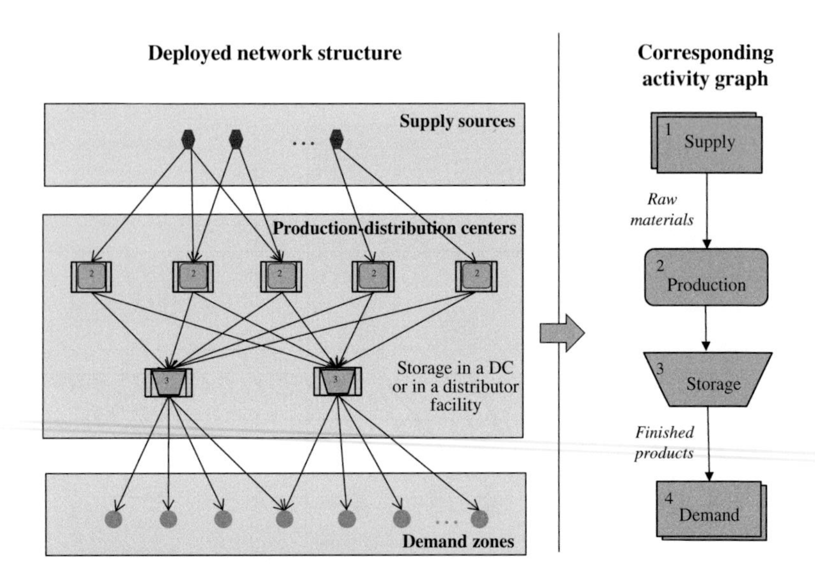

Fig. 7.4 Deployment with intermediate DCs

costs, and the decentralization of inventories increases holding costs. This SCN variant is often found in a spare parts distribution context.

A third deployment strategy frequently used in a distribution context is presented in Fig. 7.5. External suppliers ship products to storage points via transshipment centers. Transshipment sites split loads coming from suppliers and reconsolidate them in store shipments to promote economies of scale in transportation. Walmart was one of the pioneers of this strategy. The client in this context collects the goods at the store, whether it is a sales point or a pick-up point. The latter is common for online sales. For example, items purchased online from Sears (www.sears.com) in the United States can be picked up at the closest Sears outlet; in France, Cdiscount (www.cdiscount.fr) uses Casino supermarkets as picking points. One then takes advantage of the resources of a third-party network. This type of structure is also often seen for returned merchandise.

The three strategies just described distinguish themselves by the structure of their distribution network. For complex products, it is often production that is the core business and deployment strategies then focus on the form of the manufacturing network. The design of the SCN must then consider the BOM of the products manufactured as well as the technical specificities of production technologies and processes. The activity graph in Fig. 7.6 represents the case of a MTO company engaged in several component-fabrication and finished-product assembly activities. The figure shows how the activity graph relates to the BOM. Node 3 on the graph is not shaded to indicate that, because the fabrication technology of component 5 provides a competitive advantage to the company, the activity cannot be externalized. The deployed network indicates that the company currently operates a factory focused on the fabrication of component 5 (activity 3), a factory covering

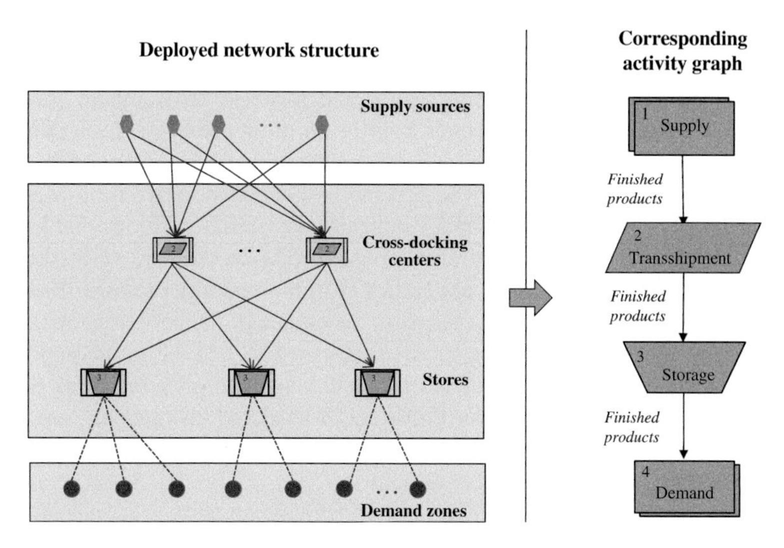

Fig. 7.5 Deployment with transshipment centers and pick-up stores

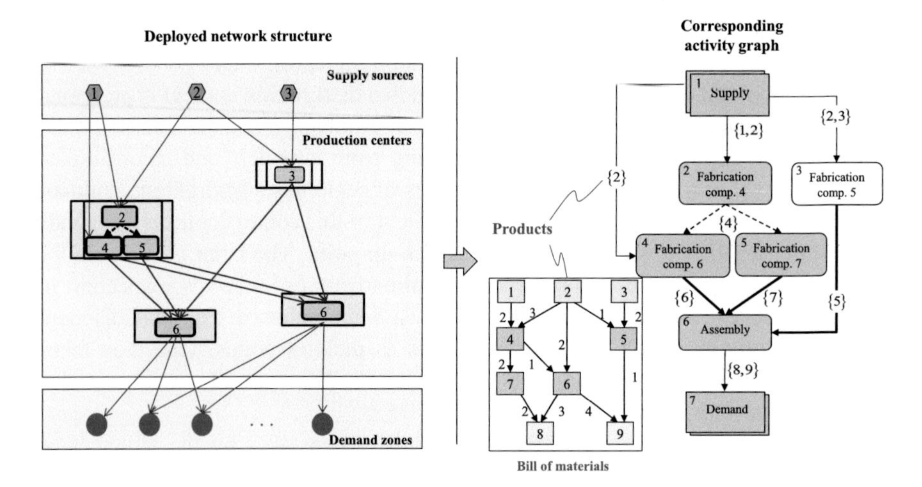

Fig. 7.6 Deployment of a manufacturing network

activities 2, 4, and 5, and two assembly plants from which products are shipped to customers. Generally, the activity graph of a manufacturing company incorporates a hierarchy of production activities, organized by product or technological process, and it involves inter- and intra-site movements. These activity graphs give rise to multi-echelon networks incorporating several focused or flexible manufacturing facilities and several finished products assembly–distribution centers. It is clear that the supply chain strategy of multinational companies leads to complex activity graphs and deployed networks. In such contexts, the SCN can be much more elaborate than in the described examples and may include raw materials storage and/or transshipment echelons.

It should now be clear that, although the physical structure of a SCN is contingent on the company's industrial sector and business environment (markets, competition, partners, etc.), it depends heavily on its SC strategy. Each company has its own specificities, and examining some generic deployment strategies, as we did in this section, is not sufficient to find the best suited structure for a particular case. Previously discussed deployment strategies are partial solutions that may be appropriate for a subset of products, a given market or country, or a particular competitive context, but they are not sufficient in the majority of cases. Based on the SC strategy developed by the company, the choice of the best structure should rather rely on a SCN optimization model formulated to find the best compromise among all desired performance indicators (i.e., to maximize value creation). Should we select a single raw material source or a set of local and international suppliers? Should we have a large centralized plant to manufacture all products or a network of small factories close to markets? Should we have a large DC to serve all customers or a network of local warehouses? To provide an answer to such strategic questions we address the modeling of SCN design problems in the following sections.

7.3 Supply Chain Network Design Models

As previously discussed, location–allocation decisions are key issues addressed in SCN reengineering projects. These decisions lay down the SCN physical structure and specify the flow of goods through the deployed network during the planning period considered in the study. This section starts by exploring the design of a distribution network, focusing on the minimization of relevant costs. We then study the design of a production–distribution network and examine variants of these two location–allocation problems.

7.3.1 Designing Distribution Networks

To start with, consider the case in which facility-location decisions concern a single echelon of intermediate sites responsible for the delivery of finished goods to customers. Three fundamental facets of this problem are displayed in Fig. 7.7: the activity graph, the potential resources available, and the resulting potential network. As seen in this figure, a family of products provided by external suppliers must be shipped to demand zones (customers) through intermediate DCs. We must therefore determine what DCs to open and specify their mission, taking into account available capacity and total customer demand so as to minimize the total cost incurred for the planning period considered (usually a typical year).

More specifically, the following is assumed:

- The products distributed are similar. They can thus be clustered in a single family and we can proceed as if we had a single product.
- The network must be designed from scratch and the platform (see Chap. 4) of the DC at the proposed sites is predetermined.
- Operating costs are linear.

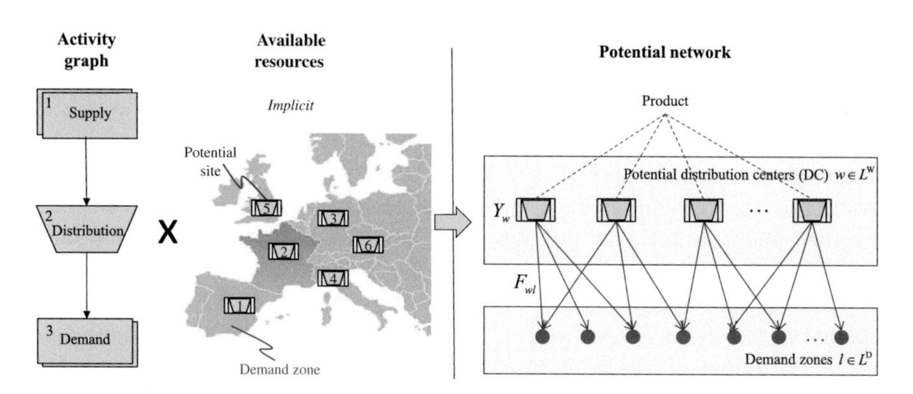

Fig. 7.7 Single-echelon distribution network design problem

To formulate the problem, the following sets must be defined:

L^{W} Set of distribution sites (warehouses) considered $(w \in L^{\mathrm{W}})$

L^{D} Set of demand zones (ship-to point clusters) to provision $(l \in L^{\mathrm{D}})$

Relevant problem data are denoted as follows:

d_l Yearly demand of zone l ship-to points

b_w^{S} Maximum storage space available in site w (in pallets, ft^2, m^3... as the case may be)

φ Turnover ratio of the company for cycle and safety stocks (see Chap. 2)

e Storage space needed for a product (in pallets, ft^2, m^3... as the case may be)

η DCs order cycle and safety stock (maximum level)/(average level) ratio for the year considered

The following relevant costs must also be taken into account:

v_w Average value of a product kept in stock in DC $w \in L^{\mathrm{W}}$ (including its purchase price, upstream transportation, reception and handling charges, as well as in-transit capital costs)

r_w^{I} Inventory holding cost in \$/\$/year (see Chap. 2), plus the rate per inventory-\$ paid for the space occupied on average during the year when DC w is owned by a third party

c_{wl} Unit picking, shipping, transport, and in-transit inventory holding cost incurred when servicing demand zone l from DC w

f_{wl} Total cost incurred per unit flow for servicing demand zone l from site w

y_w Fixed cost (lease—or annuity—plus fixed operating costs) incurred when opening and operating a platform on site w for the year considered (see Chap. 2)

To specify a solution, the following decision variables are required:

Y_w Binary variable set to 1 if site w is opened, and 0 otherwise

F_{wl} Quantity of products provided by DC w to demand zone l during the year considered.

The annual throughput of product in a DC w is given by the sum of outbound flows $X_w = \sum_{l \in L^{\mathrm{D}}} F_{wl}$. The average level of inventory required during the year to provide an adequate service is obtained by dividing this throughput by the company inventory turnover rate, giving: $\bar{I}_w = X_w / \varphi$. In practice, the inventory level is not always equal to its annual average. Given that delivery trucks carry product batches, at certain points in times the inventory in hand (cycle stock) may be significantly higher than the annual average. Obviously, the storage space of the DC must be sufficient to accommodate these higher levels of inventory. The storage space required in the DC is, therefore, $e\eta\bar{I}_w = (e\eta/\varphi)X_w$, and its throughput must thus respect the following constraint: $X_w \leq b_w$ with $b_w = (\varphi/e\eta)b_w^{\mathrm{S}}$.

Concerning relevant flow costs, to the value v_w of products shipped from DC w one must add inventory holding and customer delivery costs. As seen in Chap. 2, the inventory holding cost is given by the expression $(r_w^{\mathrm{I}} v_w)\bar{I}_w = (r_w^{\mathrm{I}} v_w/\varphi)X_w$. Therefore, the unit flow cost incurred for servicing demand zone l from site w is $f_{wl} = v_w + (r_w^{\mathrm{I}} v_w/\varphi) + c_{wl}$. Considering, in addition, fixed DC opening costs (y_w) and customer demands (d_l), the optimal structure of the distribution network is obtained by solving the following optimization model:

$$C^{\mathrm{SCN}} = \min \sum_{w \in L^{\mathrm{W}}} y_w Y_w + \sum_{w \in L^{\mathrm{W}}} \sum_{l \in L^{\mathrm{D}}} f_{wl} F_{wl} \tag{7.1}$$

subject to

- Demand constraints

$$\sum_{w \in L^{\mathrm{W}}} F_{wl} = d_l, \quad l \in L^{\mathrm{D}} \tag{7.2}$$

- DC capacity constraints

$$\sum_{l \in L^{\mathrm{D}}} F_{wl} \leq b_w Y_w, \quad w \in L^{\mathrm{W}} \tag{7.3}$$

- Nonnegativity constraints

$$Y_w \in \{0,1\}, w \in L^{\mathrm{W}}, F_{wl} \geq 0, w \in L^{\mathrm{W}}, l \in L^{\mathrm{D}} \tag{7.4}$$

This is a classical mixed-integer programming model that can be solved with the Excel Solver or with other commercial solvers. Because location variables are binary ({0,1}), the inclusion of a very large number of potential sites in the mathematical model makes the problem harder to solve.

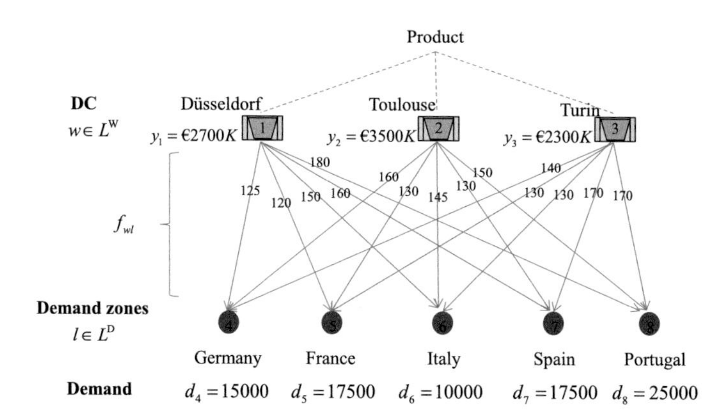

Fig. 7.8 Potential distribution network for Example 7.1

Example 7.1 The potential network in Fig. 7.8 concerns the distribution of a product family to several product markets through intermediate DCs. Five country-based markets are involved: Germany, France, Italy, Spain, and Portugal. Products could be distributed to these five markets from three potential DCs located in Düsseldorf, Turin, or Toulouse. These potential DCs are owned by 3PLs the company could sign an agreement with, based on a maximum throughput of 50,000 standard shipping loads per year for each site. Projected unit flow costs are indicated on the network arcs, and the fixed costs of operating facilities are shown beside DC nodes. We want to determine the DCs to open as well as their mission.

The Excel spreadsheet used to find the optimal solution of this problem and the Solver[1] parameters specified are found in Fig. 7.9. The top table specifies the parameters of the problem in terms of costs, capacity, and demand. The second table contains decision variables as well as DC capacity and customer demand constraints. It gives the optimal flows from DCs to demand zones, as well as binary DC opening decisions. Note that the excess capacity $(b_w Y_w - \Sigma_{l \in L^D} F_{wl})$ for each site depends on the facility-opening decisions, that is, on the 0–1 variables associated with sites. Note also that it is necessary to specify that these variables are binary in the solver.

The optimal solution obtained in terms of site openings and flows to demand zones is presented in the second table in Fig. 7.9. It suggests not using the site in Toulouse. The DC in Düsseldorf covers the demand from Germany and France as well as a portion of the orders from Spain. The mission of the DC in Turin is to service Italy and Portugal as well as the remaining customers from Spain. The total SCN cost for the typical year considered is €17,325,000.

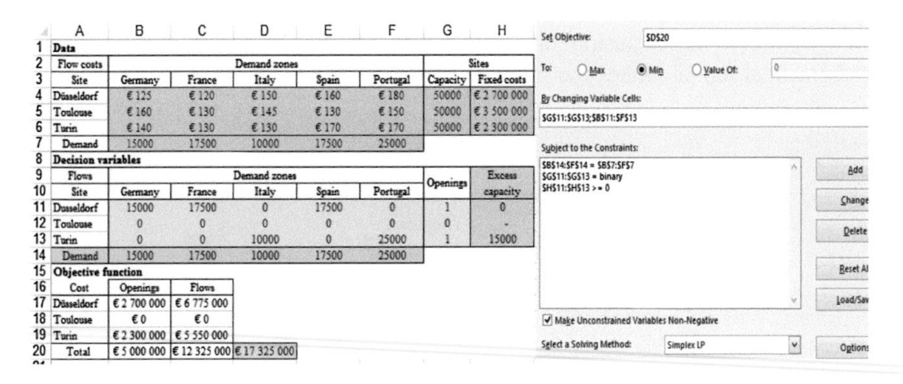

	A	B	C	D	E	F	G	H
1	**Data**							
2	Flow costs	Demand zones					Sites	
3	Site	Germany	France	Italy	Spain	Portugal	Capacity	Fixed costs
4	Düsseldorf	€ 125	€ 120	€ 150	€ 160	€ 180	50000	€ 2 700 000
5	Toulouse	€ 160	€ 130	€ 145	€ 130	€ 150	50000	€ 3 500 000
6	Turin	€ 140	€ 130	€ 130	€ 170	€ 170	50000	€ 2 300 000
7	Demand	15000	17500	10000	17500	25000		
8	**Decision variables**							
9	Flows	Demand zones					Openings	Excess capacity
10	Site	Germany	France	Italy	Spain	Portugal		
11	Dusseldorf	15000	17500	0	17500	0	1	0
12	Toulouse	0	0	0	0	0	0	-
13	Turin	0	0	10000	0	25000	1	15000
14	Demand	15000	17500	10000	17500	25000		
15	**Objective function**							
16	Cost	Openings	Flows					
17	Düsseldorf	€ 2 700 000	€ 6 775 000					
18	Toulouse	€ 0	€ 0					
19	Turin	€ 2 300 000	€ 5 550 000					
20	Total	€ 5 000 000	€ 12 325 000	€ 17 325 000				

Fig. 7.9 Excel spreadsheet and solver parameters for the example. One should ensure in the Solver window that "Make Unconstrained Variables NonNegative" is checked, and that the solving method is set to *Simplex LP*

[1]For a primer on the use of the Excel Solver, see Harmon (2013).

7.3.2 Designing Production–Distribution Networks

The basic model we just solved considers SCNs with a single echelon of distribution sites and a single product family (Fig. 7.7). We now study the case of a two-echelon network including production and distribution facilities and selling several product families. As Fig. 7.10 shows, the activity graph associated with this new problem states that the product families are manufactured using distinct technologies and are then stored, either in a PDC or a DC, before being sold to customers. The company is considering the implementation of a few PDCs to make all or any of the products and distribute them directly to demand zones in the vicinity. It is also considering the operation of DCs to facilitate access to more distant markets. This gives rise to two levels of intermediate sites in the potential network in Fig. 7.10.

The main features of the problem are the following:

- Finished products are made to stock, and are shipped to customers either directly from a PDC or via a DC. DC inventories are replenished directly from the production activities at the PDCs and not from the PDC storage points. All the products of a family require the same production, handling, and transportation technology, and use the same production–distribution planning and control methods.
- The network must be designed from scratch and the DC–PDC platform for proposed sites is predetermined. The platform of some PDCs can manufacture only a subset of the product families.
- Operating costs are linear, and we want to design the network that minimizes total relevant costs for a base year.

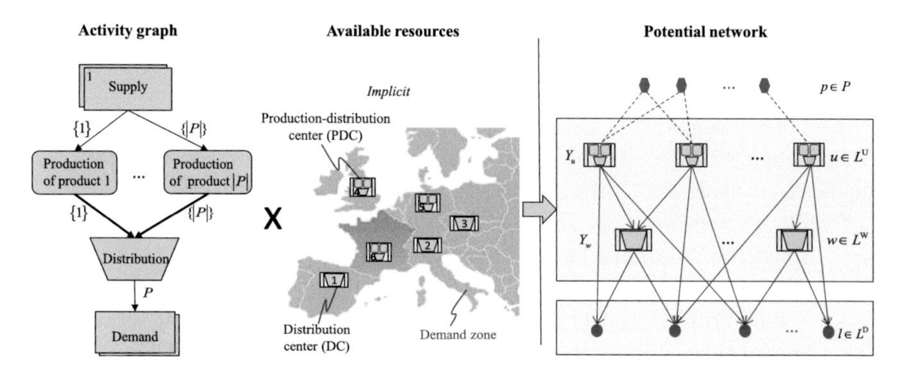

Fig. 7.10 Two-echelon production–distribution network design problem

To formulate the design model, the following sets are required:

P Set of the product families considered $(p \in P)$

L_p^{D} Set of demand zones requiring product p (with $L^{\mathrm{D}} = \cup_p L_p^{\mathrm{D}}$)

L^{U} Set of potential PDCs

$P_u \subset P$ Subset of product families that can be produced in PDC $u \in L^{\mathrm{U}}$

L^{W} Set of potential DCs

L^{S} $L^{\mathrm{U}} \cup L^{\mathrm{W}}$

The data required to complete the analysis are the following:

d_{pl} Demand for product p in demand zone l during the base year

b_s Maximum annual outflow from site $s \in L^{\mathrm{S}}$ (in standard unit loads—drums, pallets, ft^2, m^3…) to demand zones imposed by available storage space $\left(b_s = (\varphi/\eta)b_s^{\mathrm{S}}\right)$

e_p Space used by a family p product (in standard unit loads)

b_{pu} Product p production capacity for PDC u

The following relevant costs must be taken into account:

c_{pu}^{P} Average unit production cost of product p in PDC u (including the value of raw materials purchased, reception costs, and cost of manufacture)

c_{ps}^{S} Average storage cost for product p per unit of flow in site s ($c_{ps}^{\mathrm{S}} = r_s^{\mathrm{I}} v_{ps}/\varphi$, v_{ps} being the value of product p in site s)

c_{psl} Unit picking, shipping, transportation, and in-transit holding cost associated with the delivery of product p to node $l \in L^W \cup L^D$ from site $s \in L^{\mathrm{S}}$

$$
\begin{aligned}
f_{pul} &= c_{pu}^{P} + c_{pu}^{S} + c_{pul}, && u \in L^{U}, l \in L^{D} \\
f_{puw} &= c_{pu}^{\mathrm{P}} + c_{puw}, && u \in L^{\mathrm{U}}, w \in L^{\mathrm{W}} \\
f_{pwl} &= c_{pw}^{\mathrm{S}} + c_{pwl}, && w \in L^{\mathrm{W}}, l \in L^{\mathrm{D}}
\end{aligned}
$$

Note that transportation costs on a specific arc of the network depend on the transportation mode used for this arc, on whether full-load or less than full-load shipments are made, and, in the latter case, on the average shipment size (see Chap. 5).

Finally, the following decision variables are required:

Y_s Binary variable equal to 1 if a platform is implemented on site $s \in L^{\mathrm{S}}$, and 0 otherwise

F_{pwl} Annual flow of product p delivered to demand zone l from DC w

F_{pul} Annual flow of product p delivered to demand zone l from PDC u

F_{puw} Annual flow of product p shipped by PDC u to DC w

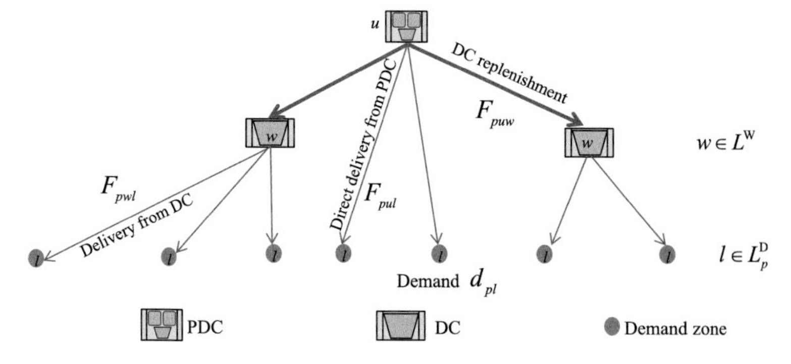

Fig. 7.11 Production-distribution chains for a product p made in PDC u

The three types of product flow variables defined are illustrated in Fig. 7.11 for a product p made in PDC u.

The following optimization model can be solved to find the optimal structure of the SCN:

$$C^{\mathrm{SCN}} = \min \sum_{s \in L^{\mathrm{S}}} y_s Y_s + \sum_{p \in P} \sum_{u \in L^{\mathrm{U}}} \sum_{w \in L^{\mathrm{W}}} f_{puw} F_{puw} + \sum_{p \in P} \sum_{s \in L^{\mathrm{S}}} \sum_{l \in L_p^{\mathrm{D}}} f_{psl} F_{psl} \tag{7.5}$$

subject to

- Demand constraints

$$\sum_{s \in L^{\mathrm{S}}} F_{psl} = d_{pl}, \quad p \in P, l \in L_p^{\mathrm{D}} \tag{7.6}$$

- Production capacity constraints

$$\sum_{l \in L_p^D} F_{pul} + \sum_{w \in L^W} F_{puw} \leq b_{pu} Y_u, \quad u \in L^U, p \in P_u \tag{7.7}$$

- Distribution capacity constraints

$$\sum_{p \in P} e_p \Big(\sum_{l \in L_p^{\mathrm{D}}} F_{psl} \Big) \leq b_s Y_s, \quad s \in L^{\mathrm{S}} \tag{7.8}$$

- Flow equilibrium constraints for intermediate sites

$$\sum_{u \in L^U} F_{puw} = \sum_{l \in L_p^D} F_{pwl}, \quad p \in P, w \in L^W \tag{7.9}$$

- Nonnegativity constraints

$$Y_s \in \{0,1\}, s \in L^S; \quad F_{puw}, F_{psl} \geq 0, p \in P, s \in L^S, l \in L^D \tag{7.10}$$

This mixed-integer network optimization model is solved easily with commercial solvers for most practical contexts.

Example 7.2 Continuing the previous example, it is assumed now that there are two product families to distribute and that they can be manufactured in one or two plants. The potential PDC sites considered are Berlin and The Hague. The potential DC sites and the demand zones considered are the same as before. Figure 7.12 illustrates the structure of the potential production–distribution network for each product. In addition to the (PDC → DC → Demand zones) chains shown in the figure, direct (PDC → Demand zones) deliveries are possible. All relevant data in terms of costs, capacity, and demand are provided in the two Excel tables in Fig. 7.13.

The mixed-integer linear program to solve with the Excel Solver is shown in Fig. 7.14 as well as its optimal solution. This solution can be interpreted as follows:

- The Hague PDC should be opened and used almost to its full capacity; its mission will be to manufacture the two products, to resupply the two opened DCs and also to deliver directly a part of the product orders in France, Spain, and Portugal. The potential PDC in Berlin should not be opened.
- The DCs in Düsseldorf and Turin should be opened and used to meet part of product 1 demands for all countries, as well as part of product 2 demands for Germany, Italy, Spain, and Portugal. The Toulouse DC should not be operated.

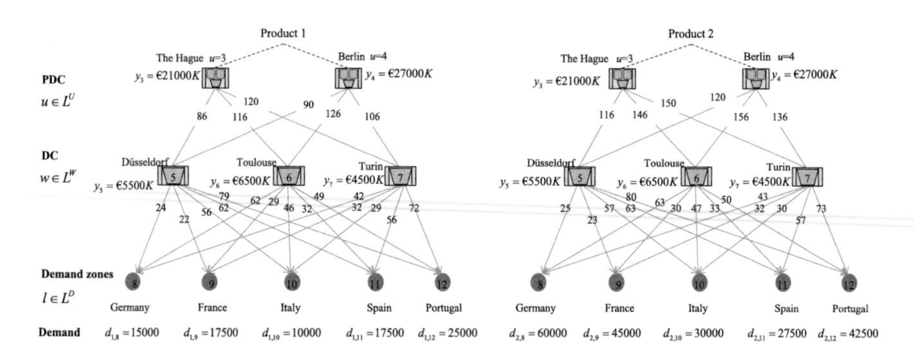

Fig. 7.12 Structure of the production–distribution network for the example

Data

Site	Flow costs: Product-markets for $p=1$: Germany	France	Italy	Spain	Portugal	Product-markets for $p=2$: Germany	France	Italy	Spain	Portugal	Fixed costs	Site parameters: Distribution capacity	Conversion e_1	e_2
The Hague	€ 108	€ 101	€ 138	€ 141	€ 158	€ 139	€ 132	€ 169	€ 172	€ 189	€ 21 000 000	100000	1	1
Berlin	€ 76	€ 107	€ 124	€ 151	€ 164	€ 107	€ 138	€ 155	€ 182	€ 195	€ 27 000 000	150000	1	1
Düsseldorf	€ 24	€ 22	€ 56	€ 62	€ 79	€ 25	€ 23	€ 57	€ 63	€ 80	€ 5 500 000	100000	1	1
Toulouse	€ 62	€ 29	€ 46	€ 32	€ 49	€ 63	€ 30	€ 47	€ 33	€ 50	€ 6 500 000	100000	1	1
Turin	€ 42	€ 32	€ 29	€ 56	€ 72	€ 43	€ 32	€ 30	€ 57	€ 73	€ 4 500 000	100000	1	1
Demand	15000	17500	10000	17500	25000	60000	45000	30000	27500	42500				

Site	Flow costs: DC (product $p=1$): Düsseldorf	Toulouse	Turin	DC (product $p=2$): Düsseldorf	Toulouse	Turin	Site parameters: Production capacity $p=1$	$p=2$
The Hague	€ 86	€ 116	€ 120	€ 116	€ 146	€ 150	90000	210000
Berlin	€ 90	€ 126	€ 106	€ 120	€ 156	€ 136	150000	250000

Fig. 7.13 Data of the two-echelon two-product example

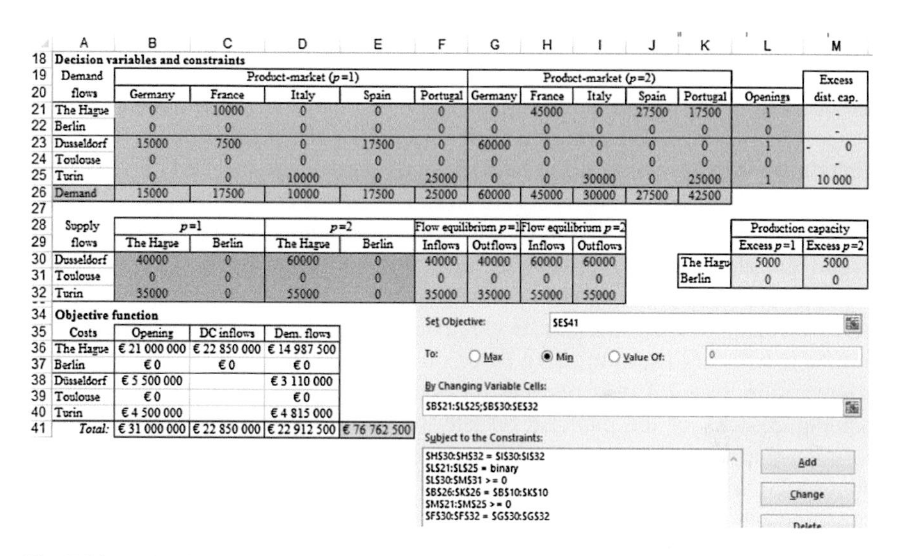

Decision variables and constraints

Demand flows	Product-market ($p=1$): Germany	France	Italy	Spain	Portugal	Product-market ($p=2$): Germany	France	Italy	Spain	Portugal	Openings	Excess dist. cap.
The Hague	0	10000	0	0	0	0	45000	0	27500	17500	1	-
Berlin	0	0	0	0	0	0	0	0	0	0	0	-
Dusseldorf	15000	7500	0	17500	0	60000	0	0	0	0	1	- 0
Toulouse	0	0	0	0	0	0	0	0	0	0	0	-
Turin	0	0	10000	0	25000	0	0	30000	0	25000	1	10 000
Demand	15000	17500	10000	17500	25000	60000	45000	30000	27500	42500		

Supply flows	$p=1$: The Hague	Berlin	$p=2$: The Hague	Berlin	Flow equilibrium $p=1$: Inflows	Outflows	Flow equilibrium $p=2$: Inflows	Outflows
Dusseldorf	40000	0	60000	0	40000	40000	60000	60000
Toulouse	0	0	0	0	0	0	0	0
Turin	35000	0	55000	0	35000	35000	55000	55000

Production capacity	Excess $p=1$	Excess $p=2$
The Hague	5000	5000
Berlin	0	0

Objective function

Costs	Opening	DC inflows	Dem. flows	
The Hague	€ 21 000 000	€ 22 850 000	€ 14 987 500	
Berlin	€ 0	€ 0	€ 0	
Düsseldorf	€ 5 500 000		€ 3 110 000	
Toulouse	€ 0		€ 0	
Turin	€ 4 500 000		€ 4 815 000	
Total:	€ 31 000 000	€ 22 850 000	€ 22 912 500	€ 76 762 500

Fig. 7.14 Optimal solution of the two-echelon two-product example

7.3.3 *Location–Allocation Problem Variants*

Several variants of the location–allocation problems studied are found in practice. Some of them are discussed in the following sections.

7.3.3.1 Service Constraints

We may require that all DCs servicing a client (or demand zone) are located within a maximum distance (or delivery time) of this client to comply with a service level policy. Rather than including these constraints explicitly in the model, it is more efficient to eliminate non-feasible arcs a priori. When the service criterion is based on response times, for example, arc (s, l) is eliminated if $RT_{sl} > RT_l^{\max}$ with

RT_{sl} Average response time when a product is delivered to client $l \in L^D$ from site $s \in L^S$

$RT_l^{\max}$ Maximum acceptable response time for customer $l \in L^D$

As shown in Fig. 7.15a, the sites that could be used to service a given client are limited based on the service constraint. In doing so, however, one must ensure that each customer can be served by at least one DC. Otherwise, there will be no feasible solution.

7.3.3.2 Transshipment Sites

The distribution strategy described in the activity graph of Fig. 7.7 involves storing products in DCs and delivering them to customers from these DCs. In other distribution contexts, as discussed in Sect. 7.2, no intermediate inventory is kept, but, to reduce transport costs, shipments are consolidated in transshipment points (using cross-docking). In addition to in-transit times, customer response times then also depend on supplier order processing times and on cross-docking times in the transshipment point. To adapt the optimization model (7.1)–(7.4) to this context, the maximum annual throughputs, $b_w, w \in L^w$, on the right-hand side of constraint (7.3) must be redefined. Instead of being based on available storage space and inventory turnover as in Sect. 7.3.1, b_w depends on the reception, cross-docking, staging, and shipping capacity of the platform on site w. Data on the number and layout of reception-shipping docks and on platform handling equipment must then be used to estimate it. Obviously, the storage activity in Fig. 7.10 could also be replaced by a transshipment activity. Constraint (7.8) in the production–distribution network design model presented previously would then have to be adapted accordingly.

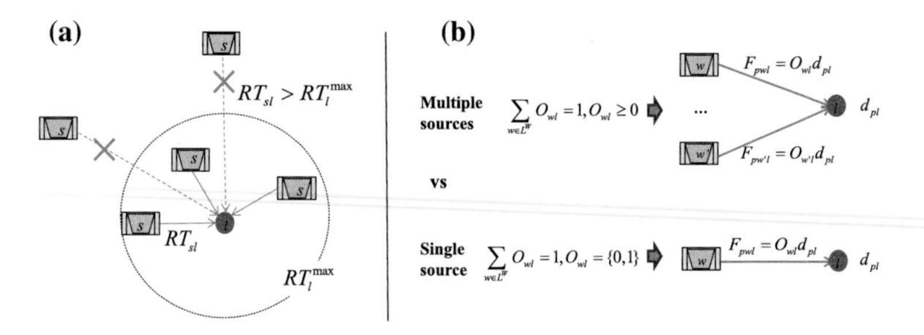

Fig. 7.15 Illustration of two location–allocation problem variants

7.3.3.3 Supply Sources

In distribution network design model (7.1)–(7.4), supplier selection decisions and procurement flows are implicit. In certain contexts, these decisions are crucial and they must be explicitly considered. This is typically the case for the supply network of service companies, such as Orange in France or Hydro-Québec (see Chap. 1). In this context, the set L^{V} of potential vendors and the sets $P_l, l \in L^{\mathrm{V}}$, of the product families they can supply must be considered. In addition to the parameters previously defined, the following data are required:

$\overline{b}_{pl}$ Upper bound on the annual quantity of product $p \in P_l$ vendor $l \in L^{\mathrm{V}}$ can supply

f_{plw} Unit flow cost of product $p \in P_l$ between vendor $l \in L^{\mathrm{V}}$ and DC $w \in L^{\mathrm{W}}$ (includes the purchase price paid to supplier l, the transportation cost on arc (l, w), and the reception cost at DC w)

y_l^{V} Fixed cost incurred when signing a contract with vendor $l \in L^{\mathrm{V}}$

The following additional decision variables are also needed:

Y_l^{V} Binary variable equal to 1 if vendor $l \in L^{\mathrm{V}}$ is selected, and 0 otherwise

F_{plw} Annual flow of product p between source $l \in L^{\mathrm{V}}$ and DC $w \in L^{\mathrm{W}}$.

The following optimization model can be solved to find the optimal structure of the SCN with supplier selection:

$$C^{\mathrm{SCN}} = \min \sum_{l \in L^{\mathrm{V}}} y_l^{\mathrm{V}} Y_l^{\mathrm{V}} + \sum_{l \in L^{\mathrm{W}}} y_l Y_l + \sum_{p \in P} \sum_{w \in L^{\mathrm{W}}} \sum_{l \in L_p^{\mathrm{D}}} f_{pwl} F_{pwl} + \sum_{p \in P} \sum_{l \in L^{\mathrm{V}}} \sum_{w \in L^{\mathrm{W}}} f_{plw} F_{plw} \tag{7.11}$$

subject to

- Demand constraints

$$\sum_{w \in L^{\mathrm{W}}} F_{pwl} = d_{pl}, \quad p \in P, l \in L_p^{\mathrm{D}} \tag{7.12}$$

- DC capacity constraints

$$\sum_{p \in P} \sum_{l \in L_p^{\mathrm{D}}} F_{pwl} \leq b_w Y_w, \quad w \in L^{\mathrm{W}} \tag{7.13}$$

- Vendor capacity constraints

$$\sum_{w \in L^{W}} F_{plw} \leq \overline{b}_{pl} Y_l^{\mathrm{V}}, \quad l \in L^{\mathrm{V}}, \quad p \in P_v \tag{7.14}$$

- DC flow equilibrium constraints

$$\sum_{l \in L^{\mathrm{V}}} F_{plw} = \sum_{l \in L_p^{\mathrm{D}}} F_{pwl}, \quad p \in P, \quad w \in L^{\mathrm{W}} \tag{7.15}$$

- Nonnegativity constraints

$$\begin{aligned} &Y_l \in \{0,1\}, l \in L^{\mathrm{W}}; \quad Y_l^{\mathrm{V}} \in \{0,1\}, l \in L^{\mathrm{V}}; \\ &F_{plw}, F_{pwl} \geq 0, p \in P, w \in L^{\mathrm{W}}, l \in L^{\mathrm{D}} \text{ or } L^{\mathrm{V}} \end{aligned} \tag{7.16}$$

7.3.3.4 Site-Selection Restrictions

If we want to force the inclusion of some sites in the design, it is sufficient to fix the value of the corresponding Y_s a priori. For example, if DC 1 is already opened and must not be closed, one sets $Y_1 = 1$ a priori. One may also want to limit the total number of DC, PDC, or sites ($s^{\max}$) in the network. The following upper bound is then added to the model:

$$\sum_{s \in L^{\mathrm{S}}} Y_s \leq \mathrm{s}^{\max}$$

For a subset of sites $L^{\mathrm{S}'} \subset L^{\mathrm{S}}$ having a specific characteristic, we make sure that *at most* one site in this subset is selected by adding the restriction $\sum_{s \in L^{\mathrm{S}'}} Y_s \leq 1$. However, by adding the constraint $\sum_{s \in L^{\mathrm{S}'}} Y_s \geq 1$ we require that *at least* one site in the subset must be selected.

In some cases, a product p may be incompatible with the platform of a subset of the potential sites considered because of a lack of space or adequate handling equipment, for example. In this case, we must simply define subsets of eligible sites by product ($L_p^{\mathrm{S}}, p \in P$), or of admissible products by site ($P_l, l \in L^{\mathrm{S}}$), and use these subsets instead of the complete set in the formulation.

7.3.3.5 Bounds on Platform Throughputs

Constraint (7.8) imposes upper bounds on product throughputs in the sites. If, in practice, there is no capacity constraints for a DC (because it is implemented in a large public warehouse, for example), we must set the corresponding b_w equal to the maximum demand that can go through the DC and $e_p = 1, p \in P$. Constraint (7.8) must be included in the model even if there's no capacity restriction because it is required to ensure that there will be no throughput through the sites that are not selected.

If we want a DC to be opened only if its annual throughput exceeds a predetermined lower bound $\underline{b}_w$, the following constraint must be added:

$$\sum_{p \in P} e_p \Big(\sum_{l \in L_p^{\mathrm{D}}} F_{pwl} \Big) \geq \underline{b}_w Y_w, \quad w \in L^{\mathrm{W}} \tag{7.17}$$

Similarly, when there are no direct deliveries, if we want a plant to be operated only if its annual activity level exceeds a predetermined lower bound $\underline{b}_u$, the following constraint must be added:

$$\sum_{p \in P} \sum_{w \in L^{\mathrm{W}}} F_{puw} \geq \underline{b}_u Y_u, \quad u \in L^{\mathrm{U}}. \tag{7.18}$$

7.3.3.6 Single Sourcing

In practice, for service motives, we may want shipments to a given demand zone to always be made by the same DC. To enforce this restriction, we need the following binary variables:

$$Z_{sl}^{\mathrm{S}} = \begin{cases} 1 & \text{if demand zone } l \text{ is assigned to site } s \\ 0 & \text{otherwise} \end{cases}, \quad s \in L^{\mathrm{S}}, l \in L^{\mathrm{D}}$$

The schemas in Fig. 7.15b illustrate how these binary variables can be used to impose single sourcing in a distribution network. When using these variables, flows between sites and clients are given by the expression $F_{psl} = d_{pl} Z_{sl}^{\mathrm{S}}$, $p \in P, s \in L^{\mathrm{S}}, l \in L_p^{\mathrm{D}}$. To enforce single sourcing, constraint (7.6) in the production–distribution network design model must be replaced by the following constraints:

$$\sum_{s \in L^{\mathrm{S}}} Z_{sl}^{\mathrm{S}} = 1, \quad l \in L^{\mathrm{D}} \tag{7.19}$$

$$F_{psl} - d_{pl} Z_{sl}^{\mathrm{S}} = 0, \quad p \in P, \quad s \in L^{\mathrm{S}}, \quad l \in L_p^{\mathrm{D}} \tag{7.20}$$

$$Z_{sl}^{\mathrm{S}} \in \{0, 1\}, \quad s \in L^{\mathrm{S}}, \quad l \in L^{\mathrm{D}} \tag{7.21}$$

When constraint (7.21) is relaxed, that is, when it is replace by $Z_{sl}^{\mathrm{S}} \geq 0$, $s \in L^{\mathrm{S}}, l \in L^{\mathrm{D}}$, the revised model is equivalent to the original one. The decision variable Z_{sl}^{S} can then be interpreted as the proportion of client l demand serviced from site s. However, when the Z_{sl}^{S} are binary variables, the model ensures that each client is serviced by a single site.[2]

7.4 Maximizing Value Added

In the previous section, we initiated our study of SCN design problems by formulating location–allocation models minimizing relevant investment and operation costs. As seen in Chap. 2, it is, however, not sufficient to minimize costs to create value; one also needs to maximize revenues. The objective of this section is thus to transform the previous basic models to maximize economic value added. Clearly, this assumes that revenues are affected by the order winners provided by the SCN, so this section examines different approaches to model this. For the moment, the impact of taxes is neglected (i.e., we assume that $\tau = 0$).

The easiest way to address this problem is to calculate the profits of the company, that is, the total sales income minus the total system cost during the base year considered. To maximize profits, objective function (7.5) in the previous SCN design model must be replaced by the function

$$EVA^{\mathrm{SCN}} = \max \sum_{p \in P} \sum_{l \in L_p^{\mathrm{D}}} R_{pl}(d_{pl}) - \left[\sum_{p \in P} \sum_{u \in L^{\mathrm{U}}} \sum_{w \in L^{\mathrm{W}}} f_{puw} F_{puw} + \sum_{p \in P} \sum_{s \in L^{\mathrm{S}}} \sum_{l \in L_p^{\mathrm{D}}} f_{psl} F_{psl} + \sum_{s \in L^{\mathrm{S}}} y_s Y_s \right] \tag{7.22}$$

where d_{pl} is the quantity of product p sold in demand zone l and $R_{pl}(d_{pl})$ is the total product p sales revenue for zone l. The complexity of the model thus obtained depends on the nature of the demand and revenue functions. If d_{pl} is predetermined and $R_{pl}(d_{pl})$ does not depend on SCN design variables, then the first term in (7.22) is a constant and it is sufficient to minimize costs (C^{SCN}), as we did in the previous section, to obtain the optimal design. Otherwise, d_{pl} becomes a decision variable, or a function of some of the decision variables in the design model, and it is bounded as follows:

$$\underline{d}_{pl} \leq d_{pl} \leq \bar{d}_{pl}, \quad p \in P, \quad l \in L_p^{\mathrm{D}} \tag{7.23}$$

[2]In this Internet era, sales and delivery processes are usually decoupled, the former often taking place at the head office and the latter in a DC. The requirement for single sourcing is therefore less frequent than it used to be.

where $\underline{d}_{pl}$ and $\bar{d}_{pl}$ are, respectively, a lower bound given by market penetration targets and an upper bound corresponding to the maximum market share the company can obtain under its current marketing strategy and given the position held by competitors. In the following sections, we examine three modeling approaches reflecting different facets of the problem.

7.4.1 Demand as a Decision Variable

A basic approach is simply to consider the demand d_{pl} as a decision variable. The complexity of the problem then depends on the nature of the revenue function $R_{pl}(d_{pl})$. If revenues are computed with the relation $R_{pl}(d_{pl}) = \pi_{pl}(d_{pl})d_{pl}$, and the unit price $\pi_{pl}(d_{pl})$ is a linear function of the demand d_{pl}, the objective function is quadratic. The model is then more complex than before but it can be solved with most commercial solvers. If, however, the sales price π_{pl} of product p in demand zone l is independent of the demand d_{pl}, the resulting model is linear and it is not more difficult to solve than the cost-minimization model.

Example 7.3 This approach can be illustrated using a variation of Example 7.1. The Excel tables in Fig. 7.16 provide the data required, including the sale prices and the demand bounds. The optimal solution of this value maximization model indicates that it is best to sell as much as possible in the most lucrative demand zones (Germany, France, Italy, and Spain) and to sell as much as capacity allows in the remaining zone (Portugal). The solution obtained is not the same as before: it recommends to open the three warehouses and to use them at their full capacity.

	A	B	C	D	E	F	G	H
1	**Data**							
2	Flow costs	Demand zones					Sites	
3	Site	Germany	France	Italy	Spain	Portugal	Capacity	Fixed costs
4	Düsseldorf	€ 125	€ 120	€ 150	€ 160	€ 180	50000	€ 2 700 000
5	Toulouse	€ 160	€ 130	€ 145	€ 130	€ 150	50000	€ 3 500 000
6	Turin	€ 140	€ 130	€ 130	€ 170	€ 170	50000	€ 2 300 000
7	Min demand	10000	12500	7500	12500	20000		
8	Max demand	30000	32500	27500	32500	40000		
9	Sales price	€ 240	€ 230	€ 220	€ 200	€ 190		
10	**Decision variables**							
11	Flows	Demand zones					Openings	Excess capacity
12	Site	Germany	France	Italy	Spain	Portugal		
13	Düsseldorf	30000	20000	0	0	0	1	0
14	Toulouse	0	0	0	32500	17500	1	- 0
15	Turin	0	12500	27500	0	10000	1	-
16	Demand	30000	32500	27500	32500	27500		
17	**Objective function**							
18		Costs		Revenues				
19		Openings	Flows					
20	Düsseldorf	€ 2 700 000	€ 6 150 000	€ 11 800 000				
21	Toulouse	€ 3 500 000	€ 6 850 000	€ 9 825 000				
22	Turin	€ 2 300 000	€ 6 900 000	€ 10 825 000				
23	Total	€ 8 500 000	€ 19 900 000	€ 32 450 000	€ 4 050 000	Profit		

Solver Parameters
Set Objective: $E$23
To: (•) Max ( ) Min ( ) Value Of: 0
By Changing Variable Cells: $G$13:$G$15;$B$13:$F$15
Subject to the Constraints:
$H$13:$H$15 <= 0
$B$16:$F$16 >= $B$7:$F$7
$B$16:$F$16 <= $B$8:$F$8
$G$13:$G$15 = binary
[x] Make Unconstrained Variables Non-Negative
Select a Solving Method: Simplex LP

Fig. 7.16 Data and solution of a value-added maximization model

7.4.2 Sensitivity of Demand to Market Offers

Assume now that demand d_{pl} depends on two order-winners: response time and sales price. More specifically, assume that a relationship

$$d_{pl} = \phi_{pl}(\pi_{pl}, RT_l), \quad p \in P, \quad l \in L_p^{\mathrm{D}} \tag{7.24}$$

can be established between the dependent variable d_{pl} and the explanatory variables:

π_{pl} Sales price for product p in demand zone l

RT_l Response time (or maximum distance between ship-to points and DCs) provided to demand zone l customers

The easiest way to address this problem is to perform a sensitivity analysis, that is, to examine the results obtained for a range of possible values of variables $\pi_{pl}, p \in P, l \in L_p^{\mathrm{D}}$ and $RT_l, l \in L^{\mathrm{D}}$. If the problem includes a large number of sites or products, the number of cases to explore can be extremely high. To circumvent this difficulty, it is generally assumed that the company's offer is the same for all demand zones $l \in L^{\mathrm{D}}$, and the price variations are usually expressed in terms of a percentage increase or decrease of the product's base price. For given base prices, π_p^o, $p \in P$, and response times RT^o, the demand is estimated using (7.24), revenues are calculated with

$$R^{\mathrm{SCN}} = \sum_{p \in P} \sum_{l \in L_p^{\mathrm{D}}} R_{pl}(d_{pl}) = \sum_{p \in P} \pi_p^o \Big(\sum_{l \in L_p^{\mathrm{D}}} d_{pl}\Big) \tag{7.25}$$

and, after eliminating all arcs not respecting response time RT^o, the cost-minimization model (7.5)–(7.10) is solved. For a given price, SCN revenue (R^{SCN}) and cost (C^{SCN}) curves can then be plotted to show how they depend on response time, as illustrated in Fig. 2.15 (Chap. 2). By analyzing these curves and the associated value added $EVA^{\mathrm{SCN}} = R^{\mathrm{SCN}} - C^{\mathrm{SCN}}$ for some price vectors, one can select the most suitable SCN design among the solutions obtained.

7.4.3 Optimizing a Company's Market Offer

The previous sensitivity analysis quickly becomes laborious and does not take into account the fact that company offers are typically modulated by product-market. We can generalize this approach by modeling the order-winners offered by a company using binary variables. A product-market $k \in K$ is defined as a geographical region covering a set of demand zones $L_k^{\mathrm{D}} \subset L^{\mathrm{D}}$ in which a set of products $P_k \subset P$ is marketed in the same manner. Consider, for example, two clients located close to each other. One of them buys MTS products that he wants delivered quickly, say, standard printing paper. The other buys specialty printing paper manufactured on order. Although the two clients are neighbors, they do not belong to the same product-market because they do not value the same order-winners. We assume in what follows that demand zones $l \in L^{\mathrm{D}}$ are defined so that they belong to a unique product-market denoted $k(l)$.

In order to get orders in these product-markets, the company must develop different offers to satisfy customers better than its competitors. Suppose, as discussed, that offers are specified in terms of response time and products prices. An offer takes the form of a market policy quantifying these order-winners. For each product market k, a set J_k of potential policies can be considered. The product-market for which a policy j is considered is denoted $k(j)$. A policy $j \in J_k$ for product-market $k \in K$ is characterized by

- A price π_{jp} for each product $p \in P_k$
- A maximum response time RT_j
- A fixed implementation cost y_j^{M} (promotion, documentation, consignment inventory, etc.)
- An upper bound $\bar{d}_{jpl}$ on the demand for products $p \in P_k$ in zones $l \in L_k^{\mathrm{D}}$ corresponding to the market share the company could obtain with this policy

The response time RT_j limits the DCs that can be used to service customers in demand zone $l \in L_{k(j)}^{\mathrm{D}}$. The example in Fig. 7.17 illustrates the shipping points that, given their location in the network, can be used to send product p to demand zone l for three potential policies. The subset of sites that can be used to ship product p to demand zone $l \in L_{k(j)}^{\mathrm{D}}$ under policy j is denoted $S_{jpl} \subseteq L^{\mathrm{S}}$.

To select the best offer, the following decision variables are required:

Y_j^{M} Binary variable equal to 1 if policy j is selected for product-market $k(j)$, and 0 otherwise

F_{jpsl} Annual flow of product p shipped to demand zone l from site $s \in S_{jpl}$ when policy $j \in J_{k(l)}$ is selected (this decision variable replaces the previous flow variable F_{psl})

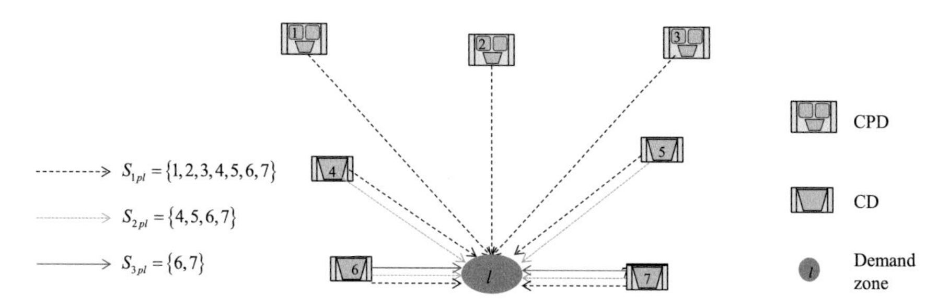

Fig. 7.17 Set of feasible shipping sites for different market policies

When this approach is used to choose the network structure and market offers that maximize value added, the minimization model (7.5)–(7.10) studied in the previous section is replaced by the following mixed-integer program (MIP):

$$\begin{aligned} EVA^{\mathrm{SCN}} = \max \sum_{p \in P} \sum_{l \in L_p^{\mathrm{D}}} \sum_{j \in J_{k(l)}} \sum_{s \in S_{jpl}} (\pi_{jp} - f_{psl}) F_{jpsl} \\ - \sum_{p \in P} \sum_{u \in L^{\mathrm{U}}} \sum_{w \in L^{\mathrm{W}}} f_{puw} F_{puw} - \sum_{s \in L^{\mathrm{S}}} y_s Y_s - \sum_{j \in J} y_j^{\mathrm{M}} Y_j^{\mathrm{M}} \end{aligned} \tag{7.26}$$

subject to

- Demand constraints

$$Y_j^{\mathrm{M}} \underline{d}_{pl} \leq \sum_{s \in S_{jpl}} F_{jpsl} \leq Y_j^{\mathrm{M}} \bar{d}_{jpl}, \quad p \in P, \quad l \in L_p^{\mathrm{D}}, \quad j \in J_{k(l)} \tag{7.27}$$

$$\sum_{j \in J_k} Y_j^{\mathrm{M}} \leq 1, \quad k \in K \tag{7.28}$$

- Production capacity constraints

$$\sum_{l \in L_p^{\mathrm{D}}} \sum_{j \in J_{k(l)}} F_{jpul} + \sum_{w \in L^{\mathrm{W}}} F_{puw} \leq b_{pu} Y_u, \quad u \in L^{\mathrm{U}}, \quad p \in P_u \tag{7.29}$$

- Distribution capacity constraints

$$\sum_{p \in P} e_p \Big(\sum_{l \in L_p^{\mathrm{D}}} \sum_{j \in J_{k(l)}} F_{jpsl} \Big) \leq b_s Y_s, \quad s \in L^{\mathrm{S}} \tag{7.30}$$

- Intermediate sites flow equilibrium constraints

$$\sum_{u \in L^{\mathrm{U}}} F_{puw} = \sum_{l \in L_p^{\mathrm{D}}} \sum_{j \in J_{k(l)}} F_{jpwl}, \quad p \in P, \quad w \in L^{\mathrm{W}} \tag{7.31}$$

- Nonnegativity constraints

$$\begin{aligned} & Y_s \in \{0,1\}, s \in L^{\mathrm{S}}; \quad Y_j^{\mathrm{M}} \in \{0,1\}, j \in J; \\ & F_{puw}, F_{jpsl} \geq 0, p \in P, s \in L^{\mathrm{S}}, l \in L^{\mathrm{D}}, j \in J_{k(l)} \end{aligned} \tag{7.32}$$

In this model, the original demand constraints (7.6) are replaced by (7.27) and (7.28). This last constraint states that at most one policy can be selected for a product-market k. One could decide not to service a product-market, however, which occurs when the sum of Y_j^{M} is zero. Constraint (7.27) states that product

p shipments to demand zone l must comply with the bounds on demand imposed by the chosen policy. The other constraints in the model were simply adapted to account for the fact that the flow variables (F_{jpsl}) now depend on the policy j selected. Objective function (7.26) was also modified to account for sales revenue and market-policy fixed costs. Of course, as for classical location–allocation problems, all sorts of variants of this value-added maximization problem may be encountered and the formulation must be adapted accordingly.

7.5 General SCN Optimization Model

In the previous optimization models, activities and sources of supply were taken into account implicitly to simplify the formulation. However, when we want to design a SCN embedding complex manufacturing processes, such as illustrated in Figs. 7.18 and 7.19, one must explicitly consider suppliers as well as activities and associated recipes (BOMs).

Consider the case of a company with the following characteristics:

- The set P of product families under consideration includes raw materials, intermediate components, and finished products. The role of each product in the manufacturing process is specified by an activity graph, as shown in Fig. 7.18. This activity graph could take any form imposed by the SCN strategy of the

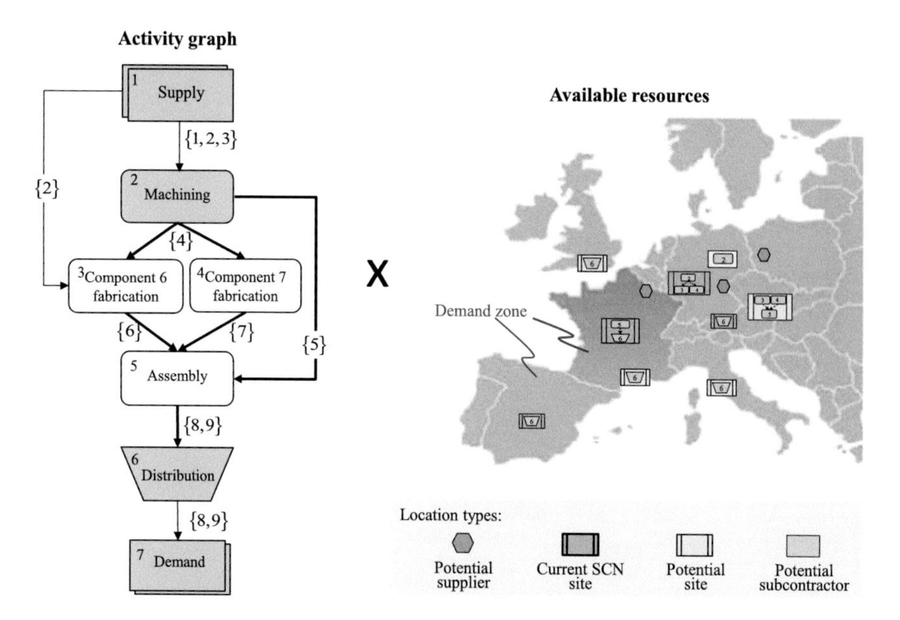

Fig. 7.18 Activity graph and potential resources for a manufacturing company

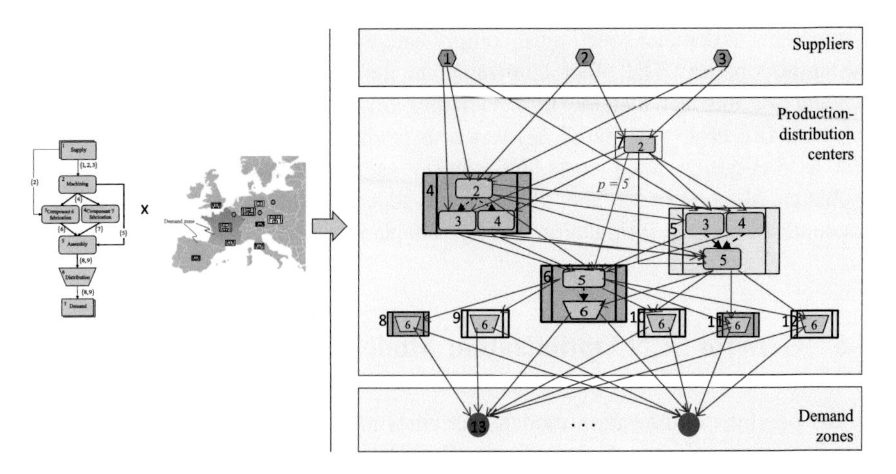

Fig. 7.19 Potential supply chain network

company. Recipes are associated to manufacturing activities to specify the quantity of input products required to manufacture each output product. As shown in Fig. 7.20, this is equivalent to defining a complete BOM.

- Some of the activities can be subcontracted.
- Vendors must be selected among a set of potential supply sources $l \in L^{V}$. The price of the products $P_l \subset P$ sold by source l is specified in a supply contract as well as a minimum purchase amount for the year. The maximum annual quantity of products that could be provided by each supplier is also known.

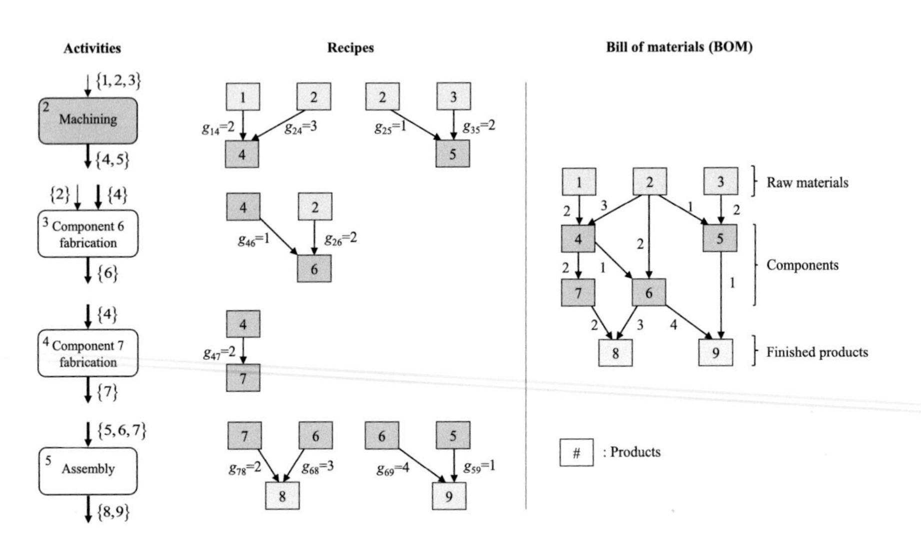

Fig. 7.20 Recipes associated with production activities (bill of materials)

- We assume each demand zone $l \in L^{D}$ is geographically located in a single product-market $k \in K$ (state, postal codes cluster, etc.), that it orders a subset of products $P_{k(l)} \subset P$, and that products sales price depends on the product-markets.
- We want to reengineer an existing SCN, that is, some platforms are already implemented in given locations, but additional sites are also considered. The platforms to implement on potential or existing production–distribution sites are predetermined, that is, the activities they can perform and the available capacity are fixed. Existing sites can be closed.
- Operating costs are linear, and we want to find a SCN design that maximizes economic value added for the base year considered.

Supply sources, production–distribution sites and demand zones define a set L of physical locations, and, as before, to distinguish them, we partition L into three subsets L^{V}, L^{S}, and L^{D}. The capability of a site $l \in L^{S}$ is specified by the set of activities $A_l \subset A^{S}$ its platform can perform. Recall that the set A^{S} can contain production (A^{F}), warehousing-storage (A^{E}), and consolidation-transshipment (A^{G}) activities. Some sites may also be partner facilities (contract manufacturers, 3PLs, etc.) engaged in production or distribution activities. In the resource and network representations in Figs. 7.18 and 7.19, the site icons indicate what activities can be done in the platforms.

The type of the potential network obtained for the design problem considered is illustrated in Fig. 7.19. As with all networks, it contains nodes and arcs. Note, however, that in this case, the arcs do not go from one site to another, as previously, but rather from an activity performed in a site to an activity carried out in the same site or another site. The nodes of the network are (location, activity) pairs and the arcs link an origin (l, a) to a destination (l, a') in site l or to a destination (l', a') in another site. An arc represents the flow of a product p between an origin and a destination. If we examine arc (7,2) $\rightarrow$ (6,5) in Fig. 7.19, for example, because it goes from activity $a = 2$ to activity $a' = 5$, it is clear from the activity graph that the associated product is $p = 5$. In some cases, however, more than one product is moved between two activities. For example, movement 1 $\rightarrow$ 2 involves products $\{1, 2, 3\}$. Therefore, there could be as many as three parallel arcs, one for each product, between suppliers and the nodes incorporating activity 2 in the network. These parallel arcs and product numbers are not all shown in Fig. 7.19 to avoid cluttering the diagram.

A generic representation of the resulting potential network is provided in Fig. 7.21. To simplify the notation, a unique label can be associated to each network node. In what follows, a label n is associated to node (l, a), and pointers $l(n)$ and $a(n)$ are used to indicate, respectively, the location and the activity of node n. We use one or the other of these identifiers interchangeably, depending on what is most convenient. An arc in the network is thus denoted by the triplet (p, n, n'). The set of all network nodes is denoted N. As we did for locations, to distinguish node types we define the following subsets:

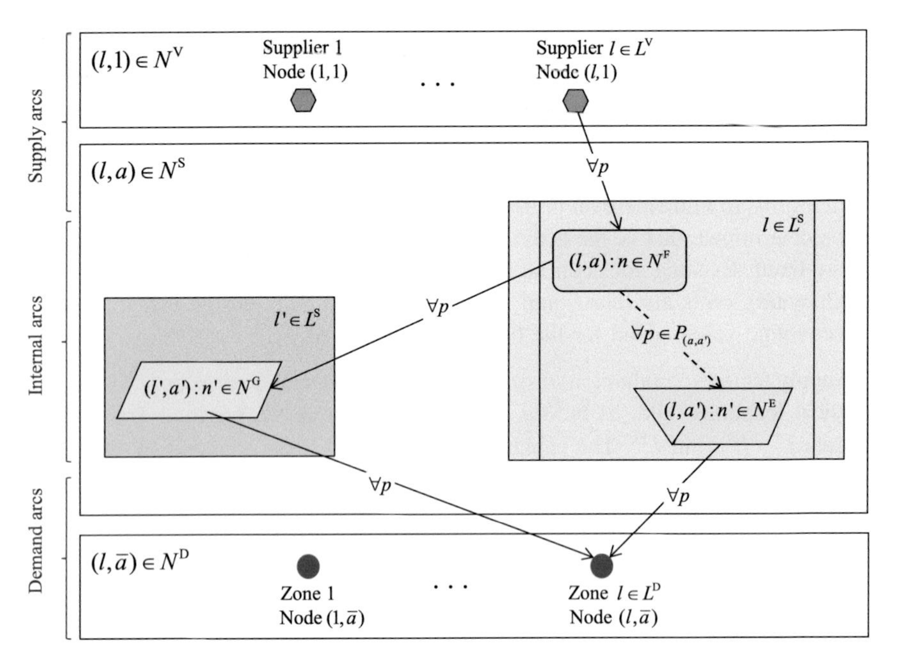

Fig. 7.21 Generic representation of a potential network

N^V Set of supply nodes $((l,1), l \in L^V)$
N^S Set of production–distribution nodes $((l,a), l \in L^S, a \in A^S)$
N^D Set of demand nodes $((l,\bar{a}), l \in L^D)$

The pointer $k(n)$ designates the product-market including demand node $n \in N^D$. The nodes in N^S are partitioned into the following subsets:
N^F Set of fabrication-assembly (production) nodes $((l,a), l \in L^S, a \in A^F)$
N^E Set of warehousing-storage nodes $((l,a), l \in L^S, a \in A^E)$
N^G Set of consolidation-transshipment nodes $((l,a), l \in L^S, a \in A^G)$

Finally, to navigate more easily in the network, for each node n and product p, we define sets of predecessor and successor nodes:
$N^{\rightarrow}_{pn}$ Destinations of outbound arcs for node n and product p
$N^{\leftarrow}_{pn}$ Sources of inbound arcs for node n and product p

Similarly, for each activity a, the following input (output) product subsets are defined:
$P^{\rightarrow}_{a}$ Set of output products for activity $a \in A \backslash \{\bar{a}\}$
$P^{\leftarrow}_{a}$ Set of input products for activity $a \in A \backslash \{1\}$

The specifications of the platform on a site determine fixed and operating costs as well as the production or distribution capacity of its activities. To model these costs and capacities the following notation is needed:

$g_{pp'}$ Quantity of product p required to make one unit of product p'

q_{pa} Capacity consumed in activity $a \in A^{\mathrm{S}}$ per unit of product $p \in P_{\vec{a}}$

$\varphi_{pn'n}$ Inventory turnover ratio for product p cycle and safety stocks stored in node $n \in N^{\mathrm{E}}$ when supplied from node n'

η_{pa} Order cycle and safety stock (maximum level)/(average level) ratio for product p stored in activity $a \in A^{\mathrm{E}}$

$\overline{b}_n$ Capacity of node n (production capacity when $n \in N^{\mathrm{F}}$ and maximum throughput when $n \in N^{\mathrm{E}} \cup N^{\mathrm{G}}$)

$\underline{b}_n$ Minimum activity level in node n required to open site $l(n)$

b_n^{S} Storage space available in node $n \in N^{\mathrm{E}}$ (in pallets, ft^2, m^3...)

c_{pn}^{X} Unit activity cost (production for $n \in N^{\mathrm{F}}$ and throughput for $n \in N^{\mathrm{E}} \cup N^{\mathrm{G}}$) for product p in node n

c_{pn}^{I} Average annual inventory holding cost per unit of product p kept in node $n \in N^{\mathrm{E}}$ ($c_{pn}^{\mathrm{I}} = r_{l(n)}^{\mathrm{I}} v_{pn}$, v_{pn} being the value of product p in node n)

y_l Fixed operating cost (lease or annuity, plus fixed exploitation costs) of site $l \in L^{\mathrm{S}}$ platform during the base year

y_l^+ Opening cost for site $l \in L^{\mathrm{S}}$ (initial provisioning of safety stocks, hiring of personnel, installation of management systems, etc.)

y_l^- Closing cost for site $l \in L^{\mathrm{S}}$ (cash inflows (outflows) for the repositioning or disposal of material, equipment, and personnel)

Y_l^0 Binary parameter giving the state of site $l \in L^{\mathrm{S}}$ when the design decisions are made ($Y_l^0 = 1$ if the site is already opened and $Y_l^0 = 0$ if it is a potential site)

It should be noted that the costs incurred for internal and external sites (contract manufacturers, public warehouses, etc.) are modeled in the same way. However, for external sites, variable costs are generally much higher and fixed costs much lower.

To model potential sourcing contracts, the following notation is required:

$\overline{b}_{pl}$ Maximum annual quantity of product $p \in P_l$ which can be supplied by source $l \in L^{\mathrm{V}}$

$\underline{b}_l$ Lower bound on the annual purchase amount from vendor $l \in L^{\mathrm{V}}$ when a contract is signed

c_{pl}^{V} Unit price of product $p \in P_l$ purchased from vendor $l \in L^{\mathrm{V}}$

y_l^{V} Fixed cost incurred when signing a contract with vendor $l \in L^{\mathrm{V}}$

To capture the revenues and variable costs associated with the network arcs, the following notation is used:

π_{kp} Sales price of product p in product-market k

$f_{pnn'}$ Unit flow cost for arc (p, n, n'), including relevant order, reception, shipping, transportation and in-transit inventory holding costs for inter-sites arcs, and handling costs for intra-site arcs

Finally, to optimize the structure of the supply chain network, the following decision variables are needed:

Y_l 1 if site $l \in L^S$ is operated during the base year considered, and 0 otherwise

Y_l^+ 1 if site $l \in L^S$ is opened at the beginning of the year, and 0 otherwise

Y_l^- 1 if site $l \in L^S$ is closed at the beginning of the year, and 0 otherwise

Y_l^V 1 if a contract is signed with vendor $l \in L^V$, and 0 otherwise

X_{pn} Activity level of node n for output product $p \in P_a^{\rightarrow}$ (production quantity when $n \in N^F$ and throughput when $n \in N^E \cup N^G$)

$F_{pnn'}$ Flow of product p between node n and node n' (transportation if $l(n) \neq l(n')$ or handling if $l(n) = l(n')$)

Several constraints must be formulated to obtain a feasible design. First, supply contracts impose the following restrictions:

$$\sum_{n \in N^{\rightarrow}_{p(l,1)}} F_{p(l,1)n} \leq Y_l^V \overline{b}_{pl}, \quad l \in L^V, \quad p \in P_l \tag{7.33}$$

$$Y_l^V \underline{b}_l \leq \sum_{p \in P_l} \sum_{n \in N^{\rightarrow}_{p(l,1)}} c_{pl}^V F_{p(l,1)n}, \quad l \in L^V \tag{7.34}$$

Constraint (7.33) states that vendor capacity must not be exceeded. Constraint (7.34) ensures that the total amount purchased from each vendor during the year is sufficient to respect contract lower bounds.

Several constraints also apply to production–distribution sites. First, we must take the state of sites into account correctly, that is, make sure that a platform can be closed only if it is already in place and opened only if it is on a potential site:

$$Y_l + Y_l^- - Y_l^+ = Y_l^0, \ l \in L^S \tag{7.35}$$

We must then make sure that flow equilibrium and production recipes are respected. This is illustrated in Fig. 7.22 for assembly node (6,5). The recipes of outbound products (8 and 9) are illustrated in the node. For each product, the quantity assembled must equal the quantity shipped to storage nodes and, therefore,

$$X_{8(6,5)} = F_{8(6,5)(8,6)} + F_{8(6,5)(9,6)} + F_{8(6,5)(6,6)} + F_{8(6,5)(10,6)} + F_{8(6,5)(11,6)} + F_{8(6,5)(12,6)}$$
$$X_{9(6,5)} = F_{9(6,5)(8,6)} + F_{9(6,5)(9,6)} + F_{9(6,5)(6,6)} + F_{9(6,5)(10,6)} + F_{9(6,5)(11,6)} + F_{9(6,5)(12,6)}$$

However, to manufacture finished products 8 and 9, components 5, 6, and 7 are required. These components come from production activities 2, 3, and 4 in

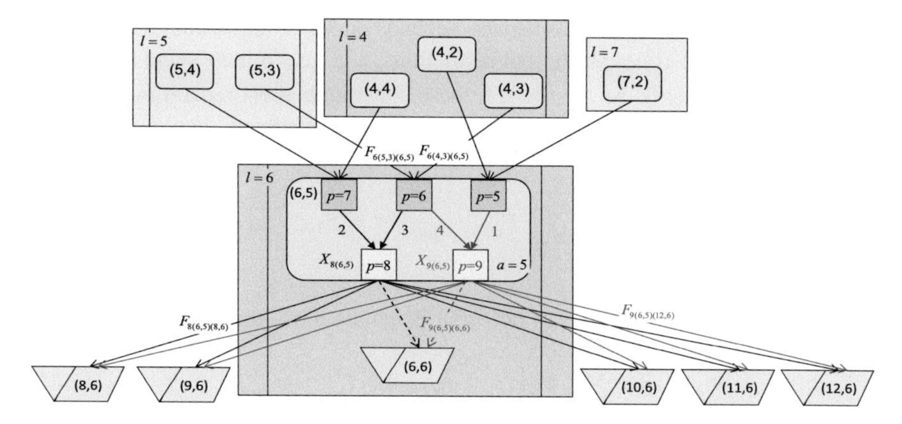

Fig. 7.22 Subnetwork associated with assembly node (6,5)

platforms 4, 5, and 7. Components received from these nodes must be sufficient to produce the quantity of finished products required, taking associated recipes into account, and therefore it is necessary that

$$
\begin{aligned}
&2X_{8(6,5)} \leq F_{7(5,4)(6,5)} + F_{7(4,4)(6,5)} \\
&3X_{8(6,5)} + 4X_{9(6,5)} \leq F_{6(5,3)(6,5)} + F_{6(4,3)(6,5)} \\
&X_{9(6,5)} \leq F_{5(4,2)(6,5)} + F_{5(7,2)(6,5)}
\end{aligned}
$$

The following general constraints must thus be imposed:

$$X_{pn} = \sum_{n' \in N_{pn}^{\rightarrow}} F_{pnn'}, \quad n \in N^{\mathrm{S}}, \quad p \in P_{a(n)}^{\rightarrow} \tag{7.36}$$

$$\sum_{p' \in P_{a(n)}^{\rightarrow}} g_{pp'} X_{p'n} \leq \sum_{n' \in N_{pn}^{\leftarrow}} F_{pn'n}, n \in N^{\mathrm{F}}, p \in P_{a(n)}^{\leftarrow} \tag{7.37}$$

For distribution nodes, constraint (7.37) is replaced by the following inflow equilibrium relation:

$$X_{pn} = \sum_{n' \in N_{pn}^{\leftarrow}} F_{pn'n}, \quad n \in N^{\mathrm{G}} \cup N^{\mathrm{E}}, \quad p \in P_{a(n)}^{\leftarrow} \tag{7.38}$$

Clearly, platform capacity and minimum flow constraints must also be respected:

$$\underline{b}_n Y_{l(n)} \leq \sum_{p \in P_{a(n)}^{\rightarrow}} q_{pa(n)} X_{pn} \leq \bar{b}_n Y_{l(n)}, \quad n \in N^{\mathrm{S}} \tag{7.39}$$

The capacity of storage nodes usually depends on space rather than the flow. For these nodes, if there is no throughput restrictions, a large arbitrary value is assigned to capacity $\overline{b}_n$ in (7.39). As mentioned, these constraints must remain in the model to preserve its validity. The following storage space constraints, however, must be added:

$$\sum_{p \in P^{\rightarrow}_{a(n)}} e_p \eta_{pa(n)} \Big(\sum_{n' \in N^{\leftarrow}_{pn}} F_{pn'n} / \varphi_{pn'n} \Big) \leq b^{\mathrm{S}}_n Y_{l(n)}, \quad n \in N^{\mathrm{E}} \tag{7.40}$$

Market penetration targets and potential demands must also be respected:

$$\underline{d}_{pl} \leq \sum_{n \in N^{\leftarrow}_{p(l,\bar{a})}} F_{pn(l,\bar{a})} \leq \overline{d}_{pl}, \quad l \in L^{\mathrm{D}}, \quad p \in P_l \tag{7.41}$$

It remains only to formulate an objective function to obtain the SCN design model. As noted, the company wants to maximize economic value added for the base year considered. When taxes are neglected, the function to maximize is the following:

$$\begin{aligned}
&\max \sum_{p \in P} \sum_{n \in N^{\mathrm{D}}} \sum_{n' \in N^{\leftarrow}_{pn}} \pi_{k(n)p} F_{pn'n} && \text{Sales revenues} \\
&- \sum_{l \in L^{\mathrm{S}}} (y^{+}_l Y^{+}_l + y_l Y_l + y^{-}_l Y^{-}_l) && \text{Fixed platform costs} \\
&- \sum_{p \in P} \sum_{n \in N^{\mathrm{V}}} \sum_{n' \in N^{\rightarrow}_{pn}} (c^{\mathrm{V}}_{pl(n)} + f_{pnn'}) F_{pnn'} && \text{Raw material costs} \\
&- \sum_{n \in N^{\mathrm{S}}} \sum_{p \in P^{\rightarrow}_{a(n)}} c^{\mathrm{X}}_{pn} X_{pn} && \text{Activity costs} \\
&- \sum_{p \in P} \sum_{n \in N^{\mathrm{E}}} c^{\mathrm{I}}_{pn} \Big(\sum_{n' \in N^{\leftarrow}_{pn}} F_{pn'n} \Big/ \varphi_{pn'n} \Big) && \text{Inventory holding costs} \\
&- \sum_{p \in P} \sum_{n \in N^{\mathrm{S}}} \sum_{n' \in N^{\rightarrow}_{pn}} f_{pnn'} F_{pnn'} && \text{Outflow costs} \\
&- \sum_{l \in L^{\mathrm{V}}} y^{\mathrm{V}}_l Y^{\mathrm{V}}_l && \text{Supply contracts fixed costs}
\end{aligned} \tag{7.42}$$

In sum, we want to maximize (7.42) subject to constraints (7.33)–(7.41) and to usual decision variable nonnegativity and binary value restrictions. For real SCN reengineering projects, this MIP can include thousands of variables and constraints. In the majority of cases, it can however be solved effectively with commercial solvers such as CPLEX, Xpress, or Gurobi.

7.6 Modeling Economies of Scale

It was assumed previously that all relevant variable costs are linear. Chapter 2 stressed that this is not always the case in practice and that several supply chain activities involve economies of scale. More specifically, we saw that inventory holding costs, facility-throughput costs, and raw material costs are often concave because marginal costs decrease when activity levels increase. With respect to the purchase of raw materials, economies of scale usually come from quantity discounts that give rise to piecewise linear functions. The approach proposed in the previous section to model supply contracts allows this type of discounts. Economies of scale in facility-throughput costs typically result from the use of more efficient technologies when volumes increase. We will examine this issue in detail in the next chapter on capacity planning. Finally, as illustrated in Fig. 2.9, the average level of inventory needed to provide adequate service decreases marginally when annual flows increase, resulting in economies of scale for inventory holding costs. It is the modeling of these costs that we address in this section.

In the models formulated in Sect. 7.3, the shipments $X_{ps} = \sum_{l \in L_p^D} F_{psl}$ of product p to demand zones and the inventory turnover ratio φ are implicitly used to compute average inventory levels for site s, $\bar{I}_{ps} = X_{ps}/\varphi$, and unit storage costs $c_{ps}^S = r_s^I v_{ps}/\varphi$, v_{ps} being the value of a product p held in inventory on site s. The inventory holding costs $C_{ps}^I(X_{ps}) = c_{ps}^S X_{ps}$ were consequently linear. As seen in Chap. 2, the average inventory of product p in site s is represented better using the concave inventory-throughput function $\bar{I}_{ps} = a_p X_{ps}^{b_p}$, a_p and b_p being regression parameters. The inventory holding cost is then given by the power function $C_{ps}^I(X_{ps}) = (r_s^I v_{ps} a_p) X_{ps}^{b_p}$. When economies of scale are modeled like this, the unit storage costs c_{ps}^S must be removed from the unit flow costs f_{psl} used in model (7.5)–(7.10), and the inventory holding costs $\sum_{s \in L^S} \sum_{p \in P_s} C_{ps}^I(X_{ps})$ must be added to objective function (7.5).

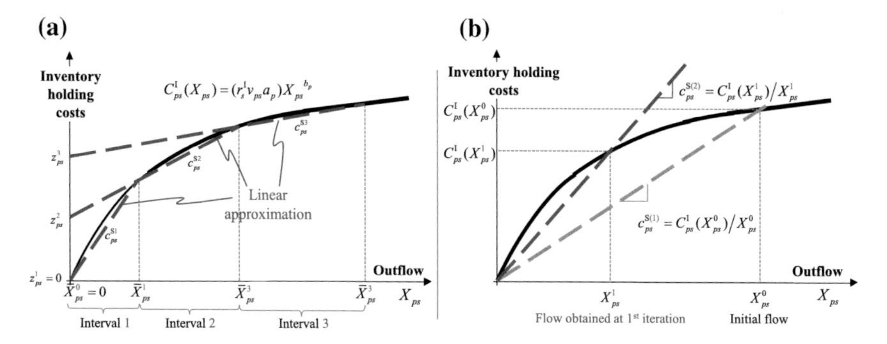

Fig. 7.23 Concave costs linearization methods

The easiest way to take these economies of scale into account is to approximate $C^{\mathrm{I}}_{ps}(X_{ps})$ by a piecewise linear function, as shown in Fig. 7.23a. The X_{ps} axis is divided into intervals and the points on the curve at the beginning and at the end of these intervals are joined by straight lines. To embed resulting polygonal functions in the model, the following data and variables are required:

m_{ps} Number of intervals specified for variable X_{ps}

$\bar{X}^i_{ps}$ Upper bound of interval i for variable X_{ps} ($\bar{X}^0_{ps} = 0$ denotes the lower bound of the first interval)

$c^{\mathrm{S}i}_{ps}$ Slope of the ith segment of the piecewise linear approximation of $C^{\mathrm{I}}_{ps}(X_{ps})$

z^i_{ps} Origin value of the ith linear segment for $C^{\mathrm{I}}_{ps}(X_{ps})$

Z^i_{ps} Binary variable equal to 1 if $\bar{X}^{i-1}_{ps} < X_{ps} \leq \bar{X}^i_{ps}$, and 0 otherwise

X^i_{ps} Continuous variable giving the shipments of product p to demand zones from storage point s when in interval i.

In the model, the variables X_{ps} are not used explicitly; rather they are replaced by copies X^i_{ps} for each interval $i = 1, \ldots, m_{ps}$. Because the optimal throughput value is necessarily in a single interval, only one of the $X^i_{ps}, i = 1, \ldots, m_{ps}$, can be non-negative and the others take a zero value. Consequently, the sites throughput are given by $X_{ps} = \sum_{i=1}^{m_{ps}} X^i_{ps}$. To ensure that the linear function associated with each interval is correctly selected and to specify the relationship between throughput (X_{ps}) and flow (F_{psl}) variables, the following constraints must be added to the original model:

$$\sum_{i=1}^{m_{ps}} X^i_{ps} = \sum_{l \in L^{\mathrm{D}}_p} F_{psl}, \quad s \in L^{\mathrm{S}}, \ p \in P_s$$

$$\bar{X}^{i-1}_{ps} Z^i_{ps} < X^i_{ps} \leq \bar{X}^i_{ps} Z^i_{ps}, \quad s \in L^{\mathrm{S}}, \ p \in P_s, \ i = 1, \ldots, m_{ps}$$

$$\sum_{i=1}^{m_{ps}} Z^i_{ps} = 1, s \in L^{\mathrm{S}}, p \in P_s; Z^i_{ps} \in \{0, 1\}, s \in L^{\mathrm{S}}, p \in P_s, i = 1, \ldots, m_{ps}$$

Finally, the concave functions $C^{\mathrm{I}}_{ps}(X_{ps})$ added to the objective function must be replaced by the following piecewise linear approximation: $\widehat{C}^{\mathrm{I}}_{ps}(X_{ps}) = z^i_{ps} Z^i_{ps} + c^{\mathrm{S}i}_{ps} X^i_{ps}$, $i = 1, \ldots, m_{ps}$. The resulting MIP can often be solved with commercial solvers; however, it incorporates a very large number of binary variables and it can be difficult to solve.

Another method, not requiring the addition of binary variables, can be used to approximate economies of scale. This method, as illustrated in Fig. 7.23b, relies on the calculation of unit costs c^{S}_{ps} for a given throughput X_{ps} of product p in storage point s. The difficulty with this method is that we do not know the optimal throughputs when unit costs are calculated. We must therefore solve model

(7.5)–(7.10) repeatedly using, at each iteration i, the unit costs $c_{ps}^{S(i)}$ estimated with the throughputs $X_{ps}^{(i-1)} = \Sigma_{l\in L_p^D} F_{psl}^{(i-1)}$ which were optimal for iteration (i-1), that is, using $c_{ps}^{S(i)} = C_{ps}^{I}(X_{ps}^{(i-1)}) / X_{ps}^{(i-1)}$. The only change required to model (7.5)–(7.10) at each iteration is to replace the unit flow costs f_{psl} by their revised value

$$f_{pul}^{(i)} = c_{pu}^{P} + c_{pu}^{S(i)} + c_{pul}, u \in L^{U}, l \in L^{D}; \quad f_{pwl}^{(i)} = c_{pu}^{S(i)} + c_{pwl}, w \in L^{W}, l \in L^{D}$$

The costs thus calculated usually stabilize after a few iterations. The details necessary to implement this approach are found in Martel (2005).

We examined the impact of concave inventory-throughput functions $\bar{I}_{ps}(X_{ps})$ on inventory holding costs and we saw how to model related economies of scale. It should be clear, however, that these functions also have an impact on the required storage space and therefore on constraints (7.8). When $\bar{I}_{ps}(X_{ps})$ is concave, these constraints are nonlinear. Fortunately, the two methods studied to linearize costs can also be used to linearize the storage space constraints. Note finally that these methods can also be used to introduce economies of scale in the value-added maximization model studied in Sect. 7.4, as well as in the general model proposed in Sect. 7.5.

Review Questions

1. Why is the optimization of a supply chain network's structure a complex problem? What are the factors to be considered when addressing this problem?
2. How can a company's supply chain strategy be formalized using an activity graph?
3. What is the difference between a location problem and an allocation problem?
4. What is the use of 0–1 variables in supply chain network modeling?
5. What is a flow equilibrium constraint? Are these constraints always necessary in SCN design models?
6. What kinds of additional difficulties are encountered when modeling the SCN of a company with complex manufacturing processes?
7. How can we take account of service constraints without complicating the formulation of SCN design models?
8. What can we do to ensure that each client will be serviced by a single source when formulating a SCN design model?
9. What can we do to account for inventory holding cost economies of scale while preserving the linearity of a SCN optimization model?
10. What is the difference between throughput capacity and storage capacity?
11. What can we do to optimize the market offer of a company?

Table 7.1 Exercise 1 data

	Transportation costs ($ per ton)		Production cost	Capacity	Fixed cost
	Market 1	Market 2	($ per ton)	(ton)	($1000)
Quebec plant	600	400	$3800	1000	$4000
Rimouski platform	1000	200	$3500	1000	$5900
Longueuil platform	300	700	$3600	1000	$6200
Potential market (ton)	800	700			

Exercises

Exercise 7.1

A Canadian company was able until now to meet the demand of its customers from its factory located in the Quebec region. Strong sales growth, however, is forecasted for next year in the two markets served by the company (1: Eastern Quebec, 2: Western Quebec). The company, therefore, is considering the construction of another plant for which two potential sites have been identified: one in Rimouski and the other in Longueuil. The collected data are provided in Table 7.1.

Products sold can be grouped in a single family and the average sale price of the products of this family is $50,000 per ton on both markets.

(a) Draw an activity graph that corresponds to this context.
(b) Propose an optimization model to solve this problem.
(c) Implement your model using the Excel Solver; determine whether to build a new plant and, if so, on which site?

Exercise 7.2 Expro has developed five (5) new markets involving annual demands for 150, 175, 100, 175, and 250 tons of goods respectively. Currently, the company has no DC covering these territories. It therefore has the intention of negotiating an agreement with a 3PL for the operation of two or three DCs, each having a maximum capacity of 500 tons, to serve these markets. Three potential sites have been identified: Charlotte, Philadelphia, and Columbus. Transportation costs per ton between each of these sites and the various markets are provided in Table 7.2. The DC set-up cost, the average value of products in stock, and the storage rate charged by the 3PL for each of the sites are given in Table 7.3.

Knowing that the inventory turnover ratio of the company is 4, and that it uses an inventory holding cost rate of 10 %, propose a mixed-integer programming model to solve this problem. Which sites should be selected and what is the cost of this decision?

Table 7.2 Transportation costs for exercise 2

Markets					
Site	1	2	3	4	5
Charlotte	$3	$7	$12	$5	$8
Philadelphia	$12	$5	$5	$6	$4
Columbia	$7	$8	$2	$3	$2

Table 7.3 Set-up and inventory holding costs for exercise 2

Site	Set-up costs	Product value	Storage cost rate
Charlotte	$9500	$110	20 %
Philadelphia	$10,000	$130	18 %
Columbia	$9600	$115	19 %

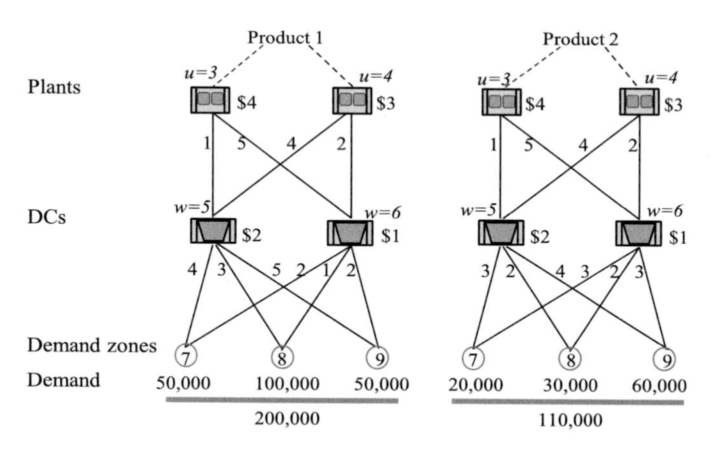

Fig. 7.24 Exercise 7.3 SCN

Table 7.4 Exercise 7.3 fixed costs and capacities

CD	Fixed costs	Distribution	Production capacity	
		Capacity	$p = 1$	$p = 2$
$w = 5$	$100,000	150,000		
$w = 6$	$200,000	–		
$u = 3$	$400,000		210,000	120,000
$u = 4$	$500,000		230,000	150,000

Exercise 7.3[3]

The potential supply chain network shown in Fig. 7.24 involves the production and distribution of two product families. Two potential plants can produce the two products. The products could be distributed to three demand zones through two DCs. The first DC corresponds to a potential contract with a 3PL that limits the

[3]This exercise is based on an example given in Ballou (1992).

annual throughput to 150,000 products. The second DC corresponds to an alternative contract with another 3PL. The fixed cost of this alternative is higher but variable costs are lower, and for all practical purposes, there is no capacity limit. Unit transportation costs are shown on the arcs of the network and unit production (storage) costs on its nodes. The fixed costs of operating facilities and capacity levels are found in Table 7.4. Determine the plants and DCs to use as well as the mission of selected facilities.

Exercise 7.4

Having solved the model formulated in Exercise 7.3, you realize that it ignores the DC inventory holding costs. Based on historical data on flows and inventories for the two product families considered, you are able to establish relationships between the average level of inventory of each product $\bar{I}_p$ in a DC and the annual product throughput X_{pw} in this DC, namely,

$$\bar{I}_1 = 100X_{1w}^{0,7} \quad \text{and} \quad \bar{I}_2 = 10X_{2w}^{0,9}, w = 5, 6$$

Knowing that products 1 and 2 in inventory are valued at \$10 and \$5 per unit, respectively, and that the inventory holding cost rate used by the company is 10 %:

(a) Calculate the total inventory holding cost for the solution found in Exercise 3.
(b) Given the magnitude of these costs, you decide to change your design model to account for inventory holding costs. The model thus obtained being non-linear, you choose to use successive linear programming (SLP) to solve it approximately. To initiate the process, suppose that the total demand for each product is distributed in equal proportion by each DC. What is the unit inventory holding cost thus obtained for each product in each DC? How are these holding costs changing the unit flow costs of your original model?
(c) Find the optimal solution of the MIP to solve for the first iteration of the SLP method.
d) From the solution obtained, calculate the unit inventory holding cost to use for each product and each DC in the second iteration of the SLP method.

Exercise 7.5

Suppose that a company must solve a SCN design problem in the context described in Fig. 7.10, but that in addition it has the choice among several transportation means to ship products from PDCs to DCs. Given that all products are sold in pallets, it is possible to deliver several product families in each shipment, and for each transportation mean considered, the size of shipments and delivery times are known. Let

M_{uw} = Set of transportation options for arc (u, w)

z_{muw}^{T} = Fixed costs incurred when using transportation mean $m \in M_{uw}$ for arc (u, w)

f_{mpuw} = Unit cost incurred to ship product p on arc (u, w) with transportation mean $m \in M_{uw}$

Table 7.5 Exercise 7.7 demand and costs per market

Markets						
Site	Germany	France	Italy	Spain	Portugal	Fixed costs
Düsseldorf	\$121.25	\$125.25	\$130.25	\$123.25	\$126.25	\$9500
Toulouse	\$151.10	\$144.10	\$144.10	\$145.10	\$143.10	\$10,000
Turin	\$130.34	\$131.34	\$125.34	\$126.34	\$125.34	\$9600
Min. demand	150	175	100	175	250	

Table 7.6 Exercise 7.7 data per policy

Maximum demand						Price	Max RT (days)	Implementation cost
Policy	Germany	France	Italy	Spain	Portugal			
ND	250	175	100	175	250	200	3	500
FD	350	275	200	275	350	225	2	1000
PD	450	375	300	375	450	250	1	2000

Table 7.7 Exercise 7.7 response time in days per site for all markets

Site	Germany	France	Italy	Spain	Portugal
Düsseldorf	1	2	2	3	3
Toulouse	3	1	3	2	2
Turin	2	2	1	3	3

Transportation costs were estimated by regression from the shipping history or rates offered by carriers (see Chap. 5). Based on the formulation presented in Sect. 7.3.2, propose a SCN design model including the choice of transportation means.

Exercise 7.6

Suppose that the manufacturing company studied in Sect. 7.5 has a seasonal demand, that is, that when the planning horizon considered (say a year) is decomposed into a set $T = \{t\}$ of seasons, it appears that the demand $\bar{d}_{plt}, t \in T$, for a zone l is not the same in every season. To cope with these fluctuations in demand, the firm must modulate its production and keep seasonal inventories. Let

c^{I}_{pnt} Inventory holding cost of product p in node $n \in N^{\mathrm{E}}$ for season t

X_{pnt} Activity level of node n for product p in season t

$F_{pnn't}$ Flow of product p between nodes n and n' in season t

I_{pnt} Seasonal inventory of product p in node $n \in N^{\mathrm{E}}$ at the end of season t

Use this additional notation to propose a model for the optimization of the structure of the manufacturing company's network that takes into account the seasonality of demand. Note that seasonal inventories must be included in the model to smooth production during the planning horizon. Seasonal inventories at the beginning and at the end of the year can be assumed to be equal. In other words, one must require that $I_{pn0} = I_{pn|T|}$, for all p and n.

Exercise 7.7

The American company *Franexpo*, is considering the possibility to expand its activities to five European countries: Germany, France, Italy, Spain and Portugal. The minimum market penetration target for these counties expressed in annual demands is 150, 175, 100, 175, and 250 tons of goods respectively. To cover these territories, the company is investigating the opportunity to operate DCs with a maximum capacity of 1500 tons in Dusseldorf, Toulouse, and Turin. Transportation costs per ton between each of these sites and the various markets, as well as fixed DC costs per site, are provided in Table 7.5. In addition, the company wants to determine the best offer for these markets among the three following potential policies: Premium Delivery (PD), Fast Delivery (FD), and Normal Delivery (ND). For each policy, the maximum response time (RT), the price specified, and the implementation costs incurred are given in Table 7.6. Finally, in order to find the set of feasible shipping sites for each policy, response times in days between sites and the various markets are provided in Table 7.7. Given that, you are asked to optimize Franexpo's market offer based on the modeling approach presented in Sect. 7.4.3.

Bibliography

Aikens C (1985) Facility location models for distribution planning. Eur J Oper Res 22:263–279

Ambrosino D, Scutellà M (2005) Distribution network design: new problems and related models. Eur J Oper Res 165:610–624

Amiri A (2006) Designing a distribution network in a supply chain system: Formulation and efficient solution procedure. Eur J Oper Res 171:567–576

Ballou R (1992) Business logistics management, 3rd edn. Prentice-Hall

Bidhandi H, Yusuff R, Ahmad M, Bakar M (2009) Development of a new approach for deterministic supply chain network design. Eur J Oper Res 198(1):121–128

Brandeau M, Chiu S (1989) An overview of representative problems in location research. Manage Sci 35(6):645–674

Camm J et al (1997) Blending OR/MS judgement and GIS: restructuring P&G's supply chain. Interfaces 27(1):128–142

Carle M-A, Martel A, Zufferey N (2012) The CAT metaheuristic for the solution of multi-period activity-based supply chain network design problems. Int J Prod Econ 139:664–677

Chopra S, Meindl P (2010) Supply chain management, 4th edn. Prentice-Hall

Cohen M, Moon S (1990) Impact of production scale economies, manufacturing complexity, and transportation costs on supply chain facility networks. J Manuf Oper Manage 3:269–292

Cooke J (2007) Weaving 2 supply chains together. CSCMP's Supply Chain Q 3:34–38

Denton B, Forrest J, Milne R (2006) IBM solves a mixed-integer program to optimize its semiconductor supply chain. Interfaces 36(5):386–399

Ferrio J, Wassick J (2008) Chemical supply chain network optimization. Comput Chem Eng 32:2481–2504

Fleischmann B (1993) Designing distribution systems with transport economies of scale. Eur J Oper Res 70:31–42

Francis R, McGinnis L, White J (1992) Facility layout and location, 2nd edn. Prentice-Hall

Geoffrion A, Graves G (1974) Multicommodity distribution system design by Benders decomposition. Manage Sci 20(5):822–844

Geoffrion A, Powers R (1995) 20 years of strategic distribution system design: an evolutionary perspective. Interfaces 25(5):105–127

Harmon M (2013) Step-by-step optimization with Excel solver. Amazon Digital Services

Hax A, Candea D (1984) Production and inventory management. Prentice-Hall

Holmberg K (1994) Solving the staircase cost facility location problem with decomposition and piecewise linearization. Eur J Oper Res 75:41–61

Klose A, Drexl A (2005) Facility location models for distribution system design. Eur J Oper Res 162:4–29

Kooksalan M, Sural H (1999) Efes Beverage Group makes location and distribution decisions for its malt plants. Interfaces 29(2):89–103

Lakhal S, Martel A, Kettani O, Oral M (2001) On the optimization of supply chain networking decisions. Eur J Oper Res 129:59–270

Lambin J (1991) Le marketing stratégique. McGraw-Hill

Love R et al (1988) Facilities location: models and methods. North-Holland

Martel A (2005) The design of production-distribution networks: A mathematical programming approach. In: Geunes J, Pardalos P (eds) Supply chain optimization. Springer, Berlin, pp 265–306

Melachrinoudis E, Min H (2007) Redesigning a warehouse network. Eur J Oper Res 176:210–229

Melo M, Nickel S, Saldanha-da-Gama F (2009) Facility location and supply chain management—A review. Eur J Oper Res 196:401–412

Perl J, Sirisoponsilp S (1988) Distribution networks: facility location, transportation and inventory. Int J Phys Distrib Mater Manage 18(6):18–26

Pirkul H, Jayaraman V (1996) Production, transportation, and distribution planning in a multi-commodity tri-echelon system. Transp Sci 30(4):291–302

Robinson P Jr, Swink M (1994) Reason based solutions and the complexity of distribution network design problems. Eur J Oper Res 76:394–409

Shapiro J (2008) Modeling the supply chain, 2nd edn. Brooks/Cole Publishing Co

Shapiro J (2004) Challenges of strategic supply chain planning and modeling. Comput Chem Eng 28:855–861

Shapiro J, Wagner S (2009) Strategic inventory optimization. J Bus Logistics 30(2):161–173

Shen Z (2007) Integrated supply chain design models: a survey and future research direction. J Ind Manage Optim 3(1):1–27

Vila D, Martel A, Beauregard R (2006) Designing logistics networks in divergent process industries: a methodology and its application to the lumber industry. Int J Prod Econ 102: 358–378

Chapter 8
Strategic Capacity Planning

When should a business add production or distribution capacity to its SCN and how much? This is the main issue addressed in this chapter, which also examines the selection of appropriate technology to provide capacity. The chapter starts by clarifying the nature of strategic capacity planning decisions. It then studies the measurement of the capacity provided by a technological system. It finally shows how to formulate mathematical programming models to help solve capacity planning problems. Capacity plans typically consider the production–distribution needs of a company over several years. But the future is necessarily uncertain, which complicates the elaboration of robust plans significantly. For simplicity, in this chapter, it is assumed that the company is able to forecast its future needs, which leads to deterministic optimization models. We will see how to take uncertainty into account explicitly in Chap. 11. The chapter examines three modeling contexts of increasing complexity. It focuses first on timing decisions regarding the addition of a single type of capacity for a given technological system. Then, assuming that demand and capacity requirements are stationary, it examines the selection of the facility platforms to implement at SCN sites to cover needs during a typical year. Finally, it shows how to model dynamic capacity decisions for a generic SCN over a multiyear planning horizon.

8.1 Problem Context

Given the expected growth and variability of capacity requirements, the investments and expenditures incurred to build or acquire and operate different types of facilities, the deterioration of existing facilities with age, the progress of relevant technologies, and the anticipated behavior of competitors and partners, a *capacity strategy* is a long-term plan that specifies *how much* and *when* various *types* of capacity should be added or reduced to improve the competitive position of a company, that is, to nurture value creation.

A. Martel and W. Klibi, *Designing Value-Creating Supply Chain Networks*,
DOI 10.1007/978-3-319-28146-9_8

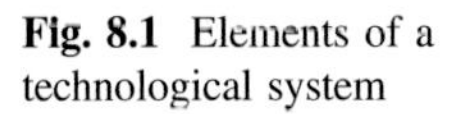
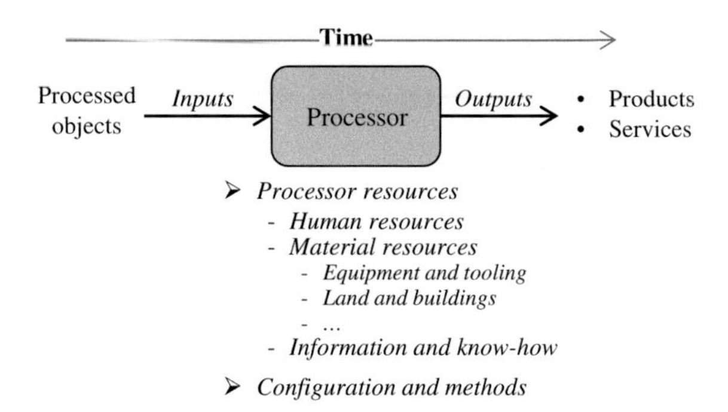

Fig. 8.1 Elements of a technological system

This definition does not clarify what is meant exactly by capacity. Roughly speaking, it is a limit on the activity that can be performed with a technological system, whether it is equipment, a workstation, an activity center, a factory, or a distribution center. In Chap. 4, technological systems are described as hierarchical resources–configuration–methods triplets. Two types of resources are distinguished: processors, which are themselves technological subsystems, and processed objects, that is, physical and symbolic inputs transformed into products and services (see Fig. 4.4). Chapter 4 also explains that the productivity of a technological system is the ratio of outputs produced over resources used, and that it depends on interrelations among its resources, configuration, and methods. The capacity of a technological system can thus be defined as the quantity of products or services that it can produce with its resources per *unit of time* under its current configuration and methods. The main elements of this definition are summarized in Fig. 8.1.

The capacity of a technological system is expressed using a standard measure of the quantity of produced outputs (number of products, kilograms, liters, etc.) during a standard time unit (hour, day, month, or year). The number of cars that can be manufactured per year and the number of liters of milk that can be produced per day are examples of capacity measures. In practice, it may be very difficult to find a standard output measure for the whole of the activity of a firm. Capacity is then often expressed in terms of the availability of a critical processor resource. For example:

- The number of paper machines in a paper mill
- The number of channels in a satellite communication system
- The number of vehicles in a transportation fleet
- The number of processing hours available for a machine shop.

Note that most companies need to manage several types of capacity, depending on the technological system they encompass: assembly capacity for a family of similar products, machining capacity for a type of component, transportation capacity, storage capacity, and so on. They must therefore plan capacity for each of the distinct technological systems they operate.

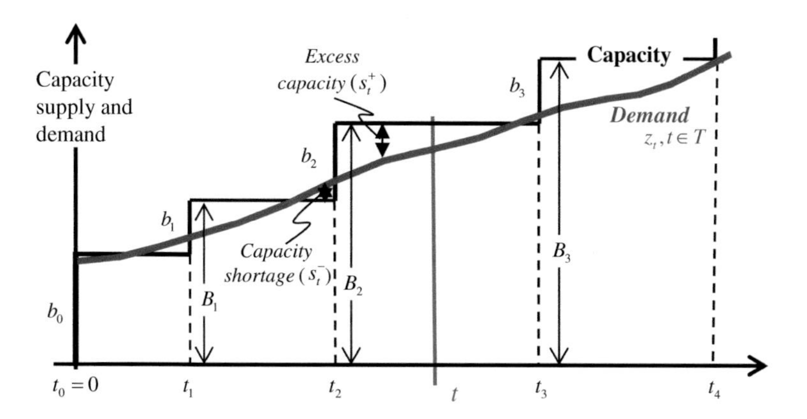

Fig. 8.2 Capacity expansion decisions to meet a growing demand

8.1.1 Capacity Planning

Having chosen an adequate measurement unit, for a given technological system, the capacity planning problem can be described more formally. For a given continuous or discrete planning horizon,[1] $T = \{t\}$, a capacity type, capacity demand forecasts $z_t, t \in T$, and the capacity b_0 available at the beginning of the planning horizon ($t_0 = 0$), a *strategic capacity plan* specifies at which points in time $t_1 \ldots t_h \ldots$ capacity amounts $b_h, h = 1, 2\ldots$, should be added (or removed) so that the capacity available at any point in time, B_t, is close to the capacity required, that is, so that

$$B_t \cong z_t, \quad t \in T; \quad B_t = \sum_{h|t_h \le t} b_h \tag{8.1}$$

In general, as Fig. 8.2 illustrates, it is not possible to ensure that available capacity (B_t) is equal to demand (z_t) at any time $t \in T$, which would anyhow not necessarily be interesting for several reasons:

- Even if demand z_t changes continuously, capacity can be added (or removed) only in *discrete* steps b_h of heights depending on the nature of the resources providing capacity (people, machines, trucks, multi-resource workstations, etc.)
- Because it is very difficult to forecast future demand accurately, it is often good to keep safety capacity to deal with the unexpected
- Short-term recourses, such as overtime, worker reallocations, subcontracting, product substitution, or safety stocks, can provide the flexibility required to alleviate temporary capacity shortages.

[1] A discrete planning horizon is of a set of subsequent planning periods of a predetermined length (e.g., a month or a year). The number of periods can be finite or infinite.

It is therefore desirable to replace relationship (8.1) by

$$B_t + s_t^- - s_t^+ = z_t, \quad t \in T; \quad B_t = \sum_{h|t_h \leq t} b_h \tag{8.2}$$

where

s_t^- Additional capacity obtained at time t using short-term recourses

s_t^+ Excess capacity available at the time t

In practice, when planning capacity expansions or contractions, managers often predict upcoming capacity needs, $z_t, t \in T$, and then assume that these are known with certainty, that is, they are content to consider a single plausible future scenario. In fact, the demand for the products of a company during a long-term planning horizon cannot be forecasted with certainty, and the capacity needed to produce and ship them is a stochastic process. Therefore, investments in capacity involve certain cash outlays to receive uncertain future benefits, which is an important source of complexity.

8.1.2 Generic Capacity Strategies

Relationship (8.2) implies that several types of capacity strategies can be considered. Three archetypal policies are illustrated in Fig. 8.3 under the assumption that demand grows linearly at a rate γ. Under policy A, capacity *precedes* demand. With this type of strategy, there is always a capacity surplus. One then has the flexibility necessary to grasp new opportunities as they arise. However, this policy may be very expensive because it only uses a portion of the capacity available. In some cases (for example, if no recourse is possible when there is a capacity shortage), however, a company may be forced to adopt this kind of policy to meet all incoming demand.

Policy C is exactly the reverse of policy A: capacity *follows* demand. There will therefore never be any nonproductive capacity. However, the company does not have any flexibility and, at times, it may have difficulty finding recourses, such as outsourcing, overtime, or resource reallocation, to ease capacity shortages. Response times can be lengthened because of congestion, which can affect market shares. Policy B is a compromise: capacity *overlaps* demand. There are a multitude of possible intermediate positions and our problem is to find the best one. To do this, one needs to examine the relationship between different capacity adjustment options and the costs they generate.

These generic capacity strategies assume an expansion context. However, in times of recession or when a company has decided to withdraw from a sector of activity, it may be necessary to make downward capacity adjustments. The generic strategies presented in Fig. 8.3 are easily adapted to a resource reduction context. Because capacity decisions have long-term effects, it is vital to base them on

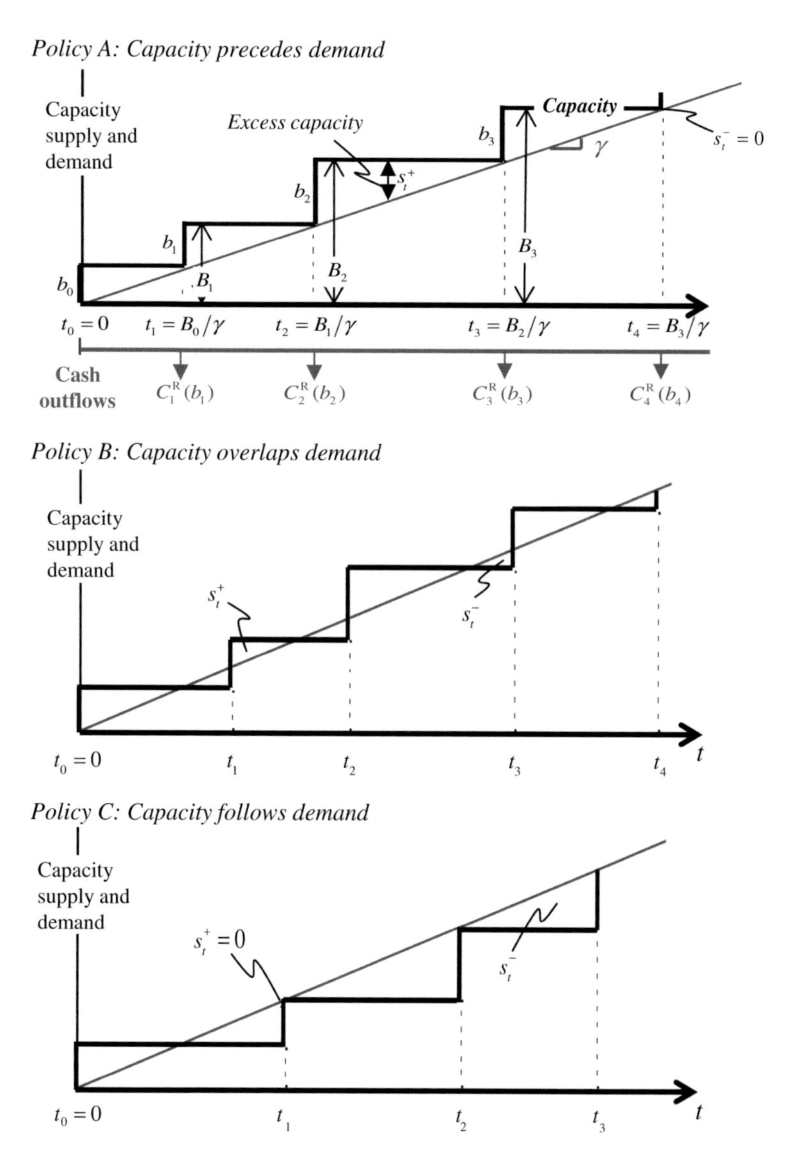

Fig. 8.3 Generic capacity strategies

accurate capacity requirement forecasts. A company that invests in capacity to deal with forecasted continued growth may be in serious trouble if an economic downturn occurs. This happened recently in the ocean shipping industry. During the growth period that preceded the recent economic decline, a very large number of ships were ordered. Because demand has slowed down considerably, there is now a substantial overcapacity, and the shipbuilding industry is currently confronted with an unprecedented slowdown.

8.1.3 Capacity Costs

The relationship between a quantity, b, of new capacity and its cost, $C^{R}(b)$, is not always straightforward. The cost of an addition in capacity mainly depends on the price of acquired resources, on fixed installation and usage costs, as well as on funding modalities. However, it may be difficult to establish a relation between the units used to measure capacity and the resources procured to provide it, especially when capacity is expressed in outputs per unit time. The relation must reflect the discrete nature of resources (persons, machines, etc.) as well as the methods and configuration of the technological system considered. If capacity is measured directly in terms of a critical resource (number of trucks, number of machines, number of hours worked, etc.), the relation is easier to characterize, but the difficulty noted previously occurs when forecasting capacity requirements. We come back to this issue in the next section.

In what follows, $C_h^{R}(b)$ denotes the net present value (NPV) of b units of additional capacity at the date, t_h, of its implementation. This amount includes investment and installation costs at time t_h as well as fixed operating costs (insurance, maintenance, security, etc.) until the end of the planning horizon. It also takes into account the tax benefits obtained and the market value of the resources installed at the end of their useful lives (see Sect. 2.2.2). In the simplest case, $C_h^{R}(b)$ comprises a fixed cost c^{+} and a variable cost c^{R} per unit of acquired capacity, that is,

$$C_h^{R}(b) = c^{+} + c^{R}b, \quad b > 0 \text{ with } C_h^{R}(0) = 0 \tag{8.3}$$

For example, the addition of b trucks to an existing fleet could cost \$300,000 per truck (c^{R}) for the horizon considered, plus an amount estimated at \$50,000 (c^{+}) in administrative expenses related to choosing the truck model and the supplier, to road tests, and so on, which is independent of the number of trucks purchased.

In most practical situations, economies of scale apply when large quantities of capacity are acquired. These economies of scale can often be modeled by the power function:

$$C_h^{R}(b) = Ab^{\beta} \tag{8.4}$$

where A is a base cost and $\beta \in [0, 1]$ is a coefficient that reflects the magnitude of economies of scale (usually between 0.6 and 1 in practice). Processors with a cylindrical or cubical shape (refineries, boilers, mixers, reactors, vessels, dryers, warehouses, etc.) usually have a capacity proportional to their volume (i.e., to the cube of their side if it has a cubic form) and acquisition and operating costs proportional to their area (i.e., to the square of their side if it has a cubic form). For example, suppose a company wants to build a cubic warehouse as shown in Fig. 8.4. One sees immediately that, when the dimension of the building sides increases, the available storage space (side^3) grows much faster than the walls' area

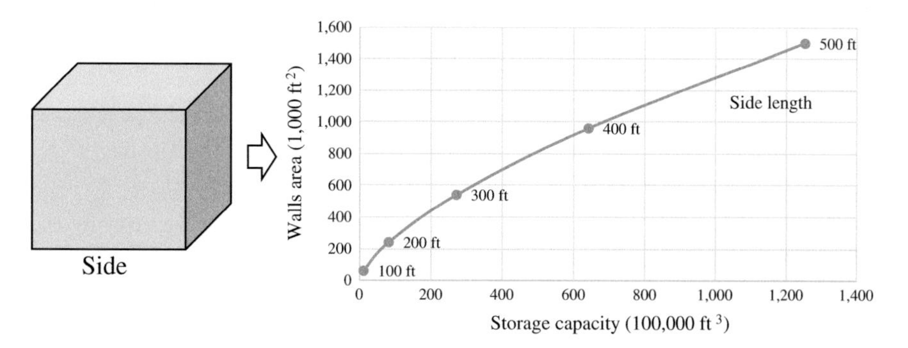

Fig. 8.4 Capacity of a cubic warehouse

(6 * side²). For this reason, these processors typically have a cost function with coefficient $\beta \cong 2/3$.

In some cases, as the required capacity increases, the technology used is changed for a more performing one, layouts are reconfigured, and methods are improved (see Chap. 4), which gives rise to polygonal cost functions close to power function (8.4). Take the function illustrated in Fig. 8.5. It could be a storage capacity cost function obtained as follows:

- When throughputs (b) are low, it is better to use the services of a public warehouse (Technology 1) with a low fixed cost but a substantial variable cost.
- When volumes increase, it becomes more economical to build a distribution center using conventional storage and handling technologies (Technology 2).
- Finally, if throughputs are very high, it is worth building a high-volume DC with automated storage and retrieval systems (Technology 3). Fixed costs are then high, but the cost per additional unit of flow is very low.

This example also shows the intimate link between the issue of capacity planning and the choice of technologies. In many cases, the problem boils down to a choice between a finite set O of technological options (types of technological systems). Each option $o \in O$ can be relatively complex and the only way to determine the capacity it provides and the costs it generates is to engage in a

Fig. 8.5 Economies made through more efficient high-volume technologies

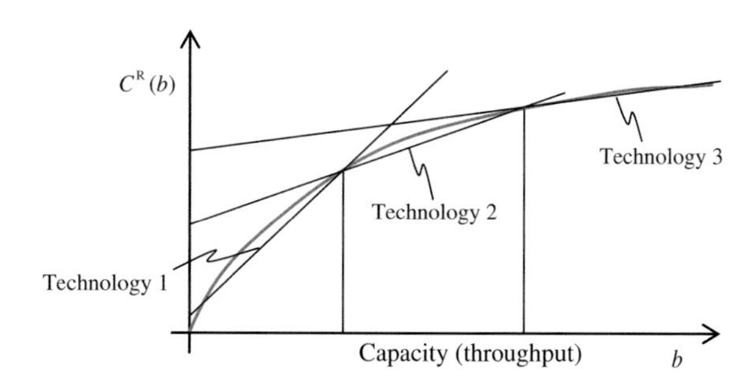

Table 8.1 Discrete cost function

Option (o)	1	2	3
b_o	3000	5000	10,000
C_o^R	\$5M	\$7M	\$12M

detailed engineering study. In this context, the capacity cost function is so complicated that the only possible approach is to use a table specifying the capacity and the cost of the options considered. Table 8.1 provides an example of a discrete cost function obtained from the detailed study of three technological options.

8.1.4 Capacity Strategy Optimization

Given the cost and the additional capacity provided by a set of technological options, as well as capacity requirements for the planning horizon considered, we want to determine which options to implement and when in order to maximize the value added to the company. The concepts of NVP and cost of capital required to calculate economic value added were studied in Chap. 2. Assume for now that capacity requirements z_t are known with certainty for the duration of the planning horizon $t \in T$, and that they were forecasted so as to comply with a predetermined value proposition made to customers by the company. Under these restrictions, the capacity strategy of the company has no influence on prices or demand. When this is the case, as explained in Chap. 7, it is not necessary to consider sales revenues explicitly to find an optimal solution, and the strategic capacity planning problem can be stated as follows:

$$\begin{array}{llcl} \min & \sum\limits_{h} \mathrm{NPV}\{C_h^{\mathrm{R}}(b_h)\} & + & \sum\limits_{t} \mathrm{NPV}\{Q_t(s_t^-)\} \\ & \text{Capacity investment costs} & + & \text{Recourse costs} \end{array} \tag{8.5}$$

subject to

$$\begin{array}{ccccc} \sum\limits_{h|t_h \le t} b_h & + & s_t^- & \ge & z_t, \quad t \in T \\ \text{Capacity} & + & \text{Recourses} & \ge & \text{Demand} \end{array} \tag{8.6}$$

where function $Q_t()$ gives the cost of recourses applied at time t. Economic function (8.5) calculates the NVP of all cash outflows related to capacity investments and recourses used to provide short-term capacity when the installed capacity is not sufficient. Constraint (8.6) ensures that the capacity needed is always available. Of course, in addition to this constraint, one must consider restrictions imposed by the nature of technological options (upper and lower bounds on the capacity provided by each option, compatibility of the options with the technology in place, etc.), by the type of recourses available (overtime restrictions specified in collective agreements, limits on subcontracting imposed by technology or by business strategy, etc.) as well as by funding opportunities available to the company. This model cannot be solved as is. However, it provides a conceptual framework that facilitates the elaboration of rigorous strategic capacity planning methods.

Assuming that demand is known in advance and that the capacity strategy of a company has no influence on prices and demand is not very realistic. If prices, demand, and capacity are interdependent, the problem is much more difficult: an explicit relationship among prices, demand, and capacity must then be established, and value added must be maximized by taking sales revenues into account. We address this problem in Sect. 8.5. In some cases, the replacement, reorganization, or disposal of current facilities must also be taken into account, which complicates things. Finally, if the assumption that the future is known with certainty is relaxed, the value-risk efficient frontier defined in Sect. 2.4 must be explored. The attitude toward risk of decision makers must then also be taken into account. Chapter 11 shows how to address strategic decisions under uncertainty.

The described conceptual framework implicitly assumes that the company considered is using a single type of capacity and that the technological options available are mutually exclusive. Under these conditions, when variable operating costs are the same for all technological options, it is not necessary to consider them explicitly. However, if they are not the same for all options, they must be added to investment and fixed operating costs in the analysis.

8.1.5 Capacity Portfolio Deployment

When a company engages in interrelated activities, that is, when its SCN activity graph is not trivial, it usually needs several types of capacity. Take the sawmill activity graph illustrated in Fig. 4.2. It is clear for that case that distinct technological systems are required for the main activities, namely bucking, sawing, drying, planning/grading, chipping, and storage. This company must therefore invest in a portfolio of various types of capacity that are not necessarily mutually exclusive. In this context, capacity types are naturally associated with the internal activities $a \in A^S$ of a company's SCN activity graph. A capacity plan is elaborated by programming the implementation of technological options, that is, by choosing among a set O of options available during a planning horizon. The selection of some of these options may depend on the choices made previously. For example, the option to upgrade a technological system can be selected only if this system has been installed previously.

To understand how various types of capacity can overlap, go back to the manufacturing company example presented in Fig. 7.19. One alternative could be to implement separate *dedicated* capacities (specialized equipment, for example) for activity 3 (Component 6 fabrication) and for activity 4 (Component 7 fabrication), which correspond to technological options $o = 1, 2$, respectively. Another option, $o = 3$, could be to implement capacity sufficiently *flexible* (a flexible manufacturing system, for example) to fabricate the two components. Among these three options,

$O = \{1, 2, 3\}$, onc could decide to install dedicated capacity (o = 1, 2), flexible capacity (o = 3), or a combination of both, depending on forecasted demand for finished products 8 and 9 and on the investments required. In this context, it is clear that variable operating costs for each activity depend on selected technology options. Therefore, they must be considered explicitly in the analysis. This also shows that, besides specifying the amount and timing of capacity additions and removals, the capacity strategy of a company must prescribe the type of technology to implement.

The previous discussion emphasizes the temporal and technological dimensions of a capacity strategy. This is not to say that the *spatial* dimension (geographic location of platforms) studied in Chap. 7 must not be taken into account. Indeed, we have seen that the production–distribution capacity of a company that serves broad markets and has a diverse supply base should be optimally deployed in its theater of operation. The conceptual framework presented does not necessarily restrict itself to technological options for a single site, but rather it considers all the capacity requirements of a company's supply chain network. A capacity strategy must therefore specify not only *how much* and *when* to modify each *type* of capacity required but also *where*. For a given site, technological options span a set of alternative platforms. The strategic capacity planning problem therefore boils down to determining which site to open, maintain, or close during the planning horizon considered and what platform to implement for each site selected.

8.2 Capacity Provided by a Technological System

The SCN design models presented in Chap. 7 were static, that is, they considered only a year of operation. Also, demand, d_{pl}, was specified by product family $p \in P$ and demand zone $l \in L_p^{\mathrm{D}}$. When considering a discrete planning horizon T covering several years,[2] the demand for planning period $t \in T$ can be denoted d_{plt}. In order to calculate a company's capacity need, z_{at}, for an activity $a \in A^{\mathrm{S}}$ during a planning period $t \in T$, a direct relationship must be established between the units used to forecast demand d_{plt} for each product-market and those used to express the capacity provided by the technological system performing activity a. When concerned with dedicated capacity measured in *throughput units* (number of units processed during a planning period), the correspondence is direct. When possible, it is always simpler to express capacity in throughput units. However, when dealing with flexible technologies or when the capacity measure used is associated with a critical resource, one has to develop a capacity–demand relationship specifying how much

[2]In what follows, we assume capacity is planned over a discrete planning horizon covering a finite number of planning periods (say years). The index t is used to identify planning periods. We also assume that capacity changes are always planned for the beginning of one of the periods in the planning horizon.

outputs can be produced per period with available resources, which is not always easy. To clarify this important issue, we examine two particular systems: a warehouse and a workstation. Then, we show how congestion caused by random demands can make things even less obvious.

8.2.1 *Storage Capacity*

Although relatively simple, the case of a warehouse can be used to illustrate many of the nuances just discussed. What unit should be used to measure the capacity of a warehouse? Because a typical warehouse keeps all kinds of products in stock, it must use flexible technologies. Suppose that the products are stored and handled in pallets. We indicated that it is desirable to express capacity in throughput units, that is, in output quantity per period. When using yearly planning periods, storage capacity should thus be expressed in terms of annual throughput, that is, the sum of annual pallet shipments over all products.

Now, what limits the capacity of a warehouse? Shipping and receiving resources (number and type of docks, staging area, etc.) may be limiting factors, but, in most cases, the dominant factor is the storage space available. Storage space can also be measured in pallets, but what is the relationship between available storage space and the maximum pallet throughput during a year? To answer this question, we need to know how much inventory the company requires to provide an adequate customer service and how much storage space is provided by the warehousing platform used. In other words, the answer depends on the resources–configuration–methods triplets of the platform considered or, roughly speaking, on the storage technology available.

As discussed in Sect. 2.2.3.2, an inventory–throughput relationship of the type $\bar{I}(X,\tau) = \alpha'(\tau^{\chi})X^{\beta}$ can usually be established empirically among the annual pallet throughput X in a warehouse, the average procurement lead times τ, and the average inventory $\bar{I}(X,\tau)$ necessary to provide required service levels. We also saw in Sect. 7.3.1 that, because suppliers deliver replenishment shipments in lots, at certain times the sum of cycle and safety stocks is necessarily higher than $\bar{I}(X,\tau)$. The ratio $\eta =$ (maximum inventory)/(average inventory) was then introduced to be able to convert average inventory levels into maximum inventory levels. Based on this, it should be clear that storage space requirements, in pallets, are given by $z(X,\tau) = \eta\alpha'(\tau^{\chi})X^{\beta}$. The storage space required depends on the inventory management methods used (DRP, order point, etc.) and on average delivery times, that is, on the transportation means used, as discussed in Chap. 4. As illustrated in Fig. 8.6, space requirements (maximum inventory levels) are higher when lead times increase. When τ is fixed, that is, for given inbound transportation means, storage space requirements reduce to $z(X) = \alpha X^{\beta}$, $\alpha = \eta\alpha'(\tau^{\chi})$. If the warehouse platform provides enough space to keep b^{S} pallets in inventory, then the following nonlinear storage capacity constraint applies: $z(X) \leq b^{\mathrm{S}}$. In order to express available capacity

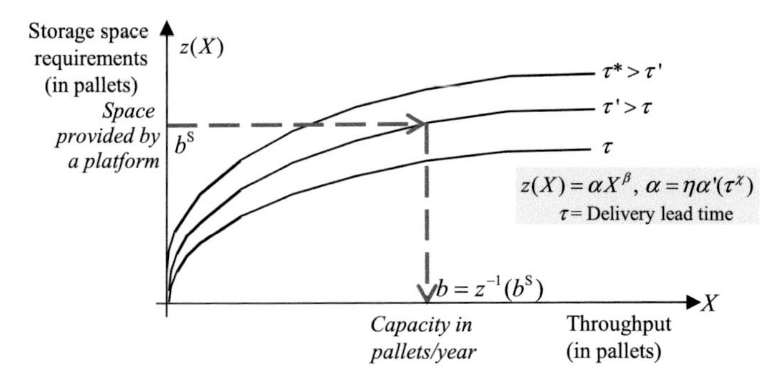

Fig. 8.6 Relationship between storage space and maximum annual throughput

in terms of maximum pallet throughput, the available space, b^{S}, must be projected onto the X-axis of the inventory–throughput function shown in Fig. 8.6. The capacity constraint then becomes $X \leq b, b = z^{-1}(b^{\mathrm{S}})$, where $z^{-1}(b^{\mathrm{S}})$ is the required storage capacity expressed in terms of annual throughput, and it is linear.

Example 8.1 From recent data on customer shipments (in pallets), maximum inventory levels (in pallets), and average delivery lead times (in days) to its warehouses, a company was able, using regression analysis, to estimate the following empirical relationship among storage space used, lead times, and annual throughputs:

$$z(X, \tau) = 3\tau^{0,6}X^{0,7} \tag{8.7}$$

The company considers the possibility of adding storage capacity to its distribution network by implementing a platform of the same type as existing warehouses on a new site. This platform would be capable of storing 100,000 pallets. Inventory management methods are not expected to change and it is estimated, considering current site locations and transportation means, that average delivery lead times will be 10 days. Given that several other technological options (different sites and different storage technologies) are available, the company wants to evaluate them. To do this, the capacity provided by each option must be expressed in terms of annual maximum throughput, that is, as the maximum number of pallets that can be shipped per year. What is the capacity of the platform considered?

To answer, one must find the value of X in relation (8.7) when $\tau = 10$ days and $z(X, 10) = b^{\mathrm{S}} = 100,000$ pallets, that is, solve equation $3(10)^{0.6}X^{0.7} = 100,000$. This gives $b = z^{-1}(100,000) = (8,372.95)^{1/0.7} = 401,895$. The capacity of the platform considered under current conditions would therefore be of 401,895 pallets per year. Suppose now that the company is renegotiating its current supplier and carrier contracts and that it is anticipating that, when the new platform will be implemented, average lead times will have been reduced to 5 days. The annual capacity provided by the platform under these new conditions is given by

$$(X,5) = 3(5)^{0.6}X^{0.7} \Rightarrow b = z^{-1}(100,000) = (12,691.03)^{1/0.7}$$
$$= 728,011 \text{ pallets}$$

This illustrates the impact that procurement and transportation means can have on storage capacity.

Suppose now that the company manages three major families of products, p = 1, 2, 3, and that annual SCN flows must be optimized. Flows are expressed in product boxes and there are 80 boxes of product 1 per pallet, 75 boxes of product 2 per pallet, and 100 boxes of product 3 per pallet. The annual pallet throughput is then given by $X = F_1/80 + F_2/75 + F_3/100$, where F_p denotes the annual outflow of product p from the warehouse, and the capacity constraint to enforce during optimization, for the platform considered, would be

$$F_1/80 + F_2/75 + F_3/100 \leq 728,011$$

Note that the capacity in this constraint is based on an empirical relationship among inventory levels, average lead time and throughput, and on an average lead time estimation. This type of constraint therefore should not be interpreted as a firm upper bound but rather as an order of magnitude to be respected. Also, one should ensure that the platform receiving and shipping resources are able to handle this annual throughput level.

8.2.2 Workstation Capacity

Now consider the case of a dedicated technology used for the assembly of a product family. To be more specific, suppose that the assembly is done in predesigned workstations involving three types of resources: two *machines* and three *operators* working in a fixed plant area (*space*). The configuration of a workstation and the work methods used are illustrated in Fig. 8.7. The stems in the center of the schema represent the working time of the operators and the shaded rectangles the working time of the machines.

As in the previous case, the processor capacity should be expressed in terms of annual throughput, that is, the number of pallets of products the workstation can assemble per year. Clearly, the capacity is limited by the resources, configuration, and methods of the assembly station, as well as by the number of hours of operation during a year. Assume that the company's policy is to work two 8-h shifts per day,

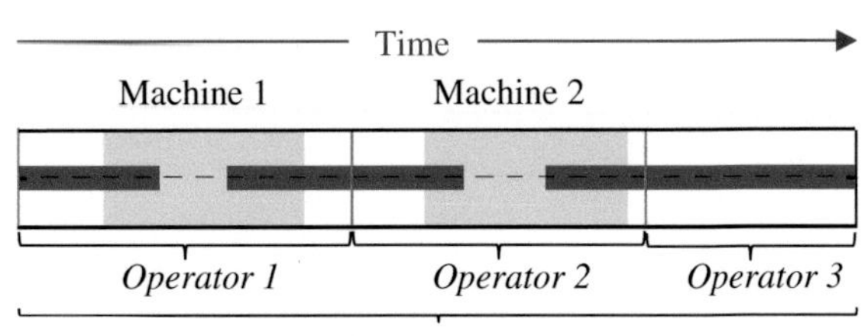

Fig. 8.7 Assembly station for a product family

that break periods summing to 1 h per shift are imposed by the collective agreement for the personal needs of workers, and that there are 240 days per year. This gives a potential of $WH = 2 * 240 * (8 - 1) = 3360$ work hours per year per workstation. However, suppose that the standard assembly time (including setup, execution, handling, and inspection) for a product pallet is T^p hours. Under these conditions, the station should be able to produce WH/T^p product pallets per year. If T^p is 2 h per pallet, for example, it should be possible to assemble (3360)/2 = 1680 product pallets per year per workstation. This does not account for unproductive times and defects, however.

Available working hours always incorporates unproductive times attributable, in part, to poor operation managements (bad scheduling, raw material shortages, poorly planned maintenance, breakdowns, accidents, etc.) and to worker inefficiencies (absenteeism, delays, indolence, work done without care, accidents, etc.). If company statistics on work hours indicate that UP % of available time is unproductive, then it is not possible to make more than $(1 - UP)WH/T^p$ product pallets per year. If $UP = 20$ %, for example, the maximum throughput is $(1 - 0.2)(1680) = 1344$ pallets per year. Defects must also be considered. If $q = 3.27$ % of the assembled products are rejected, then the annual capacity, b, per assembly station is given by

$$b = (1 - q)(1 - UP)WH/T^p = (1 - 0.0327)(1344) = 1300 \text{ pallets/year}$$

Once again, this illustrates the impact of the resources–configuration–methods triplet on capacity.

If the required capacity is high, rather than implementing several assembly stations of the type illustrated in Fig. 8.7, one could design a more performant assembly line using different resources, configurations, and methods. Proceeding as in the previous paragraph, suppose that the capacity calculated for this new type of technological system is 3000 pallets per year. The capacity obtained as a function of the number of processors implemented is plotted in Fig. 8.8 for the two technologies considered. Note that it is not possible to install a fractional number of processors. However, one could design an activity center including a few processors of each type. If the technological option considered involves the implementation of two processors of type 1 and six processors of type 2, for example, the capacity available would be 2(1300) + 6(3000) = 20,600 pallets per year.

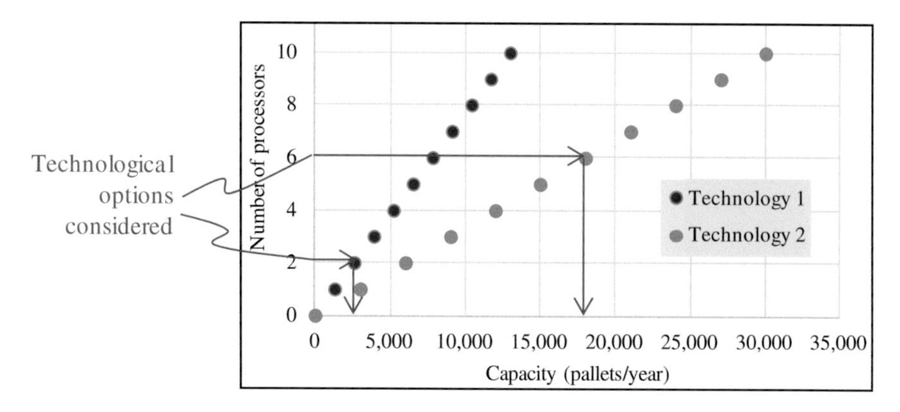

Fig. 8.8 Capacity provided by different technologies

8.2.3 *Impact of Congestion*

The case studied in the previous section implicitly assumes that production is continuous and that all products made have fixed processing times. When products are made to order, customer orders arrive randomly and processing times may vary from one product (order) to the next. We then have a production system with queues as illustrated in Fig. 8.9. In this kind of system, processors can have dead time, but they can also be congested with orders waiting to be processed. This must be taken into account in the calculation of capacity. When demand and/or processing times are random, customer response time (*RT*) and service level policies must be predetermined to establish a relationship similar to that of Fig. 8.8 between *number of processors* and *capacity*. In this context, the number of processors required to serve customers in less than 3 days 98 % of the time is higher than the number needed to complete these orders in less than 5 days 98 % of the time. The type of processor–capacity relationship obtained in such a context is shown in Fig. 8.10. Note that in this case, the relationship between resources and capacity is not linear. A good discussion of these phenomena is found in Hopp and Spearman (2008).

The preceding discussion shows that there may be a substantial difference between the absolute capacity of a technological system and the capacity it can really provide. This discrepancy may become even greater when capacity is affected by unforeseeable events such as strikes and industrial accidents. It is clearly important to take this into account when developing long-term capacity plans.

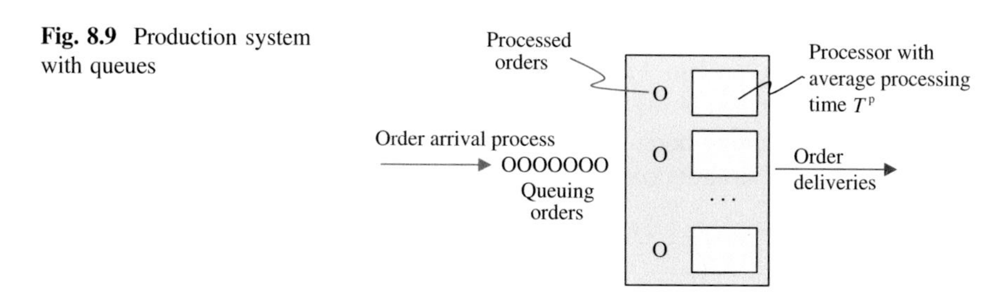

Fig. 8.9 Production system with queues

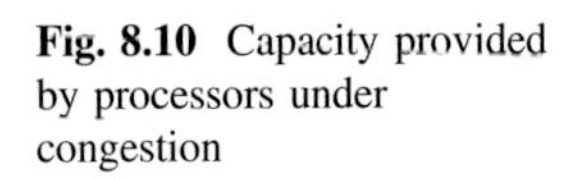
Fig. 8.10 Capacity provided by processors under congestion

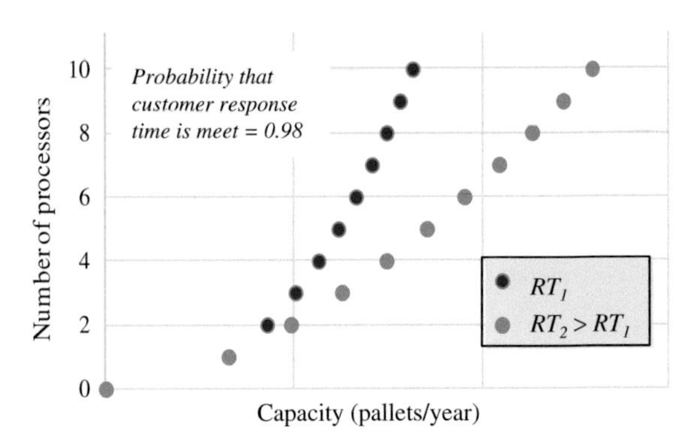

8.3 Single-Technology Optimal Capacity Plan

We now turn to the study of decision models to support the development of strategic capacity plans. This section shows how to find an optimal program for the addition of a given type of capacity under growing demand. It concentrates on the temporal dimension of the problem. The spatial deployment of technological options is examined in the next sections. The problem studied has the following features:

- The technological system considered needs a single type of capacity, and the technology used to provide it is predetermined
- Capacity is planned over a finite planning horizon, $t \in T$, in a growing demand context, that is, the capacity required increases (or stays the same) from one planning period to the next
- We need to determine when and how much capacity to add to ensure that capacity precedes demand (Policy A in Fig. 8.3).

The problem is illustrated in Fig. 8.11 and the following notation is required to formulate the optimization model:

n_t — Starting date of planning period $t \in T$, that is, number of years between the start of the planning horizon and the beginning of period t

z_t — Capacity required during period t (z_0 is the capacity available at the beginning of the planning horizon)

$Z_t = z_t - z_{t-1}$ — Additional capacity required during period t

b_t — Number of units of capacity added at the beginning of period t (decision variable)

$C_t^{\mathrm{R}}(b)$ — Net present value at time n_t of b units of additional capacity installed at the beginning of period t and used until the end of the planning horizon

r — Opportunity cost of capital of the company in \$/\$/year (see Chap. 2)

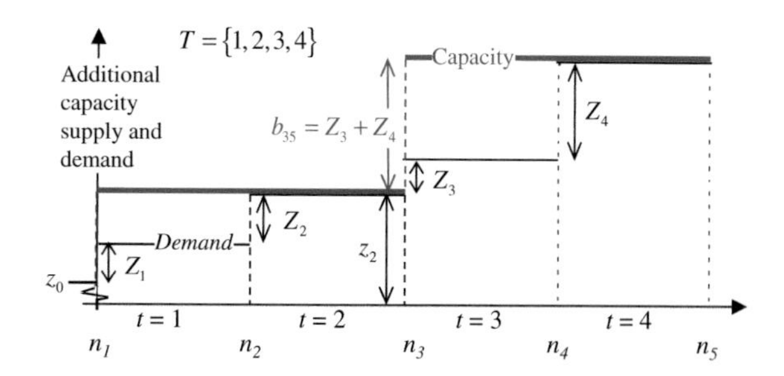

Fig. 8.11 Capacity supply and demand for the planning horizon

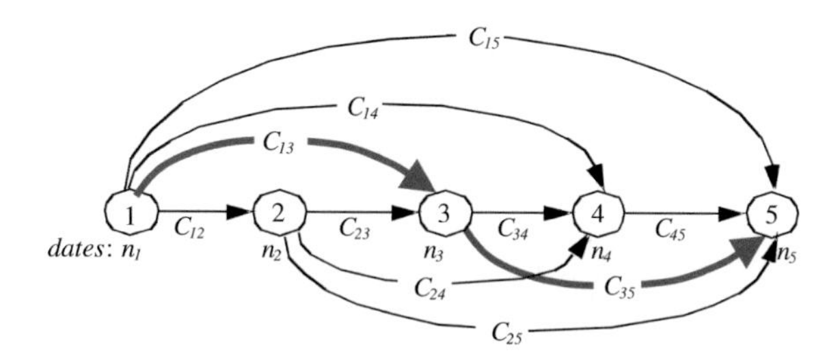

Fig. 8.12 Directed graph of the capacity planning problem

An optimal capacity strategy is obtained in this case by finding the shortest path in a directed graph of the type shown in Fig. 8.12. In this graph, node t is associated with the start *date* n_t of period t, and arc (t, m) represents a decision to add, at date n_t, enough capacity to cover needs during interval (n_t, n_m), that is, to add

$$b_{tm} = Z_t + \cdots + Z_{m-1} \tag{8.8}$$

units of capacity, where b_{tm} is the additional capacity needed between dates n_t and n_m. Each path in this graph corresponds to a feasible capacity strategy.

To understand this, examine the problem specified in Fig. 8.11 and the directed graph in Fig. 8.12 simultaneously. Note, for example, that arc (1, 2) corresponds to planning period 1, and arc (2, 5) covers periods 2, 3, and 4. The capacity required to meet additional needs during these last three periods is $b_{25} = Z_2 + Z_3 + Z_4$. Now, compare the capacity expansion strategy in bold in Fig. 8.11 with the path in bold in Fig. 8.12. The strategy mapped out in Fig. 8.11 corresponds to the addition of $b_1 = Z_1 + Z_2$ units of capacity at date n_1 and of $b_3 = Z_3 + Z_4$ units of capacity at date n_3. When looking at the path in bold in Fig. 8.12, one sees that it corresponds exactly to this solution.

It should be clear that an optimal capacity strategy is given by the shortest (minimum cost) path in the graph. To find this path, one must first compute the cost associated with each arc in the graph. The cost C_{tm} of the decision associated with arc (t, m) is the NVP at the beginning of the planning horizon of b_{tm} units of capacity installed at date n_t, that is,

$$C_{tm} = C_t^{\mathrm{R}}(b_{tm})e^{-rn_t} \tag{8.9}$$

In this expression, the present value is calculated assuming that interest is compounded continuously (see Chap. 2).

This shortest path problem is easily solved using *dynamic programming*, that is, by proceeding period by period to calculate, for each successive period m, the value of the best solution to cover needs until this period, V_m^*, using the following *recurrence equation*:

$$V_m^* = \min_{1 \le t \le m-1} \left[C_{tm} + V_t^* \right], \quad V_1^* = 0 \tag{8.10}$$

Solving the capacity planning problem with this method[3] is very simple, and the best way to explain how it works is to use an example.

Example 8.2 [Based on Freidenfelds (1981)]

It is forecasted that future demand for channels in a satellite communication system will increase at the rate of 1.5 units of 1000 channels per year. A new satellite that can provide b additional channels has a NPV, at the time of its installation, of

$$C^{\mathrm{R}}(b) = 16 + 2b \quad \text{(in \$100,000 units)}$$

If a capacity plan must be elaborated over a planning horizon including four planning periods, each lasting 4 years, and the weighted average cost of capital (WACC) is 10 %, what size of satellite should be launched and when?

Because capacity requirements grow linearly, the additional capacity needed in each planning period is 4(1.5) = 6 units of 1000 channels, that is, $Z_t = 6, t = 1, \ldots, 4$. Moreover, the costs on the arcs of the graph in Fig. 8.12 can be calculated using the relation

$$C_{tm} = C_t^{\mathrm{R}}(b_{tm})\mathrm{e}^{-rn_t} = (16 + 2b_{tm})e^{-(0.1)n_t}$$

Therefore, for arcs (1, 2) and (2, 5), for example, we have

$$b_{12} = 6, \quad C_{12} = (16 + 12)\mathrm{e}^{-(0.1)0} = 28$$

$$b_{25} = 18, \quad C_{25} = (16 + 36)\mathrm{e}^{-(0.1)4} = 34.9$$

The costs calculated for all arcs of the decisions graph are provided in Fig. 8.13. The optimal solution of the problem is calculated in Fig. 8.14.

[3] For those familiar with inventory theory, note the similarity between this method and the Wagner and Whitin (1958) algorithm developed to solve dynamic lot-sizing problems.

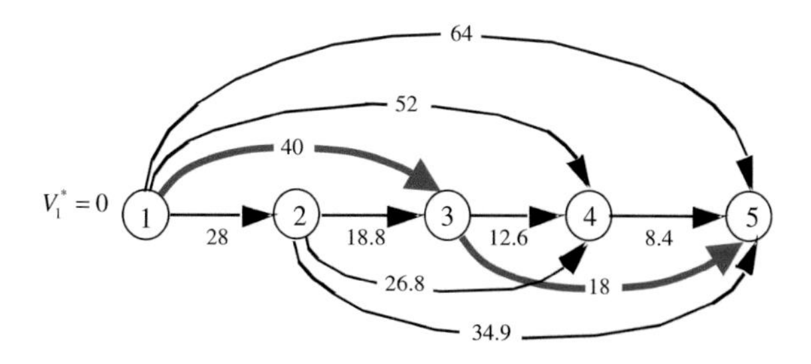

Fig. 8.13 Decisions graph for the example

	End-of-period node (m)							
	2		3		4		5	
Start node (t)	t	V_{t2}	t	V_{t3}	t	V_{t4}	t	V_{t5}
1	*1*	*28 + 0 = 28*	*1*	*40 + 0 = 40*	*1*	*52 + 0 = 52*	1	64 + 0 = 64
2			2	18.8 + 28 = 46.8	2	26.8 + 28 = 54.8	2	34.9 + 28 = 62.9
3					3	12.6 + 40 = 52.6	*3*	*18 + 40 = 58*
4							4	8.4 + 52 = 60.4

Fig. 8.14 Dynamic programming solution worksheet

The left column in the dynamic programming table in Fig. 8.14 lists all nodes t corresponding to the start of a period (arc). The top row lists all nodes m corresponding to the end of a period (arc). A cell (t, m) in the table corresponds to an arc of the decisions graph. When there is no corresponding arc, the cell is empty. Cell (t, m) is split into two parts: the left part identifies the origin t of the arc considered; the right part is used to calculate the value V_{tm} of the best path, including arc (t, m), which can be used to go to node m. V_{tm} is calculated by adding cost C_{tm} to the value V_t^* of the best path available to go to node t. When this calculation is completed for all the arcs arriving at node m (i.e., when all relevant cells in column m are filled), the best path going to node m is identified by selecting the arc (t, m) with the smallest value. In the table, this arc is shaded. This boils down to applying recurrence Eq. (8.10) for a given m. To find the optimal solution, these calculations are done column by column, starting on the left.

Example 8.2 (continued) To solve our example problem, begin with column 2 in Fig. 8.14. The only arc available to go to node 2 is (1, 2) and its value is 28. Column 3 indicates that arcs (1, 3) or (2, 3) can be used to go to node $m = 3$. Their respective costs are 40 and 18.8, which in the latter case must be added to the value of the best path found to go to node 2, that is, 28. This gives the value $V_{23} = 18.8 + 28 = 46.8$ of the best path including (2, 3). Comparing these two values, we see that the first is smaller, that is, that $V_3^* = \min[40; 46.8] = 40$, meaning that the best path to go to node 3 includes arc (1, 3). Column $m = 4$ can now be considered. Three arcs are going to node 4: (3, 4), (2, 4), and (1, 4). The value of arc (3, 4) is calculated by adding its cost, 12.6, to the value of the best path to get to node 3. From the calculations in column 3, it is seen that the best path reaching 3 has a value of 40. The value of arc (3, 4) is thus $V_{34} = V_{34} + V_3^* = 12.6 + 40 = 52.6$. After doing the calculations for arcs (1, 4) and (2, 4), we see that the best way to reach node 4 is through arc (1, 4) and that this path has a value $V_4^* = \min[52; 54.8; 52.6] = 52$. Finally, we reach column $m = 5$. Four arcs go to node 5 and, after doing the calculations, we conclude that it is preferable to use arc (3, 5). The shortest path in the decisions graph, therefore, has a value $V_5^* = 58$.

To retrace the best path from the origin, examine the choices made starting from the end. Column 5 indicates that the best way to go to node 5 is to transit through node 3. Column 3 specifies that the best way to get to node 3 is to start from node 1. The optimal path thus includes arcs (1, 3) and (3, 5), as shown in bold in Fig. 8.13. This means that the optimal capacity investment plan is to launch a satellite with a capacity of 12 one-thousand-channel units at the beginning of the horizon ($b_1 = 12$) and another one of the same size 8 years later ($b_3 = 12$). The NVP of this plan is \$5.8 million.

8.4 SCN Platform Selection Model

The modeling approach studied in the previous section focuses on the temporal dimension of strategic capacity planning, but it does not consider location or technology choices. This section studies platform selection decisions for SCN sites. As explained in Sect. 4.3.2, platforms are assemblages of technological systems supporting a set of primary activities. To address these design decisions, we concentrate initially on the formulation of a single-period capacity planning model for the two-echelon SCN design problem described in Fig. 7.10. This problem involves the distribution of several product families from a set of distribution centers (DCs) or production–distribution centers (PDCs) and the replenishment of the DCs by the PDCs. In Chap. 7, it was assumed that the SCN must be designed from scratch and that the DC–PDC platform of each potential site is predetermined. These unrealistic assumptions are relaxed in this section. The general case, for a multi-period planning horizon, is treated in the next section.

Henceforth, unlike in the previous section, capacity plans are not developed directly from a forecast of future capacity needs. Instead, demand forecasts d_{pl} for products $p \in P$ in demand zones $l \in L_p^{\mathrm{D}}$ are used, and capacity is provided by the platforms' technological systems. Capacity planning therefore involves the selection of

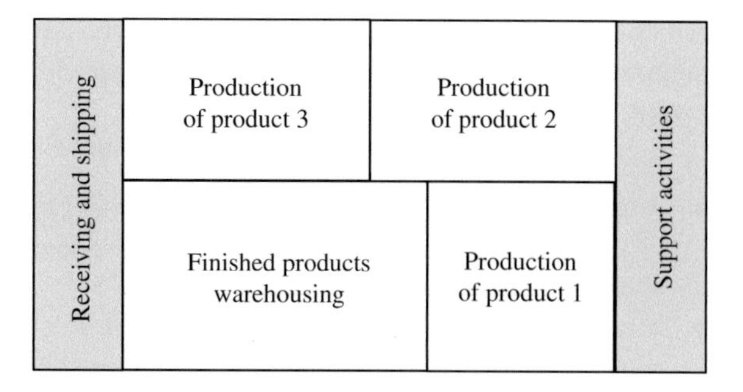

Fig. 8.15 Block layout of a production–distribution platform

(platform, site) pairs meeting product-market demands during the planning period considered. Because each product family $p \in P$ requires *dedicated* production capacity, the p-index in this case simultaneously designates a product family and a type of production capacity. Conversely, storage resources provide *flexible* capacity shared by all products. The block layout (see Sect. 4.3.2) of a typical production–distribution platform is shown in Fig. 8.15. In the capacity planning approach proposed, essential primary activities only are taken into account explicitly. Thus, although the platform in Fig. 8.15 includes a generic block for support activities (management, engineering, quality assurance, sales, maintenance, heating, cafeteria, etc.) on the right-end side, and one for receiving and shipping on the left-end side, these activities are not taken into account explicitly in the following optimization model.

In Chap. 7, we assumed that the production capacity b_{pu} for each product family $p \in P$ on each site $u \in L^{\mathrm{U}}$ and the distribution capacity b_s on each site $s \in L^{\mathrm{S}}$ are predetermined even if the site is vacant, that is, that site platforms are selected before solving the SCN design problem and that no capacity alternative is available. This approach can lead to poor designs. Take the example described in Fig. 7.12 and solved in Fig. 7.14. In this case, potential production–distribution sites in The Hague and Berlin are considered. The platform on each site is large enough to accommodate all the network demand, and the associated fixed cost reflect this. It would be in this context much too expensive to open the two PDCs, and the decision implicitly reduced to choosing one of the two. However, if smaller platforms, with necessarily lower fixed costs, were also considered for each site, it could be more profitable to open the two PDCs.

Consequently, for each site $s \in L^{\mathrm{S}}$, it is now assumed that a set O_s of technological options (platforms) is considered and that selecting the most appropriate option for each site is part of the design problem. If a site is already open when the reengineering decisions are made, the status quo platform is necessarily in the options set, and an alternative could be an expansion or a reorganization of this platform. If one considers leasing a facility or a technological system (in a public warehouse, for example) or outsourcing activities, instead of buying or building a facility, platforms then correspond to potential contracts with one or more plant owners or service providers. A site can then be a city and the platforms considered can be facilities or contracts offered by third parties in that city. In a leasing context, the fixed costs associated with a platform are typically lower, but variable costs are higher.

A potential platform $o \in O_s$ for site s is characterized by the following:

- A detailed layout, such as the one illustrated in Fig. 4.16, and a binary parameter Y_{so}^{0} set to 1 if the platform is already in operation at the beginning of the planning period and 0 otherwise
- A capacity for each supported activity, that is, a maximum annual production quantity by product family, $\bar{b}_{pso}, p \in P$, if $s \in L^{\mathrm{U}}$, and a maximum annual distribution throughput $\bar{b}_{so}$ in standard load units (pallets, cwt, etc.)
- An identifier for the new platform obtained if the current platform is renovated (the index of this renovated platform is denoted $re(o)$)
- The fixed cost y_{so} (annuity or rent, plus fixed operating costs) incurred when it is operated during the year considered
- A change-of-state cost if it is closed (y_{so}^{-}) or opened (y_{so}^{+}) at the beginning of the planning period
- A unit production cost for each product, $c_{pso}^{\mathrm{P}}, p \in P$, if $s \in L^{\mathrm{U}}$
- A storage cost $c_{pso}^{\mathrm{S}} = r_o^{\mathrm{I}} v_{ps}/\varphi$ per *unit of flow* for products $p \in P$ (parameter r_o^{I} is platform dependent because when contracting space in a public warehouse, it includes the storage costs paid by \$ of inventory held, in addition to the cost of capital).

The nature of fixed costs (y_{so}) and of change-of-state costs (y_{so}^{+}, y_{so}^{-}) was discussed in detail in Sect. 2.2.3.1.

In what follows, we show how to modify MIP (7.5)–(7.10) presented in Chap. 7 to account for capacity options. To formulate the model, in addition to the flow variables F_{puw} and F_{psl} defined previously, the following decision variables are needed:

Y_{so} Binary variable with value 1 if platform $o \in O_s$ is used on site $s \in L^{\mathrm{S}}$ during the planning period, and 0 otherwise

Y_{so}^{+} 1 if platform o is implemented on site $s \in L^{\mathrm{S}}$ at the beginning of the planning period, and 0 otherwise

Y_{so}^{-} 1 if platform o on site $s \in L^{\mathrm{S}}$ is closed at the beginning of the planning period, and 0 otherwise

X_{puo}^{P} Annual quantity of product p manufactured at site $u \in L^{\mathrm{U}}$ when platform $o \in O_u$ is selected

X_{pso}^{S} Annual throughput of product p in the storage area of site $s \in L^{\mathrm{S}}$ when platform $o \in O_s$ is selected.

Reconsider, first, the production capacity constraints (7.7). The left-hand side of these constraints gives the quantity of product p produced in PDC u, expressed in terms of outflows to DCs and to demand zones. Given that we now have several capacity options, we need to establish an explicit relationship between these outflows and the quantities produced with optional platforms, that is,

$$\sum_{o \in O_u} X^{\mathrm{P}}_{puo} = \sum_{l \in L^{\mathrm{D}}_p} F_{pul} + \sum_{w \in L^{\mathrm{W}}} F_{puw}, \quad u \in L^{\mathrm{U}}, \ p \in P_u \tag{8.11}$$

If platform o is selected, the production capacity it provides for product p must be respected and therefore the capacity constraints become

$$X^{\mathrm{P}}_{puo} \leq \bar{b}_{puo} Y_{uo}, \quad u \in L^{\mathrm{U}}, \quad o \in O_u, \quad p \in P_u \tag{8.12}$$

Clearly, we cannot choose more than one platform for a site. Also, the state of the network at the beginning of the planning period must be taken into account, that is, platforms can be upgraded or closed only if they are already in place, and a platform must be implemented on new selected sites. Therefore, the following constraints are required in the model:

$$\sum_{o \in O_s} Y_{so} \leq 1, \quad s \in L^{\mathrm{U}} \tag{8.13}$$

$$\left.\begin{array}{ll} Y_{so} + Y^{-}_{so} + Y^{+}_{s,re(o)} = 1 & \text{if } Y^{0}_{so} = 1 \\ Y_{so} - Y^{+}_{so} = 0 & \text{if } Y^{0}_{so} = 0 \end{array}\right\} \quad s \in L^{\mathrm{U}}, \ o \in O_s \tag{8.14}$$

The situation for distribution capacity constraints (7.8) is similar. Explicit relationships between outbound flows to demand zones and platform-specific storage activity throughputs must be specified:

$$\sum_{o \in O_s} X^{\mathrm{S}}_{pso} = \sum_{l \in L^{\mathrm{D}}_p} F_{psl}, \quad s \in L^{\mathrm{S}}, \ p \in P \tag{8.15}$$

If platform o is chosen, the available capacity in standard load units must be respected, and therefore the distribution capacity constraints become

$$\sum_{p \in P} e_p X^{\mathrm{S}}_{pso} \leq \bar{b}_{so} Y_{so}, \quad s \in L^{\mathrm{S}}, o \in O_s \tag{8.16}$$

To choose a single platform for the selected sites and to consider the initial state of the network, one must also add constraints similar to (8.13) and (8.14) for all DC sites $s \in L^{\mathrm{W}}$.

To complete the formulation, the platform fixed and variable costs must be added in the objective function. Given that the variable production costs (c^{P}_{puo}) and storage costs (c^{S}_{pso}) now depend on the platform selected, they cannot be included in the unit flow costs (f_{psl}) as previously. Flow costs now include only variable transportation costs c_{psl}. Consequently, the static capacity deployment model obtained is the following:

$$\begin{aligned} C^{\mathrm{SCN}} = \min & \sum_{s\in L^{\mathrm{S}}}\sum_{o\in O_s} y^{+}_{so}Y^{+}_{so} + y_{so}Y_{so} + y^{-}_{so}Y^{-}_{so} && \text{Platform fixed costs} \\ + & \sum_{u\in L^{\mathrm{U}}}\sum_{o\in O_u}\sum_{p\in P_u} c^{\mathrm{P}}_{puo}X^{P}_{puo} && \text{Production costs} \\ + & \sum_{s\in L^{\mathrm{S}}}\sum_{o\in O_s}\sum_{p\in P} c^{\mathrm{S}}_{pso}X^{S}_{pso} && \text{Storage costs} \\ + & \sum_{p\in P}\sum_{u\in L^{\mathrm{U}}}\sum_{w\in L^{\mathrm{W}}} c_{puw}F_{puw} + \sum_{p\in P}\sum_{s\in L^{\mathrm{S}}}\sum_{l\in L^{\mathrm{D}}_p} c_{psl}F_{psl} && \text{Transportation costs} \end{aligned} \quad (8.17)$$

subject to

- Demand constraints

$$\sum_{s\in L^{\mathrm{S}}} F_{psl} = d_{pl}, \quad p\in P, \quad l\in L^{\mathrm{D}}_p \quad (8.18)$$

- Platform selection constraints

$$\sum_{o\in O_s} Y_{so} \le 1, \quad s\in L^{\mathrm{S}}, \quad (8.19)$$

$$\left.\begin{aligned} Y_{so} + Y^{-}_{so} + Y^{+}_{s,re(o)} &= 1 \quad \text{if} \quad Y^{0}_{so} = 1 \\ Y_{so} - Y^{+}_{so} &= 0 \quad \text{if} \quad Y^{0}_{so} = 0 \end{aligned}\right\} \quad s\in L^{\mathrm{S}}, \quad o\in O_s \quad (8.20)$$

- Platform activity-level definition constraints

$$\sum_{o\in O_u} X^{\mathrm{P}}_{puo} = \sum_{l\in L^{\mathrm{D}}_p} F_{pul} + \sum_{w\in L^{\mathrm{W}}} F_{puw}, \quad u\in L^{\mathrm{U}}, \quad p\in P_u \quad (8.21)$$

$$\sum_{o\in O_s} X^{\mathrm{S}}_{pso} = \sum_{l\in L^{\mathrm{D}}_p} F_{psl}, \quad s\in L^{\mathrm{S}}, \quad p\in P \quad (8.22)$$

- Production capacity constraints

$$X^{\mathrm{P}}_{puo} \le \bar{b}_{puo}Y_{uo}, \quad u\in L^{\mathrm{U}}, \quad o\in O_u, \quad p\in P_u \quad (8.23)$$

- Distribution capacity constraints

$$\sum_{p\in P} e_p X^{\mathrm{S}}_{pso} \le \bar{b}_{so}Y_{so}, \quad s\in L^{\mathrm{S}}, \quad o\in O_s \quad (8.24)$$

– Flow equilibrium constraints for intermediate sites

$$\sum_{u \in L^{\mathrm{U}}} F_{puw} = \sum_{o \in O_w} X^{\mathrm{S}}_{pwo}, \quad w \in L^{\mathrm{W}}, \; p \in P \tag{8.25}$$

The usual restrictions on the binary or nonnegative value of the decision variables must also be included. This mixed-integer program is larger than the one presented in Sect. 7.3.2, but it can also be solved with commercial solvers (CPLEX, Xpress, Gurobi, etc.) for most practical situations.

Example 8.3 To see how capacity plans can be elaborated using this model, return to Example 7.2. The state of the SCN following the optimization made in Fig. 7.14 is shown in Fig. 8.16. Suppose that the company forecasts sales increases of 10 to 12 % in its traditional markets for next year and that in addition it wishes to distribute its products in Switzerland. Because the current production capacity is not sufficient to satisfy this additional demand, the company is considering either an enlargement of the PDC in The Hague or the construction of a PDC with substantial storage capacity in Barcelona. If the first option is retained, it might be interesting to open a DC in Bordeaux, and the company spotted two public warehouses with different cost structures that could fill the need. However, if a PDC is opened in Barcelona, the storage capacity currently available in Turin could be excessive and the company is considering the renegotiation of this contract. The new contract would ensure sufficient capacity to accommodate an annual throughput of 50,000 standard shipping loads instead of 100,000. Relevant data to evaluate these restructuring opportunities are provided in Fig. 8.17.

The decision variables, objective function, and constraints of the optimization model to solve to get the best SCN design are provided in Fig. 8.18.

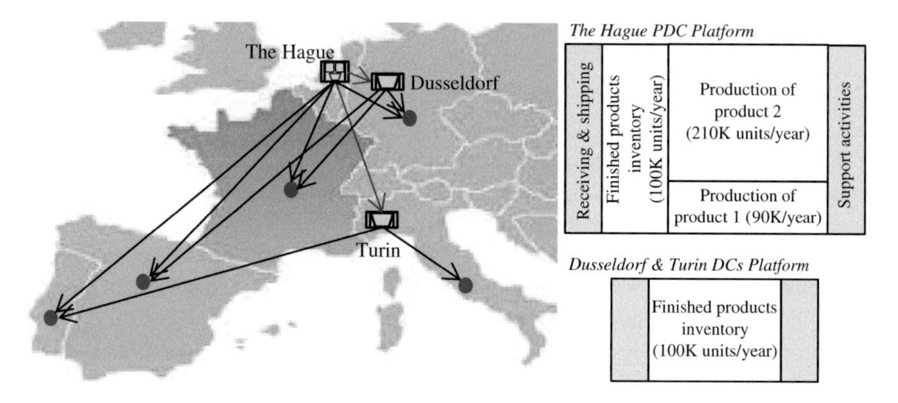

Fig. 8.16 State of the SCN at the beginning of the planning period

	A	B	C	D	E	F	G	H	I	J	K	L	M
1	Platforms data:												
2				Fixed costs			Production costs		Storage costs		Production capacity		Distrib.
3	Site	Plateform	State	Open	Use	Close	$p=1$	$p=2$	$p=1$	$p=2$	$p=1$	$p=2$	capacity
4	The Hague	1 (status quo)	1		€ 21 000 000	€ 10 000 000	€ 80,00	€ 110,00	€ 5	€ 7	90000	210000	100000
5	The Hague	2 (expansion)	0	€ 1 500 000	€ 30 000 000		€ 75,00	€ 105,00	€ 5	€ 7	125000	300000	100000
6	Barcelona	1 (addition)	0	€ 2 000 000	€ 12 000 000		€ 70,00	€ 100,00	€ 5	€ 6	50000	100000	130000
7	Dusseldorf	1 (status quo)	1		€ 5 500 000	€ 500 000			€ 6	€ 7			100000
8	Bordeaux	1 (contract)	0	€ 1 000 000	€ 5 000 000				€ 9	€ 10			120000
9	Bordeaux	2 (contract)	0	€ 1 000 000	€ 4 000 000				€ 11	€ 12			100000
10	Turin	1 (status quo)	1		€ 4 500 000	€ 500 000			€ 7	€ 8			100000
11	Turin	2 (contract)	0	€ 500 000	€ 2 500 000				€ 7	€ 8			50000
12													
13	Unit transportation costs to markets:												
14		Product $p=1$ markets						Product $p=2$ markets					
15	Site	Germany	France	Italy	Spain	Portugal	Switzerl.	Germany	France	Italy	Spain	Portugal	Switzerl.
16	The Hague	€ 23	€ 16	€ 53	€ 56	€ 73	€ 26	€ 25	€ 17	€ 58	€ 61	€ 80	€ 28
17	Barcelona	€ 60	€ 33	€ 43	€ 20	€ 40	€ 33	€ 66	€ 36	€ 47	€ 22	€ 44	€ 36
18	Dusseldorf	€ 18	€ 16	€ 50	€ 56	€ 73	€ 20	€ 19	€ 17	€ 55	€ 61	€ 80	€ 22
19	Bordeaux	€ 53	€ 20	€ 50	€ 23	€ 36	€ 33	€ 58	€ 22	€ 55	€ 25	€ 39	€ 36
20	Turin	€ 36	€ 26	€ 23	€ 50	€ 66	€ 13	€ 39	€ 28	€ 25	€ 55	€ 72	€ 14
21	Demand	16500	20000	11000	19500	28000	10000	67500	50000	34000	31000	47500	30000
22	Sum:						105000						260000
23	Unit transportation costs to CDs:												
24		CD (product $p=1$)			CD (product $p=2$)					Unit loads/product			
25	Site	Dusseldorf	Bordeaux	Turin	Dusseldorf	Bordeaux	Turin						
26	The Hague	€ 6	€ 36	€ 40	€ 6	€ 39	€ 44			e_1	e_2		
27	Barcelona	€ 46	€ 20	€ 30	€ 50	€ 22	€ 33			1	1		

Fig. 8.17 Spreadsheet with capacity planning data

The optimal solution obtained with the Excel Solver suggests keeping the PDC in The Hague as is but to build a new factory in Barcelona. Also, it recommends to renegotiate the storage contract in Turin, which reduces the capacity available from 100,000 to 50,000. However, it is not necessary to open a DC in Bordeaux. The mission of The Hague PDC is to replenish Düsseldorf and Turin DCs for both products and also to deliver product 1 directly to Germany, France, Italy, and Switzerland, as well as product 2 to France. The mission of the PDC in Barcelona is to replenish the Turin depot and to deliver both products directly in Spain and Portugal as well as product 2 to Italy.

The SCN design thus obtained is much more appropriate than that provided by the model formulated in Sect. 7.3.2 because alternative capacity options are considered. However, remember that the static model studied considers capacity requirements for a single year. Even when demand is stable, this poses fixed cost estimation challenges and it does not account for the capacity lost and the delays incurred when engaging in facility construction or renovation projects. When using a static model of this type to develop a capacity strategy in a demand growth or decline context, the model must be solved sequentially for several years and the interpretation of the results may be difficult. Recall also that one should ideally maximize value creation instead of

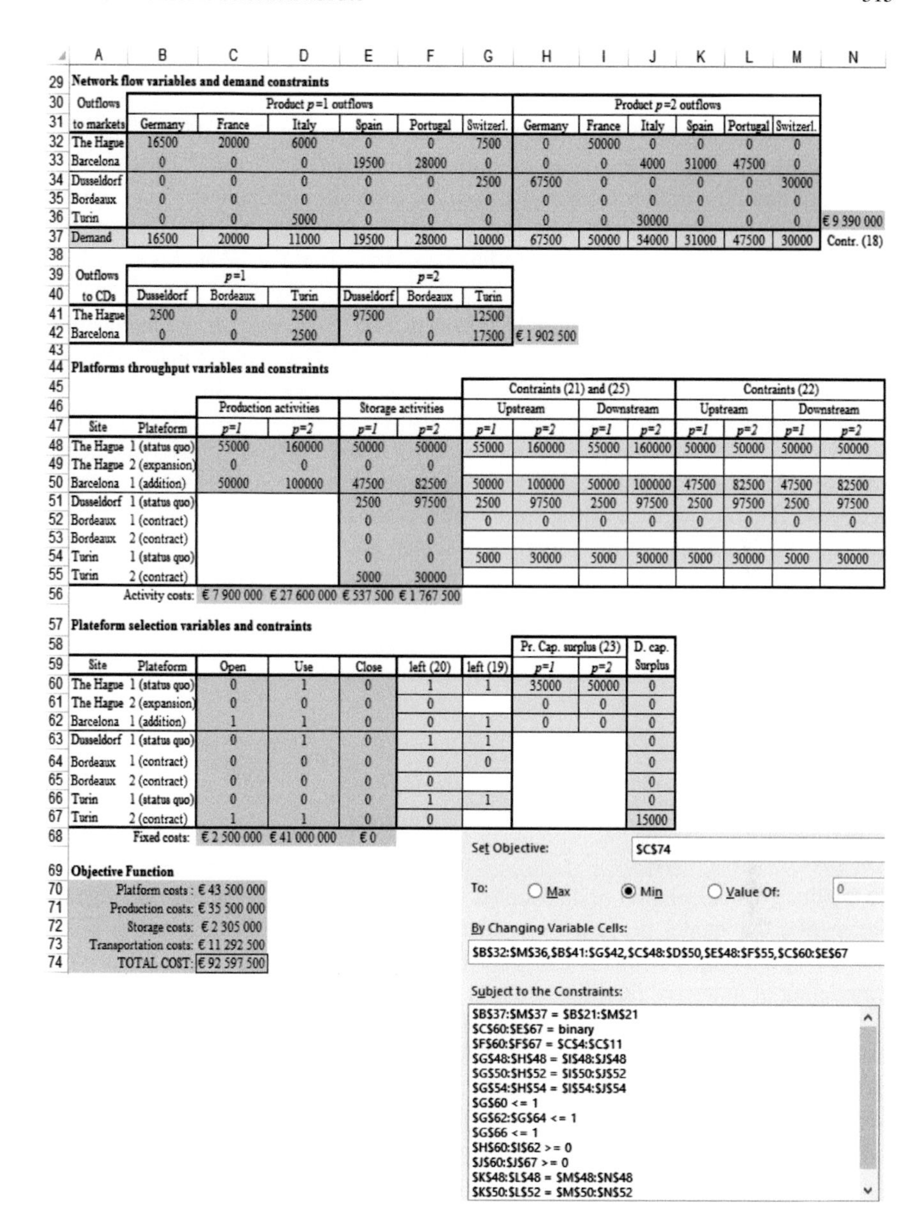

Network flow variables and demand constraints

Outflows to markets	Product p=1 outflows: Germany	France	Italy	Spain	Portugal	Switzerl.	Product p=2 outflows: Germany	France	Italy	Spain	Portugal	Switzerl.	
The Hague	16500	20000	6000	0	0	7500	0	50000	0	0	0	0	
Barcelona	0	0	0	19500	28000	0	0	0	4000	31000	47500	0	
Dusseldorf	0	0	0	0	0	2500	67500	0	0	0	0	30000	
Bordeaux	0	0	0	0	0	0	0	0	0	0	0	0	
Turin	0	0	5000	0	0	0	0	0	30000	0	0	0	€ 9 390 000
Demand	16500	20000	11000	19500	28000	10000	67500	50000	34000	31000	47500	30000	Contr. (18)

Outflows to CDs	p=1: Dusseldorf	Bordeaux	Turin	p=2: Dusseldorf	Bordeaux	Turin	
The Hague	2500	0	2500	97500	0	12500	
Barcelona	0	0	2500	0	0	17500	€ 1 902 500

Platforms throughput variables and constraints

		Production activities		Storage activities		Contraints (21) and (25) Upstream		Downstream		Contraints (22) Upstream		Downstream	
Site	Plateform	p=1	p=2	p=1	p=2	p=1	p=2	p=1	p=2	p=1	p=2	p=1	p=2
The Hague	1 (status quo)	55000	160000	50000	50000	55000	160000	55000	160000	50000	50000	50000	50000
The Hague	2 (expansion)	0	0	0	0								
Barcelona	1 (addition)	50000	100000	47500	82500	50000	100000	50000	100000	47500	82500	47500	82500
Dusseldorf	1 (status quo)			2500	97500	2500	97500	2500	97500	2500	97500	2500	97500
Bordeaux	1 (contract)			0	0	0	0	0	0	0	0	0	0
Bordeaux	2 (contract)			0	0								
Turin	1 (status quo)			0	0	5000	30000	5000	30000	5000	30000	5000	30000
Turin	2 (contract)			5000	30000								
	Activity costs:	€ 7 900 000	€ 27 600 000	€ 537 500	€ 1 767 500								

Plateform selection variables and contraints

Site	Plateform	Open	Use	Close	left (20)	left (19)	Pr. Cap. surplus (23) p=1	p=2	D. cap. Surplus
The Hague	1 (status quo)	0	1	0	1	1	35000	50000	0
The Hague	2 (expansion)	0	0	0	0		0	0	0
Barcelona	1 (addition)	1	1	0	0	1	0	0	0
Dusseldorf	1 (status quo)	0	1	0	1	1			0
Bordeaux	1 (contract)	0	0	0	0	0			0
Bordeaux	2 (contract)	0	0	0	0				0
Turin	1 (status quo)	0	0	0	1	1			0
Turin	2 (contract)	1	1	0	0				15000
	Fixed costs:	€ 2 500 000	€ 41 000 000	€ 0					

Objective Function

Platform costs : € 43 500 000
Production costs: € 35 500 000
Storage costs: € 2 305 000
Transportation costs: € 11 292 500
TOTAL COST: € 92 597 500

Fig. 8.18 Spreadsheet with the optimal solution of the reengineering model

minimizing costs. To alleviate these shortcomings, one must rely on an optimization model defined over a multiyear planning horizon. A general model of this type is presented in the next section.

8.5 Dynamic Capacity Deployment Model

In this section, we return to the business context described in Sect. 7.5. The structure of the SCNs considered is described in Figs. 7.18 and 7.19. However, instead of having a predetermined platform for each site, multiple capacity options are considered. Also, the capacity required depends on the company's product-market offer. Instead of expanding capacity, a company can choose to limit its offer. The case dealt with in this section therefore covers the elements of the market offer optimization problem discussed in Sect. 7.4. Activity internalization–externalization decisions (see Chap. 6) also have an impact on capacity. Because some components can be purchased or outsourced instead of being manufactured in-house, supplier selection decisions must also be considered to address capacity deployment issues adequately. It is assumed that future demand varies dynamically from year to year and that capacity options can be selected as needed during a multiyear planning horizon. More specifically, plans on the platforms to implement, renovate, or withdraw; on the suppliers to select; and on the offers to make during the planning horizon are optimized. Reengineering plans must maximize value added during the planning horizon and take into account constraints on the availability of capital. The model presented in this section is therefore an integration and a multi-period extension of the static models presented in Chap. 7 and in the previous section.

8.5.1 Decision Variables and Constraints

To facilitate the calculation of discounted cash flows and the preparation of financial statements, the model proposed considers a planning horizon covering a set of one-year *planning periods* $t \in T$. However, insofar as the construction and remodeling of industrial facilities can take a long time, it is not necessarily desirable to revise the design of an SCN every year. For this reason, the planning horizon is divided into *reengineering cycles* $h \in H$ (this being congruent with the SCN reengineering methodology summarized in Fig. 1.17). These cycles may last a few years and they are not necessarily all of the same length. The relationship between the period-based and the cycle-based views of the planning horizon is illustrated in Fig. 8.19. A cycle $h \in H$ includes a set of periods $t \in T_h$ and the notation $h(t)$ is

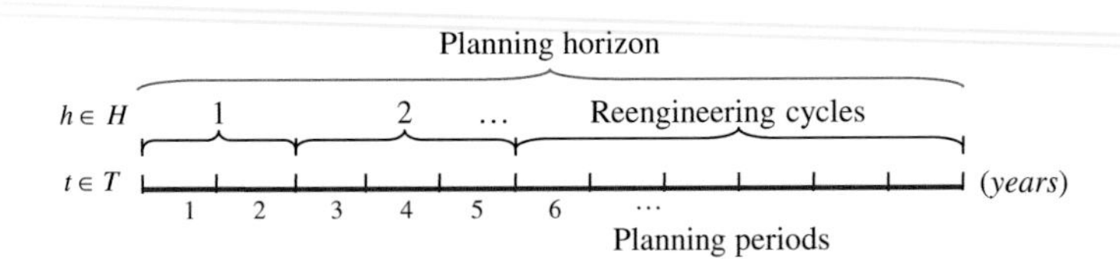

Fig. 8.19 Planning horizon

used to indicate the cycle including period t. The first cycle of the planning horizon has a profound impact on short-term restructuration plans, and its length rarely exceeds a year or two. However, as we move farther in the horizon, the cycle's length can be stretched. This is appropriate because subsequent design decisions will be revised before they are implemented. Also, cash flows being discounted, they have less impact than the irreversible decisions made at the beginning of horizon. To facilitate the solution of the model, the horizon rarely includes more than three or four reengineering cycles.

To formulate a dynamic capacity planning model, the reengineering cycle index h must be added to the subscripts of previously defined design variables. The model's binary variables then become

- $Y^{+}_{loh}, Y^{-}_{loh}, Y_{loh}, l \in L^{S}, o \in O_l, h \in H$ for platform implementation, removal, and usage decisions
- $Y^{V}_{lh}, l \in L^{V}, h \in H$, for supplier selection decisions
- $Y^{M}_{jh}, j \in J_k, h \in H$, for product-market $k \in K$ offer selection decisions.

Similarly, the planning periods index t must be added to the subscripts of the activity-level variables defined previously, giving the following continuous decision variables:

- $X_{pnot}, n \in N^{\mathbf{S}}, p \in P^{\rightarrow}_{a(n)}, o \in O_{l(n)}, t \in T$, for platform's activity levels
- $F_{pnn't}, p \in P, n' \in N^{S}, n \in N^{\leftarrow}_{pn'}, t \in T$, for internal nodes inbound product flows
- $F_{jpnn't}, j \in J_k, p \in P_k, n' \in N^{D}, n \in N^{\leftarrow}_{pn'}, t \in T$, for product flows servicing product-market $k \in K$ demand zones.

For the general case studied here, the platforms $o \in O_l$ considered for site $l \in L^{S}$ can perform a set of activities $A_{lo} \subset A^{S} = A^{F} \cup A^{E} \cup A^{G}$. The nodes $n \in N$ of the potential SCN (see Fig. 7.21) correspond to location–activity pairs (l, a). This is why the throughput variables X^{P}_{puo} and X^{S}_{puo} defined in the previous section are replaced by X_{pnot}. This also explains why flow variables are associated to potential network arcs (n, n').

Suppose that the subset $O_{lh} \subseteq O_l$ of platforms could be used on site $l \in L^{S}$ during cycle $h \in H$. Each of these platforms is characterized by

- A detailed layout specifying the set of activities A_{lo} it can perform
- A capacity $\bar{b}_{(l,a)o}$, for each activity $a \in A_{lo}$, expressed in terms of an upper bound on a standard annual throughput measure (production time, number of pallets, etc.)—it is assumed that all the output products $p \in P^{\rightarrow}_{a}$ of activity $a \in A_{lo}$ share the capacity provided by the platform for this activity
- A maximum storage space $b^{S}_{(l,a)o}$ for storage activities $a \in A_{lo} \cap A^{E}$ (in pallets, m^2, m^3, etc.)
- A capacity loss $\delta_{(l,a)ot} \leq \bar{b}_{(l,a)o}$ incurred during period $t \in T_h$, for each activity $a \in A_{lo}$, when the platform is implemented at the beginning of cycle $h \in H$

- A minimum throughput $\underline{b}_{(l,a)oh}$ required during cycle $h \in H$ for each activity $a \in A_{lo}$ to implement the platform
- The index $re(o)$ of the new platform obtained when platform o is remodeled
- A fixed exploitation cost y_{lot} for periods $t \in T_h$, this cost including fixed operating costs as well as a *rent* paid for using the platform (the rent charged by the owner if the platform is rented or leased, or calculated by the company to cover financial charges, market value depreciation, and opportunity costs for year t)
- Fixed change-of-state costs y^+_{lot} and y^-_{lot} for periods $t \in T_h$, these one-time costs being generally positive for the first period of cycle T_h, when the platform is opened or closed, and near zero otherwise
- A variable throughput cost, $c^{\mathrm{X}}_{p(l,a)ot}, p \in P, t \in T_h$, for each activity $a \in A_{lo}$
- A variable inventory holding cost, $c^{\mathrm{I}}_{p(l,a)ot}, p \in P, t \in T_h$, for each storage activity $a \in A_{lo} \cap A^{\mathrm{E}}$.

When a platform is closed on a site, this site cannot be reused during the planning horizon. Platform implementation constraints (see Sect. 8.4) must be adapted to the broader context of the problem. First, it is clearly not possible to implement more than one platform on a site during a reengineering cycle:

$$\sum_{o \in O_{lh}} Y_{loh} \leq 1, \quad l \in L^{\mathrm{S}}, \quad h \in H \tag{8.26}$$

Also, a site cannot be opened or closed more than once during the planning horizon:

$$\sum_{h \in H} \sum_{o \in O_{lh}} Y^+_{loh} \leq 1 - Y_{lo0} \quad l \in L^{\mathrm{S}} \tag{8.27}$$

$$\sum_{h \in H} \sum_{o \in O_{lh}} Y^-_{loh} \leq 1 \quad l \in L^{\mathrm{S}} \tag{8.28}$$

Moreover, a platform can be remodeled at the beginning of a cycle only if it is already in operation and not closed:

$$Y^+_{l,re(o),h} \leq Y_{l,o,h-1} - Y^-_{l,o,h-1} \quad l \in L^{\mathrm{S}}, \; h \in H, \; o \in O_{lh} \tag{8.29}$$

Finally, network state changes from one cycle to the other must be accounted for properly:

$$Y_{loh} + Y^+_{l,re(o),h} + Y^-_{loh} - Y^+_{loh} - Y_{l,o,h-1} = 0 \quad l \in L^{\mathrm{S}}, h \in H, o \in O_{lh} \tag{8.30}$$

Data and constraints on potential product-market offers (see Sect. 7.4.3) must also be adapted to the problem context. A policy $j \in J_k$ for product-market $k \in K$ is now characterized by the following:

- A sales price π_{jpt} for products $p \in P_k$ during each period $t \in T$ of the horizon
- A subset $N^{\leftarrow}_{jp(l,\bar{a})} \subseteq N^{\mathrm{S}}$ of internal nodes which can be used to ship product p to demand zone $l \in L^{\mathrm{D}}_k$ under policy j
- A fixed implementation cost y^{M}_{jt} for periods $t \in T$
- A lower bound $\underline{d}_{jplt}$ and an upper bound $\bar{d}_{jplt}$ on product $p \in P_k$ demand in zone $l \in L^{\mathrm{D}}_k$ during periods $t \in T$

Subset $N^{\leftarrow}_{jp(l,\bar{a})}$ is defined from the standpoint of product-markets, and it specifies the internal nodes that can be used to deliver products to demand zones. Conversely, taking the viewpoint of internal nodes, for a product p one can specify the set $NJ^{\rightarrow}_{pn}$ of (demand zone, policy) pairs $((l,\bar{a}),j), l \in L^{\mathrm{D}}, j \in J_{k(l)}$, which can be served from node $n \in N^{\mathrm{S}}$. As we shall see, the use of this set simplifies the formulation of flow equilibrium constraints significantly.

Because product offers can be revised only at the beginning of a reengineering cycle, and because a single policy can be selected for a cycle, it is required that

$$\sum_{j \in J_k} Y^{\mathrm{M}}_{jh} \leq 1, \quad h \in H, \; k \in K \tag{8.31}$$

For supply contracts, product availabilities $\bar{b}_{plt}$, minimum purchase amounts $\underline{b}_{lt}$, prices c^{V}_{plt}, and fixed contract management costs y^{V}_{lt} must now be specified for each period $t \in T$ of the planning horizon. Supply constraints (7.33) and (7.34) thus become

$$\sum_{n \in N^{\rightarrow}_{p(l,1)}} F_{p(l,1)nt} \leq Y^{\mathrm{V}}_{lh(t)} \bar{b}_{plt}, \quad l \in L^{\mathrm{V}}, \; p \in P_l, \; t \in T \tag{8.32}$$

$$Y^{\mathrm{V}}_{lh(t)} \underline{b}_{lt} \leq \sum_{p \in P_l} \sum_{n \in N^{\rightarrow}_{p(l,1)}} c^{\mathrm{V}}_{plt} F_{p(l,1)nt}, \quad l \in L^{\mathrm{V}}, \; t \in T \tag{8.33}$$

As in Sect. 8.4, the throughput in the network nodes can be defined in terms of outflows to other internal nodes and to demand zones:

$$\begin{gathered} \sum_{o \in O_{nh(t)}} X_{pnot} = \sum_{n' \in N^{\rightarrow}_{pn} \cap N^{\mathrm{S}}} F_{pnn't} + \sum_{(n',j) \in NJ^{\rightarrow}_{pn}} F_{jpnn't}, \\ n \in N^{\mathrm{E}}, \; p \in P^{\rightarrow}_{a(n)}, \; t \in T, \; n \in N^{\mathrm{S}}, \; p \in P^{\rightarrow}_{a(n)}, \; t \in T \end{gathered} \tag{8.34}$$

Average inventory levels in storage nodes $n \in N^{\mathrm{E}}$ must also be modeled. Because a single platform can be implemented on site $l(n)$ during a reengineering cycle, the average level of product p cycle and safety stocks $\bar{I}_{pnot}$ in period t when using platform $o \in O_{l(n)h(t)}$ can be calculated as in Sect. 7.5, giving

$$\sum_{o \in O_{l(n)h(t)}} \bar{I}_{pnot} = \sum_{n' \in N_{pn}^{\leftarrow}} F_{pn'nt} \Big/ \varphi_{pn'n} \quad n \in N^{E}, \; p \in P_{a(n)}^{\rightarrow}, \; t \in T \tag{8.35}$$

In this case, because we consider several reengineering cycles, it can be interesting to transfer inventory from one cycle to another, to compensate for capacity lost during the setup or remodeling of a platform on a site, for example. The following additional notation is necessary to model these *strategic* inventories:

I_{pnot} Level of strategic inventory of product $p \in P$ held at the end of period $t \in T$ in storage node $n \in N^{E}$ when using platform $o \in O_{l(n)h(t)}$

For these inventory levels to be computed properly, the following flow equilibrium constraints must be met:

$$\sum_{o \in O_{l(n)h(t)}} \left(I_{pnot} + X_{pnot} - I_{pnot-1} \right) = \sum_{n' \in N_{pn}^{\leftarrow}} F_{pn'nt}, \quad n \in N^{E}, \; p \in P_{a(n)}^{\leftarrow}, \; t \in T \tag{8.36}$$

Flow equilibrium relations must also be specified for production and transshipment nodes:

$$\sum_{o \in O_{l(n)h(t)}} \sum_{p' \in P_{a(n)}^{\rightarrow}} g_{pp'} X_{p'not} \leq \sum_{n' \in N_{pn}^{\leftarrow}} F_{pn'nt}, \quad n \in N^{F}, \; p \in P_{a(n)}^{\leftarrow}, \; t \in T \tag{8.37}$$

$$\sum_{o \in O_{l(n)h(t)}} X_{pnot} = \sum_{n' \in N_{pn}^{\leftarrow}} F_{pn'nt}, \quad n \in N^{G}, \; p \in P_{a(n)}^{\leftarrow}, \; t \in T \tag{8.38}$$

Minimum activity levels and maximum platform throughputs must also be respected:

$$\underline{b}_{(l,a)oh} Y_{loh} \leq \sum_{t \in T_h} \sum_{p \in P_a^{\rightarrow}} q_{pa} X_{p(l,a)ot}, \quad (l,a) \in N^{S}, \; h \in H, \; o \in O_{lh} \tag{8.39}$$

$$\sum_{p \in P_a^{\rightarrow}} q_{pa} X_{p(l,a)ot} \leq \bar{b}_{(l,a)o} Y_{loh(t)} - \delta_{(l,a)ot} Y_{loh(t)}^{+}, \quad (l,a) \in N^{S}, \; t \in T, \; o \in O_{lh(t)} \tag{8.40}$$

Note that (8.40) accounts for capacity loss caused by the implementation of new platforms. For storage nodes, capacity can also be limited by available storage space:

$$\sum_{p \in P_a^{\rightarrow}} e_p \left(\eta_{pa} \bar{I}_{p(l,a)ot} + I_{p(l,a)ot} \right) \leq b_{(l,a)o}^{S} Y_{olh(t)}, \quad (l,a) \in N^{E}, \; t \in T, \; o \in O_{lh(t)} \tag{8.41}$$

The left side of (8.41) provides an approximation of the storage space required at the end of periods. $\bar{I}_{pnot}$ being an average based on node inventory turnover ratios [estimated via (8.35)], the physical cycle and safety stock level during a period may sometime be higher than this average. Moreover, when new platforms are implemented, the activity in the site stops during the duration of the construction or remodeling work. Consequently, flows during these periods do not cover a full year, and the average inventory level calculated with (8.35) is underestimated. Factor η_{pa} in (8.41) is used to inflate cycle and safety stock levels in an attempt to make the constraint more realistic. However, the accuracy of this constraint should never be taken for granted.

Finally, potential demand and market penetration policies must also be taken into account:

$$Y^{\mathrm{M}}_{jh(t)}\underline{d}_{jplt} \leq \sum_{n \in N^{\leftarrow}_{jp(l,\bar{a})}} F_{jpn(l,\bar{a})t} \leq Y^{\mathrm{M}}_{jh(t)}\overline{d}_{jplt}, \quad t \in T, \ \ l \in L^{\mathrm{D}}_t, \ \ p \in P_l, \ \ j \in J_{k(l)} \tag{8.42}$$

8.5.2 *Objective Function*

The objective of the model formulated is to maximize the value added by the SCN during the planning horizon. To facilitate the presentation and interpretation of results, its objective function is expressed in a form matching company annual financial statements. Revenues and expenses calculations are based on the following simplifying assumptions:

- All variable transportation costs on network arcs, except those coming from vendors, are paid by the shipper, and variable ordering, receiving, and shipping costs are independent of the platform used. Therefore, the variable costs associated with network arcs for period $t \in T$ can be defined as follows:

$f_{pnn't}$ Product p handling cost between node $n = (l, a)$ and $n' = (l, a')$ in site l

$f^{\rightarrow}_{pnn't}$ Unit product p flow cost between nodes n and n' paid by *origin* n (includes relevant customer order, shipping, transportation, and in-transit inventory holding costs)

$f^{\leftarrow}_{pn'nt}$ Unit product p flow cost between nodes n' and n paid by *destination* n (includes relevant supplier–order and reception costs for $n \in N^{\mathrm{S}}$, as well as transportation costs when the origin is a vendor, that is, when $l(n') \in L^{\mathrm{V}}$

- Income taxes are all paid within the same jurisdiction, and the tax rate is the same for all network facilities and for all periods. Consequently, the effect of taxation can be considered indirectly in the calculation of annual rent y_{lot}. As we

Table 8.2 Fixed network costs for period $t \in T$

		Period $t \in T$
Expenses	(a) Market policies	$\sum_{j \in J} y_{jt}^{\mathrm{M}} Y_{jh(t)}^{\mathrm{M}}$
	(b) Supply contracts	$\sum_{l \in L^{\mathrm{V}}} y_{lt}^{\mathrm{V}} Y_{lh(t)}^{\mathrm{V}}$

shall see in the next chapter, this assumption is not valid for multinational SCNs and the explicit consideration of tax rates then becomes important.

Operating expenses can be separated into two categories: (1) fixed costs paid across the network, such as market policy and vendor contract management costs, and (2) expenses incurred by individual network sites. Table 8.2 lists the fixed network costs for each period $t \in T$. Table 8.3 gives the revenues and expenses of each site $l \in L^{\mathrm{S}}$ for each period $t \in T$. Each revenue or expense element is identified by a label, from (a) to (j).

To facilitate the formulation of the objective function, the following financial statement elements are defined:

C_t Corporate network expenses for period t

C_{lt} Total site l expenses (including asset ownership costs) for period t

R_{lt} Total site l revenues for period t.

From the expenses and revenues listed in Tables 8.2 and 8.3, one sees that the value of these financial variables is given by

$$C_t = (a) + (b), \quad t \in T \tag{8.43}$$

$$C_{lt} = (c) + (d) + (e) + (f) + (g) + (h) + (i), \quad l \in L^{\mathrm{S}}, \quad t \in T \tag{8.44}$$

Table 8.3 Site $l \in L^{\mathrm{S}}$ revenues and expenses for period $t \in T$

		Site $l \in L^{\mathrm{S}}$, period $t \in T$
	(c) Raw material procurement	$\sum_{p \in P_l} \sum_{a \in A_l} \sum_{n \in N_{p(l,a)}^{\leftarrow} \cap N^{\mathrm{V}}} c_{pl(n)t}^{\mathrm{V}} F_{pn(l,a)t}$
	(d) Inflows	$\sum_{p \in P} \sum_{a \in A_l} \sum_{n \in N_{p(l,a)}^{\leftarrow}} f_{pn(l,a)t}^{\leftarrow} F_{pn(l,a)t}$
	(e) Platforms	$\sum_{o \in O_{lh(t)}} (y_{lot}^{+} Y_{loh(t)}^{+} + y_{lot} Y_{loh(t)} + y_{lot}^{-} Y_{loh(t)}^{-})$
	(f) Activity	$\sum_{a \in A_l} \sum_{o \in O_{(l,a)h(t)}} \sum_{p \in P_a^{\rightarrow}} c_{p(l,a)ot}^{\mathrm{X}} X_{p(l,a)ot}$
Expenses	(g) Handling	$\sum_{p \in P} \sum_{a \in A_l} \sum_{n \in N_{p(l,a)}^{\rightarrow} \mid l(n)=l} f_{p(l,a)nt} F_{p(l,a)nt}$
	(h) Inventory holding	$\sum_{a \in A_l \cap A^{\mathrm{E}}} \sum_{o \in O_{(l,a)h(t)}} \sum_{p \in P_a^{\rightarrow}} c_{p(l,a)ot}^{\mathrm{I}} [\bar{I}_{p(l,a)ot} + I_{p(l,a)ot}]$
	(i) Outflows	$\sum_{p \in P} \sum_{a \in A_l} [\sum_{n \in N_{p(l,a)}^{\rightarrow} \cap N^{\mathrm{S}}} f_{p(l,a)nt}^{\rightarrow} F_{p(l,a)nt}$ $+ \sum_{(n,j) \in NJ_{p(l,a)}^{\rightarrow}} f_{p(l,a)nt}^{\rightarrow} F_{jp(l,a)nt}]$
Revenues	(j) Sales	$\sum_{a \in A_l} \sum_{p \in P_a^{\rightarrow}} \sum_{(n,j) \in NJ_{p(l,a)}^{\rightarrow}} \pi_{jpt} F_{jp(l,a)nt}$

$$R_{lt} = (j), \quad l \in L^{S}, \ t \in T \tag{8.45}$$

Under the assumptions made earlier, the economic value added by the SCN during planning period t is given by

$$EVA_{t}^{SCN} = \sum_{l \in L^{S}} R_{lt} - (\sum_{l \in L^{S}} C_{lt} + C_{t}), \quad t \in T \tag{8.46}$$

The objective of the company being to maximize the sum of discounted value added over the planning horizon, the mathematical programming model to solve is the following:

$$V^{SCN} = Max \sum_{t \in T} \left[\frac{\text{EVA}_{t}^{\text{SCN}}}{(1+r)^{t}} \right]$$

subject to

- Platform selection constraints (8.26) to (8.30)
- Vendor contract and capability constraints (8.32) and (8.33)
- Flow equilibrium and inventory accounting constraints (8.34) to (8.38)
- Platform implementation and capacity constraints (8.39) to (8.41)
- Supply and demand constraints (8.31) and (8.42)
- Financial variables definition constraints (8.43) to (8.46)
- Nonnegative and binary decision variables domain constraints.

As before, coefficient r in the objective function denotes the weighted average cost of the capital (WACC) of the company.

This mathematical program is more elaborate than the previous ones and it may be difficult to solve even with current commercial solvers. An efficient meta-heuristic to solve it was proposed by Carle et al. (2012). Decision support systems developed to facilitate the formulation and the resolution of this kind of model, as well as to ease the exploration of the solutions obtained, are generally used in practice in SCN design projects. These specialized tools are examined in the last chapter of the book.

8.5.3 Financial Constraints

The SCN design model formulated in the previous sections implicitly assumes that the company concerned can always raise the capital required to invest in capacity. This is not necessarily the case. For example, despite the fact that Canada has been one of the largest exporters of pulp and paper in the world since the beginning of the twentieth century, Canadian pulp and paper companies have had difficulty raising new capital in recent years because their ROCE is substantially lower than the cost of capital and because of their high debt-equity ratio. The equipment used by several companies is old and productivity is suffering from this. The efforts of the companies to modernize their SCN in order to improve profitability are hampered by financial constraints. Taking the financial situation of a company into account precisely requires a detailed modeling of balance sheet elements. Nonetheless, capital availability constraints can be added to our current model without too much effort.

In our model, the capital financing means used by a company are hidden in the calculation of the annual *rent* included in the fixed platform costs y_{lot}, which also include fixed operating costs. To keep things simple, assume that fixed operating costs can be covered by the operating revenues generated with SCN assets but that all capacity investments must be financed with new capital (long-term debt or stock) or with capital recovered from the disposal of current assets. Recall, however, that some resources may be leased and some activities may be outsourced, which does not require any new capital. In our current model, all interests paid for the amounts borrowed, dividends paid for the stocks sold, and ensuing tax effects are factored in the calculation of annual rents. However, there is a limit to the amount of capital the company is able (or wants) to raise on the market.

To take this into account, the following notations are introduced:

R^+_{loh} Additional capital required to implement capacity option o on site $l \in L^S$ at the beginning of reengineering cycle h

R^-_{loh} Capital recovered at the beginning of cycle h when platform o on site $l \in L^S$ is shut down

CA_h Limit on the capital available for investments in capacity during cycle h.

Note that R^+_{loh} and/or R^-_{loh} can be null for some of the options considered, for example, when storage space is rented in a public warehouse or when the production of a component is outsourced to a contract manufacturer. The following financial constraints can be added to our model:

$$\sum_{l \in L^S} \sum_{o \in O_{lh}} R^+_{loh} Y^+_{loh} - \sum_{l \in L^S} \sum_{o \in O_{lh}} R^-_{loh} Y^+_{loh} \leq CA_h, \quad h \in H$$

Because the financial position of the company and of capital markets may be difficult to anticipate for long planning horizons, the constraint may be included only for the first reengineering cycles.

Another approach may be to include a debt-to-equity ratio constraint or a debt service constraint for each planning period. This, however, requires an explicit modeling of loan reimbursements, dividend payments, and asset depreciation, which complicates the models significantly. The modeling of long-term financing decisions is discussed in detail in Peterson (1969).

Review Questions

8.1. What is meant by capacity?
8.2. What is a capacity strategy?
8.3. Capacity can often be expressed in terms of a critical production resource. How would you measure capacity for the following food industry companies?

- A food processor in the dairy industry
- A wine producer
- A slaughterhouse
- A food products distributor
- A retail store
- A cattle ranch
- A restaurant

8.4. Propose a capacity strategy that could apply to a company of your choice. To better answer this question, refer to Fig. 8.3 on generic capacity strategies. Justify your answer.
8.5. Why should net present value (NPV) be considered in the assessment of a capacity option?
8.6. Why do capacity strategies focus on three dimensions: technological, spatial, and temporal?
8.7. How does the resources–configuration–methods triplet have an impact on capacity? Give an example, other than the one presented in the text, to support your explanations.
8.8. In this chapter, static and dynamic strategic capacity planning models were presented. What is the fundamental difference between these two approaches?

Exercises

Exercise 8.1 Bonbiscuit Co. began to export its cookies over the last 2 years and has to expand to meet demand. In light of its development strategy and of the results obtained to date, it forecasts that demand will significantly increase during the next 5 years, as shown in Table 8.4. However, Bonbiscuit mixers are already used at full capacity, and the company needs to invest in additional capacity to meet demand. An engineering study concluded that the price of mixers depends on the quantity of dough it may stir, as shown by the following cost function:

$$C(b) = 50b^{0.7}(\text{in } \$1000)$$

where b is the volume of the mixer in kiloliters. Note that commercially available mixers have volumes of 4, 5, 6, 7, 8, 9, 10, 11, 12, 13, or 14 kl.

Table 8.4 Demand for the next 5 years

Year	1	2	3	4	5
Additional demand (kiloliters/day)	3	5	4	7	3

If the weighted average cost capital (WACC) of the company is 12 %, what should be its capacity expansion strategy?

Exercise 8.2 Topclim, a Canadian manufacturer of household air conditioning units, saw its demand south of the border increase significantly during the last year, which is forcing the company to develop a manufacturing network in the United States. For the coming years, the company forecasts an annual demand of 180,000 units in the southern United States, 120,000 units in the Midwest, 110,000 units in the East, and 100,000 units in the West. Four potential sites have been identified for the construction of the company's new plants: New York, Atlanta, Chicago, and San Diego. Platforms for producing 200,000 units or 400,000 units have been designed by the company engineers. The annual fixed cost for the implementation of each platform at each potential site is given in Table 8.5. The table also provides variable production and transportation costs for servicing targeted markets from each potential site.

Propose an optimization model that could be used to solve this problem. What manufacturing network should Topclim implement?

Exercise 8.3 After studying the optimum solution found for Exercise 8.2, Topclim management realizes that, because the company already has a plant in Quebec City to service the Canadian market, and because it plans to develop the Mexican market in the near future, it would be more appropriate to optimize their manufacturing network across North America. Management also comes to realize that in such a context it would be preferable to maximize net operating profits in Canadian dollars rather than minimize costs.

(a) What additional data must be compiled by Topclim to design a manufacturing network maximizing net operating profits?
(b) Modify the model proposed in Exercise 8.2 to adapt it to this new context.
(c) In real SCN design projects, a much larger number of demand zones and more realistic capacity options are considered, and the models obtained incorporate

Table 8.5 Topclim data

	New York	Atlanta	Chicago	San Diego
Annual fixed cost				
200,000 units platform	$6 million	$5.5 million	$5.6 million	$6.1 million
400,000 units platform	$10 million	$9.2 million	$9.3 million	$10.2 million
Prod.-tranp. cost/unit				
East	$211	$232	$238	$299
South	$232	$212	$230	$280
Midwest	$240	$230	$215	$270
West	$300	$280	$270	$225

thousands of variables and constraints. What tools would you use to formulate and solve a problem of this size?

Exercise 8.4 Soapcan manufactures industrial detergents, packaged in metal drums, for the northeastern North-American market. Currently, the company has a production–distribution center (PDC) with an annual capacity of 200,000 cwt in the Montreal area. Manufactured products are shipped directly to customers who are close to the PDC or, for more distant clients, via a distribution center (DC) in the Toronto area. The Toronto DC is located in a public warehouse and it has ample capacity to meet the needs of the company. Customers buy products in truck loads (TL) and they expect deliveries in less than 2 days.

Exports to the United States recently have gained momentum and the projected sales for next year in the main demand zones of the company are as follows:

- Eastern Quebec (Demand Zone 5): 15,000 cwt
- Western Quebec (Demand Zone 6): 65,000 cwt
- Ontario (Demand Zone 7): 75,000 cwt
- NE USA (Demand Zone 8): 100,000 cwt
- MW USA (Demand Zone 9): 85,000 cwt

To cope with this demand, Soapcan needs to increase its production capacity, and it considers expanding the Montreal PDC to be able to produce an additional 200,000 cwt. Alternatively, it could implement a new PDC with a capacity of 200,000 or 400,000 cwt in the Pittsburgh area, which could lead to a shutdown of the Montreal plant. To maintain a high service level, Soapcan does not want the centroid of its demand zones to be more than 500 miles from the DC or PDC that serves it. To be able to do this, the company also considers the possibility of signing a contract with a large public warehouse in the Detroit area.

Currently, variable warehousing costs in Montreal and Toronto are approximately \$2/cwt and it is expected that they will be of the same order of magnitude in Pittsburgh and Detroit. The annual fixed cost for the management of a warehousing contract is estimated at \$10,000. Second, from historical data, it was estimated that transportation costs by cwt as a function of distance (in miles) follow the following linear relationship:

$$\$/\text{cwt} = 2.00 + 0.01 * \text{distance}$$

Distances in miles between the different locations of the potential SCN are provided in Table 8.6. Finally, the capacities as well as the annual fixed costs and variable production costs associated with PDCs are provided in Table 8.7.

(a) Formulate an optimization model that could be used to solve this problem.
(b) Find the model's optimal solution using the Excel Solver or any another optimization tool.

Table 8.6 Inter-location network distances in miles

	DZ 5	DZ 6	DZ 7	DZ 8	DZ 9	Toronto	Detroit
1-Montreal	100	150	550	550	1000	350	700
2-Pittsburgh	800	600	300	400	450	500	300
3-Toronto	450	200	200	400	650		
4-Detroit	800	550	250	600	300		

Table 8.7 Platform capacity and costs

Site	Platform	Capacity (cwt)	Fixed cost	Prod. cost
1-Montreal	1 (status quo)	200,000	\$2000,000	\$20
1-Montreal	2 (expansion)	400,000	\$5500,000	\$18
2-Pittsburgh	1 (addition)	200,000	\$4000,000	\$18
2-Pittsburgh	2 (addition)	400,000	\$7,000,000	\$16

Exercise 8.5 Having solved the model formulated for Exercise 8.4, you realize it does not account for inventory holding costs in DCs and PDCs. Based on historical data on flows and inventories, you are able to establish the following relationship between the average level of inventory to keep in a storage point (in cwt) $\bar{I}(X)$ and the annual flow of products X in this storage point (in cwt):

$$\bar{I}(X) = 35X^{0.6}$$

Knowing that the detergent sold is worth \$40 per cwt and the inventory holding cost rate used by the company is 20 %, do the following:

1. Calculate the inventory holding costs of the solution you proposed for Exercise 8.4.
2. Given the magnitude of these costs, you decide to change your model to account for inventory holding costs. The model thus obtained being nonlinear, you decide to solve it approximately using successive linear programming (see Sect. 7.6). To initiate the solution process, divide the total demand equally among all DCs and PDCs. What is the storage cost per cwt thus obtained for each PDC and DC? How does this cost alter the unit variable flow costs in your original model?
3. The optimal solution of the linear program solved for this first iteration gives the following flows:

$$F_{15} = 15{,}000 \quad F_{27} = 15{,}000$$
$$F_{16} = 65{,}000 \quad F_{28} = 100{,}000$$
$$F_{13} = 60{,}000 \quad F_{29} = 25{,}000$$
$$F_{37} = 60{,}000$$

Knowing this, calculate the storage cost per cwt to use for each DC and PDC in the second iteration of the solution algorithm.

Exercice 8.6 The capacity planning approach described in Sect. 8.4 requires the a priori definition of a number of potential platforms $o \in O_s$ for each site s with *given* production capacity $\bar{b}_{pso}, p \in P$, for the product families manufactured, and *given* storage capacity $\bar{b}_{so}$, which is not always ideal because it is difficult to determine the capacity required on each site before the model is solved. An alternative approach would be to derive capacity cost functions reflecting potential economies of scale (see Fig. 8.5 for an example) for each activity and to use them instead of optional platforms to formulate an optimization model. Assuming that economies of scale for an activity can be modeled by a power function similar to (8.4), and that the amount of capacity that can be added to a site may not exceed a predetermined limit, propose a model for the optimization of the capacity strategy of an SCN. Suggest an approach to solve your model.

Exercice 8.7 As illustrated in Fig. 8.15, receiving and shipping activities are not considered explicitly in the models proposed in Sects. 8.4 and 8.5. A simple way to take these activities into account explicitly is to assume that the annual inbound and outbound flows in and out of a given platform are limited by an upper bound reflecting the capacity provided by its receiving and shipping resources (docks, staging area, sorting equipment, etc.). Formulate a constraint to add to the models in Sects. 8.4 and 8.5 to ensure that the receiving and shipping capacity of selected platforms is respected.

Exercice 8.8 In the model presented in Sect. 8.5, it was implicitly assumed that the transportation means used for network arcs are predetermined. As seen in Chap. 5, several transportation options, however, could be considered: alternative potential contracts with 3PLs, for example. Basing yourself on the approach used to model platform selection decisions, suggest an extension of the model developed in Sect. 8.5 to take into account transportation options.

Bibliography

Amrani H, Martel A, Zufferey N, Makeeva P (2011) A variable neighborhood search heuristic for the design of multicommodity production-distribution networks with alternative facility configurations. OR Spectrum 33(4):989–1007

Ballou R (1992) Business logistics management, 3rd edn. Prentice-Hall, Englewood Cliffs

Carle M-A, Martel A, Zufferey N (2012) The CAT metaheuristic for the solution of multi-period activity-based supply chain network design problems. Int J Prod Econ 139(2):664–677

Correira I, Melo T, Saldanha da Gama F (2012) Comparing classical performance measures for multi-period, two-echelon supply chain network design problem with sizing decisions. Technical report on logistics of the Saarland Business School

Couillard J, Martel A (1990) Vehicle fleet planning in the road transportation industry. IEEE Trans Eng Manage 37(1):31–36

Driver C (2000) Capacity utilisation and excess capacity: theory, evidence, and policy. Rev Ind Organ 16:69–87

Eppen G, Kipp Martin R, Schrage L (1989) A scenario approach to capacity planning. Oper Res 37(4):517–527

Freidenfelds J (1981) Capacity expansion. North-Holland
Hayes R, Wheelwright S (1984) Restoring our competitive edge: competing through manufacturing. Wiley, New York
Hodder J, Dincer M (1986) A multifactor model for international plant location and financing under uncertainty. Comput Oper Res 13(5):601–609
Hopp W, Spearman M (2008) Factory physics, 3rd edn. Waveland Pr Inc
Julka N, Bainesb T, Tjahjonob B, Lendermanna P, Vitanov V (2007) A review of multi-factor capacity expansion models for manufacturing plants: searching for a holistic decision aid. Int J Prod Econ 106:607–621
Lee S, Luss H (1987) Multifacility-type capacity expansion planning: algorithms and complexities. Oper Res 35(2):249–253
Li S, Tirupati D (1994) Dynamic capacity expansion problem with multiple products: technology selection and timing of capacity additions. Oper Res 42(5):958–976
Martel A (2005) The design of production-distribution networks: a mathematical programming approach. In: Geunes J, Pardalos P (eds) Supply chain optimization. Springer, Berlin, pp 265–306
Mazzola J, Schantz R (1997) Multiple-facility loading under capacity-based economies of scope. Naval Res Logistics 44:229–256
Montreuil B, Boctor F, Martel A (1995) La maîtrise des technologies de production. In: Martel A, Oral M (eds) Les défis de la compétitivité: Vision et stratégies. Publi-Relais, Montreal
Olhager J, Rudberg M, Wikner J (2001) Long-term capacity management: Linking the perspectives from manufacturing strategy and sales and operations planning. Int J Prod Econ 69:215–225
Paquet M, Martel A, Desaulniers G (2004) Including technology selection decisions in manufacturing network design models. Int J Comput Integr Manuf 17(2):117–125
Paquet M, Martel A, Montreuil B (2008) A manufacturing network design model based on processor and worker capabilities. Int J Prod Res 46(7):2009–2030
Peterson D (1969) A quantitative framework for financial management. Irwin
Porter M (1980) Competitive strategy. Free Press
Rajagopalan S, Soteriou A (1994) Capacity acquisition and disposal with discrete facility sizes. Manage Sci 40(7):903–917
Sahinidis N, Grossmann I, Fornani R, Chathrathi M (1989) Optimization model for long range planning in the chemical industry. Comput Chem Eng 13(9):1049–1063
Shapiro J (2008) Modeling the supply chain, 2nd ed., Brooks/Cole Publishing
Van Mieghem J (2003) Capacity management, investment and hedging: Review and recent developments. Manuf Serv Oper Manag 5(4):269–302
Verter V, Dincer C (1992) An integrated evaluation of facility location, capacity acquisition, and technology selection for designing global manufacturing strategies. Eur J Oper Res 60:1–18
Verter V, Dincer C (1995) Facility location and capacity acquisition: an integrated approach. Naval Res Logistics 42:1141–1160
Vila D, Martel A, Beauregard R (2006) Designing logistics networks in divergent process industries: a methodology and its application to the lumber industry. Int J Prod Econ 102: 358–378
Wagner H, Whitin T (1958) Dynamic version of the economic lot size model. Manage Sci 5:89–96
Weigel G, D'Amours S, Martel A, Watson P (2009) A modeling framework for maximizing value creation in pulp and paper mills. INFOR 47(3):247–260

Chapter 9
Designing Multinational SC Networks

Impacts of doing business globally, or at least in several countries, have been discussed gradually in several preceding chapters. The forces that drive globalization are examined in Chap. 1, which looks at preferential trade agreements (PTAs), global demographic changes, industrial delocalization–relocalization, worldwide resources availability, country technology development–adaptation trends, mergers leading to the creation of multinational giants, and increased risks (and security measures) ensuing from international commerce. It explains that the current global two-speed economy requires the implementation of complex supply chains. Chapter 4 surveys country factors that are crucial for the location of business units as well as sources of information available to make multinational comparative studies. Chapter 5 looks at international trade flows, and it identifies the steps and terms of sale necessary for international shipping. Finally, Chap. 6 discusses partnerships in global supply chains. This provides necessary background for the multinational SCN design problem studied in this chapter.

The majority of the SCN design modeling elements introduced in Chaps. 7 and 8 apply to multinational networks but, beyond the explosion of the problem size, several additional complexity factors must be considered. In this chapter, we start by completing our review of the international business context. We then look at the structure of multinational supply chain networks and we stress the necessity of considering total landed costs and lead times to make adequate SCN design decisions. Additional complexity factors related to exchange rates fluctuations, transfer pricing, taxation, tariffs, government subsidies, and various barriers to trade are examined. Finally, we present a typical multinational SCN design model taking these factors into account.

A. Martel and W. Klibi, *Designing Value-Creating Supply Chain Networks*,
DOI 10.1007/978-3-319-28146-9_9

9.1 International Business Context

International business can take many forms: the exportation of locally manufactured products or services to foreign markets; the importation of raw materials, components, or services from foreign sources; manufacturing abroad; or a combination of all these. Obviously, what is exported by one country is necessarily imported by other countries. Doing business internationally adds complexity because of the increased geographical, cultural, institutional, and economic distances between business partners. Despite this, the development of international supply chain networks provides many advantages. This section examines the context, opportunities, and challenges of international business.

9.1.1 *The Gains from International Trade*

Common sense suggests that at least some international trade is necessary. Canadians would have a very restricted choice of fruits and vegetables during winter without it, and OPEC (Organization of the Petroleum Exporting Countries) countries would not benefit from petrodollars. Economic theory (Smith, Ricardo, Heckscher-Ohlin, and Krugman) has long been advocating unrestricted free trade. It suggests that a country should concentrate on the production and export of the goods it can manufacture more efficiently and import goods that can be produced elsewhere more efficiently. The comparative advantage of a country comes from factors such as climate, location, natural resources endowment, availability of capital, technological know-how, research facilities, transportation and telecommunication infrastructures, and labor productivity. Some of these factors are inherited but others can be improved by people, businesses, and governments. Countries export goods that require locally abundant factors and import goods needing locally scarce factors. This partly explains why Brazil exports coffee, Saudi Arabia oil, and Japan automobiles. Exports lead to higher production volumes, which prompt economies of scale. The price of products thus decreases and demand increases. All this leads to higher levels of prosperity.

Economic theory also points out that some countries are front-runners for some products not because they have better factors but simply because their firms were the first to make that product (first-mover advantage). This probably explains, for example, why Boeing led the aircraft industry for several decades. The number of firms competing in several global industries is relatively small and barriers to entry are high. After studying 100 industries in 10 countries, Porter (1990) stresses that, in addition to factor endowments, the creation of a national competitive advantage in an industry depends on demand conditions, efficient related and supporting industries, and intense domestic rivalry. Rivalry is also important at the international level. As the Nobel laureate, Robert Solow puts it (Baily and Comes 2014), exposure to competition from whoever in the world has the best practice is necessary to spur local productivity improvements: "…international trade serves a purpose beyond exploiting comparative advantage. It exposes high-level managers in various countries to a little fright. And fright turns out to be an important motivation."

Exports account for about 30 % of the world GDP. Moreover, there is a strong positive correlation between countries' GDP per capita and exports, which suggest that international trade contributes to prosperity (Ghemawa and Altman 2012). At an industry level, the ADDING framework proposed by Ghemawa (2007) provides a rationale for explaining why deploying abroad may create value. It stresses that by becoming international, companies can accomplish the following:

- **A**dd volume (growth) to obtain economies of scale and scope
- **D**ecrease costs by getting access to better factors (raw materials, energy, capital, proximity, talent, labor, etc.)
- **D**ifferentiate themselves by offering better order winners and by building on brand reputation to increase willingness to pay (and thus revenues)
- **I**mprove industry attractiveness by moving to high profitability countries
- **N**ormalize risks (supply and demand variability, prices and exchange rates volatility, etc.) by doing business in an efficient portfolio of countries
- **G**enerate knowledge to improve products and processes

All these potential advantages do not necessarily apply to all industries in all countries, and a serious multinational strategy analysis must be performed to assess potential gains.

9.1.2 Country Differences and Similarities

The DHL Global Connectedness Index (Ghemawa and Altman 2012, 2014), a measure assessing the connectedness of 140 countries covering 99 % of the world GDP based on trade, capital, information, and people flows, indicates that the world is less globalized than is often presumed, and that borders still matter. It shows that most international flows (even online) take place within rather than between continental regions and that distance continues to be a serious barrier. It is much easier to do business with countries that are relatively closer than with those that are distant. And the notion of distance alluded to here is not purely geographical; it also involves cultural, administrative (institutions, governance), and economic dimensions. A successful business model at home is not necessarily profitable elsewhere. Take the case of Walmart. Figure 9.1 gives an estimate of Walmart's profits in different countries in 2004. Observe that profitable countries are neighbors of the United States sharing key similarities (common language, NAFTA partners). Differences between the US and nonprofitable countries are much stronger. Note, however, that Walmart is very successful at procuring low-cost products from China for its domestic market, so distance is not an unsurmountable difficulty. Substantial negative distance effects, however, have been detected for trade flows,

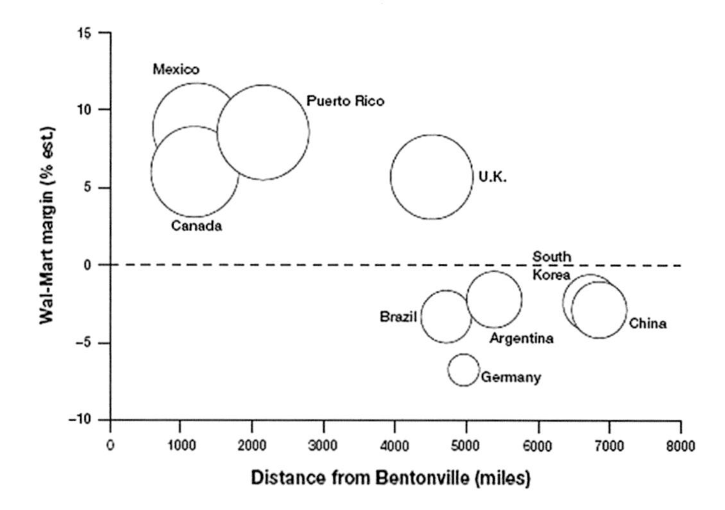

Fig. 9.1 2004 Walmart international's operating margin by country (estimated)—*Source* Ghemawa (2007)

foreign direct investment (FDI), equity trading, patent citations, and e-commerce, and they do not seem to decrease significantly with time (Ghemawa 2007).

The world can be seen as a network of interrelated countries. Assuming a company has a home country, then a multi-attribute distance vector can be attached to each link with foreign countries. Some of these attributes are characterized by quantifiable metrics, for example, the distance in miles between the capitals of two countries. Others reveal a difference or similarity between a pair of countries, for example, the fact that they speak the same language or use different currencies. Table 9.1 lists a number of attributes that should be examined when evaluating the attractiveness of a foreign market, supply source, or manufacturing solution. According to Ghemawa and Altman (2012), "countries that share a common language trade 42 % more than countries that don't, countries in the same trade bloc trade 47 % more, and if you double the [geographic] distance between a pair of countries their trade will drop by half." It should be noted that distance attributes are conditioned by industry characteristics and by the type of business relationship considered. For example, geography is more critical for products having a low value-to-weight ratio and for fragile or perishable products; the localization of a new plant overseas requires large investments, so it is very sensitive to administrative distance. Making differences visible is a prerequisite for making sound international business decisions.

Table 9.1 International distance attributes

Geographic	Cultural	Administrative	Economic
Physical distance	Language	Colonial ties	Economic freedom
Physical area	Ethnicity	Trading block	Economic development
Ease of access	Religion	Currency	Per-capita income
Common border	Education	Political system	Factor endowment
Time zone	Social structures	Legal environment	Industry concentration
Climate	Values	Regulations	
Inter-country transportation	Norms	Home bias	
Inter-country communications	Business customs	International organizations membership	
	Criminality		
	Ethics		

See Ghemawa (2007) and Hill (2014) for a detailed discussion of these attributes

9.1.3 Industry SC Structures

Supply chain structures vary widely from one industry to another. In some industries, production and consumption are still relatively local, but in others, they are clearly global. For multinational industries, dominant product flows do not always follow the same pattern. For example, in the pharmaceutical industry, both production and demand are concentrated in advanced economies. However, in the mobile phone industry, a large part of production and consumption is taking place in developing countries. In the car industry, production and demand are spread across the world. Finally, numerous labor-intensive products are now produced essentially in emerging economies. In their 2012 global connectedness report, Ghemawa and Altman (2012) examine the depth and breadth of 22 industrial sectors. Depth measures how much the flows of an industry are international by comparing finished product exports to domestic production. Breadth looks at how broadly exports are distributed across countries based on interregional flows, average distance traversed by products, and value-to-weight ratios. Industry depth and breadth are compared in Fig. 9.2 using, respectively, finished product exports to production quantity ratios and average distance traversed by exported products.

As can be seen, the *depth* metric varies widely among industries. It shows, at one end, that production and consumption of products such as electricity and milk remain essentially local, and at the other end, integrated circuits and microwaves are traded globally. Note that the depth ratio exceeds 100 % for some industries because the flows of intermediate and finished products are not distinguished and the industry SC may cross borders more than once. The *breadth* metric also displays a wide variation among industries. It indicates, for example, that the small quantity of electricity exported does not travel very far. Also, although a large part

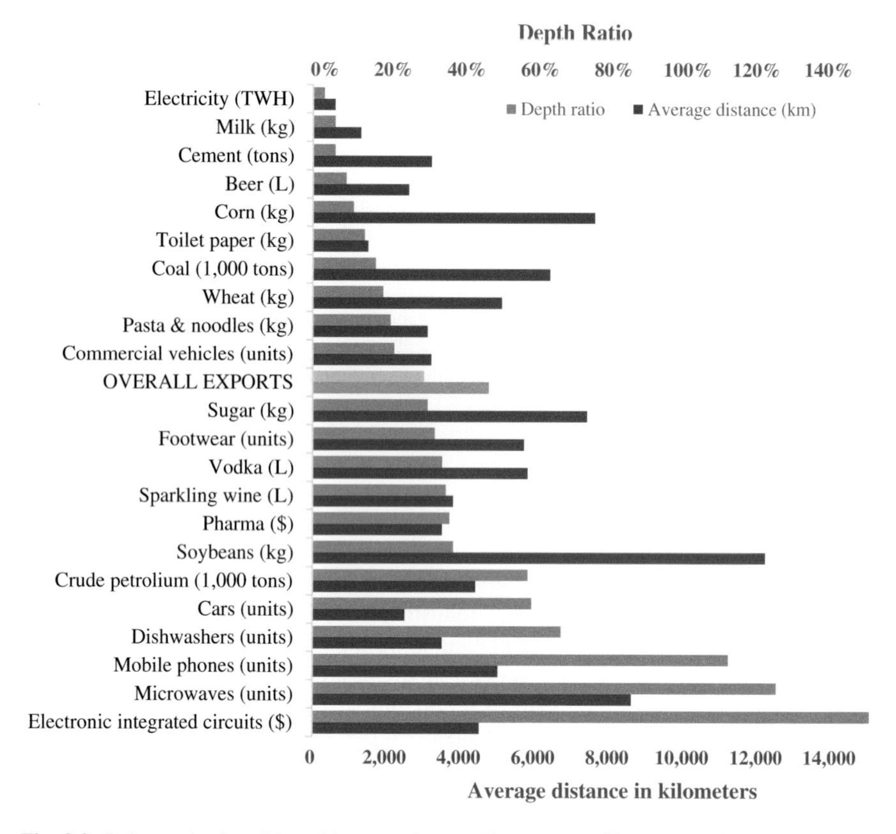

Fig. 9.2 Industry depth and breadth comparisons—*Data source* Ghemawa and Altman (2012)

of the car and mobile phones produced are exported, the moderate distance they travel (2500 km for cars and 5000 km for phones) indicate that a significant part of these exports occur within continental regions. At the other extreme, soybeans, largely produced in the Americas, travel more than 12,000 km to Asian countries where a significant proportion of derived products is exported. Another measure examined in the global connectedness report is exported products value-to-weight ratio. The study confirms that products with high ratios are more likely to be traded internationally because their high value offsets their comparatively low transportation costs.

Another relevant factor for the development of multinational SCs is that the maturity of an industry may be quite different from country to country. For example, although the Canadian pulp and paper industry is large, its typical mill at the beginning of the twenty-first century was relatively old and small, with the consequence that its manufacturing costs were comparatively high. The global position of Canadian bleached softwood kraft pulp (BSKP) mills at that time is

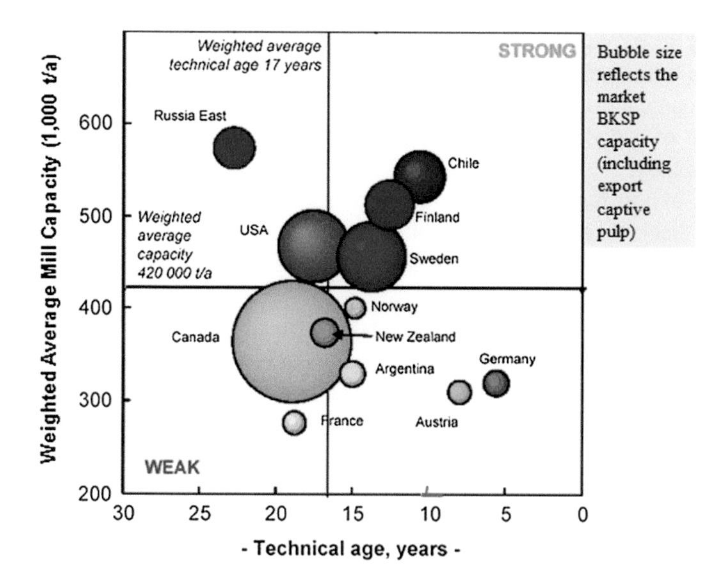

Fig. 9.3 Bleached softwood kraft pulp capacity by country—*Source* Jaakko Poyry, quoted from Roberts (2005)

depicted in Fig. 9.3. This, coupled with a revaluation of the Canadian dollar, a demand slowdown, and increased raw material costs, drove the Canadian pulp and paper industry into an unprecedented crisis. In order to get out of this storm, North American SCNs had to be completely revamped: older inefficient mills were closed, mills dedicated to declining markets, such as newsprint, were transformed to produce value-added papers, new technologies were adopted, and the industry reorganized itself through mergers and acquisitions.

9.1.4 Government Interventions

Given the potential gains of international commerce, it is natural for governments to try to improve the attractiveness of their country. They can do this via policies that target trade but also by improving their domestic business environment. Several instruments can be used to influence trade and foreign direct investments (FDI):

- Tariffs, that is, fix-charge or value-percentage taxes collected on imported goods
- Import quotas, that is, limitations on the quantity of imports (set globally or against specific countries) during a specified period of time; limits on the quantity of imports at a reduced tariff may also be set
- Voluntary export restraints, that is, restrictions on export quantities to a specific country during a given period of time; these may be offered to an importing

country to avoid the effects of possible trade restraints following a surge of exports to that country
- Local content requirements, that is, constraints specifying that some specific fraction of a good (in terms of number of components or value) be produced domestically
- Import–export regulations that relate to customs clearance procedures, import–export license requirements, health and safety standards, and so on; some countries use these procedures and norms to artificially restrain imports or exports
- Subsidies (government payments made to local producers to help compete against low-cost foreign imports or to encourage exports) taking the form of cash grants, low-interest loans, tax breaks, and government equity participation
- National procurement programs, that is, policies requiring that a specified percentage of purchases by federal or state governments be made from domestic firms
- Antidumping measures, that is, restraining actions against foreign companies exporting a product at a price lower than its production cost or than the price it charges in its home market
- FDI incentives (auspicious conditions offered to foreign firms to invest in the country) taking the form of tax concessions, low-interest loans, subsidies, or access to state-controlled resources at a low price (e.g., electricity)—home countries may encourage FDI by eliminating double taxation of foreign income or by providing insurance programs to cover risks of expropriation, war losses, and so on
- FDI restrictions, that is, ownership or performance policies applied to control foreign direct investments; the former may require that a significant proportion of the equity of a foreign subsidiary be owned by local investors; performance requirements typically are related to local content, exports, technology transfer, and local participation in top management

Policy choices must be made to fit country structural conditions and in particular its geographic position. As mentioned, the tendency over the last decades has been to eliminate barriers through the negotiation of preferred trade agreements. Several nontariff barriers are still present, however. On the other end, governments are increasingly competing to attract companies to their country. A company can therefore typically negotiate its presence abroad and benefit from substantial advantages.

The attractiveness of a country does not depend only on its trade and FDI policies but also on the quality of its domestic business environment. The World Bank compiles a logistics performance index (Arvis et al. 2014) every 2 years since 2007 to measure the efficiency of international trade supply chains and in particular to assess how governments are successful at improving logistics performance in their country. The index analyzes six crucial global business factors:

- The efficiency of customs and border management clearance
- The quality of trade and transport infrastructure

- The ease of arranging competitively priced shipments
- The competence and quality of logistics services
- The ability to track and trace consignments
- The frequency with which shipments reach consignees within scheduled or expected delivery times

Long-term commitments to the improvement of these performance factors from policymakers are essential for the development of international commerce. Leaders in Europe (e.g., Germany and the Netherlands), in Asia (e.g., Singapore and Japan), or in the Americas (e.g., the United States and Canada) see seamless logistics as an engine of growth and integration with global supply chains.

9.1.5 Import–Export Transactions

Import–export transactions are much more complex than domestic buying–selling transactions because two different business environments, two legal systems, and two currencies are involved and because distance complicates everything, as discussed previously. The terms of sale applying to international trade, the tasks required to ship goods overseas, and the service providers able to help facilitate the process were discussed in Chap. 5. In sum, the steps involved in a typical import–export transaction are the following:

1. After some negotiation between parties, the importer places an order with the exporter in which the items purchased, the terms of sale (incoterms), and the payment method are specified.
2. The exporter confirms acceptance of the terms of sale and payment method, and the order then becomes a binding contract.
3. The importer selects a custom broker to clear goods when they arrive at the border.
4. The exporter asks a freight forwarder to arrange transportation (by container, ship, air, rail, and/or truck) between the shipper location and the port of debarkation or buyer premises (depending on incoterms). This may involve freight consolidation at the port of embarkation.
5. The exporter prepares the merchandise (picking, packing, labeling, and export documents) for transportation.
6. The initial carrier loads the cargo at the exporter's dock and transports it to the next transshipment point in the multimodal transportation route setup. Subsequent carriers take charge of the cargo as it arrives at transshipment points.
7. After the goods are shipped, the forwarder sends the necessary documentation (commercial invoice, custom invoice, packing list, bill of lading or air waybill, certificate of origin) to the customs broker clearing the goods for the importer at the port of entry.

8. The customs broker submits documents to customs to obtain release of the goods. Some taxes, duties, and/or penalties may have to be paid before the release.
9. The merchandise is transported to the importer's premises under the control of the forwarder or the importer depending on incoterms. The importer accepts the goods, after inspection, by signing the bill of lading or the waybill.

Clearly, when frequent shipments are made, long-term contracts may be signed between the importer and the exporter, between the exporter and the freight forwarder, and between the importer and the custom broker to streamline operations. Also, the importer and exporter may be two divisions of a multinational company, which alters the process to a certain extent. When a preferred trade agreement exists between the two countries, some steps may be simplified. However, additional validations are required to ensure that the transaction satisfies PTA terms.

The payment method used may also be quite complex because two currencies are involved and because of perceived trust or risk issues. Shipping before payment is generally viewed as risky for the exporter and advanced payment as risky for the importer. An intermediate solution involving banks is thus often adopted. Despite the risks involved, it is estimated that more than one-third of US and UK exports are made on an open account (payment within an agreed time) or consignment (payment after the goods are sold by the importer) basis (Seyoum 2013). These approaches should be used only with foreign customers who have good credit ratings and are well known by the exporter. Note, however, that exporters can protect themselves to some extent by buying foreign credit risk insurance. When payments are delayed, there is also the risk that the currency specified in the sales contract devaluates in the meantime. Protection against foreign exchange rate variations is discussed at length later on in the chapter.

More secure modes of payment involve the use of a *draft* or of a *letter of credit* (L/C). In both cases, the payment is made through the parties' banks. Using a draft is simpler and less costly than using an L/C, but it is not as secure. After an exporter has arranged for shipment and prepared the necessary documents, he or she forwards the documents to his or her bank with payment instructions. The bank then forwards the documents to the importer's bank (collecting bank) with the instructions for their release against payment (or other terms). The collecting bank then gives the documents to the importer in exchange for payment and sends the proceeds to the other bank for payment to the exporter. The L/C is similar but it involves an additional a priori step to ensure, before the merchandise is shipped by the exporter, that the money required to make the payment has already been received by the importer's bank. All sorts of variants exist to comply with specific needs. In some cases, countertrade may also be used. These are import–export transactions in which products or services are traded in exchange for other products or services.

9.1.6 Ethical Issues

The complexity of global supply chains is such that it is often difficult to know what is going on beyond first-tier suppliers. Nonetheless, multinationals are increasingly expected to ensure that all their suppliers respect human rights, safety standards, and ecosystems, and that they adopt honest business practices. Their reputation is at stake each time the integrity of their SC is compromised. Our tightly connected world makes it impossible to hide unethical sourcing practices. In international commerce, ethical issues and dilemmas arise from the distance between the political system, laws, economic development, and culture of the countries involved and the low-bid-wins mentality, which, fortunately, seems to be fading. Managers may find themselves in situations in which none of the available alternatives seems ethically acceptable. This is complicated by the fact that what is considered ethically acceptable by managers from one country may be perceived as unethical by managers from another country. Ethical dilemmas are often related to employment practices, environmental regulations, corruption, and the moral obligations of multinationals. These are important and difficult issues, and their study goes beyond the scope of this book. A good discussion of ethics in international business is found in Hill (2014).

9.2 Multinational Supply Chains

Multinationals must take advantage of the production factors offered by globally dispersed facilities and of the attractiveness of offshore markets to create value for their stakeholders. They must be able to design and manage huge SC networks capable of adapting to market changes, offering high-quality product assortments at competitive prices, responding quickly to customer demands, being resilient when faced with adversity, and avoiding unethical practices. This is a tremendous challenge. This section looks at possible strategies and network structures that can be adopted to do this. It also discusses additional factors to consider when evaluating total landed costs and delays for multinational deployment strategies.

9.2.1 Generic SCN Strategies and Structures

The generic structure of multitier supply chains was portrayed in Fig. 6.5. Although all SCNs do not cover all production–distribution stages depicted in this figure, it provides a good starting point for the analysis of multinational SCs. The structure of a multinational SC depends on several factors, such as the maturity (age) and size of the company, its industrial sector, and its SC strategy (Sect. 1.3.1). These strategies usually aim for sustainable value creation. As discussed in previous chapters, this

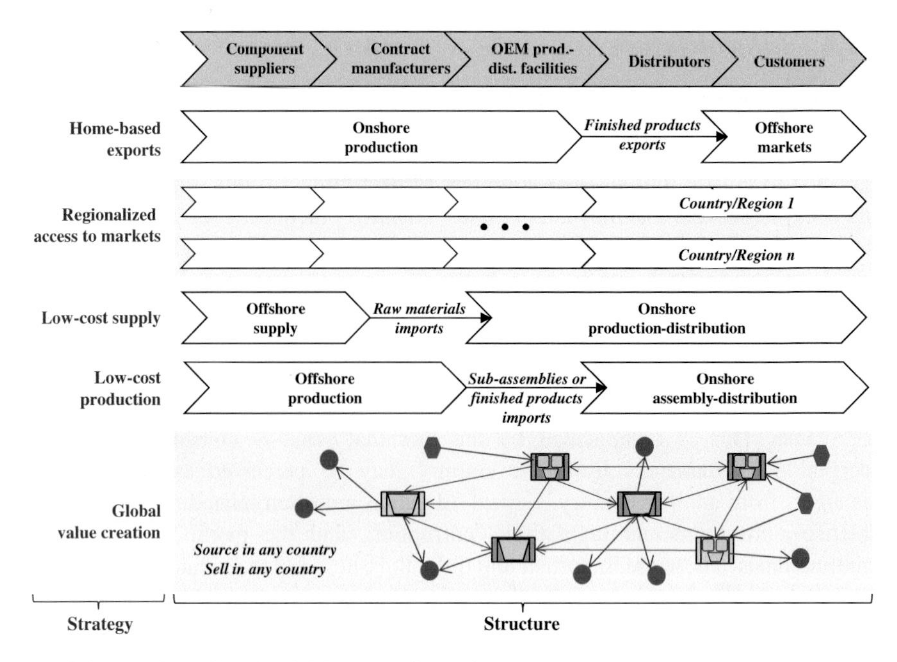

Fig. 9.4 Generic multinational SCN strategies and structures

involves the long-term minimization of operating costs and current and fixed assets, the maximization of revenues through adequate order-winning offers and product-market expansion, and risk minimization. It is thus natural that multinational deployment strategies be based on at least some of these value drivers. A number of typical strategies and SCN structures are illustrated in Fig. 9.4. They are used in the next paragraphs to support our discussion of important multinational SCN design issues.

The two top strategies in Fig. 9.4 essentially aim to provide access to foreign markets. A *home-based exports* strategy is adopted by domestic companies wanting to export their products abroad to expand their markets. These companies are typically relatively small, production and raw material supply remains onshore, but they want to start selling their products overseas. This usually occurs progressively. The company often begins using the services of a *trading house* (business intermediary between manufacturers and foreign buyers or consumers) with expertise in the target country or by making an alliance (through a license agreement, franchise, or joint venture) with a company already well positioned in the foreign market. As sales expand, to have more control on foreign operations, it may open oversea offices and eventually develop its own distribution network.

The *regionalized access to market* strategy is usually adopted by mature successful companies wanting to expand geographically, often through the acquisitions of foreign companies in the same sector or by exporting successful home solutions. Full SCs are then managed in parallel in each targeted region or country. This approach is common in sectors producing bulky products, such as cement, with a low value-to-weight ratio, making them difficult and/or expensive to transport over long distances. Cemex, a Mexico-based company, and Lafarge, a France-based company, used this strategy to become global leaders in the cement industry. With this approach, each domestic SCN can be optimized independently of the others. The global gains come from sharing know-how, technology, and brand name. Coca-Cola, for example, implements *server* bottling factories around the world, each of them serving a relatively small regional market.

The next two approaches illustrated in Fig. 9.4 are low-cost sourcing strategies. *Low-cost supply* involves the importation of raw materials or components from foreign countries where they can be procured at a low cost. This has become a very common practice. *Low-cost production* goes one step further, either by subcontracting the production of major subassemblies or finished products to foreign contract manufacturers or by implementing offshore factories. The autonomy of these factories can vary. In some cases, local managers may simply follow instructions from the head office, but in others they may be responsible for procurement, production planning, outbound logistics, product customization, process innovation, and R&D activities. The factory may be seen by the company as an outpost to gain needed knowledge or skills through proximity with key competitors or suppliers. For example, several non-US companies implement outpost factories in Silicon Valley. With time, these facilities may become lead factories responsible for the development of new products and processes for the entire company.

Clearly, a given company can combine several of the strategies examined previously. For example, midsize enterprises often combine *home-based exports* with *low-cost supply*. Zara, a Spanish company, manages production and logistic flows among its 14 factories and 200-plus suppliers and distributes products directly to its 650 stores on four continents from a facility in northwest Spain (Ferdows 2003). For multinationals that can source in any country and sell in any country, the strategic options available are extremely diverse and the simple generic strategies discussed previously are too limited. These companies usually pursue a *global value-creation* strategy, which requires the solution of complex SCN optimization problems. Geographical dispersion favors knowledge development, learning, and technology transfer. In addition, it provides a hedge against possible catastrophic events (e.g., strike, fire, earthquake, supplier bankruptcy, political unrest, etc.) that may occur in a country or region. We show how to formulate multinational SCN design models in Sect. 9.4.

9.2.2 Global Logistics Infrastructures

The previous discussion concentrates on the strategic selection of a company's facility locations, supply sources, and markets. However, when doing business internationally, oceans must often be crossed, which requires multimodal transportation solutions. In addition to selecting facility sites, suppliers, and product-markets, this usually requires the selection of ports of embarkation and debarkation. Moreover, to simplify import–export formalities and to reduce tariffs and taxes, most counties have now developed free trade zones (FTZ) around sea ports, airports, or multimodal hubs. It is crucial to take these opportunities into account when designing a multinational SCN. From a SC modeling point of view, this translates into the addition of a significant number of nodes and arcs to the potential SCN considered.

Ocean shipping is a dominant link in multinational SCNs, and significant handling (staging and loading) costs and delays may be incurred at port terminals. Also, transportation costs and transit times between ports and shipper or customer premises depend on the quality of transportation infrastructures. Hence the choice of inbound and outbound ports may have a significant impact on SC performance. As discussed in Chap. 5, the size of ships, and in particular containerships, has increased significantly in the last two decades and some ports are not equipped to handle larger vessels. This has led to the consolidation of shipments in larger sea ports and to increased competition among ports. In some basic import–export contexts, the shipper may use a third-party door-to-door transportation solution. The choice of the ports is then made by the carrier without direct input from the shipper. However, because large ports are now often coupled with FTZs, it is usually advantageous to include port-FTZ selection decisions in multinational SCN design projects and also to consider these FTZs as potential locations for production–distribution facilities.

Free trade zones (called foreign trade zones in the United States, *maquiladoras* in Mexico, *special economic zones* in China, etc.) are geographical areas in a country where foreign goods can be received, stocked, transformed (or repacked), and shipped to other countries without paying import–export tariffs. Bureaucratic requirements are usually also lowered, and companies operating within an FTZ may be granted certain host country income tax breaks or holidays. FTZs were initially introduced to attract foreign direct investments and to foster local employment. Duties on products are paid only if they are shipped elsewhere in the country. FTZs thus also can be used to delay tariff payments until a product is sold, which has some obvious cash flow advantages. To serve their domestic market, several companies now find that they do better if they use a low-cost supply strategy rather than a low-cost production approach (see Fig. 9.4), provided that their final assembly plant is located in a domestic FTZ. In addition to duty and tax advantages, they can then be much more responsive to customer needs. When final assembly is done locally, delays caused by overseas shipments and custom clearance can be avoided and customization is easier.

9.2.3 *Offshore Sourcing Decisions*

The previous discussion makes it clear that when offshore supply lines are present in an SCN, additional costs, delays, and risks are incurred. To show this, consider the *low-cost sourcing* strategies schematized in Fig. 9.4. When importing raw materials, components, subassemblies, or finished products for a domestic production–distribution facility, supply lines must be set up between the offshore sources considered and the onshore destination. More specifically, shipping and customs clearance solutions must be elaborated for the lanes considered. Assuming that the required products can be procured from different foreign suppliers, and that alternative transportation means are available for each (supplier, home facility) lane, it should be clear that the offshore sourcing problem combines elements of the transportation means selection problem examined in Sect. 5.5, and of the supplier selection problem examined in Sect. 6.4.Selecting a global sourcing solution is thus a multi-criteria decision problem in which total costs are usually a preponderant criterion. We saw in Sect. 5.5 that several apparently noneconomic criteria such as transit times duration and variability can be taken into account indirectly in *total costs analysis*. The costs, delays, and risks of international trade were only alluded to in these sections, but they can have a significant impact and they must be included in the analysis. In an international context, this type of study is usually referred to as a *total landed costs* analysis, but it is essentially the same approach as total cost analysis.

When considering a specific offshore sourcing solution for a given product (family of products), the following costs must be evaluated:

- *Value at the origin*, that is, the purchase price of the products or their value (see the discussion on transfer prices in the next section) when shipped from an offshore factory of the company
- *Physical movement costs*, that is, transportation costs and transshipment (loading, unloading, staging, sorting, boarding, debarkation, etc.) costs in intermediate port and frontier areas
- *Inventory holding costs*, including the cost of the inventory in transit in transportation equipment, the inventory waiting at transshipment points or at borders, as well as the cost of cycle and safety stocks stemming from transportation lot sizes and lead time uncertainty at the origin and destination
- *Investment costs*, that is, the fixed cost of setting up the offshore supply solution, which may include learning costs to bridge cultural and administrative gaps, system enhancement costs to be able to coordinate material flows from a distance and avoid nonethical behaviors at the supply end, as well as investments in capacity to set up an FTZ-bounded warehouse, for example
- *Operating costs* incurred when processing products in an intermediate location: shipment consolidation in a port or inspecting, repacking, and relabeling products in an FTZ facility, for example
- *Border crossing costs*, including custom duties, taxes, currency conversion fees, and fines
- *Clerical and legal support costs*, that is, freight and custom brokerage fees, and/or customs codification and documentation costs, trade and security compliance costs, dispute resolution costs, and so on

- *Risk mitigation and penalty costs* such as insurances, currency fluctuation protection instruments, ocean shipping crating costs, and disruption or damage costs
- *Reverse logistics costs* for the return, repair, or disposal of defective or non-compliant products

In practice, several of these costs are typically overlooked, which often leads to unjustified offshoring decisions.

9.3 Key Multinational SCN Design Features

Each company has its particularities, and to be able to design multinational value-creating SCNs, one must go beyond the conceptual discussions of the previous sections. The modeling approach introduced in Chaps. 7 and 8 can be generalized to cope with multinational SCN design problems. However, before we show how this is done, a number of international complexity factors previously introduced must be examined more carefully.

9.3.1 Currency Exchange

Without the *foreign exchange market* (forex or FX for short), a collection of currency trading products and services, international trade and investments would not be possible. Currency trading is conducted electronically (over the counter) between traders around the world rather than on a centralized exchange. The market is open 24 h a day, 5 days a week, and currencies are traded worldwide in London, New York, Tokyo, Zurich, Frankfurt, Hong Kong, Singapore, Paris, and Sydney. Today, the global FX market trades more than $4 trillion per day. The rate at which a currency is converted into another changes continuously in time, depending on supply and demand, and can make seemingly profitable international transactions unprofitable or vice versa. The exchange rate of the US dollar against the euro and the Canadian dollar over the last 10 years is plotted in Fig. 9.5. As can be seen, variations in time are significant, which largely contribute to international business risks. Some of these risks can be attenuated through forward exchange contracts, futures contracts, or currency swaps. The currencies of some developing or dependent countries are not convertible on the FX market. These are exceptional situations, however, and some multinationals use countertrade to circumvent the problem.

The *spot* exchange rate is the amount of domestic currency required to buy or sell one unit of foreign currency at a particular moment. A transaction fee is usually charged by the selling or buying institution (e.g., bank, currency broker) and thus buying and selling rates are not the same. Also, the selling and buying rates depend

Fig. 9.5 Weekly US dollar exchange rate from 2005 to 2015—*Data source* GVI forex database

on who is selling or buying and on the amount converted. For example, a European consumer pays more euros per dollar to obtain $10,000 from his bank, than a European corporation buying $10,000,000 from a trader. A *forward exchange* occurs when two parties engage in a contract to exchange currencies at a specific date in the future. *Futures* are essentially standardized forward contracts. Forward exchanges are individually negotiated and traded, whereas futures are traded on organized exchanges. For major currencies, *futures exchange rates* are quoted for 30 days, 90 days, and 180 days into the future. Suppose a US importer is buying products costing € 10,000 from Germany, and that these products are payable in 30 days. At the time the order is placed, the spot rate is 1.2291, meaning that each product costs $12,291. The importer knows that he can sell these products for $15,000, which yields a reasonable profit. However, if in the next month the dollar depreciates, he could be in trouble. To avoid this, knowing that the 30-day-forward rates are quoted at 1.2379, he can conclude a contract with a foreign exchange dealer at this rate and be certain that he will pay $12,379 for the products. This can be seen as paying an $88 insurance to protect the buying price. Forward exchange contracts are not available for all currencies.

A *foreign exchange swap* is a contract in which one party (a company) borrows an amount in foreign currency from another party (a bank) and simultaneously lends an equivalent amount in domestic currency to that party for a predetermined period of time. The repayment obligation of the parties serves as collateral. Swaps can be seen as sophisticated forward exchanges and they are often used by multinationals producing and selling in several countries. Consider a French company assembling its finished products locally and exporting products to Great Britain. Suppose also that these products contain components manufactured in Britain. So the French company both buys from and sells to Britain. The company knows that in 30 days, it will be paid at least £ 1,100,000 by its British customers, but it needs £ 1,000,000 today to pay its British component supplier. The company could buy pounds on the spot market today to pay its supplier, and convert the pounds received in a month to euros, but that would expose the company to FX risk. To eliminate this risk, it can instead set up a euro–pound swap with its bank. Suppose the spot rate is € 1 = £ 0.78 (S = 0.7800) and the 30-day-forward rate is F = 0.7850. To pay its supplier, the company simultaneously borrows £ 1,000,000 from its bank and lends € 1,282,051.28 (£ 1,000,000 converted at the spot rate S = 0.7800) to its bank. When the contract expires in 30 days, the company returns £ 1,006,410.25 to its bank (€ 1,282,051.28 converted at the forward rate F = 0.7850) and the bank returns € 1,282,051.28 to the company. This amount offsets the amount originally landed by the company. However, the amount paid to the bank after 30 days is £ 6,410.25 (£ 1,006,410.25− £ 1,000,000) higher than the original loan. This is the cost of the insurance paid to eliminate FX risk.

Exchange rates are extremely difficult to forecast, especially for short horizons (1 year or less). Consequently, it is hard to anticipate potential future problems. Several methods have been proposed to forecast exchange rate movements but without great success. These methods belong to two broad families: *fundamental analysis* and *technical analysis*. The former are explanatory models based on fundamental economic variables such as inflation rates and interest rates differentials, relative purchasing power parity, balance of payment position, and so on. The latter are time series analysis models. A third option favored by many is the use forward exchange rates to forecast future spot rates. Several studies (Meese and Rogoff 1983; Cheung et al. 2005) show that it is extremely difficult to elaborate exchange rate forecasting models that can consistently beat a random walk model (i.e., using the current spot rate as a predictor of all future spot rates). They also show that random walk models do not predict well. Consequently, using point estimates of exchange rates in SCN design models has its limitations. It is preferable to consider exchange rates as random variables.

All companies trading and/or manufacturing internationally are thus exposed to exchange rate risk. Two types of exposure are usually distinguished: transaction (or contractual) and operating (or competitive, economic, strategic) exposure. *Transaction exposure* comes from contracts denominated in a foreign currency, and it is relatively easy to identify. It can be controlled using forward exchange rate contracts or swaps as discussed previously. This protects the firm against downside risk, but the firm also loses the upside potential. For receivables or payables in foreign currency, a company may also be able to use leading (paying early to profit from a strong local currency) or lagging (delaying payment until the local currency strengthen) tactics. This is easier when the foreign party is a division or subsidiary. Another possibility may be to transfer the risk to the other party by invoicing in one's own currency. This may be dangerous, however, because it weakens the company's market offer. Finally, contracts with foreign partners can be negotiated to include explicit risk-sharing agreements.

Operating exposure is the risk that future exchange rate movements will affect the value of the company, which, as indicated in Chap. 1, is the sum of all discounted future residual cash flows. This is not easy to measure, but it is clear that it is directly related to the global footprint of a company's SCN as well as to the structure of its competitor's SCN. Operating exposure cannot be managed solely with financial hedging tactics; it needs a long-term perspective. In fact, it is a fundamental multinational SCN design issues. It is widely held that a way to limit operating exposure is to create a “natural hedge,” that is, to match the currency footprint by locating plants in the countries where sales are generated (or in countries using the same currency). Several car manufacturers have recently adopted this strategy (Rosemain 2013). It has been shown, however, that this approach results in inflexible national or regional networks and that firms pursuing it are less capable of improving corporate profits and controlling downside risk (Dong et al. 2014).

Generally speaking, multinational companies can adopt three types of SCN strategies to mitigate their operating exposure: speculation, portfolio, or operational hedging. A *speculation* strategy leads to the concentration of production in a small number of countries hoping that their currency will support other comparative advantages. If the currencies of these countries weaken, one may suddenly have an enormous advantage over competitors, but this is a gamble. A *portfolio* strategy, by contrast, involves the dispersion of the plants, contract manufacturers, and suppliers of the enterprise in various countries to ensure that unpredictable exchange rate movements cancel out. The natural hedge approach discussed previously can be seen as a variant of this strategy. The approach is attractive but it lacks flexibility. The *operational hedging* strategy involves the design of a flexible SCN that can be adjusted as exchange rates change. The required flexibility is obtained by investing in additional capacity or negotiating flexible multisourcing contracts (on volumes and timing of orders) in several countries so as to be able to move production or supply in response to exchange rate fluctuations and modify product-market allocations. This added flexibility is not free, however, and one must find the optimal balance.

Dornier et al. (1998) provide an excellent discussion of operational exposure and of operational hedging strategies. Multinational SCNs providing adequate protection against exchange rate risks can be designed using a stochastic programming model. A deterministic model to support the design of multinational SCNs is proposed in Sect. 9.4. The two next chapters show how it can be transformed into a stochastic programming model.

9.3.2 Transfer Pricing

Transfer prices are amounts charged by the responsibility centers of a company for cross-border transfers of tangible goods, intangible property (e.g., technology, brand names), services, or financing to affiliated foreign legal entities such as divisions or subsidiaries. Tax authorities are interested in the methods used to set transfer prices because, as we shall see, they directly affect the taxable profits of the entities involved. To avoid opportunistic practices, most countries have adopted stringent transfer pricing regulations, often based on the OECD transfer pricing guidelines for multinational enterprises and tax administrations (OECD 2010). The purpose of these guidelines is to propose methods for establishing (or testing) transfer prices that adhere to the arm's length principle. This principle requires that related parties price controlled transactions as if they were taking place under exactly the same conditions between unrelated parties on the open market (uncontrolled transactions). Although many countries have aligned their regulations with OECD guidelines, each has local features. Also, some countries (e.g., Brazil, Russia, and India) do not follow OECD guidelines. The arm's length principle is also applied by many customs administrations for the calculation of tariffs on goods exchanged between related parties, but the methods used are not necessarily aligned with OECD guidelines.

Five transfer pricing methods designed to comply with the arm's length principle are recommended by the OECD. The choice of a method depends on the business context, the strength and weaknesses of the methods, the availability of the information required, and the degree of compatibility between controlled and uncontrolled transactions. These include the following *traditional transaction methods*:

- *Comparable uncontrolled price (CUP) method.* This involves using the price of identical products or services sold by or to a third party in comparable circumstances. If slight differences exist, adjustments can be made. This method should be selected if it can be reliably applied. Unfortunately, comparable products rarely can be found for proprietary components or subassemblies crossing borders in multinational SCNs.
- *Resale price method.* If an internally transferred product is subsequently sold to an independent reseller, the reseller sale price can be reduced by an appropriate gross margin to calculate the transfer price. This approach applies mainly to finished products or spare parts when reliable information on resale margins can be obtained.
- *Cost-plus method.* This involves adding an appropriate markup to the costs incurred by the internal shipper in light of the functions performed and market conditions. The markup used must be established by reference to the markup earned in comparable uncontrolled transactions. To apply this method, the company must have a very good cost accounting system.

When these methods cannot be applied, one can adopt one of the following two *transactional profit methods:*

- *Net margin method.* This approach is similar to the cost-plus and resale price methods, except that the margin used is based on the net profit that the company earns in comparable *internal* uncontrolled transactions.
- *Profit split method.* This approach differs again in the way the margin is calculated. It first identifies the combined profits (or losses) of the associated enterprises for controlled transactions. It then splits combined profits using an economically valid basis that approximates the division of profits that would have occurred in an arm's length agreement. The split can be based on projected or actual profits.

These methods are easier to apply because they can rely on aggregate data analysis rather than on specific transactions.

Multinationals can apply methods not described in the OECD guidelines provided the prices obtained satisfy the arm's length principle. In fact, independent of the method used, companies must demonstrate that their transfer prices comply with the arm's length principle. This requires a *comparability analysis* usually involving the comparison of related party transactions to comparable transactions between unrelated parties. Two transactions are considered comparable if none of the differences between them affect prices or profit margins. If there are minor differences it may be possible to make adjustments. Features to consider when selecting comparable transactions include the characteristics of the products or services, the functions performed by the shipper, contractual terms (e.g., incoterms), economic circumstances surrounding the transactions, and the business purpose of the transaction. Also, when a transfer price for a product or service has been accepted, any modification must be justified even if the business context has changed. Despite the limited flexibility in the choice of transfer pricing method and of comparable transactions, this often leaves enough scope to vary transfer prices within reasonable bounds. To avoid disputes, companies may be able to obtain advance pricing rulings from some tax authorities.

9.3.3 Taxation

Taxation systems worldwide are very diversified, both from the point of view of the types of taxes applied and of the sources of income to tax. Glautier and Bassinger (1987) distinguish between direct and indirect taxes. *Direct taxes* include corporate income (profit) taxes, dividend taxes, capital gains tax, social security and labor taxes paid by employers, property taxes, and so on. *Indirect taxes* include purchase taxes, sales taxes, value-added taxes, customs duties, and so on. The sources of taxable income are not the same in every nation. In some countries, such as Canada, the United States, Mexico, Germany, Brazil, Australia, and Japan, taxes are imposed on domestic and foreign revenues. Other countries, such as France, Egypt, and Singapore, tax only domestic income. In the former case, to avoid double taxation, the countries concerned generally negotiate treaties on taxes. In our context, direct taxes and customs duties are of particular interest.

Direct taxes paid by companies and available tax breaks differ significantly from one country (and even state or province) to another. PwC and the World Bank publish an annual report on government-mandated direct taxes (at any level—federal, state, or local) paid by a standardized business (PwC and World Bank 2014). It reports the taxes paid as a percentage of commercial profit in 189 economies. The 2015 total tax rates for a sample of countries are provided in Table 9.2. The table also provides a breakdown of the total taxes in profit tax, labor tax, and other taxes. Highest and lowest values are highlighted. As can be seen there is a significant variation in terms of total tax rate per country and of tax types within countries. Some countries (Bahamas, Kuwait) have 0 % profit tax but their other taxes may be substantial. Among developed economies, the breakdown can be very

Table 9.2 2015 tax rates for a sample of countries

Country	Total tax (%)	Profit tax (%)	Labor tax (%)	Other taxes (%)
Brazil	**69**	24.7	40.3	4
France	66	7.4	**51.7**	7.5
China	64.6	7.8	49.3	7.5
Germany	48.8	23.3	21.2	4.3
United States	43.8	**28.2**	9.7	5.9
Bahamas	41.1	*0*	*6.3*	**34.8**
United Kingdom	33.7	20.9	11.3	1.5
Canada	21	3.9	12.5	4.6
Singapore	18.4	2.2	15.5	1.1
Kuwait	*12.8*	0	12.8	*0*

Source PwC and World Bank (2014)

different. France, for example, has a low profit tax (7.4 %) but a very high labor tax (51.7 %). On the other end, the United States has high profit taxes (28.2 %) but a low labor tax (9.7 %). Because governments adapt their policies to the evolving economic context, tax rates also vary significantly in time. According to the PwC and World Bank report, over the last 9 years, labor taxes have on average remained relatively constant, but profit taxes were reduced from an average of 19 % to 16.7 %. When modeling SCNs, labor tax and other taxes tend to be included in the calculation of fixed and variable production–distribution costs. Taxable profits, however, must be modeled by country to take income taxes into account.

Taxation mechanisms also differ from one country to the other. First, the breakdown among federal, state, and local taxes is not the same, and a firm operating in a city within a province or state in a given country does not necessarily pay the same taxes as one located in another city or province. For example, in the United States, the total tax rate in New York City is 45.8 %, but it is 40.9 % in Los Angeles. Second, the method used to compute taxable profits is not uniform: certain types of interests (say mortgage interests) are tax deductible in some countries (e.g., the United States) but not in others (e.g., Canada), which affects the weighted average cost of capital (WACC) of a company in its SCN countries; the asset depreciation methods enforced for tax purposes are not uniform; and some countries do not have the same rules for domestic companies, foreign divisions, and controlled foreign subsidiaries (separate legal entities partially or wholly owned by the parent company). The earnings of foreign divisions are repatriated after taxes have been paid in the host country, and additional taxes on repatriated amounts may have to be paid in the home country. The repatriation of earnings from subsidiaries is achieved through the payment of royalties and dividends. Also, as mentioned, taxable profits depend on transfer prices and each country regulates them differently. In short, multinational SCN design decisions require a detailed understanding of international tax laws rather than just a consideration of tax rates.

Customs duties are based on material flows and, consequently, the associated cash flows must be taken into account when goods cross borders. The proliferation of preferential trade agreements over the last decades has had the effect of reducing tariffs but customs duties can still be perceived as barriers to trade, and they have an impact on location decisions. It is possible in certain circumstances to avoid paying customs duties or to be reimbursed for duties paid. Duty drawback and avoidance opportunities must be considered explicitly in SCN design models. There are three main types of situations that can lead to duty reliefs:

- When a company imports a product and then exports it without change, it can claim a duty drawback for *reexport in the same condition*.
- When a company imports a product (component or subassembly), adds value to it in a manufacturing process, and then exports it as part of a subassembly or finished product, it can claim a duty drawback for *reexport in a different condition*.
- When a company exports a product and then imports it back as part of a subassembly or finished product, it can avoid domestic import duties for *domestic good returned in a different condition*.

Time limits are usually imposed to benefit from these conditions. In the United States, for example, imported products must be reexported within 5 years to qualify for a duty drawback. Drawbacks can also be obtained on goods destroyed under customs supervision. Drawback claims may necessitate a lot of time and efforts. However, drawback contexts can usually be transformed into avoidance contexts by locating production–distribution activities in FTZs. For example, when goods are imported in a customs-bounded warehouse, duties are charged only if they are shipped to a domestic location. If the goods are exported, the payment of duties is avoided. In the United States, goods can remain in a bounded warehouse for a maximum of 5 years.

A Harmonized Commodity Description and Coding System (HS) was developed by the World Customs Organization (WCO) to describe products for customs purposes. HS six-digit codes are used by more than 200 countries to classify products and apply tariff rates. An online HS-code database is maintained by the WCO (www.wcoomd.org). Most countries base the calculation of duty charges on CIF values (see Fig. 5.14 on Incoterms). For customs valuation purposes, the CIF value is the price paid for the goods plus the cost of transportation, loading, unloading, handling, insurance, and associated costs incidental to delivery of the goods to the port or place of entry in the importing country. For example, suppose a British clothing retailer is importing skirts from the United States. The price paid for the products is US$1000, the transportation cost is US$150, insurances cost US$50, the British import duty rate on skirts is 12 %, and the exchange rate is 1 GBP = 1.57 USD. The CIF value in USD being US$1200 (1000 + 150 + 50), the duty paid is US$144, and the landed price in pounds is £ 856.05 (1344/1.57).

9.3.4 *Profit Shifting*

There is a relatively large consensus among governments and the public in general to the effect that the distribution of the profits of multinational corporations should be congruent with the geographical dispersion of its activities. The perception that multinationals are routinely using tax avoidance practices, that is, shifting profits to low-tax countries via transfer price manipulations, has been the source of a vibrant debate on domestic tax-base erosion and profit shifting (BEPS). Base erosion is considered by most governments to be a serious risk to tax revenues, tax sovereignty, and tax fairness (OECD 2013). To fully appreciate the impact transfer prices can have on the duties and taxes paid by corporations, let us look at a simplified example. Suppose that a US company has a manufacturing division in China that sells a product to a German division distributing its products to the European market. Relevant costs (in USD) and tax rates are as follows:

- Import duties charged by Germany on the value of incoming products: 12 %
- Annual fixed cost of the Chinese plant: $100,000
- Annual fixed cost of the German DC: $600,000
- Sales price of the product in Europe: $100/unit
- Smallest compliant transfer price (based on CIF value): $55/unit
- Largest compliant transfer price (based on CIF value): $65/unit
- Profit tax rate in China: 10 %
- Profit tax rate in Germany: 25 %
- Variable production and transport cost of the Chinese division: $35/unit
- Annual sales: 20,000 units

The net profit after tax of the company for different transfer prices, is calculated in Table 9.3.

As can be seen, when the transfer price is raised from $55 to $60, the net profit after tax of the company increases. However, when we add another $5 to the transfer price, the net profit after tax decreases. This shows clearly that the transfer price has an impact on taxable profits, and that in this case the company maximizes its net profit after tax when using a price in the vicinity of $60. This also shows that the relationship between transfer prices and net profits after tax is not linear. To simplify the example, we assumed that transfer prices were based on CIF incoterms, which then makes them consistent with the value used by customs authorities to calculate duties. This is not necessary, however. By changing the incoterms used to calculate transfer prices, the company can affect the expenses incurred between a shipper in one country and a destination in another country to either the source or the destination. For example, under CIP terms, most carriage, port handling, and insurance costs would end up in the shipper's income statements, but under EXW terms they would end up in the buyer's income statement. The incoterm selected would be accepted by tax authorities provided that its choice is defendable under the arm's length principle.

Table 9.3 Calculation of net profit after tax for the example—Based on Vidal (1998)

	Chinese division	German division	Corporate results
Transfer price = $55/unit			
Revenues (20,000 at $55/unit)	*$1,100,000*		
Revenues (20,000 at $100/unit)		$2,000,000	$2,000,000
• Variable cost ($35/unit)	$700,000		$700,000
• Purchase cost		*$1,100,000*	
• Custom duties (12 %)		$132,000	$132,000
• Fixed costs	$100,000	$600,000	$700,000
Net profit	*$300,000*	*$168,000*	*$468,000*
• Profit tax (10 and 25 %)	$30,000	$42,000	$72,000
Net profit after tax	**$270,000**	**$126,000**	**$396,000**
Transfer price = $60/unit			
Revenues (20,000 at $60/unit)	*$1,200,000*		
Revenues (20,000 at $100/unit)		$2,000,000	$2,000,000
• Variable cost ($35/unit)	$700,000		$700,000
• Purchase cost		*$1,200,000*	
• Custom duties (12 %)		$144,000	$144,000
• Fixed costs	$100,000	$600,000	$700,000
Net profit	*$400,000*	*$56,000*	*$456,000*
• Profit tax (10 and 25 %)	$40,000	$14,000	$54,000
Net profit after tax	**$360,000**	**$42,000**	**$402,000**
Transfer price = $65/unit			
Revenues (20,000 at $65/unit)	*$1,300,000*		
Revenues (20,000 at $100/unit)		$2,000,000	$2,000,000
• Variable cost ($35/unit)	$700,000		$700,000
• Purchase cost		*$1,300,000*	
• Custom duties (12 %)		$156,000	$156,000
• Fixed costs	$100,000	$600,000	$700,000
Net profit	*$500,000*	*($56,000)*	*$444,000*
• Profit tax (10 and 25 %)	$50,000	$0	$50,000
Net profit after tax	**$450,000**	**($56,000)**	**$394,000**

Intangibles activities, resources, and risks, such as R&D, global supply management, licenses, brand name, insurances, and so on, also can be used to move costs from one country to another (in particular from the parent company to foreign divisions or subsidiaries), but it is this type of practice that is severely questioned in BEPS debates. Intangible activities of a given entity generate services that may be required by other affiliated legal entities, and it is normal, to some extent, that the cost of these intangibles be shared by all beneficiaries. This is an important issue for companies delivering digital goods or using cloud services because, in that context, a company can operate in a country with virtually no local resources. According to the OECD (2013), many multinational tax structures focus on allocating significant risks and hard-to-value intangibles to low-tax jurisdictions or on the artificial splitting of asset ownership between the legal entities of the company to reduce corporate taxes. Although these practices are technically legal, they are ethically questionable. To comply with the arm's length principle, any transaction between related entities that would not usually take place between independent companies should be avoided.

9.4 Multinational SCN Optimization

The business environment of distinct multinational SCNs may differ significantly depending on the region and industry involved. Also, no SCN design model currently available in the literature captures all the international peculiarities discussed in the previous sections. The optimization model presented in this section therefore should be viewed as an example of what can be done rather than as a generic global SCN design model. Nontrivial multinational SCs are typically complex networks of geographically dispersed supply sources, plants, distribution centers, and customers. As such, they would be described best by an international extension of the generic multi-period SCN design model presented in Sect. 8.5. However, this would give a complex model, overly complicated to demonstrate the international modeling features introduced in this section. For this reason, the following mathematical programming model is based on the formulation introduced in Sect. 7.5 to solve the supply–production–distribution network design problem described in Figs. 7.18 and 7.19. In Sect. 7.5, we implicitly assumed that the company considered operates in a single country or in a free trade region with a single currency and without customs duties (e.g., European Union), and we neglect income tax effects. In this section, we relax these assumptions to consider a possibly global SCN.

9.4.1 Assumptions and Notation

The industrial context of the multinational considered in this section is the same as in Sect. 7.5. We study the case of a corporation manufacturing and distributing several finished products with nontrivial BOMs described by an activity graph and recipes. Supply contracts must be selected and some activities can be subcontracted to contract manufacturers. Demand zones are part of product-markets with a given sales price (i.e., the company is a price taker and prices depend on markets). The company is already deployed internationally and it wants to reengineer its SCN. The capacity of the platform of existing and potential sites is predetermined. The business environment is relatively stable and the new design is to be based on a typical year (i.e., static design model). However, the multinational is structured in national divisions. Each division is a profit center and it manages its local SCN (i.e., it is responsible for short-term supply, production, distribution, and sales decisions). The head office in the home country plays a coordinating role and all divisions (home and foreign) are willing to work together to achieve corporate goals, and in particular to maximize total repatriated after-tax profits.

In addition to these contextual characteristics, the following assumptions regarding international issues are made:

- The prices and costs associated with SCN nodes (including supply nodes and demand zones) are incurred in local currency. Sales prices are net of any local sales expenses. The costs associated with network arcs are, when applicable, divided according to the incoterm used. The use of incoterms is not compulsory for product flows between company sites; however, we assume that they are used for all international arcs. This subsequently facilitates transfer price compliance acceptation by tax authorities. For international arcs, the costs covered by the shipper are quoted in the currency of the shipper and those under the responsibility of the destination in the currency of the destination.
- The exchange rates used are averages over the base year considered. Because the business environment is assumed to be stable, this is sufficient to capture country differences. A sensitivity analysis subsequently can be done to explore the impacts of unexpected variations.
- Each time products cross a border, duties are charged based on CIF product values or the equivalent. With the proliferation of PTAs in recent years, tariffs have decreased and the impact of duty drawback or avoidance opportunities can be taken into account in the estimation of customs rates. This implies that they do not have to be considered explicitly in the model.
- Products shipped by sea are sold with a CIF incoterm and those shipped by other modes with a CIP term (see Fig. 5.13). This means that duties are paid by the importer. It also means that loading, transfer to port, embarkation, long-haul carriage, and insurance costs are paid by the shipper. For sea transport, the costs incurred for disembarkation and transfer to the buyer premises are paid by the importer. For other modes, they are paid by the shipper.
- Transfer prices for products shipped on international internal arcs are specified a priori to satisfy arm's length regulations of tax authorities. They cover all upstream costs, all incoterm costs under the responsibility of the shipper, plus a margin comparable to industry practices.
- The corporate tax structure of a country is approximated by a profit tax rate selected based on the tax brackets applicable to the company division of the country. The income taxes paid by a division are based on the net profits (or loss) of the country's SCN. The corporate taxes of the parent company are deferred until it pays dividends, and the decision to pay dividends is independent of the network design.

As mentioned, the multinational wants to maximize the total repatriated net profit after tax of its divisions in the currency of its home country (e.g., USD). Without loss of generality, we assume that the corporation considers each country as a product-market. Consequently, the set $K = \{k\}$ previously used to denote product markets can now be considered as the set of countries in which the company has divisions. The index $k = 1$ is used for the home country. Also, the notation $k(n)$ is now used to identify the country in which node n is located. The additional notation required to model costs and revenues is following:

$f^{\rightarrow}_{pnn'}$ Unit flow cost of product p between node $n \in N^{S}$ and node $n' \in N^{S} \cup N^{D}$ paid by shipper $l(n)$; this includes customer order processing costs, the expenses under the responsibility of the shipper according to the incoterm or domestic sales term selected, and the inventory in transit holding cost

$f_{pnn'}$ Unit handling cost for the move of product p between node n in site $l(n)$ and node n' in the same site

$f^{\leftarrow}_{pn'n}$ Unit flow cost of product p between node $n' \in N^{S} \cup N^{V}$ and node $n \in N^{S}$ paid by destination $l(n)$; this covers the vendor-order processing costs and the expenses under the responsibility of the buyer according to the incoterm (including customs duties) or sales term (for domestic vendors) selected

c^{V}_{pl} Unit price of product p purchased from vendor $l \in L^{V}$ in the currency of the vendor; this price covers the value of the product at the origin and the expenses under the responsibility of the vendor according to the incoterm or sales term selected

$\pi^{T}_{pnn'}$ Transfer price of product p shipped from node $n \in N^{S}$ to node $n' \in N^{S}$ covering all upstream costs, incoterm costs under the responsibility of shipper $l(n)$, plus an acceptable margin; for domestic transfers, $\pi^{T}_{pnn'} = 0$

$\varepsilon_{kk'}$ Average exchange rate, that is, number of units of country k currency by units of country k' currency; when $k = k'$, $\varepsilon_{kk'} = 1$

τ_{k} Corporate income tax rate of country k

R_{l} Total site $l \in L^{S}$ revenues

C_{l} Total site $l \in L^{S}$ expenses

VA^{SCN}_{k} Value added in country $k \in K$

VL^{SCN}_{k} Value lost in country $k \in K$

9.4.2 *Optimization Model*

To properly account for transfer prices and income taxes, it is necessary to elaborate a financial statement for each site. The revenues and expenses incurred in a site $l \in L^{S}$ for the year considered are outlined in Table 9.4. Expression (a) for the total transfer prices paid by site l to other sites is obtained by converting the prices charged by exporters into local currency. The same rational is used to calculate other expenses and revenues.

Table 9.4 Revenues and expenses for site l

		Site $l \in L^{\mathrm{S}}$
Expenses	(a) Foreign inbound transfers	$\sum_{p \in P} \sum_{a \in A_l} \sum_{n \in N^{\leftarrow}_{p(l,a)} \cap N^{\mathrm{S}}} \varepsilon_{k(l,a)k(n)} \pi^{\mathrm{T}}_{pn(l,a)} F_{pn(l,a)}$
	(b) Raw material procurement	$\sum_{p \in P_l} \sum_{a \in A_l} \sum_{n \in N^{\leftarrow}_{p(l,a)} \cap N^{\mathrm{V}}} \varepsilon_{k(l,a)k(n)} c^{\mathrm{V}}_{pl(n)} F_{pn(l,a)}$
	(c) Inflows	$\sum_{p \in P} \sum_{a \in A_l} \sum_{n \in N^{\leftarrow}_{p(l,a)}} f^{\leftarrow}_{pn(l,a)} F_{pn(l,a)}$
	(d) Platform	$y^{+}_{l} Y^{+}_{l} + y_l Y_l + y^{-}_{l} Y^{-}_{l}$
	(e) Activity	$\sum_{a \in A_l} \sum_{p \in P^{\rightarrow}_{a}} c^{\mathrm{X}}_{p(l,a)} X_{p(l,a)}$
	(f) Handling	$\sum_{p \in P} \sum_{a \in A_l} \sum_{n \in N^{\rightarrow}_{p(l,a)} \mid l(n)=l} f_{p(l,a)n} F_{p(l,a)n}$
	(g) Inventory holding	$\sum_{a \in A_l \cap A^{\mathrm{E}}} \sum_{p \in P^{\rightarrow}_{a}} c^{\mathrm{I}}_{p(l,a)} \left[\sum_{n \in N^{\leftarrow}_{p(l,a)}} F_{pn(l,a)} \big/ \varphi_{pn(l,a)}\right]$
	(h) Outflows	$\sum_{p \in P} \sum_{a \in A_l} \sum_{n \in N^{\rightarrow}_{p(l,a)}} f^{\rightarrow}_{p(l,a)n} F_{p(l,a)n}$
Revenues	(i) Sales	$\sum_{a \in A_l} \sum_{p \in P^{\rightarrow}_{a}} \sum_{n \in N^{\rightarrow}_{p(l,a)} \cap N^{\mathrm{D}}} \varepsilon_{k(l,a)k(n)} \pi_{k(n)p} F_{p(l,a)n}$
	(j) Foreign outbound transfers	$\sum_{p \in P} \sum_{a \in A_l} \sum_{n \in N^{\rightarrow}_{p(l,a)} \cap N^{\mathrm{S}}} \pi^{\mathrm{T}}_{p(l,a)n} F_{p(l,a)n}$

The entries of Table 9.4 are used to calculate site revenues and expenses as follows:

$$C_l = (a) + (b) + (c) + (d) + (e) + (f) + (g) + (h), \quad l \in L^{\mathrm{S}} \tag{9.1}$$

$$R_l = (i) + (j), \quad l \in L^{\mathrm{S}} \tag{9.2}$$

To calculate the economic value added by a division, we must distinguish between the case when there is a profit and the case when there is a loss because there is no tax to pay on losses. To do this, the value added or lost by a division before tax is calculated as follows:

$$\begin{aligned} & VA^{\mathrm{SCN}}_{k} - VL^{\mathrm{SCN}}_{k} = \sum_{l \in L^{\mathrm{S}}_{k}} R_l - \Big(\sum_{l \in L^{\mathrm{S}}_{k}} C_l + \sum_{l \in L^{\mathrm{V}}_{k}} y^{\mathrm{V}}_{l} Y^{\mathrm{V}}_{l}\Big), \quad k \in K \\ & VA^{\mathrm{SCN}}_{k}, VL^{\mathrm{SCN}}_{k} \geq 0 \end{aligned} \tag{9.3}$$

where L^{S}_{k} is the set of potential facility sites in the country of division k and L^{V}_{k} is the set of potential supply contracts managed by division k. When expression (9.3) is included in a MIP, the profit and loss cannot be simultaneously positive, that is, the solution obtained is always such that $VA^{\mathrm{SCN}}_{k} \cdot VL^{\mathrm{SCN}}_{k} = 0, k \in K$. Taking taxes and exchange rates into account and summing the repatriated economic value added (EVA^{SCN}_{k}) of all divisions, the corporate economic value added by the multinational SCN (EVA^{SCN}) is given by

$$EVA^{\mathrm{SCN}} = \sum_{k \in K} EVA_k^{\mathrm{SCN}}, \quad EVA_k^{\mathrm{SCN}} = \varepsilon_{1k}\left[(1-\tau_k)VA_k^{\mathrm{SCN}} - VL_k^{\mathrm{SCN}}\right] \tag{9.4}$$

We can now formulate the optimization model sought to help design multinational SCNs. It involves the following objective function and constraints:

Max EVA^{SCN}

subject to

- Vendor contract and capability constraints (7.33) and (7.34)
- Site selection constraints (7.35)
- Flow equilibrium constraints (7.36)–(7.38)
- Platform capacity constraints (7.39) and (7.40)
- Demand constraints (7.41)
- Financial variables definition constraints (9.1)–(9.4)
- Nonnegative and binary decision variables domain constraints

This optimization model is a mixed integer program similar to those formulated previously and it can often be solved using commercial solvers. For larger problem instances, it may be necessary to use decomposition or heuristic methods to solve the model (see Perron et al. 2010; de Matta and Miller 2015).

9.4.3 Model Extensions

Several extensions and variants of the model formulated in the previous section can be proposed either to relax some of the assumptions made or to adapt to particular contexts. This section introduces some of them.

9.4.3.1 International Transfer Incoterms Selection

In the previous section, we assumed that transfer prices are set a priori by the multinational considered. As shown in Sect. 9.3.4, transfer prices play an important role in global SCN design problems because they have a major impact on the income taxes paid by home and foreign divisions. For this reason, instead of setting them exogenously, they should be considered as decision variables to optimize. Several SCN design models were proposed in the literature to optimize transfer prices. When this is done, however, constraints (9.1) and (9.2) become quadratic (because transfer prices are multiplied by flow variables to calculate site revenues and expenses), which complicates the solution of the problem considerably. Some authors (e.g., Vidal and Goetschalckx 2001; Perron et al. 2010) proposed approaches to solve this type of problem under the assumption that transfer prices can be varied between an upper and a lower bound while continuing to be compliant with the arm's length principle. This was true at the time, but as discussed in Sect. 9.3.2, tax authorities are now applying more stringent regulations, and it is unlikely that transfer prices optimized this way (to profit the corporation) would be compliant. de Matta and Miller (2015) and Hammami and Frein (2014) recently proposed models to optimize transfer prices based, respectively, on the OECD *cost-plus method* and *profit split method* (see Sect. 9.3.2), but this introduces significant nonlinearities and modeling complications.

The authors cited previously also raise the issue of who should pay transportation expenses (the importer or the exporter) to maximize repatriated after-tax profits from all divisions. This is more generally treated by asking which incoterm should be used for international transfers and requiring that transfer prices be fixed to cover all charges under the responsibility of the shipper, which complies with OECD (2010) transfer pricing guidelines. This implicitly discretizes the problem of finding the best transfer prices. In addition to its practical appeal, this approach is justified by the recommendations of Villegas and Ouenniche (2008), who showed that optimal transfer prices and transportation cost assignments tend to avoid mix solutions, that is, prices are set at their lower or upper bound and transportation costs are assigned to the importer or exporter.

The model proposed thereafter to optimize the selection of incoterms assumes that the same incoterm must be used for all the product transfers between a given pair of countries to comply with tax authority regulations. It requires the following additional notation:

KK Set of origin–destination country pairs (k,k') for which alternative incoterms can be considered while complying with the countries transfer pricing regulations

$IT_{kk'}$ Set of alternative incoterms considered for origin–destination pair $(k,k') \in KK$

NN Set of international internal SCN arcs (n,n') for which alternative incoterms can be considered (i.e., such that $(k(n),k(n')) \in KK$)

$f^{\rightarrow}_{pnn'i}$ Unit flow cost of product p on arc $(n,n') \in NN$ paid by shipper $l(n)$ when incoterm $i \in IT_{k(n)k(n')}$ is used

$f^{\leftarrow}_{pn'ni}$ Unit flow cost of product p on arc $(n,n') \in NN$ paid by destination $l(n)$ when incoterm $i \in IT_{k(n)k(n')}$ is used

$\pi^{\mathrm{T}}_{pnn'i}$ Transfer price of products p shipped from node $n \in N^{\mathrm{S}}$ to node $n' \in N^{\mathrm{S}}$ under incoterm $i \in IT_{k(n)k(n')}$

$F_{pnn'i}$ Flow of product p on arc $(n,n') \in NN$ shipped under incoterm $i \in IT_{k(n)k(n')}$

$Z_{ikk'}$ Binary variable equal to 1 if incoterm $i \in IT_{kk'}$ is selected for product transfers between origin–destination country pair $(k,k') \in KK$, and zero otherwise

To ensure that a single incoterm is selected for each country pair and that flows on the arcs corresponding to nonselected incoterms are null, the following constraints must be added to the previous model:

$$\sum_{i \in IT_{kk'}} Z_{ikk'} = 1, \quad (k, k') \in KK \tag{9.5}$$

$$\sum_{p \in P} F_{pnn'i} \leq \mathrm{M} Z_{ik(n)k(n')}, \quad (n, n') \in NN, \quad i \in IT_{k(n)k(n')} \tag{9.6}$$

where M is a large number. The following substitutions must also be made in the original model:

- The terms $\pi^{\mathrm{T}}_{pnn'} F_{pnn'}$ in lines (a) and (j) of Table 9.4 must be replaced by $\sum_{i \in IT_{k(n)k(n')}} \pi^{\mathrm{T}}_{pnn'i} F_{pnn'i}$.
- The term $f^{\leftarrow}_{pnn'} F_{pnn'}$ in line (c) of Table 9.4 must be replaced by $\sum_{i \in IT_{k(n)k(n')}} f^{\leftarrow}_{pnn'i} F_{pnn'i}$.
- The term $f^{\rightarrow}_{pn'n} F_{pn'n}$ in line (h) of Table 9.4 must be replaced by $\sum_{i \in IT_{k(n)k(n')}} f^{\rightarrow}_{pn'ni} F_{pn'ni}$.
- The variable $F_{pnn'}$ in constraints (7.36)–(7.38) and in constraint (7.40) must be replaced by $\sum_{i \in IT_{k(n)k(n')}} F_{pnn'i}$.

The model thus obtained has a larger number of variables and constraints, but it is still linear.

9.4.3.2 Local Content Rules

Multinationals operating in a foreign country must often comply with local content rules. These rules differ with respect to content and noncompliance penalties from one country to another. Local content can be defined in terms of value or volume. Rules based on volumes require a fraction of the total inputs used to be of domestic origin. They apply mainly when intermediate and final products are relatively homogenous (e.g., tobacco). For complex products (e.g., automobiles), value-based local content rules are more common. The value may refer to the total costs of local raw material and component purchases or to the value added in the country (supply plus manufacturing). This may be compared to the value of all supplied raw materials and components or to the value of the goods sold by the local division. The ensuing constraints to add to the SCN design model depend on the nature of the regulations, but, in most cases, they are not too difficult to formulate, at least with the level of precision allowed by the nature of flow and throughput variables.

Suppose, to illustrate, that the rules imposed by a country k stipulate that the value added to the products manufactured in the country (including supply, production, and warehousing costs but not transportation and asset-related costs) must be at least as large as a predetermined fraction of the value of the products sold by the local division of the company. Then the constraint to add would be

$$\sum_{l\in L_k^S}\left[\sum_{p\in P_l}\sum_{a\in A_l}\sum_{n\in N_{p(l,a)}^{\leftarrow}\cap N_k^V} c_{pl(n)}^{V} F_{pn(l,a)} + \sum_{a\in A_l}\sum_{p\in P_a^{\rightarrow}} c_{p(l,a)}^{X} X_{p(l,a)} + \sum_{p\in P}\sum_{a\in A_l}\sum_{n\in N_{p(l,a)}^{\rightarrow}|l(n)=l} f_{p(l,a)n} F_{p(l,a)n}\right] \geq \beta_k \sum_{l\in L_k^S} R_l \quad (9.7)$$

where β_k is the fraction of the total value of products sold imposed as a lower bound by the local content rule of country k, and N_k^V is the set of local supply sources of division k.

9.4.3.3 Duty Drawbacks and Avoidance

The model presented in the previous section assumes that duties can be calculated a priori and included in the unit transaction costs under the responsibility of the importer for the incoterm specified. It also assumes that the impact of duty drawbacks and avoidance is negligible. Duty drawbacks and avoidance can be taken into account in SCN design models, but this requires an explicit modeling of duties and the tracking of the path followed by products in the supply chain. The model proposed earlier is an *arc-based* formulation and it is memoryless. Flow equilibrium constraints (7.36)–(7.38) ensure that the quantity of products flowing out of a node does not exceed what can be manufactured with the inputs received, but the model does not keep track of where the input products come from, which is necessary to compute duty drawbacks. *Path-based* multi-echelon location-allocation models were proposed in the literature for simpler production–distribution contexts than the one considered here. These models track the route of products in the network using path-based, as opposed to arc-based, flow variables. When this is done, duty drawbacks and avoidance can be considered explicitly in the SCN design model. A good example of this type of formulation is found in Arntzen et al. (1995), who applied their model to the reengineering of the global SCN of Digital Equipment Corporation.

9.4.3.4 Port or Freight Forwarder Selection

When a multinational is managing its own intercontinental shipments, it may want to select the best embarkation and debarkation ports in the countries or regions (e.g., Europe) where it does business. Moreover, as indicated previously, it may be advantageous to set up intermediate facilities in the FTZ often coupled with these ports. These considerations can be introduced in our SCN design model simply by adding some potential network nodes to represent the transshipment, storage, or even production activities that can be done in the port area. If the implementation of a customs-bounded warehouse is considered, for example, then duty avoidance easily can be taken into account. In such a case, a FTZ facility l supporting a set Λ_l of activities would have an associated set of nodes N_l, and its operations would involve import and export flows from and to foreign countries as well as local outbound shipments. In such a context, the unit flow cost $f_{pn'n}^{\leftarrow}, n \in N_l$, on inbound international arcs would not include any duties, but the unit flow cost $f_{pnn'}^{\rightarrow}, n \in N_l$, on local shipment arcs would include the duties to be paid on the imported part of the products shipped. The model presented in the previous section is sufficiently general to take these possibilities into account.

If the multinational is not managing its intercontinental shipments but is rather using the services of freight forwarders and customs brokers, then port selection decisions are made by its service providers. The multinational then has to select the third parties it wants to sign contracts with. This can be done using parallel international flow arcs, one for each of the freight forwarders considered for a given lane, and by associating 0–1 variables to the freight forwarders considered. Additional constraints must also be added to ensure that a single arc can be used, as was done for the selection of incoterms.

9.4.3.5 Dynamic Model with Capacity Options

The multinational SCN design model presented in Sect. 9.4.2 is static, that is, it considers a single year, and it assumes that the capacity of potential site platforms is predetermined. As indicated in Chap. 8, these are limitative assumptions and in most cases it is preferable to use a dynamic model defined to cover a finite planning horizon and to consider alternative capacity options. Such a model can be formulated easily by merging the multinational design model presented here with the dynamic capacity deployment model presented in Sect. 8.5. Several of the extensions examined in Chaps. 7 and 8, such as the inclusion of service and financial constraints, could also be adapted to this dynamic multinational design model. Real-life instances of the resulting model, however, may be extremely large and difficult to solve.

The dynamic multinational model obtained would aim to maximize long-term value creation. Based on the definition of value introduced in Chap. 2, the value of the optimal SCN over the planning horizon considered is given by

$$V^{\text{SCN}} = \text{Max} \sum_{t \in T} \sum_{k \in K} \left[\frac{EVA_{kt}^{\text{SCN}}}{(1 + r_k)^t} \right]$$

where r_k is the WACC in country k, and where EVA_{kt}^{SCN}, the economic value added by division k in year t, can be defined by a relation similar to (9.4). It should be understood, however, that (9.4) provides only an approximation of the economic value added. This can be seen by comparing (9.4) to definition (2.5) in Chap. 2 or to Fig. 2.7. The main difference comes from the fact that the tax calculation method used in (9.4) does not take the depreciation of the firm resources explicitly into account. Expression (9.4) should thus be seen as an approximation. A more precise calculation of the taxes paid can be incorporated in the model but this requires a more detailed modeling of the financial statements of the corporation explicitly taking into account the depreciation of SCN resources of each division and their salvage value at the end of the planning horizon (Guillén-Gosbalez and Grossmann 2009; M'Barek et al. 2011). The WACC is not the same for all divisions because available government subsidies and capital market opportunities differ from country to country.

The SCN design models proposed in this chapter are deterministic, that is, they assume that all parameters are known with certainty. As explained previously when discussing exchange rates, demand forecasts, market prices, capacity needs, and so on, this is not very realistic, mainly for multi-period dynamic models. Sensitivity analysis can be used to explore the impact of value variations for key parameters such as exchange rates and tax rates; however, whenever possible, stochastic programming models should be used to obtain robust designs. The next two chapters explain how to formulate stochastic SCN design models.

Review Questions

9.1. Why does the SCN design problem become more complex in an international context?
9.2. Trade policies adopted by governments play an important role in business location decisions. Why?
9.3. What are transfer prices and what is their impact on the profitability of multinational companies?
9.4. Why do we need to express the objective function of the design model presented in this chapter as the sum of the corporation divisions' financial statement revenues and expenses?
9.5. What is the benefit for a company to operate in a FTZ? How can FTZs be considered in SCN design models?

Exercises

Exercise 9.1 Go back to Exercise 8.2, which describes the SCN design problem of Topclim, a Canadian manufacturer of air conditioning units. In order to be able to satisfy its increasing demand south of the border, Topclim is considering the implementation of a manufacturing network in the United States. To do this, an optimization model using the following notation was proposed by a consulting firm:

F_{ul} Flow of products from production–distribution center $u \in L^{\mathrm{U}}$ to demand zone $l \in L^{\mathrm{D}}$

Y_{uo} 1 if platform $o \in O_u$ is used on potential site $u \in L^{\mathrm{U}}$

d_l Demand in zone $l \in L^{\mathrm{D}}$

y_{uo} Fixed cost for using platform $o \in O_u$ on site $u \in L^{\mathrm{U}}$

f_{ul} Unit production–transportation costs of products shipped to demand zone $l \in L^{\mathrm{D}}$ from production–distribution center $u \in L^{\mathrm{U}}$

b_{uo} Capacity provided by platform $o \in O_u$ on site $u \in L^{\mathrm{U}}$

The model proposed to design the network is formulated as follows:

$$\min \sum_{u \in L^{\mathrm{U}}} \sum_{o \in O_u} y_{uo} Y_{uo} + \sum_{u \in L^{\mathrm{U}}} \sum_{l \in L^{\mathrm{D}}} f_{ul} F_{ul}$$

subject to

- Demand constraints

$$\sum_{u \in L^{\mathrm{U}}} F_{ul} = d_l, \quad l \in L^{\mathrm{D}}$$

- Capacity constraints

$$\sum_{l \in L^{\mathrm{D}}} F_{ul} \leq \sum_{o \in O_u} b_{uo} Y_{uo}, \quad u \in L^{\mathrm{U}}$$

$$\sum_{o \in O_u} Y_{uo} \leq 1, \quad u \in L^{\mathrm{U}}$$

- Decision variable value restrictions

$$Y_{uo} \in \{0, 1\}, u \in L^{\mathrm{U}}, o \in O_u; \quad F_{ul} \geq 0, \quad u \in L^{\mathrm{U}}, \quad l \in L^{\mathrm{D}}$$

Topclim, however, knows that, because there is already a plant in Quebec to produce air conditioning units for the Canadian market and because the Mexican market is becoming more attractive, it would be more advantageous to optimize their manufacturing network across North America. In such a context, management also comes to the conclusion that it would be preferable to maximize the economic value added in Canadian dollars rather than minimize costs.

(a) What additional data must Topclim collect to be able to design a multinational manufacturing network maximizing repatriated after-tax profits?
(b) Modify the given optimization model to adapt it to this new context.

Exercise 9.2 Consider a Canadian company selling glass on the European market. The company has a manufacturing division in Brazil that sells the products to a French division in charge of distribution in Europe. The smallest compliant transfer price is $100/unit and the largest compliant transfer price is $115/unit. You are asked to find which one maximizes the company's net profits after tax. The profit tax rates are given in Table 9.2 and the following additional data apply

- Import duties charged by France on the value of incoming products: 10 %
- Annual fixed cost of the Brazilian plant: $250,000
- Annual fixed cost of the French DC: $2000,000
- Sales price of the product in Europe: $200/unit
- Variable production and transport cost of the Brazilian division: $50/unit
- Annual sales: 50,000 units

Exercise 9.3 Section 9.4.3 explains how to formulate a multinational SCN design model including decisions for the selection of the incoterms to use for international product transfers. Following these guidelines, propose an extension of the design model in Sect. 9.4.2 incorporating incoterm selection decisions.

Exercise 9.4 Section 9.4.3 provides guidelines on how to proceed to model freight forwarder selection decisions for intercontinental lanes. Starting from the design model presented in Sect. 9.4.2, propose a multinational SCN design model including these decisions.

Exercise 9.5 Go back to the production–distribution network of Ex. 8.3 and let us consider now that the company is based in Netherlands.

(a) Check if the solution is still optimal if the following constraints are added

  - Local contents rules in Spain limits exported flows from the plants located in the country to 20,000 and 50,000 units for products 1 and 2, respectively.
  - The Dusseldorf DC is in a FTZ and when it is used to service German clients, duties of 5 euros per unit shipped must be added to the cost.
  - The two previous additional conditions apply.

(b) Collect data on profit taxes and exchange rates for the countries considered in this network (from PwC and World Bank (2014) and OANDA (www.oanda.com), for example) and propose an approach to optimize the net profit after taxes for the company.

Bibliography

Arntzen B, Brown G, Harrison T, Trafton L (1995) Global supply chain management at Digital Equipment Corporation. Interfaces 25(1):69–93

Arvis J-F, Saslavsky D, Ojala L, Shepherd B, Busch C, Raj A (2014) Connecting to compete 2014: trade logistics in the global economy. The World Bank

Baily M, Comes F (2014) Prospects for growth: an interview with Robert Solow. McKinsey Quarterly

Cheung Y-W, Chinn M, Pascual A (2005) Empirical exchange rate models of the nineties: are any fit to survive? J Int Money Finance 24(7):1150–1175

Cohen M, Lee H (1989) Resource deployment analysis of global manufacturing and distribution networks. J Manuf Oper Manage 2:81–104

Dasu S, Li L (1997) Optimal operating policies in the presence of exchange rate variability. Manage Sci 43(5):705–722
de Matta R, Miller T (2015) Formation of strategic manufacturing and distribution network with transfer prices. Eur J Oper Res 241:435–448
Dong L, Kouvelis P, Su P (2014) Operational hedging strategies and competitive exposure to exchange rates. Int J Prod Econ 153:215–229
Dornier P-P, Ernst R, Fender M, Kouvelis P (1998) Global operations and logistics. Wiley, New York
Ferdows K (ed) (1989) Managing international manufacturing. Elsevier
Ferdows K (2003) New world manufacturing order. Ind Eng (Feb):28–33
Fine C (2013) Intelli-sourcing to replace offshoring as supply chain transparency increases. J Supply Chain Manage 49(2):6–7
Ghemawa P (2007) Redefining global strategy: crossing borders in a world where differences still matter. Harvard Business School Press, Boston, MA
Ghemawa P, Altman S (2012) DHL global connectedness index 2012. Deutsche Post DHL
Ghemawa P, Altman S (2014) DHL global connectedness index 2014. Deutsche Post DHL
Glautier M, Bassinger F (1987) A reference guide to international taxation: profiting from your international operations. Lexington Books, Lexington, MA
Guillén-Gosbalez G, Grossmann I (2009) Optimal design and planning of sustainable chemical supply chains under uncertainty. AIChE J 55(1):99–121
Hammami R, Frein Y (2014) Redesign of global supply chains with integration of transfer pricing: mathematical modeling and managerial insights. Int J Prod Econ 158:267–277
Hill C (2014) International business, 10th edn. McGraw-Hill/Irwin, New York
Hodder J, Dincer C (1986) A multifactor model for international plant location and financing under uncertainty. Comput Oper Res 13(5):601–609
Huchzermeier A, Cohen M (1996) Valuing operational flexibility under exchange rate risk. Oper Res 44(1):100–113
Kogut B (1985) Designing global strategies: profiting from operational flexibility. Sloan Manage Rev (Fall):27–38
Kogut B, Kulatilaka N (1994) Operating flexibility, global manufacturing, and the option value of a multinational network. Manage Sci 40(1):123–139
Kouvelis P (2000) Global sourcing strategies under exchange rate uncertainty. In: Tayur S, Ganeshan R, Magazine M (eds) Quantitative models for supply chain management. Kluwer, Dordrecht, pp 669–702
Kouvelis P, Su P (2007) The structure of global supply chains. Now Publishers
Krugman P, Obstfeld M (2006) International economics: theory and policy, 7th edn. Darly Fox
Lootsma F (1994) Alternative optimization strategies for large-scale production-allocation problems. Eur J Oper Res 75:13–40
M'Barek W, Martel A, D'Amours S (2011) Designing multinational value-creating supply chain networks for the process industry. WP CIRRELT-2010-51, CIRRELT, Université Laval
Meixell M, Gargeya V (2005) Global supply chain design: a literature review and critique. Trans Res E 41:531–550
Meese R, Rogoff K (1983) Empirical exchange rate models of the seventies: do they fit out of sample? J Int Econ 14:3–24
Munson C, Rosenblatt M (1997) The impact of local content rules on global sourcing decisions. Prod Oper Manage 6(3):277–290
OECD (2010) OECD transfer pricing guidelines for multinational enterprises and tax administrations. OECD
OECD (2013) Addressing base erosion and profit shifting. OECD
Perron S, Hansen P, Le Digabel S, Mladenovic N (2010) Exact and heuristic solutions of the global supply chain problem with transfer pricing. Eur J Oper Res 202:864–879
Pomper C (1976) International investment planning: an integrated approach. North-Holland
Porter M (1990) The competitive advantage of nations. Free Press
PwC, World Bank (2014) Paying taxes 2015. PricewaterhouseCoopers

Roberts D (2005) Changes in the global forest products industry: Defining the environment for Canada. Presentation at Vision 2015, CIBC World Markets, Montréal

Rosemain M (2013) Carmakers expand production reach to limit currency risk. Business Week (Mar 5)

Seyoum B (2013) Export-import theory, practices, and procedures, 3rd edn. Routledge, London

Taylor D (1997) Global cases in logistics and supply chain management. Thomson Business Press

Vidal C (1998) A global supply chain model with transfer pricing and transportation cost allocation. Doctoral thesis, Georgia Tech

Vidal C, Goetschalckx M (2001) A global supply chain model with transfer pricing and transportation cost allocation. Eur J Oper Res 129:134–158

Vila D, Martel A, Beauregard R (2006) Designing logistics networks in divergent process industries: a methodology and its application to the lumber industry. Int J Prod Econ 102:358–378

Villegas F, Ouenniche J (2008) A general unconstrained model for transfer pricing in multinational supply chains. Eur J Oper Res 187:829–856

Vos B (1997) Restructuring manufacturing and logistics in multinationals. Avebury

Wilhelm W, Liang D, Rao B, Warrier D, Zhu X, Bulusu S (2005) Design of international assembly systems and their supply chain under NAFTA. Trans Res E 41:467–493

Chapter 10
Risk Analysis and Scenario Generation

On several occasions in the previous chapters we insisted that some of the fundamental factors to consider when designing SCNs are random variables. This includes product demand—and thus capacity requirements (see Chap. 8); transportation lead times and rates—because of, among other things, the cost of fuel and cost function regression errors (see Chap. 5); production resource costs—because of, in particular, exchange rate variations (see Chap. 9); and so on. When considering multi-year planning horizons, one is thus naturally exposed to random cash inflows and outflows (see Chap. 2). Despite this, the SCN design models proposed in previous chapters assume that all required data is known with certainty. In addition, in the last decades, several companies have suffered from natural disasters, industrial accidents, supplier bankruptcies, strikes, and other major disruptive events, resulting in a significant loss of market share and revenue, the destruction of resources, and high recourse costs. Business leaders should be able to provide adequate protection against business as usual random events as well as high-impact destructive events through the design of robust and resilient SCNs. Consequently, this chapter addresses the following issues: How can the vulnerabilities of a SCN be identified? How can the ups and downs of everyday business be anticipated? How can the likelihood and the impact of catastrophic events be estimated? How can these methods be used to facilitate the design of robust SCNs? The answers to these questions suggest that better SCN designs are obtained when considering sets of plausible future scenarios instead of a single expected future. The chapter thus proposes an approach for modeling risk and for the generation of plausible future scenarios. The proposed approach is based on standard probability and statistics concepts, and we assume from now on that the reader is familiar with these notions.

A. Martel and W. Klibi, *Designing Value-Creating Supply Chain Networks*,
DOI 10.1007/978-3-319-28146-9_10

10.1 Supply Chain Networks Under Uncertainty

10.1.1 Supply Chain Network Vulnerabilities

The Centre for Research on the Epidemiology of Disasters (CRED—www.emdat.be) identified 330 disasters affecting 108 countries in 2013. Table 10.1 provides a sample of major catastrophic events observed over the past two decades and highlights their impact on company SCNs. It shows that they can affect SCN supply, production–distribution capacity, and demand. According to a recent survey (Taylor 2013), in 2012, 63 % of European, Middle East, and African companies experienced SC disruptions because of unforeseen events beyond their control linked to the economic context (24 %), natural disasters (19 %), subcontractor difficulties (16 %), and even terrorism (5 %). On average, after an incident, it has taken companies 63 days to get back to business as usual. According to a recent WEF survey (Bhatia et al. 2013), natural disasters are the main type of extreme events threatening SCNs. Conflicts, political unrests, terrorism, and demand shocks (e.g., decline in construction material demand in the United States after the 2008 financial crisis) are also major concerns. The report also underlines that cyber risk is an increasingly serious problem. In a 2007 study (van Opstal 2007), the Council on Competitiveness found that 93 % of the companies losing access to their data centers for ten days or more filed for bankruptcy in the following year. SCN disruptions thus result in significant financial losses and threaten companies' competitiveness.

The cascade effects that major disruptions may have on clusters of interconnected companies can also be significant. For example, the 2011 Japanese tsunami resulted in a 1.1 % reduction of world industrial production in the month that followed. In particular, PSA Peugeot Citroën in Europe was severely affected. The temporary shutdown of a Hitachi plant in Japan leads to the interruption of the supply of an electronic component to PSA, which in turn resulted in a 25–60 % reduction of production in eight PSA assembly plants in France, Spain, and Slovakia (Bourgin and Lenoire 2012). Similarly, following the Thailand floods in 2012, more than 1000 plants were strongly affected in the automotive and high-tech industries. Catastrophic events have been largely ignored by companies until recent years. Risk analysis is, however, now seen as an important part of any SC system design project, as confirmed by its recent addition to the SCOR model (see Sect. 3.2.4).

Three broad categories of SCN vulnerability sources are distinguished in Fig. 10.1: internal SCN assets, SC partners, and public infrastructures. Internal assets include equipment, vehicles, human resources, and inventories found in production–distribution centers, recovery–rehabilitation centers, and service centers. These assets are vulnerable when located on sites exposed to disruptions. SC partners include customers, raw material and energy vendors, contract manufacturers, and 3PLs. The SCN is vulnerable when there is a unique supply source for some materials or components, when supply lanes are long, or when sales depend

Table 10.1 Sample of catastrophic events over the last two decades

Event	Date	Where	Firm	Description	SC Impact	Loss
Supplier fire	02/97	JAP	Toyota	Fire destroys all Aisin valve production capacity	20 plants affected. 72,000 less vehicles produced	0.1 % drop of country industrial growth
Hurricane Mitch	10/98	USA	Dole	Hurricane destroys the entire banana crop	Dole loses 25 % of its banana inventory	4 % revenue loss for Dole and increase of competitor's revenues
Supplier fire	03/00	MEX	Nokia and Ericson	Fire at supplier semiconductor factory (Philips's Albuquerque, NM)	Nokia and Ericsson affected by 40 % delivery delay in mobile phone parts	Loss of USD 400 M for Ericsson, who exited mobile phones market at end of year
Terrorist attack	09/01	USA	Ford US	WTC terrorist attack	Delivery of materials by air stopped for a week	Ford factory output decreases by 13 % for the fourth quarter
Supplier bankruptcy	12/01	USA	Land Rover	Supplier bankruptcy stops chassis deliveries	Land Rover production almost stopped for 9 months	16 M GBP recourse payment
Tornado	05/03	USA	General Motors	A tornado damages the roof of a GM truck plant	Production decrease of 20,000 units (3 % of N-A production)	USD 140–200 M lost
Industrial accident	03/05	USA	British Petroleum	Hydrocarbon vapor cloud explosion	15 persons killed and 170 injured; factory rebuilt five years later	Cost of USD 1 billion
Strike	01/07	USA	Good-year	12-day strike	Loss of eight suppliers	Cost of USD 30 M/week and loss of USD 360 M in market share

(continued)

Table 10.1 (continued)

Event	Date	Where	Firm	Description	SC Impact	Loss
Earthquake/tsunami	03/11	JAP	Electronics and auto industry	Tohoku earthquake causes destruction of Fukushima nuclear plant	Country production decreased by 15.3 % and automotive production by 57.3 %	Global industrial production decline of 1.1 % in April 2011

Sources Rice private communication (2007), World Bank (2011)

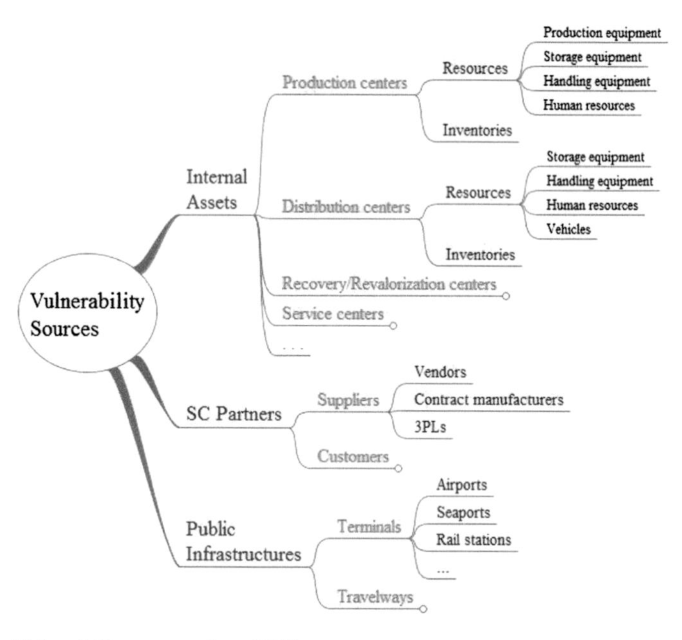

Fig. 10.1 Vulnerability sources for a SCN

heavily on an exposed major product market. The public infrastructures (transportation terminals, roads, telecommunication networks, electricity networks, utility plants, etc.) used by an SCN can also fail. Assets, partners, and infrastructures are physically located in specific geographical areas. They are thus exposed to natural disasters (cyclones, earthquakes, floods, volcanos, wild fires, drought, ice storms, infectious disease epidemic, etc.). They can also fail because of industrial accidents, fires, labor disputes, terrorist attacks, and so on. Supply and demand can also be perturbed by the bankruptcy of a partner, sudden drops in material availability, political instability, market failure, and so forth.

Day-to-day random business events are relatively easy to model using stochastic processes. However, even if they can have catastrophic consequences, major disruptive events are difficult to predict and thus difficult to model. Natural and

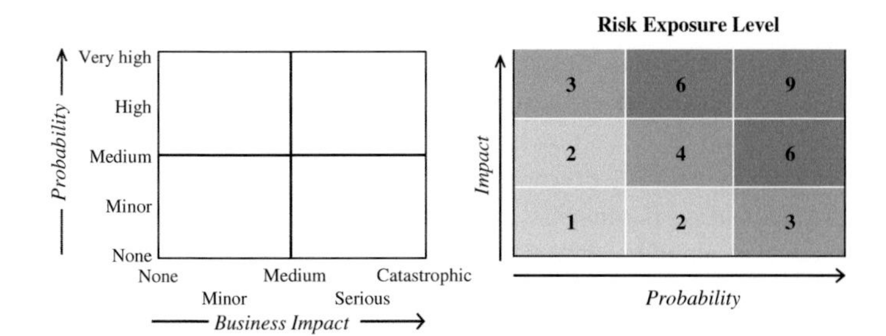

Fig. 10.2 Risk exposure matrices

accidental disasters follow Pareto's law: a small proportion of the disruptions causes a large part of the damage (Sheffi 2005). This is why the exposure to such events is typically assessed by associating its probability of occurrence with its business impact. The risk exposure matrix proposed by Norrman and Jansson (2004), and illustrated on the left side of Fig. 10.2, summarizes the two dimensions of basic risk assessment approaches. This type of risk matrix is often used in practice to categorize vulnerabilities. Matrices, such as the one shown on the right of Fig. 10.2, are often used to subjectively assign an exposure level to potential hazards. We show in this chapter that this approach is quite simplistic and that the modeling of extreme events requires a more elaborate statistical analysis approach. A workbook for the subjective assessment of vulnerabilities and for the evaluation of their impact was developed by Cranfield University for the Department for Transport in the United Kingdom (Christopher 2003).

10.1.2 Business Environment Characterization

A SCN must be designed to cope with its future environment, but at the point in time when it is reengineered the future is not known with certainty. The business as usual and extreme events mentioned are important sources of uncertainty, but the information available to characterize some of them may be lacking. Such decision-making situations can be characterized based on the quality of information: decisions are made under *certainty* when perfect information is available and under *uncertainty* when the obtainable information is partial (or imperfect) (Klibi et al. 2010). The term *uncertain* is value neutral, that is, it includes the chance of gain and, conversely, the chance of damage or loss. Uncertainty may lead to *risk* and this term refers to the possibility that undesirable outcomes could occur. The risk increases as the likelihood and negative impact of possible outcomes increases, as illustrated by the risk exposure matrices in Fig. 10.2.

The planning horizon considered in SCN design problems under uncertainty covers a set of *operational periods* $\tau \in T^{u}$ corresponding to discrete time intervals between SCN users' operational decisions such as days or weeks. These periods are thus much shorter than the aggregated *planning periods* $t \in T$ defined in Chap. 8 for the formulation of dynamic design models under certainty (see Fig. 8.19). As shown in Fig. 10.3, the business environment in which the future logistics network will operate is generally not known when design decisions are made. It becomes known gradually as daily operational decisions are made. However, to be able to make design decisions, it must be anticipated. The simplest way to do this is to incorporate several plausible future scenarios in SCN design models. Because the recovery from SCN disruptions may last only a fraction of a planning period or may overlap adjacent planning periods, plausible future scenarios must be elaborated over operational periods and then aggregated into planning periods for design purposes.

During the planning horizon, the environment evolves dynamically. To simplify we assume that changes in the state of the environment can occur only at the beginning of operational periods. An *environment* is thus defined as the internal and external conditions under which the SCN operates during a given operational period. The future is considered by specifying possible sequences of environments over the planning horizon. Each possible sequence of environment defines a plausible future *scenario*. An *event* is a measurable (i.e., having observable consequences) factor or incident influencing the business environment during a given time interval, that is, during an adjacent subset of periods in T^{u}. As shown in Fig. 10.3, single-period and multi-period events can be observed, depending on the granularity of operational periods. The environment of planning period $t \in T$ is thus a compound event, that is, the result of all the events occurring during planning period t.

Under uncertainty, different qualities of information may be available. The worst case is *total uncertainty* or complete ignorance. Three types of uncertainties may be distinguished when partial information is available: randomness, hazard, and deep uncertainty. *Randomness* is characterized by random variables related to business as usual operations; *hazard* by low-probability, high-impact, unusual events; and *deep uncertainty* by the lack of any information to assess the likelihood of plausible future extreme events. From this characterization of uncertainty, it is seen that three types of events shape SCN environments:

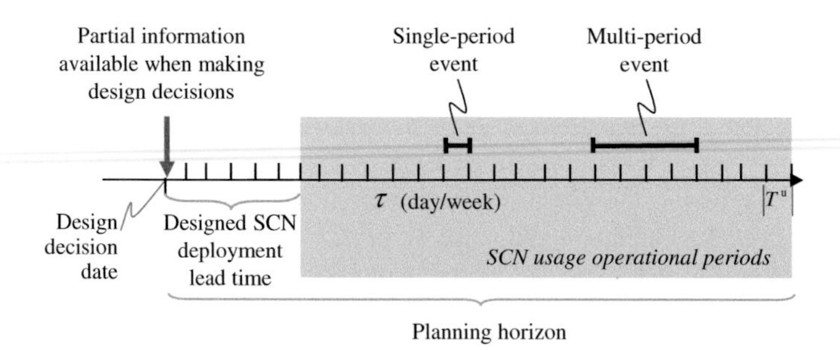

Fig. 10.3 Business environment during a planning horizon

- *Random events*, that is, possible outcomes defined over a single period and such that their probability of occurrence can be estimated. Historic information on supply, demand, costs, lead times, failure rates, exchange rates, and so on can be used to estimate the probability distribution of random variables related to the business as usual operations of a SCN. These events include the degenerate *certain events* case that occurs when perfect information is available.
- *Hazardous events* (*hazards*), that is, factors or incidents affecting a number of adjacent periods and resulting in SCN disruptions. Hazards are rare but repetitive events that may be characterized by formal location, severity, and occurrence processes. They include natural, accidental, or willful incidents affecting SCN resources. They also include disruptions arising from natural hazards affecting a geographical region, such as earthquakes, floods, windstorms, volcanic eruptions, droughts, forest fires, heat waves, freezes, and cold waves. For such events, catastrophe models have been used to provide a likelihood of occurrence and/or a likelihood of associated monetary losses based on historical data and/or professional expert opinions (Grossi and Kunreuther 2005). When data are not sufficient to estimate objective probabilities, subjective, or partially subjective, probabilities are used.
- *Deeply uncertain events*, that is, incidents affecting a number of adjacent periods for which little information exists. These events include isolated, non-repetitive, extreme events for which a likelihood of occurrence cannot be evaluated (Banks 2006). Events related to terrorism (sabotage, bombing, etc.) and political instability (sudden currency devaluation, coup, etc.), with unpredictable time of occurrence, severity and location, are considered as deeply uncertain. In the recent past, some of these disruptions, such as the 9/11 WTC attack, the SARS epidemic, and the Arab Spring, led to major business failures.

A summary of the characteristics of the different types of event affecting SCNs is provided in Fig. 10.4.

Information ↑ Impact ↓

Event type	Info.	Impact	Characteristics	Typical examples
Known events	Perfect	Normal	Structural data independent of business state	Bill of material, standard processing times,work time during a period, etc.
Random events	Partial	Normal	Stochastic processes for business as usual factors estimated from historic dataand/or expert knowledge	Supply, demand, prices, energy and raw material costs, interest rates, exchange rates,etc.
Hazardous events	Partial	Serious to catastrophic	Rare but repetitive factors or incidents which may be characterized by formal location, severity and occurrence processes	Major equipment breakdowns, strikes, discontinuities in supply, natural disasters, industrial accidents,etc.
Deeply uncertain events	Little information	Serious to catastrophic	Isolated, non repetitive extreme events, with unpredictable time of occurrence, severity and location	Sabotage, terrorist attacks, sudden currency devaluation, political coup,etc.

Fig. 10.4 Event types characteristics

10.2 Modeling Random Events

This section focuses on the modeling of business as usual random events characterized by random variables with known probability functions. The probability functions typically used to model these stochastic processes include normal, log-normal, exponential, and uniform continuous distributions, as well as discrete binomial, Poisson, and uniform distributions. As mentioned, we assume that random events occur during a single operational period τ. Before a design decision is made, each event e is modeled by a random variable ζ^e, which may depend on time and/or exogenous explanatory variables. Figure 10.5 illustrates the modeling of a stationary stochastic process (say daily or weekly demand). Historical observations are used to estimate the cumulative distribution function $F^e(\zeta)$ of the random variable ζ^e associated with the event. This distribution is then used to generate plausible future scenarios, that is, series of possible values ζ_τ^e of the random variable during the periods $\tau \in T^u$ of the planning horizon. The historical observations used to estimate the random variable should not be affected by hazards so as to capture the fundamental nature of the underlying phenomenon.

Figure 10.6 illustrates typical random variables affecting the business environment. The left side plot shows the fluctuation of crude oil prices during the last 15 years, one of the main determinants of transportation costs. Two superimposed phenomena are visible in this graph. In normal times oil prices increase gradually and they show variations of up to ±USD 15 around an upward trend. The impact of the 2008 financial crisis, however, is also clearly visible. The plot on the right shows residential construction spending in the United States. House construction influences the demand for many products, such as lumber and other construction material. The impact of the 2006–2009 economic crisis is also visible on this graph.

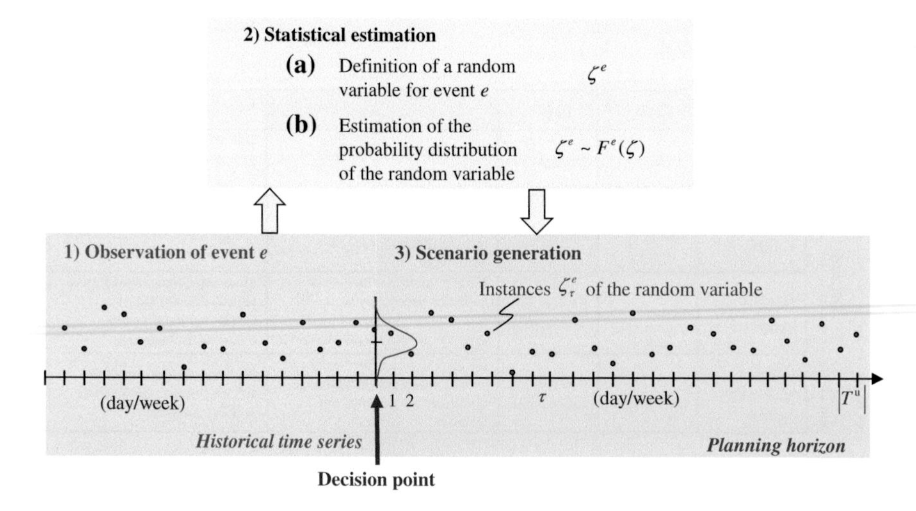

Fig. 10.5 Random event modeling and scenario generation

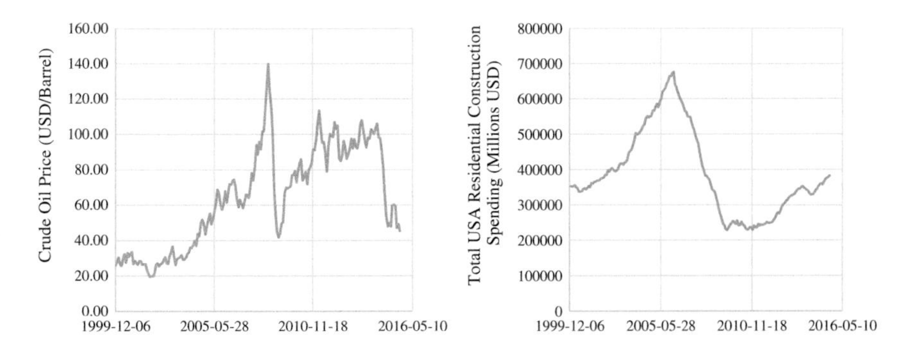

Fig. 10.6 Examples of SCN environment random variables—data retrieved from FRED, Federal Reserve Bank of St. Louis

Other examples of business as usual random variables include the Canada–USD and the Euro–USD exchange rates shown in Fig. 9.5, as well as the regression errors for the transportation cost function shown in Fig. 5.19.

Christopher and Holweg (2011) calculated an aggregate volatility index based on exchange rates, interest rates, raw material prices, and shipping costs data, and they found that SCs currently are experiencing their greatest level of volatility since the 1970s. From a statistical point of view, underlying processes are also often nonstationary, and it is clear that the approximation of their behavior by expected values is not sufficient. Nowadays, most companies maintain detailed databases to track business transactions and the state of their resources (see Sect. 3.2.3). Private or public databases on exogenous variables affecting the business environment are also easier to access via Internet. These data can be used to estimate the parameters of stochastic processes. Several modeling approaches and statistical estimation methods are available to do this, but their in-depth study goes beyond the scope of this chapter. Stochastic processes can be classified into three broad categories:

- *Simple processes* involving a single random variable, often in the form of a time series. Their model is estimated using basic statistical methods or forecasting methods such as exponential smoothing (Makridakis et al. 1998). The process can be stationary or nonstationary, that is, it can incorporate a trend, seasonal effects, or cyclical effects. It can also evolve over with time which is taken into account by giving more weight to recent data.
- *Compound processes* involving several random variables. The model formulated to describe the phenomenon is a function of the random variables specified. Several probability distributions must then be estimated.
- *Dependent processes*. In this case the phenomenon depends on explanatory variables corresponding either to decisions to make (e.g., a product demand depends on its price, delivery time, and advertising budget) or to exogenous factors (e.g., transportation costs depend on oil prices). This is the domain of regression analysis and econometric models (Wooldridge 2008).

Stochastic process models can be estimated using Excel's data analysis tools, or specialized statistical packages such as IBM-SPSS, SAS, and STAT-FIT. When examining phenomena in depth, one realizes that randomness is present virtually everywhere. Taking all sources of randomness into account in a SCN design project would quickly become overwhelming. To avoid the paralysis by analysis syndrome, stochastic modeling must be limited to processes that have a major impact on the quality of SCN designs. The type of statistical analysis required can be illustrated with the following two examples.

Example 10.1
Table 10.2 provides a series of weekly demand observations for a product family in two separate markets. Using STAT-FIT, for example, a histogram of these two samples of 50 observations can be plotted and a density function fitted. As shown in Fig. 10.7, the demand for product-market 1 takes the form of a lognormal distribution with a mean of 4.86 and standard deviation of 0.19. The figure also shows that product-market 2 demand takes the form of a uniform distribution with a lower bound of 90 units and an upper bound of 160 units. Clearly, this estimation method assumes that the underlying stochastic process is stationary.

Example 10.2
When a product's demand is presented as a time series defined with discrete time periods (days, weeks, months, years), the demand for each period is in fact an aggregate quantity giving the sum of all orders (sales) during that period. In reality, demand takes the form of a series of orders (sales) for given quantities of products taking place at different points in time. To represent this compound process adequately, the customer (order) *arrival process* must be distinguished from the customer's *order size*. Customer interarrival times can be programmed or random, and order sizes can be fixed or random. It was shown that when the arrival of customers is purely random, customer

Table 10.2 Weekly demand samples for two product markets

Weekly demand for product market 1					Weekly demand for product market 2				
193	131	129	111	120	123	94	150	111	138
140	120	132	173	125	154	99	101	107	121
173	167	117	115	119	145	109	101	135	128
105	92	107	105	114	158	103	160	141	144
124	128	132	146	145	153	149	139	108	150
139	124	85	116	114	92	138	125	139	108
109	147	93	163	95	107	104	131	148	105
97	199	166	104	146	90	143	125	110	134
122	141	137	123	129	109	109	137	122	108
125	147	147	164	176	102	148	145	94	157

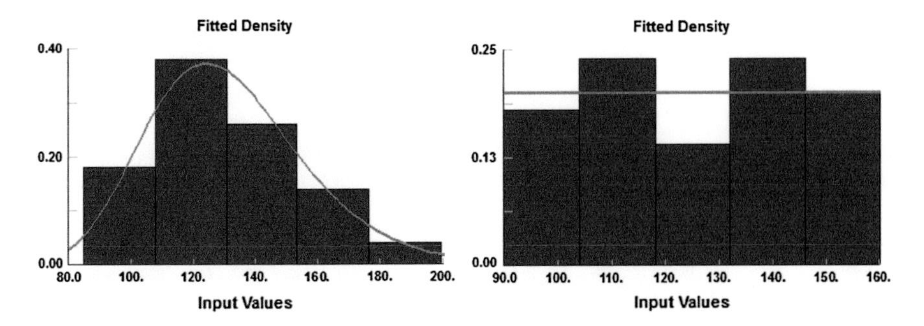

Fig. 10.7 Density functions fitted for two continuous random variables

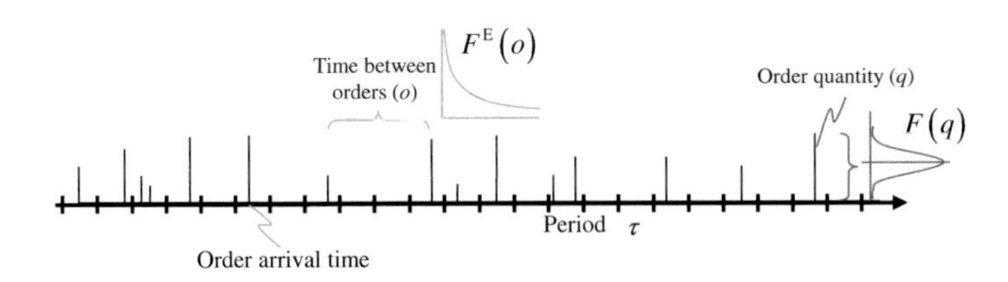

Fig. 10.8 Compound demand process

Table 10.3 Customer order sample for a product marketproduct market

Period	Quantity	Period	Quantity	Period	Quantity	Period	Quantity	Period	Quantity
5	652	66	540	99	500	135	556	183	508
20	582	69	513	102	537	139	572	184	537
21	549	71	540	104	628	139	513	186	559
26	515	76	545	110	567	141	569	195	533
27	544	77	534	111	553	142	574	197	500
33	547	82	570	112	544	151	529	235	532
44	500	84	579	115	524	167	516	237	539
47	532	86	571	118	500	168	539	256	540
53	539	94	545	135	500	174	545		

interarrival times follow an exponential distribution, and the number of customers arriving during a given time interval follows a Poisson distribution (Heyman and Sobel 1982). If the size of the orders is random, as shown in Fig. 10.8, the demand follows a *compound Poisson process*. The interarrival time is then a random variable o with an exponential distribution function $F^{E}(o)$, and the order size is a random variable q with distribution function $F(q)$. In practice, the latter is often normal or lognormal.

The estimation of this kind of process can be illustrated with the data found in Table 10.3, which provides customer arrival dates (day) and quantity

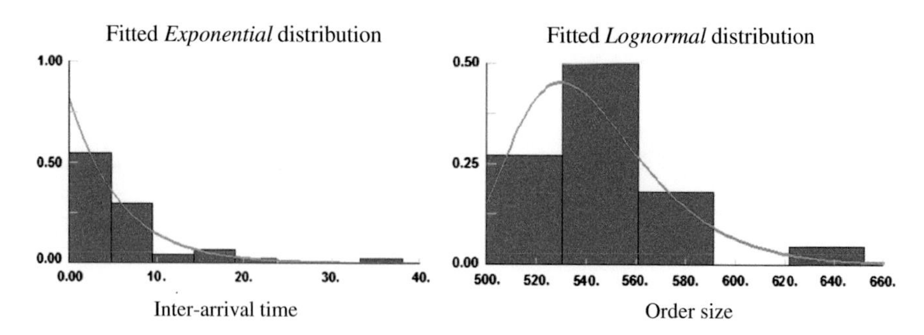

Fig. 10.9 Probability distributions fitted for example

ordered during a year. Probability distributions can be fitted to these historical data and their parameters estimated using STAT-FIT. As shown in Fig. 10.9, the time between the arrivals of successive customers takes the form of an exponential distribution with an average of 5.82 days. The customer order size is lognormal with an average of 543 and a standard deviation of 31.

10.3 Risk Analysis

This section presents a risk modeling approach to support the design of SCNs and in particular to permit the explicit consideration of hazardous and deeply uncertain events. The emphasis is not on the detailed modeling of disasters but rather on incorporating a sufficient degree of realism in SCN risk modeling to provide more robust and resilient designs. SCNs are considered to be vulnerable systems under threats. Unlike for random events, it is difficult to make direct use of historical data to estimate probability distributions for hazards. To characterize hazards adequately, three fundamental risk analyses questions (Haimes 2004) must be addressed:

1. What can go wrong?
2. What is the likelihood of that happening?
3. What are the consequences?

For hazards this leads to a three-phase analysis and modeling approach. This approach first identifies the potential extreme events and vulnerabilities to consider and it evaluates SCN exposure levels. This is done by working with *multi-hazards*, that is, meta-events having generic impacts on SCN resources, partners, and markets. Multi-hazard arrival processes are then estimated and their consequences on the SCN are assessed using adequate disruption severity metrics and recovery functions. The latter are related to key SCN design variables, such as facility and supplier capacity as well as customer demand. For deeply uncertain events, the second question cannot be answered and some adjustments to the approach are required. In what follows, the preceding questions are examined and several illustrative examples are provided.

10.3.1 Multi-hazards and Vulnerability Sources

The first phase of the risk analysis approach proposed answers the question: what can go wrong? The SCN vulnerability sources and multi-hazards to consider in the study are identified and potential threats are characterized. The territory over which the network is deployed is partitioned into hazard zones, and their exposure levels are evaluated. When the phase is completed, each SCN location is associated with a vulnerability source, a hazard zone, and an exposure level.

As mentioned, to perform its activities a SCN exploits internal resources, does business with SC partners, and uses public infrastructures. These resources and partners are associated with specific geographical locations. As in the previous chapters, L denotes the set of all considered SCN locations. When modeling a SCN, some locations may be aggregated into geographical zones with a computable centroid. For example, in a business context, ship-to points are usually aggregated into demand zones and, in a military or humanitarian context, demand is naturally associated with regions where conflicts or disasters of various types may develop.

10.3.1.1 Hazard Classes

Hazards can take many forms and our approach must find a convenient way to take them into account without getting lost in a maze of possible incident types. This is done by categorizing hazards into a small number of multi-hazards (Gogu et al. 2005; Scawthorn et al. 2006), that is, recurrent extreme event classes, such as natural, accidental, or willful hazards, which do not affect the SCN in the same way. Also, depending on the scope of the study, some hazard types may not be relevant. For example, when designing an American SCN, natural disasters are relevant, but the risk of armed conflicts resulting from a political failure is negligible. However, when designing an international SCN, potential state failures must be taken into account. Also, even if a hazard type is relevant, in some parts of the world the data required to characterize it may not be available. For all these reasons, for a given SCN reengineering project, the set M of multi-hazards to consider must be specified. This set could, for example, contain the following four elements: natural disasters, geopolitical failures, market failures, and industrial accidents. Multi-hazard sets can be elaborated from the data provided by several public sources such as CRED (*Centre for Research on the Epidemiology of Disasters*—www.cred.be), Global Risk Data Platform (www.preventionweb.net), HIIK

(*Heidelberg Institute for International Conflict*—www.hiik.de), and FEMA (US *Federal Emergency Management Agency*—www.fema.gov), as well as private sources such as large reinsurers (Swiss Re—www.swissre.com, Munich Re—www.munichre.com, etc.).

10.3.1.2 Vulnerability Sources

When considering potential SCN risks, a large number of vulnerability sources can be identified. However, the impact of hazards on these vulnerability sources can vary from catastrophic to low. At the strategic level, the number of vulnerability sources considered explicitly should be reduced to a manageable level. To help identify them, company vulnerability maps can be constructed (Sheffi 2005; WEF 2013). Figure 10.1 is an example of such a map for a typical SCN. To select the vulnerabilities to consider explicitly in a SCN design study a filtering process based on a subjective evaluation of vulnerabilities can be used (Haimes 2004). In our context, vulnerability sources are usually associated with types of internal or external network locations: production, distribution, and service facilities affecting capacity (plants, DCs, stores, intermodal transfer points, etc.), product-markets influencing demand, and vendors shaping supply (raw material and energy suppliers, contract manufacturers, etc.). It is assumed in what follows that all strategic vulnerabilities come from SCN locations $l \in L$ and not from network arcs.

Let S be the set of all relevant vulnerability sources retained for a study. The overriding criterion for the definition of a vulnerability source $s \in S$ is that all the locations $l \in L_s \subset L$ it covers have a similar behavior in terms of impact intensity, time to recovery, and recovery pattern when hit by a multi-hazard, so that they can all be described in terms of the same metrics. When an extreme event occurs, all locations are not affected in the same way. For example, a fire in a plant may decrease production capacity but an earthquake in a demand zone may increase demand for first aid products but decrease demand for luxury products. Vulnerability sources must also be defined so that the sets L_s, $s \in S$ are mutually exclusive and collectively exhaustive. This may lead to the definition of more than one location l for a same geographical point or region. For example, suppose that four vulnerability sources are a priori retained from those identified in Fig. 10.1: production centers, DCs, vendors, and customers. However, if the sales of two product categories in a given region (say first aid products and luxury products) are not affected in the same way by a multi-hazard, then the source previously labeled 'customers' must be split into two distinct vulnerability sources: first aid product-markets and luxury product-markets. Also the geographical region would then be associated with two location indexes l and l' to distinguish the product-markets. The notation $s(l)$ is used to identify the vulnerability source of location l.

The association of possible multi-hazards with selected vulnerability sources ($M \times S$) provides the spectrum of *what can go wrong*. Note that, in some SCN design contexts, a vulnerability source is not necessarily affected by all multi-hazards considered, which is accounted for by specifying the subset $M_s \subseteq M$ of multi-hazards that may hit a source $s \in S$. In 2009, researchers from MIT conducted a survey (Arntzen 2012) on experiences, attitudes, and risk management practices of 1400 SC professionals in 70 countries. The results show that the prioritization of vulnerability sources varies from one country and one industry to another and that supply chain executives do not all have the same risk perceptions and tolerance.

10.3.1.3 Hazard Zones

In order to map threats, the geographical territory in which the SCN operates must be partitioned into a set of *hazard zones* Z. Using geographical coordinates, the hazard zone $z(l) \in Z$ of a location $l \in L$ can be identified, as illustrated in Fig. 10.10. In this figure, the nodes and arcs of the SCN are represented on the *vulnerability sources layer*, and hazard zones on *multi-hazard layers*. Hazard zones delineate areas with similar geological, meteorological, political, economic, and critical infrastructure characteristics. These zones may correspond to countries, states or provinces, counties, three-digit zip codes, or a combination of those, depending on the level of precision desired and the data available. They must be constructed, however, so that the defined SCN location aggregates fit uniquely in a hazard zone. They must also be defined so that the sets $L_z \subset L$ of locations in zones $z \in Z$ are mutually exclusive and collectively exhaustive. The zonation process is a

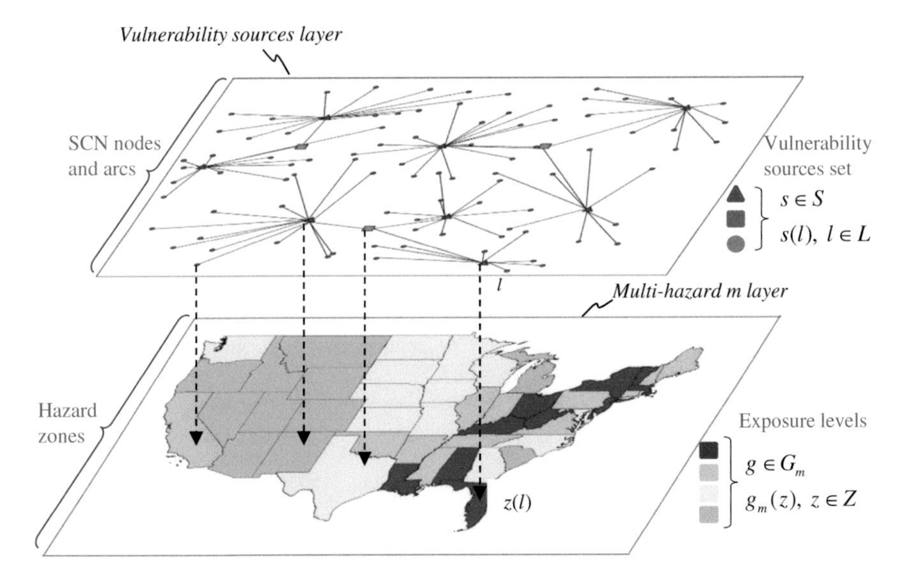

Fig. 10.10 SCN exposure modeling

key issue because the zone granularity determines the realism of the multi-hazard incidents models formulated to support the SCN design process.

10.3.1.4 Exposure Levels

As illustrated in Fig. 10.10, exposure levels must then be estimated for each multi-hazard $m \in M$ considered. In the figure, the defined exposure levels $g \in G_m$ are represented by colors. For multi-hazard $m \in M$, $g_m(z)$ denotes the exposure level of zone $z \in Z$. Exposure levels can be defined top-down or bottom-up, depending on the context. For some regions of the world and some multi-hazards, the data available are not sufficient to obtain detailed exposure estimates. This leads to a top-down approach in which exposure levels for multi-hazard $m \in M$ are associated a priori to geographical regions such as continents. The territories used as zones (countries, states, provinces, etc.) in the continent then provide the relationship $g_m(z)$ between zones and exposure levels.

Exposure levels can also be evaluated bottom-up via the estimation of relevant zone exposure indexes $(I_{mz}, m \in M, z \in Z)$. Exposure levels (G_m) are then associated with adjacent index value intervals, and zones are assigned a level $g \in G_m$ based on their index value I_{mz}. The exposure index used may be based on data provided by previously mentioned public or private data sources (CRED, FEMA, Munich Re, etc.) for natural disasters. The map on the left of Fig. 10.11, for example, illustrates indexes calculated from CRED data. A geopolitical failure index can be calculated based, for example, on HIIK and/or *Fund for Peace* FSI (*Fragile States Index*—www.ffp.statesindex.org/) data. 2015 indexes for a sample of countries are displayed on the right of Fig. 10.11. A market failure index can be calculated by combining data on economic performances provided by sources such as IMD (www.imd.ch) and WEF (www.weforum.org), on public infrastructures provided by sources such as the *CIA World Factbook* (www.cia.gov), and so on.

Independent of the approach used, the exposure level $g_m(l)$ of a location $l \in L$ is uniquely given by $g_m(z(l))$ for each multi-hazard $m \in M$. As illustrated by Fig. 10.10, this initial risk analysis phase thus leads to the specification of a

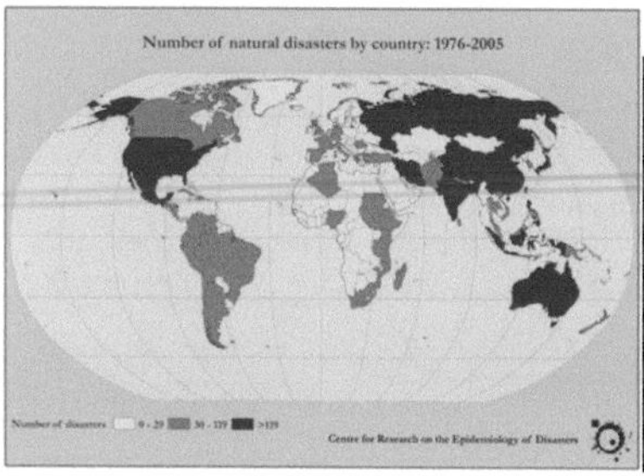

Fund for Peace Fragile States Index 2015

High Alert		*Less Stable*	
South Sudan	114.5	Trinidad	58.7
…		…	
Zimbabwe	100.0	Croatia	51.0
Alert		*Stable*	
Guinea Bissau	99.9	Barbados	49.3
…		…	
Russia	80.0	Belgium	30.4
Warning		*Sustainable*	
Lesotho	79.9	Portugal	29.7
…		…	
Albania	61.9	Finland	17.8

Fig. 10.11 Multi-hazard exposure indexes—*Sources* CRED (2015), FFP (2015)

vulnerability source $s(l)$, a hazard zone $z(l)$, and exposure levels $g_m(l)$, $m \in M$, for all SCN locations $l \in L$.

10.3.2 Multi-hazard Processes Modeling

Having determined what can go wrong, the second phase of the approach addresses the second question raised previously: what is the likelihood of that happening? To do this, it is essential to characterize the occurrence, intensity, and duration of multi-hazards by zones or exposure levels, depending on the availability of data. A compound stochastic process is defined to describe how multi-hazards occur in space and in time and to specify incident's intensity and duration. This phase is independent of the SCN considered; it relates to the multi-hazard layer depicted in Fig. 10.10. We assume that each extreme event occurs at the beginning of an operational period in a subset of adjacent hazard zones and that the event intensity and duration depend on exposure levels.

Figure 10.12 illustrates the approach proposed to model multi-hazards. It involves, first, the collection of data and expert opinions on historical incidents and, second, the formulation of an explanatory model and the estimation of the distribution function of the random variables involved using statistical estimators, when sufficient data is available, or subjective methods. The probability distributions estimated can finally be used to generate multi-hazard scenarios over the planning horizon. When the arrival process for a multi-hazard is stationary, the time between two successive hits can be modeled by a random variable. The seriousness of a hit may differ from one instance to another. It generally can be characterized by two random variables: the impact intensity and the incident duration. These variables, however, are usually highly correlated.

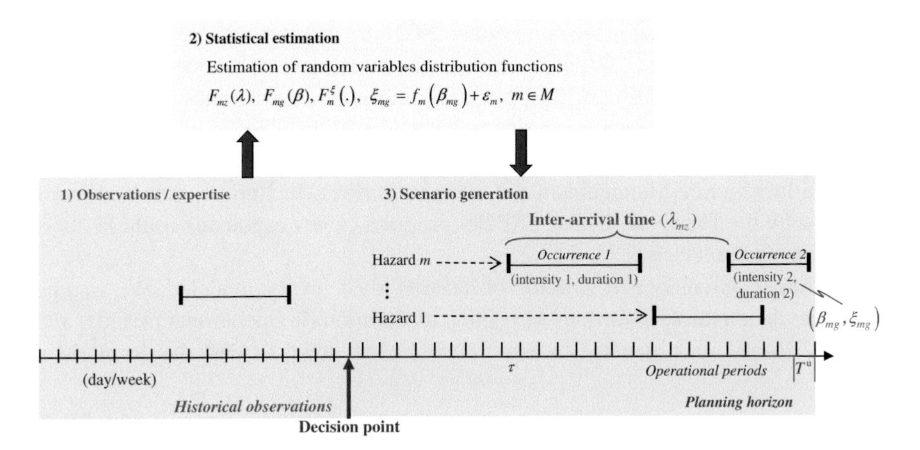

Fig. 10.12 Multi-hazard processes modeling and scenario generation

As mentioned, the proper approach to estimate spatiotemporal hazard processes strongly depends on the granularity of the hazard zones used and on available data. The simplest case can be used to illustrate the approach. When designing a national SCN for a developed country, using states, provinces, or departments as hazard zones, the data needed to estimate the process can usually be obtained relatively easily. Now, assume that multi-hazards occur independently in hazard zones and that the time between hits is a stationary stochastic process. More specifically, assume that when a multi-hazard $m \in M$ occurs in zone $z \in Z$ at a given point in time, then the time before the arrival of the next multi-hazard in the zone is a random variable λ_{mz} with cumulative distribution function $F_{mz}(\lambda)$, implying that the distribution used is estimated from hazard zone data. In practice, catastrophe models often use Poisson processes to determine the number of extreme events that can occur in a given period (Banks 2006). Accordingly, $F_{mz}(\lambda)$ is typically an exponential distribution with an expected time between multi-hazards $\bar{\lambda}_{mz}$. The estimation of this mean is straightforward. If disasters interarrival times are not stationary, an exponential distribution with a moving average estimated by regression can be used. The slope of the arrival process can then be estimated from data on the frequency of disasters provided by CRED, for example. Evolutionary trends are examined more closely in Sect. 10.4.

Variants corresponding to other contexts can be developed using a similar approach. For example, when designing a global SCN, the data provided by organizations such as CRED and HIIK are not sufficiently detailed to support a zone-based estimation approach. A hierarchical modeling approach based on exposure level arrival processes and subjective conditional hazard zone hit probabilities can then be used. The subjective zone hit probabilities can be based on indexes such as those illustrated in Fig. 10.11. This approach was used by Martel et al. (2013) for the design of a global SCN to support the overseas military and humanitarian missions of the Canadian Armed Forces. At the other end of the spectrum, another variant is required when the zones used are small (say counties or zip code groupings) and multi-hazards affect several adjacent zones. This occurs, for example, when designing a regional SCN to support disaster relief using counties as hazard zones. When multi-hazards affect several adjacent zones, one must model the propagation of the physical damage to several zones surrounding a *centroid* zone. An approach to do this was developed by Klibi et al. (2014) in the context of a project to reengineer the humanitarian support network of the North Carolina Emergency Management (NCEM) authorities. In Europe, data and maps provided by the European Union ESPON program (www.espon.eu) could be used for similar risk analysis.

The impact intensity and duration of hazards must also be modeled. We assume here that when a multi-hazard $m \in M$ occurs, its duration (in operational periods) and intensity (based on a generic measure such as loss level or casualty level[1] or a

---

[1]See for instance FEMA's methodology for estimating potential losses from disasters (www.fema.org/Hazus).

normalized scale) are characterized by two correlated random variables associated with the impact area exposure level, namely, the impact intensity β_{mg}, with cumulative distribution function $F_{mg}(\beta)$, and the duration ξ_{mg}. The duration can be related to the intensity through an incident *impact-duration function* $\xi_{mg} = f_m(\beta_{mg}) + \varepsilon_m, m \in M$, estimated by regression and with a random error term $\varepsilon_m \sim \text{Normal}(0, \sigma_m^{\varepsilon})$. These distribution functions and incident impact duration functions can be estimated from the data provided by organizations such as CRED, HIIK, and FEMA. When the data available are not sufficient to use standard statistical estimation methods, expert opinions based on partial data can be used to subjectively estimate the compound process parameters.

10.3.3 Modeling SCN Hits

In this third risk analysis phase, the consequences of hits on SCN resources are examined. This brings us back to the SCN vulnerability sources layer of Fig. 10.10. It involves the elaboration of multi-hazard incident profiles as well as the modeling of SCN damage and recovery behaviors. The objective of this phase is to ensure that extreme events with serious impacts are not treated in the same way as business as usual events. However, two important facts must first be recognized: (1) the occurrence of an extreme event in a hazard zone does not necessarily imply that all the SCN locations in the zone are hit and (2) property damages and loss of income caused by extreme events can be insured. Figure 10.13 shows insurable building, equipment, and inventory value for a metalworking plant. Insuring loss of income is more difficult, but it is also possible. Thus when evaluating the impact of a hit, insurance should be taken into account. In fact, when designing a SCN under risk, insurance options should be considered as strategic decision variables. However, in what follows, insurances are not taken into account explicitly to simplify the presentation. In other words, we assume that when designing a SCN to minimize risks, insurance costs are minimized.

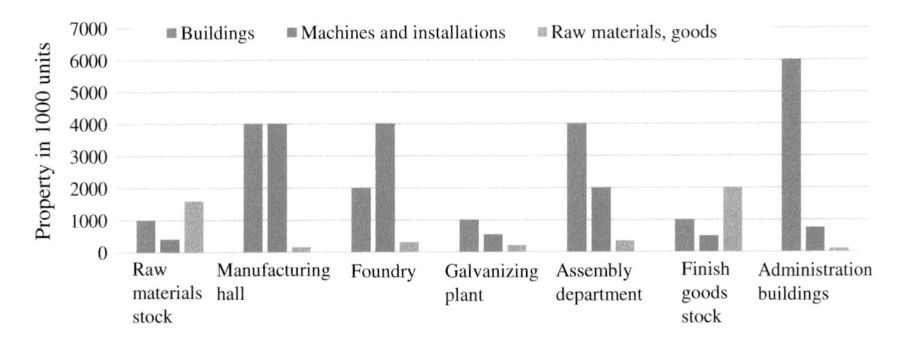

Fig. 10.13 Metalworking plant insurable property—*Source* Swiss Re (2004)

In our context, it is also crucial that the incident severity metrics used to characterize the impact of SCN hits be related to the variables and parameters of the design model formulated to optimize the network structure (see Chaps. 7–9). This is essential to be able to anticipate future activity interruptions, resource and partner failures, as well as necessary recourse actions such as the temporary relocation and rerouting of activities during recovery. As explained in Chap. 11, if the SCN design model does not consider these perturbations and recourses explicitly, it is unlikely that the SCN designs proposed will be able to avoid or minimize risks.

10.3.3.1 Attenuation Probabilities

As mentioned, the occurrence of an extreme event in hazard zone z does not necessarily imply that all the SCN locations $l \in L_z$ are hit. When the hazard zones are large (countries or states), it is likely that only a part of the zone locations will be hit. Also, when considering the impact on product-markets or service offers, the SCN does not necessarily respond to all incidents. In a global disaster relief context, for example, a relief organization's response to a natural disaster may depend on its policies, UN solicitations, and the resources available given other commitments. In such cases, a demand surge for first aid products in a hazard zone does not necessarily generate demands in the corresponding demand zones of the relief network. This leads to the estimation of *attenuation probabilities* α_{ml}, which are conditional probabilities that location l is hit when a multi-hazard $m \in M$ occurs in zone $z(l)$. It is clear that these probabilities are related to the hazard zones granularity. Large zones lead to small attenuation probabilities and vice versa. Attenuation probabilities can be estimated by experts for each SCN location, based on experience and data available. Resource constraints may also apply.

10.3.3.2 Modeling the Impact of a Hit

The impact of an incident on the capacity of a SCN and on product market demands is modeled using recovery functions. These are based on the intensity and duration variables introduced in Sect. 10.3.2. Figure 10.14 illustrates the behavior of a platform's capacity under disruptions. Let b_τ be the capacity of the platform during an operational period τ under business as usual conditions. When the platform is hit, there is a loss of capacity during a number of operational periods. The remaining capacity b'_τ in period τ during the time to recovery following a hit is thus smaller than b_τ. This dependence on the occurrence of hazards transforms the capacity into a random variable. Product-market demand can also be affected upward (surge) or downward (drop) by hazards. Consequently, a known demand under business as usual conditions also becomes a random variable under hazards.

In order to model these impacts, we partition the vulnerability source set S in two subsets: capacity-based sources S^c and demand-based sources S^d. The severity of a

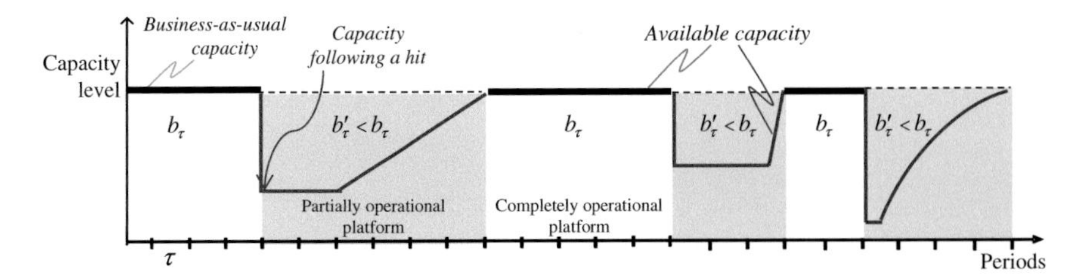

Fig. 10.14 Capacity behavior for a platform under disruptions

SCN hit is characterized on two correlated dimensions: the impact intensity and the time to recovery. These are clearly related to the multi-hazard intensity and duration variables β_{mg} and ξ_{mg} defined previously. The severity of a hit must, however, be expressed in units related to the capacity or demand of vulnerability sources S^c or S^d. Hence, incident profiles such as the ones illustrated in Table 10.4 must be elaborated for all (vulnerability source, multi-hazard) pairs. Damage affecting suppliers typically is assessed using an unfilled rate (% of material ordered not delivered) and the time required to restore supplies. Damage affecting production–distribution resources usually is expressed using a capacity loss rate and the time before production or distribution can resume. For vulnerability sources affecting demand, damage is generally stated using an inflation or deflation rate reflecting a demand surge or drop for a given period of time. Note that the evaluation of severity may also be influenced by the state of the resources or partners associated with a vulnerability source. In some cases, an engineering analysis may be required to establish the fragility of vulnerability source resources depending on the building type, age, and so on.

To model impacts more precisely, we need to define a discrete random variable θ_{ml} giving the time to recovery (in operational periods) of location $l \in L$ when hit by a multi-hazard $m \in M_{s(l)}$. We assume that this time to recovery is related to the multi-hazard duration $\xi_{mg(l)}$ using a translation function $\theta_{ml} = q_{ms(l)}\left(\xi_{mg(l)}\right)$ specified for each vulnerability source $s \in S$ and multi-hazard $m \in M_s$. This function may be based on a proportion estimated from past instances or provided by experts. Also, because ξ_{mg} is a function of the impact intensity β_{mg}, in some contexts it is simpler to relate θ_{ml} directly to $\beta_{mg(l)}$, which eliminates the need to estimate the previous impact-duration function. Instead, impact-recovery functions $\theta_{ml} = q'_{ms(l)}\left(\beta_{mg(l)}\right) + \varepsilon_{ms}$ are estimated for all $s \in S$ and $m \in M_s$, where $\varepsilon_{ms} \sim F_{ms}(\varepsilon)$ is a random deviation term. Now, suppose that a multi-hazard $m \in M$ hits location $l \in L$ at the beginning of operational period τ'. Then, as illustrated in Fig. 10.15, the impact of the hit lasts during operational periods $\tau = \tau', \ldots, \tau' + \theta_{ml} - 1$.

When a multi-hazard $m \in M_s$ hits a location l, its impact is not necessarily felt uniformly during the time to recovery θ_{ml} (Sheffi 2005). Several phases are usually observed, depending on the nature of the multi-hazard and of the vulnerability

Table 10.4 Multi-hazard incident profiles example

			Capacity-based vulnerability sources $S^c = \{1, 2, 3\}$			Demand-based vulnerability sources $S^d = \{4, 5, 6\}$		
			1	2	3	4	5	6
			Suppliers	Plants	DCs	First-aid markets	Sustainment markets	Luxury markets
Impact intensity	Multi-hazards	Natural disaster	Unfilled supply rate	Capacity loss rate	Capacity loss rate	Demand inflation rate		Demand deflation rate
		Market failures	Unfilled supply rate				Demand deflation rate	Demand deflation rate
		Industrial accidents		Capacity loss rate				
Time to recovery	Multi-hazards	Natural disaster	Time to restore	Time to restore	Time to restore	Surge duration		Drop duration
		Market failures	Time to restore				Drop duration	Drop duration
		Industrial accidents		Time to restore				

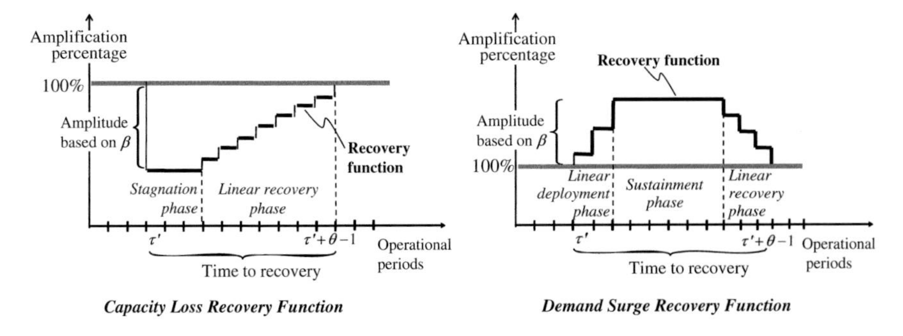

Fig. 10.15 Recovery functions for an incident at the beginning of period τ'

source. For example, when a manufacturing plant is hit by a natural disaster, production capacity drops quickly during a first phase, then there may be a stagnation period while recovery measures are organized, and during a third phase the capacity is gradually restored. However, when a disaster relief organization initiates a support mission, it typically involves the three following phases: deployment, sustainment, and recovery. Such phase-dependent impacts can be characterized by defining discrete recovery functions $(r_{ms\tau}(\beta, \theta), \tau = 1, \ldots, \theta)$, $m \in M$, $s \in S$, providing capacity or demand amplification percentages for the θ operational periods affected by the multi-hazard. As illustrated in Fig. 10.15, the amplification percentages depend on the multi-hazard impact intensity measure β. Multi-hazard recovery functions are defined by experts for each vulnerability source and product family, based on experience and data available.

Using these recovery functions, the capacity available or the demand under disruptions can be calculated for specific operational periods, products, and locations. More specifically, consider the production–distribution network context described in Sect. 7.3.2. The behavior of the production capacity $b'_{pl\tau}$, the storage capacity $b'_{l\tau}$, and/or the demand $d'_{pl\tau}$ resulting from a multi-hazard m hitting a location l at the beginning of period τ' is described by the following relations:

$$b'_{pl\tau} = r_{m,s(l),\tau-\tau'+1}\left(\beta_{mg(l)}, \theta_{ml}\right) b_{pl\tau}, \tau = \tau', \ldots, \tau' + \theta_{ml} - 1, l \in L^{\mathrm{U}}, p \in P_l \quad (10.1)$$

$$b'_{l\tau} = r_{m,s(l),\tau-\tau'+1}\left(\beta_{mg(l)}, \theta_{ml}\right) b_{l\tau}, \tau = \tau', \ldots, \tau' + \theta_{ml} - 1,\ l \in L^{\mathrm{S}} \quad (10.2)$$

$$d'_{pl\tau} = r_{m,s(l),\tau-\tau'+1}\left(\beta_{mg(l)}, \theta_{ml}\right) d_{pl\tau}, \tau = \tau', \ldots, \tau' + \theta_{ml} - 1, l \in L^{\mathrm{D}}, p \in P_l \quad (10.3)$$

In these expressions, $b_{pl\tau}$, $b_{l\tau}$, and $d_{pl\tau}$ are, respectively, the production capacity, the distribution capacity, and the random demand that would prevail at location l in periods $\tau = \tau', \ldots, \tau' + \theta_{ml} - 1$ if there was no hit in period τ'. Note that a hit could occur before a location has completely recovered from previous hits. For this reason, it is necessary to make these computations in chronological order. Note, also that the previous expressions implicitly assume that the shape of recovery functions depends purely on the vulnerability source and not on the type of capacity or demand (i.e., product family) associated with a location. Modifying this would be straightforward.

Example 10.3
Reconsider the single-product European distribution network example (Ex. 7.1) presented in Chap. 7. The capacity of the Düsseldorf site in this example was stated in terms of annual throughput and is equal to 50,000 standard shipping loads. This aggregate throughput capacity was computed using the storage space available at the DC, the storage space required by a standard shipping load, and a *yearly* inventory turnover ratio. Suppose that when this capacity is calculated in terms of maximum daily throughput (instead of yearly throughput) using the same approach, the value obtained is 150 standard loads per day. Suppose also that the only type of multi-hazard considered is natural disasters, that their intensity β is measured using a normalized scale defined over the interval [1, 10], and that the only capacity vulnerability source considered is DCs.[2] Suppose finally that the capacity loss recovery function estimated has the shape of the function illustrated on the left of Fig. 10.15, and that it is given by:

$$r_\tau(\beta,\theta_l)=\begin{cases} 1-0.1\beta & \text{if } 1\le\tau\le\lfloor 0.25\theta_l\rfloor \\ 1-0.1\beta[(\theta_l+1-\tau)/(\theta_l+1-\lfloor 0.25\theta_l\rfloor)] & \text{if } \lfloor 0.25\theta_l\rfloor+1\le\tau\le\theta_l \end{cases}$$

If a natural disaster of intensity $\beta = 7.5$ occurs on April 7 (at the beginning of the day so that $\tau' = 7$) and the time to recovery for the Düsseldorf DC ($l = 1$ in Ex. 7.1) is $\theta_1 = 20$ days, then, based on relation (10.2), the capacity available during the time to recovery is given by $b'_{1\tau} = r_{\tau-\tau'+1}(\beta,\theta_l)b_{1\tau}$, $\tau = 7,\ldots,26$, where $b_{1\tau} = 150$, $\tau = 7,\ldots 26$, is the DC capacity under normal operating conditions. The resulting capacity available during the month of April is plotted in Fig. 10.16. As can be seen, during the stagnation phase the available capacity is only 25 % of the total capacity (37 loads per day). During the recovery phase, the available capacity increases by 4.7 % every day until full recovery on day 27.

This SCN impact modeling approach is based on a simplified representation of SCN resources and demands, but it should be relatively easy to adapt to the specificities of real-life cases. Also, expressions (10.1), (10.2), and (10.3) model multiplicative impacts, which is appropriate in most business contexts. However, for humanitarian relief or military organizations, this is inadequate because the

[2] The indexes m and s are dropped in this example, because there is a single multi-hazard and a single capacity vulnerability source.

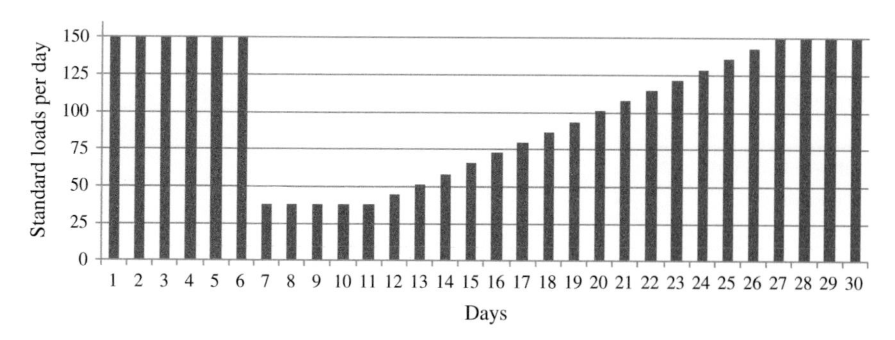

Fig. 10.16 Daily capacity of Düsseldorf's distribution platform during April

demand is usually zero when there is no incident. The demand recovery functions must then be expressed in absolute terms, that is, they must provide a demand level for periods $\tau = \tau', \ldots, \tau' + \theta_{ml} - 1$. For disaster relief networks, for example, the impact intensity (β_{mg}) is often measured in terms of the proportion of the population requiring assistance. In such cases the daily demand during the deployment, sustainment, and recovery phases can be expressed in terms of demand zone population and daily needs per habitant. Standards for humanitarian support, in terms of water and food supplies, health care needs, and so on are provided, for example, in the SPHERE project handbook (www.spherehandbook.org).

10.4 Evolutionary Paths

The two previous sections explain how historical data on the internal and external environment of a company can be used to model business as usual and extreme events affecting its resources and operations. These models can be used to develop plausible future scenarios by projecting historical trends into the future. However, businesses operate in a complex world and, when looking far away, as required when designing a SCN, the projection of historical trends is not sufficient. When developing their strategies, world-class companies such as Shell (www.shell.com) and DHL (www.dhl.com) explore what the world might look like over the next decades, and they anticipate the effect alternative plausible futures may have on their business (DHL 2012; Shell 2011). In other words, they study the impact of plausible economic, social, political, and environmental *evolutionary paths* resulting from an in-depth study of several of the contextual factors examined in previous chapters (see Sects. 1.1, 5.1, and 9.1).

An evolutionary path is a plausible vision of the future capturing current world transitions and potential development alternatives with a focus on opportunities and threats. An illustrative example of a set of evolutionary paths suitable for strategic planning is provided in the following box. To be useful in our context, evolutionary paths must be quantified and related to key SCN design parameters or stochastic

Table 10.5 Examples of links between SCN design parameters and evolutionary paths

Typical SCN design parameters	Evolutionary path indicators				
	Inflation	Oil price	Total consumption	Interest rate	Disaster frequency
Platform fixed costs				X	
Production costs	X				
Inventory holding costs				X	
Transportation costs		X			
Raw material costs	X				
Product sales price	X				
Product market demand			X		
Disaster interarrival times					X

processes. Evolutionary paths are characterized by trends for a number of socioeconomic, political, and environmental indicators such as inflation, interest rates, GDP, energy prices, natural resources reserves, government deficits, city sizes, CO_2 emissions, disaster frequencies, and so on. These trends must then be related to key SCN cost, price, supply, and demand parameters. An example of potential links between evolutionary path indicators and SCN parameters is given in Table 10.5. When data permit, causal relationships can be established using econometric models (Wooldridge 2008) or simulation models (Thomopoulos 2013) in which indicators are considered as explanatory variables. Shell, for example, has developed a simulation model to study the impact of socioeconomic indicators on future energy needs.

Three Evolutionary Paths to Characterize Tomorrow's World

In view of overlapping trends on the evolution of the world in terms of economic development, available energy sources, technological innovations, foreign policies, demographic changes, corporate social responsibility, and so on, the following three plausible evolutionary paths could be considered in a SCN design project: (1) a carefree world, (2) a world in search of sustainability, and (3) a world committed to global sustainability.

- *A carefree world* ($\kappa = 3$). This evolutionary path, inspired by current economic and political instabilities in several countries, by the "Scramble World" energy scenario developed by Shell (2011) and by the "Untamed Economy" scenario elaborated by DHL (2012), envisions a world dominated by protectionism, nationalism, absence of collaboration, and deterioration in weather conditions. As a result, global logistics networks deployment capabilities are impaired, access to some markets is limited, availability of energy and raw materials is significantly reduced, and so on. This leads to a significant increase in several logistics costs (interest rates, energy costs, raw material prices, third-party resources usage costs, etc.), increased SCN vulnerabilities, and a drop in demand for several product markets.

- *A world in search of sustainability* ($\kappa = 2$). This evolutionary path describes a world aware of the importance of social and ecological issues. Sustainability initiatives are gradually adopted, mainly in developed countries, but their implementation is slow. This plausible future is congruent with the "Blueprint" scenario developed by Shell (2011) and with the vision of a world moving toward efficiency in megacities elaborated by DHL (2012). It results in a slight increase or stagnation of logistics costs because of better collaboration and efficient use of energy. In addition, it foresees a moderate economic growth, a slight improvement in the purchasing power of consumers, and thus an increase in demand and the emergence of new markets in developing countries. It also anticipates technological advances and investments fostering improved availability of some resources and production–distribution capabilities. It nurtures economies of scale, partnerships, and the sharing of resources in industrial and logistics clusters.
- *A world committed to global sustainability* ($\kappa = 1$). This evolutionary path corresponds to a more optimistic vision of the future. It relies on major initiatives to foster economic, ecological, and social sustainability such as those imagined in the Physical Internet (PI) manifesto (Montreuil 2011), the Shell (2011) "Wide Oceans" scenario, and the DHL (2012) "Global Resilience—Local Adaptation" scenario. It is characterized by economic prosperity coupled with an improvement of social and environmental factors. For example, the implementation of logistic systems promoting interconnectivity through the standardization of logistics interfaces and operations would enable companies to supply, make, store, and move products more effectively. This would have a positive impact on the availability of resources, the capabilities of production–distribution networks, as well as the growth of existing and emerging markets. This would also dampen inflation and stabilize logistics costs.

In what follows, it is assumed that a set K of evolutionary paths with probability of occurrence p_κ, $\kappa \in K$, is elaborated to support the design of a SCN, and that the key parameters or stochastic processes used to generate plausible future scenarios depend on these evolutionary paths. When a design parameter or process (say transportation costs) depends on a single indicator (say oil prices) or simply on time, and the anticipated indicator trend is linear, then its future values may be calculated simply through the use of an inflation–deflation factor. As in Sect. 10.2, consider a design parameter modeled by a random variable ζ_τ^e, which is now assumed to depend on time. Let $F_0^e(\zeta)$ be a recent estimate of the probability distribution of the random variable for the period preceding the planning horizon, that is, for period $\tau = 0$, so that $\zeta_0^e \sim F_0^e(\zeta)$ with an expected value $\mathrm{E}\left[\zeta_0^e\right] = \mu_0^e$. Then, the random variable to use for period $\tau \in T^{\mathrm{u}}$ under evolutionary path $\kappa \in \mathrm{K}$ could be $\zeta_{\kappa\tau}^e = \zeta_0^e\left(1 + \delta_\kappa^e \tau\right)$, where δ_κ^e is the inflation–deflation factor estimated for the random variable considered under evolutionary path $\kappa \in \mathrm{K}$, or $\zeta_{\kappa\tau}^e \sim F_{\kappa\tau}^e(\zeta)$, where the parameters of $F_{\kappa\tau}^e(\zeta)$ are functions of δ_κ^e. If $F_{\kappa\tau}^e(\zeta)$ is normally distributed, for example, its mean and standard deviation could be given by

$$\mu_{\kappa\tau}^e = \mu_0^e\left(1 + \delta_\kappa^e \tau\right) \text{ and } \sigma_{\kappa\tau}^e = CV_0^e \mu_{\kappa\tau}^e$$

where CV_0^e is the coefficient of variation of the random variable estimated at time $\tau = 0$. Inflation–deflation factor values are based on historical data trends, if any, modified to take into account the evolutionary path effects either subjectively or using projections of the indicator value. Depending on the context, to simplify, one may want to base projections directly on planning periods $t \in T$ instead of operational periods $\tau \in T^{\mathrm{u}}$. The two following examples illustrate this modeling approach.

Example 10.4
Return to Ex. 8.3 in which a company sells its product in six European countries. Suppose that the demands given in the example were averages estimated for year 2014 ($\mu_{l0}, l \in L^{\mathrm{D}}$) and that we want to convert the static design model used into a dynamic model such as the one presented in Sect. 8.5 with a five-year planning horizon $T = \{1, 2, 3, 4, 5\}$, these periods corresponding to year 2015–2019. Assume also that the annual demand for location l is modeled by a Normal$(\mu_l, CV\mu_l)$ distribution with a coefficient of variation $CV = 0.25$. Based on the three evolutionary path $\mathrm{K} = \{1, 2, 3\}$ described in the previous box, the company wants to project the probability distributions of its product-market demand over the next 5 years. For the evolutionary path $\kappa = 1$ (*A world committed to global sustainability*), an optimistic 40 % yearly demand increase is anticipated in every country ($\delta_{1l} = 0.4, l \in L^{\mathrm{D}}$). However, for path $\kappa = 2$ (*A world in search of sustainability*), a moderate 10 % increase is forecasted for all countries ($\delta_{2l} = 0.1, l \in L^{\mathrm{D}}$). Finally for path $\kappa = 3$ (*A carefree world*), a gradual yearly decrease of 10 % is anticipated in each country ($\delta_{3l} = -0.1, l \in L^{\mathrm{D}}$). Using this data, the average demand for each country, year, and evolutionary path is calculated with relation $\mu_{l\kappa t} = \mu_{l0}(1 + \delta_{\kappa l} t)$. The results obtained are given in Table 10.6. Based on this, the average projected demand for Germany in year 2018 under evolutionary path 2 would be $\mu_{124} = 23,100$, its standard deviation $CV\mu_{124} = 0.25(23,100) = 5775$, and its probability distribution would be a Normal$(23,100; 5775)$.

Table 10.6 Projected average demand over the planning horizon

	Factor ($\delta_{\kappa l}$)	Dem. 2014 (μ_{l0})	Location l					
			Germ.	France	Italy	Spain	Port.	Switz.
			16,500	20,000	11,000	19,500	28,000	10,000
Path $\kappa = 1$	0.4	Dem. 2015 ($t = 1$)	23,100	28,000	15,400	27,300	39,200	14,000
	0.4	Dem. 2016 ($t = 2$)	29,700	36,000	19,800	35,100	50,400	18,000
	0.4	Dem. 2017 ($t = 3$)	36,300	44,000	24,200	42,900	61,600	22,000
	0.4	Dem. 2018 ($t = 4$)	42,900	52,000	28,600	50,700	72,800	26,000
	0.4	Dem. 2019 ($t = 5$)	49,500	60,000	33,000	58,500	84,000	30,000
Path $\kappa = 2$	0.1	Dem. 2015 ($t = 1$)	18,150	22,000	12,100	21,450	30,800	11,000
	0.1	Dem. 2016 ($t = 2$)	19,800	24,000	13,200	23,400	33,600	12,000
	0.1	Dem. 2017 ($t = 3$)	21,450	26,000	14,300	25,350	36,400	13,000
	0.1	Dem. 2018 ($t = 4$)	23,100	28,000	15,400	27,300	39,200	14,000
	0.1	Dem. 2019 ($t = 5$)	24,750	30,000	16,500	29,250	42,000	15,000
Path $\kappa = 3$	-0.1	Dem. 2015 ($t = 1$)	14,850	18,000	9900	17,550	25,200	9000
	-0.1	Dem. 2016 ($t = 2$)	13,200	16,000	8800	15,600	22,400	8000
	-0.1	Dem. 2017 ($t = 3$)	11,550	14,000	7700	13,650	19,600	7000
	-0.1	Dem. 2018 ($t = 4$)	9900	12,000	6600	11,700	16,800	6000
	-0.1	Dem. 2019 ($t = 5$)	8250	10,000	5500	9750	14,000	5000

Example 10.5

We previously indicated that interarrival times between successive natural disasters are typically modeled as an exponential random variable. More specifically, we indicated that when a natural disaster occurs in a hazard zone $z \in Z$ at a given point in time, then the time before the arrival of the next disaster in the zone is a random variable λ_z with cumulative distribution function $F_z(\lambda)$ and expected value $\mathrm{E}[\lambda_z] = \bar{\lambda}_z = 1/\bar{\phi}_z$, where $\bar{\phi}_z$ is the average number of disasters per year for hazard zone $z \in Z$. However, when looking at the CRED data on the frequency of disasters in Fig. 10.17, it is

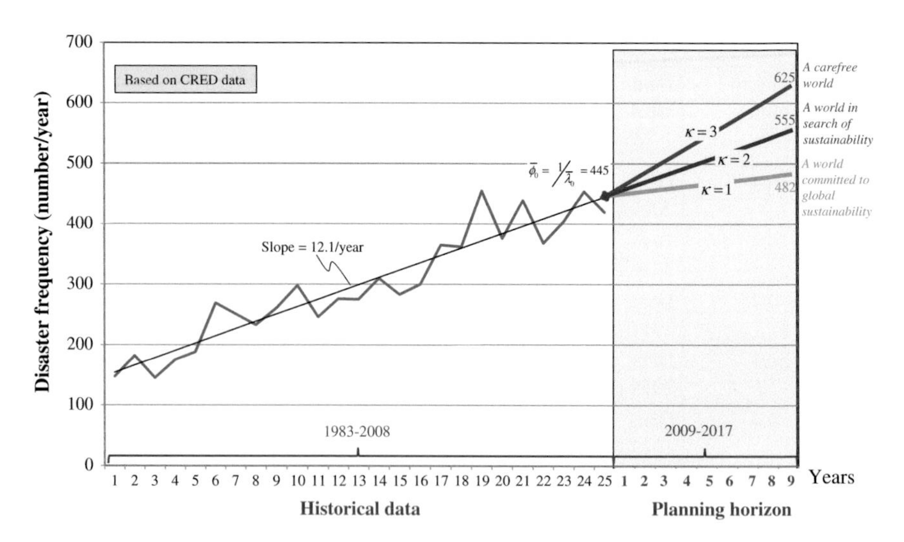

Fig. 10.17 Disaster frequency functions for three evolutionary paths

clear that they are not stationary. Moreover, the greening efforts made in coming years will influence disaster frequency. Thus, the three evolutionary paths $K = \{1, 2, 3\}$ described in the previous box are likely to yield different disaster frequency trends, as shown in the figure. As a consequence, disaster interarrival times in a zone $z \in Z$ for a planning period $t \in T$ can be modeled by an exponential random variable with a mean $\bar{\lambda}_{z\kappa t} = 1/\bar{\phi}_{z\kappa t}$, $\bar{\phi}_{z\kappa t} = \bar{\phi}_{z0}(1 + \delta_\kappa t)$, where $\bar{\phi}_{z0}$ is an estimate of the average number of disasters per year at the beginning of the planning horizon (i.e., in period $t = 0$) for hazard zone $z \in Z$ and δ_κ is the yearly disaster frequency inflation factor used for evolutionary path $\kappa \in K$.

Inflation factors and initial disaster frequencies can be based on a regression line estimated with historical data. For the annual disaster frequency data plotted in Fig. 10.17, the regression line estimated with Excel is $\bar{\phi} = 141.85 + 12.1t$. This means that from 1983 to 2008 the disaster frequency has increased on average by 12.1 per year, and that, at the beginning of the planning horizon ($t = 25$), the average disaster frequency is $\bar{\phi}_0 = 141.85 + 12.1(25) = 445$. The historical yearly inflation factor, based on planning period $t = 0$ average disaster frequency, is thus $\bar{\delta} = 12.1/445 = 0.027$. For evolutionary path $\kappa = 2$, which essentially corresponds to recent trends, an inflation factor $\delta_2 = 0.027$ (i.e., 2.7 %) could thus be used. For path $\kappa = 1$, which is very optimistic, an inflation factor $\delta_1 = 0.009$ could be specified. Finally, for path $\kappa = 3$, which is somewhat pessimistic, one could set $\delta_3 = 0.045$. The mean time between disasters (in years) anticipated in planning period (year) $t = 9$ under path $\kappa = 3$ would then be calculated as follows:

$$\bar{\phi}_9 = \bar{\phi}_9(1 + \delta_3 t) = 445(1 + 9(0.045)) = 625,$$
$$\bar{\lambda}_9 = 1/\bar{\phi}_9 = 1/625 = 0.0016$$

The three evolutionary paths just defined are illustrated in Fig. 10.17.

10.5 Plausible Future Generation

A scenario is a plausible configuration of the internal and external business environment of a SCN during a planning horizon. The construction of scenarios relies mostly on a rigorous statistical approach, but it may also involve some subjective elements. It is based on the descriptive models elaborated to characterize business as usual and extreme events as well as evolutionary paths. Thus, a scenario can be seen as a compound event; it is the result of the juxtaposition of random events, multi-hazards, and deeply uncertain events over a planning horizon. After highlighting the role of scenarios in strategic decision making, this section explains how to use Monte Carlo methods to generate plausible future scenarios useful to design robust SCNs.

10.5.1 Scenario-Based Decision-Making

Scenarios are an indispensable tool for decision-making under uncertainty. They can be used to analyze the risks associated with a given SCN, to formulate optimization models under uncertainty, or to assess and compare a set of candidate SCN designs. Scenarios help decision-makers project historical behaviors into the future according to anticipated evolutionary paths. Figure 10.18 uses a cone to illustrate a plausible futures scenario space as seen at the beginning of a planning horizon. It shows how an existing business environment can be projected over a planning horizon, yielding a possibly infinite number of progressively more detailed scenarios. Clearly, all plausible scenarios cannot be explicitly examined in a SCN design project. To reap the benefits of the approach, one has, as illustrated in the figure, to generate samples of scenarios that provide a good representation of the entire space of plausible futures.

The role of scenarios in strategic planning is increasingly recognized and its use by businesses is widening. The scenario analysis approach helps decision-making in three ways:

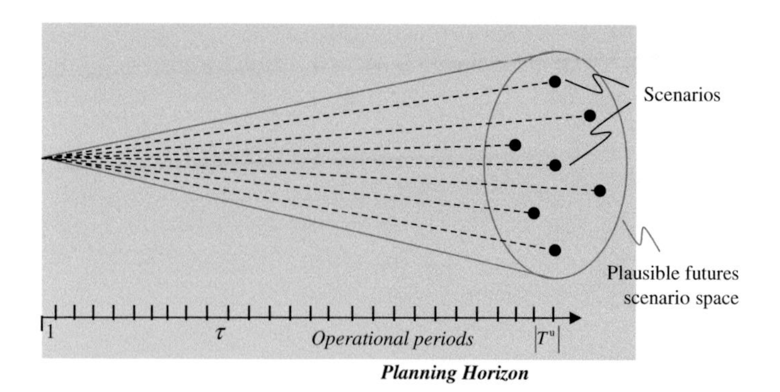

Fig. 10.18 Plausible futures scenario space

- Understanding historical behaviors through a statistical analysis of business as usual and extreme events data, as explained in previous sections
- Anticipating plausible futures through the development of evolutionary paths and the Monte Carlo generation of representative plausible future scenario samples and examining their impact on candidate solutions
- Studying the impact of unusual (surprising) or *worst case* plausible futures corresponding to significant departures from historical behaviors

Historical behavior modeling is now facilitated by the widespread availability of statistical analysis and forecasting tools such as IBM-SPSS and SAS or even the Data Analysis Module of Excel and by the all-encompassing data provided by corporate and public databases (*big data*). As explained in Chap. 3, several specialized APS are also now available to facilitate these analyses. The use of Monte Carlo methods to generate plausible future scenarios is examined in detail in the next section. Several ERP-SCM suites offered by vendors such as SAP are now incorporating Monte Carlo modules for the generation of scenarios. The construction and analysis of imaginative or worst-case scenarios is often associated with so-called *what-if* analysis (Godet 2001; van Der Heijden 2005). Unlike statistical analysis, this is a reverse engineering approach: starting from future imaginative events anticipated by a panel of experts, based on structured questionnaires (Rowe and Wright 2001) and brainstorming sessions, scenarios are gradually detailed and quantified to make them useful for decision-making. Worst-case scenarios can also be constructed by perturbing extreme Monte Carlo scenarios.

Several approaches have been used to anticipate the future in strategic decision making, each being a compromise between simplicity of use and the quality of the representation of the plausible futures scenario space. Figure 10.19 illustrates the main approaches found in practice. Approach (a) corresponds to the case where the future is assumed to be a deterministic replication of the recent past. SCN designs are often optimized for a *base period* associated with a recent year with complete data. This has obvious limitations. Approach (b) involves using a single scenario based on the expected value of relevant business as usual random variables and

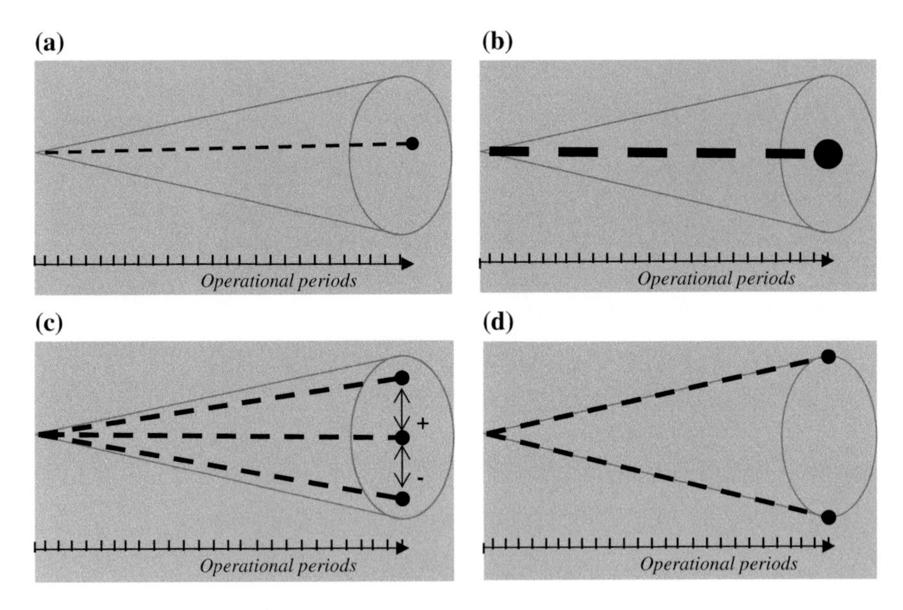

Fig. 10.19 Typical plausible future anticipation approaches

ignoring the variability of these random variables and hazards. This type of scenario is often derived using standard forecasting methods. Approach (c) is an attempt to study the robustness of the solution found when using a single scenario by doing a sensitivity analysis. Doing this properly is difficult for complex SCN design problems and it does not necessarily yield better solutions. Approach (d) is a worst-case analysis based on extreme scenarios on the boundaries of the plausible futures scenario space. This approach is useful to test the robustness of a solution, but when worst-case scenarios only are considered, the solution selected tends to be overly conservative. Another approach involving the use of a sample of representative plausible future scenarios was illustrated in Fig. 10.18. In a SCN design context, when data and problem size permits, combining the latter with a worst-case analysis is likely to yield more robust value-creating network structures. The following section shows how to build representative and worst-case scenarios using Monte Carlo methods.

10.5.2 Monte Carlo Method

Based on our definition of a scenario as a compound event, a generic approach for the generation of plausible future samples is illustrated in Fig. 10.20. Starting on the left side, this figure shows how realizations of a set of disjoint random, hazardous, and/or deeply uncertain events occurring during a planning horizon can be concatenated to obtain a plausible scenario. The generation of random events was

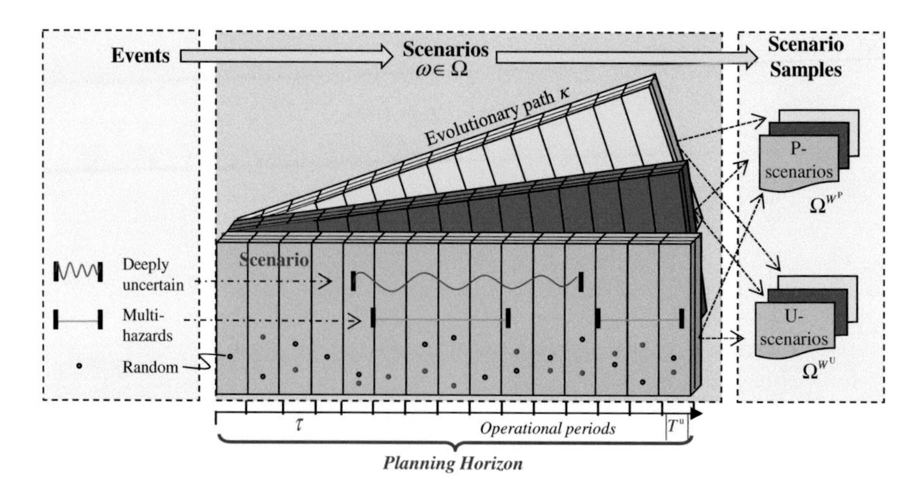

Fig. 10.20 Scenario samples generation approach

illustrated in Fig. 10.5 and the generation of multi-hazards in Fig. 10.12. Different realizations of the random variables yield distinct scenarios, and thus the set of all plausible futures, denoted Ω, is usually extremely large if not infinite. All scenarios include random events, but they do not necessarily include hazardous or deeply uncertain events. The set Ω can thus be partitioned into two mutually exclusive and collectively exhaustive subsets: Ω^{P} including all *probabilistic* scenarios without deeply uncertain events (*P*-scenarios), and its complement Ω^{U} (*U*-scenarios). In principle, the probability $\mathrm{p}(\omega)$ of scenarios $\omega \in \Omega^{\mathrm{P}}$ can be evaluated, but the probability of *U*-scenarios cannot. When evolutionary paths are also distinguished, the following subsets are obtained: $\Omega_{\kappa}^{\mathrm{P}}, \Omega_{\kappa}^{\mathrm{U}}, \kappa \in \mathrm{K}$. The plausible futures generation process involves the construction of representative samples of scenarios from each of these subsets.

The most common approach for the generation of probabilistic scenarios is the Monte Carlo method. It generates realizations of stochastic processes using pseudorandom numbers and the inverse of the distribution function of the random variables involved (Thomopoulos 2013). If some random variables are dependent, the generation process is more complicated but straightforward. This approach enables the generation of scenarios linked to key parameters of SCN design models. When a sample of scenarios is generated using this method, all the scenarios in the sample are *equiprobable*, which simplifies the calculation of their probability of occurrence significantly. In addition to its interesting statistical properties, this method is relatively simple to implement with largely available tools such as Excel, as shown in the following examples. The key steps of the Monte Carlo method, when applied to the design of SCNs, are summarized in Fig. 10.21 under the assumption that multi-hazards occur independently in hazard zones. The last step aggregates the operational period values obtained into planning period values to be used in strategic SCN design models. Note that this aggregation process does not

1. Select an evolutionary path κ randomly using p_{κ}, $\kappa \in K$.
2. For each hazard zone $z \in Z$, construct a chronological hazard list specifying the time (period) of arrival and the type of multi-hazard.
3. Generate the intensity and duration of the multi-hazards in the hazard lists and, when a network location is hit, use them to calculate recovery times and recovery function amplification factors.
4. For all operational periods $\tau \in T^{u}$, generate a realization of all the business-as-usual random variables using the inverse of their distribution functions.
5. For all operational periods $\tau \in T^{u}$, use the amplification factors to calculate demands and capacities.
6. For all $t \in T$, aggregate operational period quantities into planning period quantities.

Fig. 10.21 Monte Carlo scenario generation procedure

always involve a simple sum over all the operational periods contained in a planning period. When aggregating capacity, for example, taking congestion into account properly may require the application of a correcting factor.

The execution of the procedure in Fig. 10.21 yields a probabilistic scenario $\omega \in \Omega^{\mathrm{P}}$. In practice, because some business as usual or extreme events are usually characterized by continuous probability distributions (uniform, exponential, normal, lognormal, etc.), the number of plausible probabilistic scenarios $|\Omega^{\mathrm{P}}|$ is infinite. In this situation, or when the number of plausible scenarios is extremely large, one must limit itself to the consideration of a sample $\Omega^{W^{\mathrm{P}}} \subset \Omega^{\mathrm{P}}$ of W^{P} independent P-scenarios. In order to generate a sample $\Omega^{W^{\mathrm{P}}}$ of scenarios, the Monte Carlo procedure in Fig. 10.21 is repeated W^{P} times. Each of the scenarios generated is then equiprobable, that is, each has a probability of occurrence $1/W^{\mathrm{P}}$. To complete the process, a small sample $\Omega^{W^{\mathrm{U}}}$ of W^{U} imaginative or worst-case U-scenarios can also be elaborated as shown later on. This gives a total of $W = W^{\mathrm{P}} + W^{\mathrm{U}}$ scenarios for further analysis. In what follows, the approach is examined in more detail for two particular cases. Because we generally want to limit the analysis to relatively small scenario samples, it is desirable to use a sampling method that ensures that each evolutionary path is adequately represented. More advanced importance sampling techniques (Ducapova et al. 2000) can be used to improve the representativity of small scenario samples.

10.5.2.1 Demand Scenarios Generation Under Business as Usual Conditions

Assume that the demand for a product follows a compound Poisson process (see Ex. 10.2), and that multi-hazards and evolutionary paths are neglected. The inter-arrival time is an exponential random variable o with distribution function $F^{\mathrm{E}}(o)$ and mean time between arrivals $\mathrm{E}[o] = \mu^{o}$ days. The order size is a normally distributed random variable q with distribution function $F^{\mathrm{N}}(q)$, expected value

MonteCarlo (i/p: $F^E(o)$, $F^N(q)$, T^u; o/p: $d_\tau(\omega)$, $\tau \in T^u$)

$\eta = 0$; $d_\tau(\omega) = 0$, $\tau \in T^u$

While $\eta \leq |T^u|$, do:

a) Generate a pseudorandom number u_o and compute the next order arrival time $\eta = \eta + F^{E-1}(u_o)$ and $\tau = \lceil \eta \rceil$

b) Generate a pseudorandom number u_q and compute the period τ demand $d_\tau(\omega) = d_\tau(\omega) + F^{N-1}(u_q)$

End While

Fig. 10.22 Monte Carlo procedure for demand generation

$E[q] = \mu^q$, and standard deviation σ^q. Using the Monte Carlo method, we want to generate a demand scenario $\omega \in \Omega^P$ over a planning horizon T^u, i.e., we want to calculate $d_\tau(\omega), \tau \in T^u$. Assuming that customer orders are independent of each other, to do this we generate independent pseudorandom numbers u_o and u_q uniformly distributed on the interval $[0; 1]$, and we compute the inverse, $F^{E-1}(u_o)$ and $F^{N-1}(u_q)$, of the distributions of interarrival times and order sizes. For this simple case, the generic procedure in Fig. 10.21 reduces to the **MonteCarlo** procedure presented in Fig. 10.22. In this procedure, the continuous variable η is used to denote order arrival time realizations. Order arrivals are generated in the interval $[0, |T^u|]$ and mapped onto the corresponding operational periods $\tau \in T^u$. More than one order can arrive in a given period. The use of the procedure with Excel is illustrated in the following paragraph.

To illustrate this, suppose that the mean time between customer arrivals is $\mu^o = 5$ days, that the average and standard deviation of order sizes are $\mu^q = 500$ and $\sigma^q = 50$, and that we want to generate daily demands for 100 days. The density function and the cumulative distribution of exponential interarrival times o with mean μ^o are given by

$$f^E(o) = (1/\mu^o)e^{-(1/\mu^o)o} \text{ and } F^E(o) = 1 - e^{-(1/\mu^o)o}$$

For a given random number u_o, we have $F^E(o) = u_o = 1 - e^{-(1/\mu^o)o}$. Consequently, the inverse of the distribution function is given by $F^{E-1}(u_o) = -\ln(1 - u_o)\mu^o$. With Excel, uniformly distributed random numbers are given by the function RAND(), and the inverse of the normal distribution is directly provided by the function NORM.INV$(u_q; \mu^q; \sigma^q)$. Given this, the results of the implementation of the procedure in Fig. 10.22 with Excel are shown in Fig. 10.23.

The second column in the Excel spreadsheet calculates arrival times $\eta = \eta + F^{E-1}(u_o)$ and the third one converts them into days ($\tau = \lceil \eta \rceil$). The last column generates the order sizes with the inverse normal function. Note that more than one order is received on day 12, 63, and 94. Consequently, the demand for these days is $d_{12}(\omega) = 570 + 487 = 1057$, 1016, and 1033. Note also that the demand for the days without orders is nil (e.g., $d_1(\omega) = 0$).

Fig. 10.23 Customer order generation with Excel

Random number (u_o)	Arrival time (η)	Arrival day (τ)	Random number (u_q)	Demand (d_τ)
0.312449218	1.87309794	2	0.105788082	438
0.027692985	2.013516266	3	0.742997284	533
0.83075091	10.89543499	11	0.6274703	516
0.054282565	11.17449223	12	0.918553876	570
0.122926702	11.83031579	12	0.394970449	487
0.613589632	16.58459251	17	0.159349341	450
0.688274215	22.41274937	23	0.843786956	551
0.306498581	24.24275933	25	0.265385659	469
0.976368499	42.96863259	43	0.316959542	476
0.978029449	62.05889405	63	0.440947495	493
0.133947352	62.77794195	63	0.67933355	523
0.94846258	77.60517759	78	0.653754673	520
0.301361216	79.39828475	80	0.666212519	521
0.934641263	93.0376055	94	0.423601985	490
0.036336296	93.22267	94	0.803486803	543
0.239276116	94.5900941	95	0.657153928	520
0.505953018	98.1157174	99	0.32024126	477

10.5.2.2 Capacity Scenarios Generation Under Natural Disasters

In Sect. 10.3, we showed how to model the arrival and intensity of multi-hazards and how they affect the facilities of a SCN. This section explains how these descriptive models can be used to generate capacity scenarios for the facilities of the SCN. To simplify the presentation, we consider the case of a distribution network. Because there is a single capacity vulnerability source, the index s can be dropped. Also, the only multi-hazards considered are natural disasters, so that the index m can also be dropped. We assume that disasters occur independently in hazard zones, and that their arrival and intensity processes are not influenced by evolutionary paths. Under these assumptions, the simplified Monte Carlo procedure presented in Fig. 10.24 can be used to generate a capacity scenario $\omega \in \Omega^{\mathrm{P}}$ over planning horizon T, that is, to calculate $b_{lt}(\omega), t \in T,\ l \in L^{\mathrm{S}}$.

The procedure is based on the random variables $\lambda_z \sim F_z(\lambda)$, $\beta_z \sim F_{g(z)}(\beta)$ and θ_l, the recovery functions $r_\tau(\beta, \theta), \tau = 1, \ldots, \theta$, and the attenuation probabilities α_l defined previously. It assumes that the time to recovery is directly related to the disaster intensity by a predetermined impact-recovery function $\theta_l = \left\lceil q'\left(\beta_{z(l)}\right) + \varepsilon \right\rceil$, where $\varepsilon \sim F(\varepsilon)$ is a random deviation term. The function $q'(\beta)$ is assumed to be continuous, and we take the ceiling to obtain discrete θ_l values. Pseudorandom numbers u_λ, u_β, u_α and u_ε are used to generate random variable realizations based on the inverse of probability functions. Disaster arrivals are associated to the local continuous variable η and converted to operational periods $\tau \in T^{\mathrm{u}}$. To be able to include the scenario generated in SCN design models, daily capacities are aggregated into planning period capacities $b_{lt}(\omega),\ t \in T, l \in L^{\mathrm{S}}$. Repeating this procedure W^{P} times yield a scenario sample $\Omega^{W^{\mathrm{P}}} \subset \Omega^{\mathrm{P}}$.

To illustrate the procedure, we go back to the European distribution network examined in Exs. 7.1 and 10.3, in which three public DCs (in Toulouse, Düsseldorf,

MonteCarlo(i/p: $(F_z(\lambda), L_z, z \in Z), (F_g(\beta), g \in G), q'(\beta), (\alpha_l, l \in L^S), (T_t^u, t \in T), r(\beta,\theta)$;
o/p: $b_{lt}(\omega), l \in L^S, t \in T$)

1) For all $l \in L^S$ and $\tau \in T^u$, set the business-as-usual capacity $b_{l\tau}(\omega) = b_l$
2) For all $z \in Z$ do:
$\eta = 0$
While $\eta \leq |T^u|$ do:
Generate u_λ and compute the next hazard arrival moment $\eta = \eta + F_z^{-1}(u_\lambda)$
Add day $\lceil \eta \rceil$ to the chronological list T_z
End While
For all $\tau' \in \mathrm{T}_z$ do:
a) Generate u_β and compute the multi-hazard intensity $\beta_z = F_{g(z)}^{-1}(u_\beta)$
For all $l \in L_z$ pseudorandom number $u_\alpha \leq \alpha_l$ do: *(Hit test)*
b) Generate u_ε and compute the time to recovery $\theta_l = \lceil q'(\beta_z) + F^{-1}(u_\varepsilon) \rceil$
c) Compute the capacity $b_{l\tau}(\omega) = r_{\tau-\tau'+1}(\beta_z, \theta_l) b_{l\tau}(\omega)$, $\tau = \tau', \ldots, \tau' + \theta_l - 1$
End For
End For
End For
3) Aggregate these values over periods $\tau \in T_t^u$, $t \in T$, to get $b_{lt}(\omega), l \in L^S, t \in T$

Fig. 10.24 Monte Carlo procedure for the generation of a capacity scenario

and Turin), with a platform capacity of 150 standard shipping loads per day, are considered for implementation. Suppose that the SCN is operating 365 days per year and that the planning horizon considered includes 4 years. Suppose also that the hazard zones specified correspond to European countries. Based on historical data for natural disasters provided by CRED and ESPON, relevant information for the three countries concerned (France, Germany, and Italy) can be derived. Three exposure levels (low, moderate, and high) are distinguished, and the required exposure, zone, and DC parameters estimated are given in Table 10.7. Interarrival times in a zone z are exponential as before, and the impact β_z of natural disasters is measured on a scale from 1 to 10. For exposure level $g(z)$, β_z is assumed to be uniformly distributed between the bounds $(\underline{\beta}_g, \bar{\beta}_g)$ given in the table. The time to

Table 10.7 Estimated parameter values for the European SCN example

Country (hazard zone)	Exposure level	Yearly disaster frequency	Mean interarrival time (days)	Impact probability distribution	DC	Attenuation probability
1 France	Low	2.96	121	$U(1, 4)$	Toulouse	0.2
2 Germany	Moderate	3.18	115	$U(1, 7.5)$	Düsseldorf	0.3
3 Italy	High	3.62	99	$U(1, 10)$	Turin	0.4

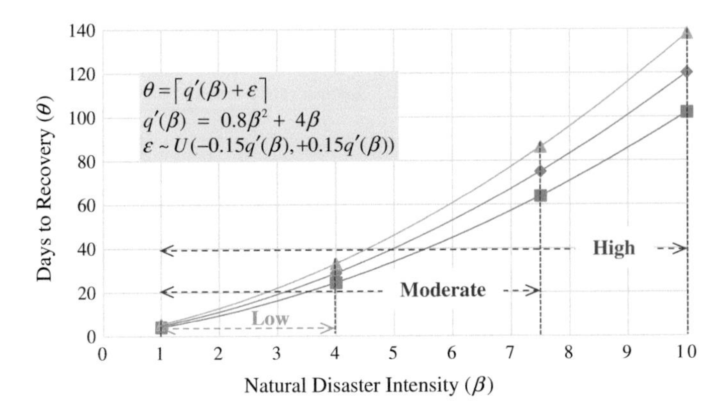

Fig. 10.25 Impact-recovery function for the European SCN example

recovery when a DC $l \in L^{S}$ is hit is estimated directly from the disaster intensity $\beta_{z(l)}$ using the impact recovery function in Fig. 10.25. Deviations around the mean time are assumed to be uniformly distributed in a ±15 % interval. The recovery function $r_{\tau}(\beta, \theta_l)$ used to compute capacity loss is the same as in Ex. 10.3. Finally, when aggregating the daily demands into yearly demands at the end of the procedure, an efficiency factor of 0.913 applies to compensate for nonoperational time because of congestion and maintenance.

For this example, the procedure **Monte Carlo** is initialized (Step 1) by setting the business as usual capacity to 150 for all days and DCs, that is, $b_{l\tau}(\omega) = 150, \forall l, \tau$. Then, a list of natural disaster arrival dates is generated for each country using the inverse of the exponential distribution, as was done in Ex. 10.3. The intensity of each incident in the list is subsequently calculated using the inverse of the uniform distribution (Step 2a). For France ($z = 1$), for example, when using Excel, this is done by computing

$$\beta_1 = \underline{\beta}_1 + \text{RAND}() * (\bar{\beta}_1 - \underline{\beta}_1) = 1 + \text{RAND}() * 3$$

One then uses Toulouse's attenuation probability to test if the DC in France is hit. A pseudorandom number is generated with RAND() and, if it is smaller than 0.2, one concludes that the Toulouse DC is hit. Next, the function in Fig. 10.25 is used to compute the time to recovery (Step 2b). Suppose that the Toulouse DC is hit on day 314 and that the intensity of the disaster is 3.87. Then, we have

$$q'(3.87) = 0.8(3.87)^2 + 4(3.87) = 27.46, \quad \varepsilon \sim U(-4.12, +4.12)$$

If the random deviation generated is 1.7, then $\theta_1 = \lceil 27.46 + 1.7 \rceil = 30$ days. When these calculations are repeated for all the incidents in the list for France, the results in Fig. 10.26 are obtained.

France				Toulouse	
Inter-arrival time	Arrival day (η)	Exposure level	Disaster intensity (β)	Hit test result	Time to recovery
8.42	8	Low	2.44	0	-
306.22	314	Low	3.87	1	30
303.38	617	Low	2.22	0	-
8.96	625	Low	2.76	0	-
95.98	720	Low	3.97	0	-
57.43	777	Low	3.37	1	21
261.47	1038	Low	1.50	0	-
66.71	1104	Low	3.16	0	-
6.03	1110	Low	1.83	0	-
51.51	1161	Low	1.75	0	-

Fig. 10.26 Disaster list generated with Excel for France and the Toulouse DC

To complete the daily capacity generation process (Step 2c), one applies the recovery functions over the time to recovery for each hit, as done in Ex. 10.3. A graphical representation of a capacity scenario generated over four years for Italy is given in Fig. 10.27. These calculations are repeated for each country. To complete the procedure, the capacity is aggregated per years (Step 3). For a year $t \in T$ and DC $l \in L^S$, this is done in this case by calculating $b_{lt}(\omega) = 0.913 \Sigma_{\tau \in T_t^u} b_{l\tau}(\omega)$. The efficiency factor (0.913) is applied to account for the fact that capacity not used on a given day cannot be recuperated in subsequent days. The aggregated yearly capacities obtained for three scenarios and the average capacity generated are given in Fig. 10.28. As can be seen, the amplitude and variation of capacity loss are more pronounced when the DC exposure level increases. Also, individual scenarios can differ significantly from the average scenario. Values in bold underline the worst-case obtained for each year and DC. They could be used to construct a worst-case scenario. Note, however, that any serious SCN design analysis would require the use of a much larger sample of scenarios.

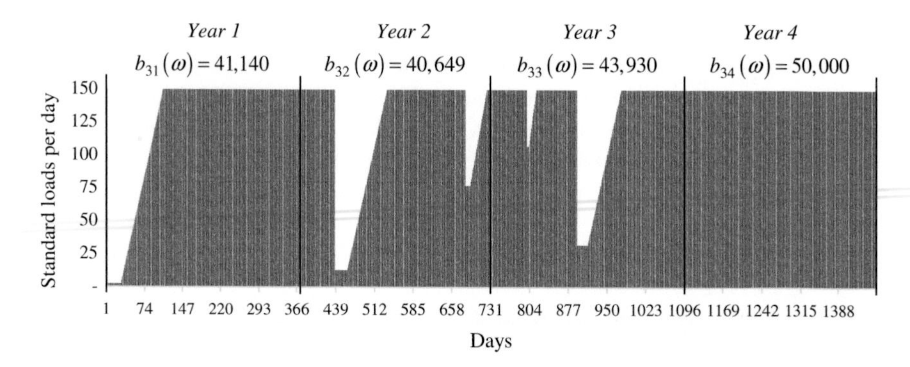

Fig. 10.27 Four-year capacity scenario for the Turin DC in Italy

Fig. 10.28 Three capacity scenarios generated with Excel

		Toulouse DC Capacity	Düsseldorf DC Capacity	TurinDC Capacity
Scenario 1	Year 1	**49,030**	49,080	**41,140**
	Year 2	50,000	**46,466**	40,649
	Year 3	49,411	**48,197**	43,930
	Year 4	50,000	**45,474**	50,000
Scenario 2	Year 1	49,161	50,000	43,746
	Year 2	49,819	48,729	**39,868**
	Year 3	50,000	45,996	**41,403**
	Year 4	49,743	46,182	**43,957**
Scenario 3	Year 1	50,000	**44,173**	49,893
	Year 2	**49,416**	50,000	41,053
	Year 3	**48,995**	50,000	40,598
	Year 4	**49,651**	50,000	49,132
Average Scenario	*Year 1*	*49,397*	*47,751*	*44,926*
	Year 2	*49,745*	*48,398*	*40,523*
	Year 3	*49,469*	*48,064*	*41,977*
	Year 4	*49,798*	*47,219*	*47,696*

10.5.2.3 Assessing Risks with Scenario Samples

The execution of the Monte Carlo procedure in Fig. 10.21 yields a probabilistic scenario $\omega \in \Omega^{P}$. Some of the plausible future scenarios generated with this procedure may involve only a few multi-hazards over the planning horizon but others may be much more chaotic, that is, involve several hits. An intuitive measure of the risk associated with a scenario is the number of hits it undergoes during the planning horizon. The left plot in Fig. 10.29 illustrates the distribution of the number of hits for a large sample of scenarios with exponential multi-hazard interarrival times. As expected, it has the shape of a Poisson distribution. An alternative measure would be the cumulative damage level during the planning horizon. For the scenario sample shown in Fig. 10.29, the right plot provides a

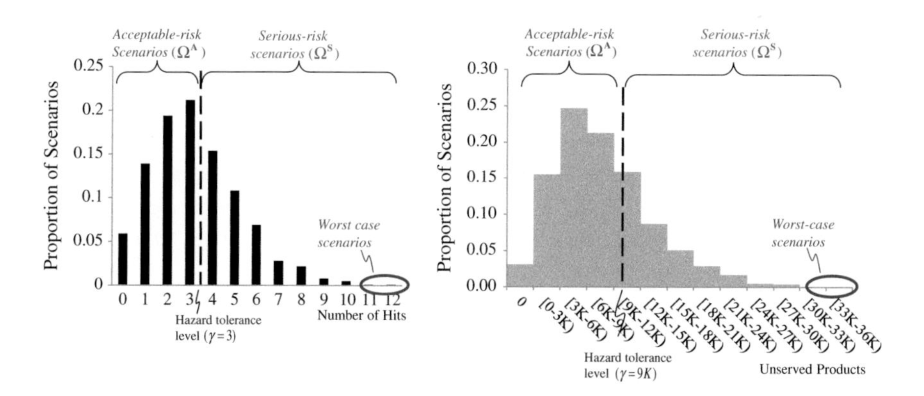

Fig. 10.29 Number of hits and unserved products for a large scenario sample

damage level distribution based on the cumulative number of products not shipped to customers from a depot following a hit. As can be seen, for these two measures, the shapes of the resulting histograms are similar. In order to distinguish between the scenarios a decision-maker would consider as *acceptable*, in terms of the risks involved, and those that would raise a *serious* concern, a hazard tolerance level γ can be defined. This level is the maximum number of hits (or the maximum cumulative damage level) the decision-maker can tolerate over the planning horizon without serious concern. This tolerance level can be used to partition the set of probabilistic scenarios Ω^{P} in two subsets, namely, the set of acceptable risk scenarios Ω^{A} and the set of serious risk scenarios Ω^{S}. These scenario subsets can then be used in the SCN design evaluation process to take into account decision-makers' aversion to risk.

The sets, measures, and functions used to characterize SCN hazards are necessarily based on the information and experience available and, consequently, they may completely overlook some potential extreme events for which no information and experience exist. It is to investigate these potential threats that deeply uncertain scenarios must be elaborated. As mentioned, some uncertain extreme events associated with *U*-scenarios can be identified through structured brainstorming sessions and/or expert interviews related to SCN threats and vulnerabilities. Recent books by Bremmer and Keat (2009) and Taleb (2007) study the development of chaotic scenarios through rational arguments based on geopolitical knowledge of capital markets, terrorism, revolutions, and so on. Lempert et al. (2006) suggest the use of narrative scenarios for such situations, and they show how these scenarios can be used to improve the quality of decision-making processes.

The scenarios generation procedure in Fig. 10.21 could be enriched to enable the manual modification of statistically generated scenarios and the introduction of imaginative scenarios. All scenarios necessarily include random events and they may include hazards so *U*-scenarios are most easily created by perturbing probabilistic scenarios or, following the structured process described previously, by replacing probability distributions and impact functions with human inputs for multi-hazards that cannot be described probabilistically. In any case, for our purposes, the resulting scenarios must be expressed quantitatively in terms of the parameters used for SCN design. Decision-makers interest in *U*-scenarios is mainly related to the desire to examine worst-case scenarios. These are typically probabilistic scenarios in the tail of the distribution of the number of hits or damage level, as illustrated in Fig. 10.29, or serious-risk scenarios coupled with deep-uncertainty events imagined by experts. When doing this, however, the equal probability property of Monte Carlo scenarios is lost, so that *U*-scenarios must be used with care. They are valuable for what-if analysis of existing or candidate solutions, but they are not easily incorporated in SCN optimization models. The next chapter shows how to use scenarios in stochastic programming models to obtain robust, value-creating SCN structures.

Review Questions

10.1 What is the difference between the concepts of risk and uncertainty?
10.2 Why is the fact that the future business environment is uncertain important in SCN design problems?
10.3 Why is it simpler to consider random business as usual events than extreme events?
10.4 Why is the elaboration of a risk exposure matrix not sufficient to design robust SCNs?
10.5 How is scenario analysis different from conventional forecasting methods?
10.6 What role can scenarios play in strategic planning?
10.7 How can *U*-scenarios be created and used in decision-making?
10.8 What is the role of evolutionary paths? How can trend functions and their parameters be estimated? Provide some examples.

Exercises

Exercise 10.1 The demand for a set of product families $P = \{1, 2, 3, 4, 5, 6\}$ is known to follow a compound Poisson processes with exponential order interarrival times $Exp(\bar{\lambda}_p)$ and a mean time between arrivals of $\bar{\lambda}_p$ days. The order size for products 1, 3, and 5 follows a normal distribution $N(\mu_p, \sigma_p)$ with a mean μ_p and standard deviation σ_p, and the order size for the other products follows a uniform distribution $U(\underline{v}_p, \bar{v}_p)$ with lower bound $\underline{v}_p$ and upper bound $\bar{v}_p$. The estimated value of the parameters of these probability distributions is given in Table 10.8. You are asked to generate five three-month demand scenarios for the six products using Excel.

Exercise 10.2 Based on the United States FEMA historical disaster database (available at www.fema.gov/disasters), assess the exposure level of a few US states. Also, using available data, estimate the mean time between natural disasters for each exposure level.

Exercise 10.3 Go back to Example 10.3 and assume that the distributor considered sells first aid products. You are asked to analyze the impact of the April 7 natural disaster on the German market demand. The business as usual demand in Germany is 40 standard loads per day. The intensity of the disaster results in a 50 % demand inflation and a 15-day time to recovery. The deployment phase (assumed to be

Table 10.8 Exercise 10.1 data

Product	Interarrival times	Order size
$p = 1$	*Exp* (4)	*N* (500, 50)
$p = 2$	*Exp* (6)	*U* (350, 400)
$p = 3$	*Exp* (12)	*N* (300, 30)
$p = 4$	*Exp* (8)	*U* (175, 250)
$p = 5$	*Exp* (21)	*N* (200, 20)
$p = 6$	*Exp* (15)	*U* (80, 100)

linear) accounts for 20 % of the time to recovery and the final recovery phase lasts only 2 days. Plot the resulting changes in demand during the month of April and compare your results to the demand surge recovery function in Fig. 10.15.

Exercise 10.4 Reconsider the Düsseldorf DC described in Example 10.3 which has a capacity of 150 standard shipping loads per day. Recall that the site is hit by a natural disaster of intensity 7.5 on April 7, resulting in a 75 % initial capacity loss and a time to recovery of 20 days, with 25 % of this time taken by the stagnation phase. Now suppose that the site is hit again on April 16 by a natural disaster of intensity 5, and that the time to recovery of this second incident is 16 days. Using the recovery function given in Ex. 10.3, estimate the cumulative capacity loss resulting from these two disasters and calculate the daily capacity levels of the Düsseldorf platform for the months of April and May. Plot these levels as in Fig. 10.16.

Exercise 10.5 Based on the SCN context described in Example 5.3, we wish to anticipate the evolution of transportation costs over a 5-year planning horizon for the evolutionary paths described in the Sect. 10.4 box. As explained in Sect. 5.4.2, a linear function $\bar{c}_0(D) = \bar{c}^{\mathrm{o}} + \bar{c}^{\mathrm{d}} D$ can be used to estimate transportation costs per shipping load as a function of distance (D), where $\bar{c}^{\mathrm{o}}$ and $\bar{c}^{\mathrm{d}}$ are regression parameters estimated from historical data. Based on the approach presented in Sect. 10.4, you are asked to anticipate transportation cost functions for the planning horizon 2015–2019 using the projection $\bar{c}_{\kappa t}(D) = \bar{c}_0(D)\left(1 + \delta_{\kappa}^{\bar{c}} t\right), \kappa \in \mathrm{K}, t \in T$, where $\delta_{\kappa}^{\bar{c}}$ is the transportation costs inflation factor estimated for evolutionary path $\kappa \in \{1, 2, 3\}$ described in the Sect. 10.4 box.

(a) Based on the oil prices data plotted in Fig. 10.6, evaluate a transportation costs inflation factor for each evolutionary path $\kappa \in \{1, 2, 3\}$.
(b) Using the transportation cost function estimated in Fig. 5.19 ($\bar{c}_0(D) = 21.937 + 0.0779D$), compute the values of the cost function parameters to use for each year of the planning horizon under each evolutionary path.
(c) As seen in Fig. 5.19, the regression errors for the transportation cost function estimated from historical data are normally distributed with mean 0 and standard deviation 20.4. Using the Monte Carlo method, and assuming that the probability of the three evolutionary paths considered are $\mathrm{p}_1 = 0.2$, $\mathrm{p}_2 = 0.5$, and $\mathrm{p}_3 = 0.3$, generate five transportation costs scenarios over the next five years for shipments on three origin-destination lanes with a 500, 1000, and 1500 miles length, respectively.

Exercise 10.6 Go back to Topclim's case described in Exercise 8.2. The Canadian manufacturer of household air conditioning units is considering four potential sites for the construction of the company's new plants: New York, Atlanta, Chicago, and San Diego. Recall that platforms for producing 200,000 units or 400,000 units are considered by the company. You are asked to take into account the risk exposure of the sites considered and to revise their maximum production capacity using a

Table 10.9 Exercise 10.6 estimated risk exposure data

US State (hazard zone)	Exposure level	Yearly disaster frequency	Mean interarrival time (days)	Impact probability distribution	Plant	Attenuation probability
1 New York	High	3.98	92	$U(1, 10)$	New York	0.4
2 Georgia	Moderate	2.46	148	$U(1, 7.5)$	Atlanta	0.1
3 Illinois	Low	1.1	331	$U(1, 4)$	Chicago	0.2
4 California	High	3.5	105	$U(1, 10)$	San Diego	0.3

scenario-based approach. Table 10.9 provides the natural disasters exposure data for each potential site. The impact-recovery function given in Fig. 10.25 can be used in the analysis.

(a) Generate three scenarios providing 1-year production capacity for the sites
(b) For each scenario, solve the optimization model of Exercise 8.2 and compare the candidate designs obtained for the scenarios considered with the one obtained in Exercise 8.2 (with fixed capacities). Discuss your findings and determine which candidate design the company should retain.

Bibliography

Arntzen B (2012) Global supply chain risk management. MIT CTL White Papers Parts 1, 2, and 3

Banks E (2006) Catastrophic risk: analysis and management. Wiley Finance, Hoboken

Bhatia G, Lane C, Wain A (2013) Building resilience in supply chains. World Economic Forum

Bourgin E, Lenoire C (2012) Risk management: Éviter l'effet domino. Supply Chain Mag 61: 106–108

Bremmer I, Keat P (2009) The fat tail: the power of political knowledge in an uncertain world. Oxford University Press, New York

Christopher M (2003) Understanding supply chain risk: a self-assessment workbook. Center for Logistics and Supply Chain Management, Cranfield University, UK

Christopher M, Holweg M (2011) Supply chain 2.0: managing supply chains in the era of turbulence. Int J Phys Distrib Logistics Manage 41(1):63–82

Cook T (2008) Managing global supply chains: compliance, security, and dealing with terrorism, Auerbach

CRED (2015) Natural disasters by country. http://www.emdat.be/world-maps. Accessed 22 Feb 2015

DHL (2012) Delivering tomorrow: logistics 2050, a scenario study. Deutsche Post AG

Ducapova J, Consigli G, Wallace S (2000) Scenarios for multistage stochastic programs. Ann Oper Res 100:25–53

FFP (2015) Fragile state index 2014. http://www.ffp.statesindex.org. Accessed 22 Feb 2015

Godet M (2001) Creating futures: scenario planning as a strategic management tool. Economica Ltd

Gogu R, Trau J, Stern B, Hurni L (2005) Development of an integrated natural hazard assessment method. Geophys Res Abstr 7:03724

Grossi P, Kunreuther H (2005) Catastrophe modeling: a new approach to managing risk. Springer, New York

Haimes Y (2004) Risk modeling, assessment, and management, 2nd edn. Wiley, Hoboken
Helferich O, Cook R (2002) Securing the supply chain. Council of Logistics Management (CLM)
Hendricks K, Singhal V (2005) Association between supply chain glitches and operating performance. Manage Sci 51(5):695–711
Heyman D, Sobel M (1982) Stochastic models in operations research, vol 1. McGraw-Hill, New York
Klibi W, Ichoua I, Martel A (2014) Prepositioning emergency supplies to support disaster relief: a stochastic programming approach. CIRRELT Working Paper. Université Laval
Klibi W, Martel A (2012) Scenario-based supply chain network risk modeling. Eur J Oper Res 223:644–658
Klibi W, Martel A, Guitouni A (2010) The design of robust value-creating supply chain networks: a critical review. Eur J Oper Res 203(2):283–293
Lempert R, Groves D, Popper S, Bankes S (2006) A general, analytic method for generating robust strategies and narrative scenarios. Manage Sci 52(4):514–528
Makridakis S, Wheelwright S, Hyndman R (1998) Forecasting: methods and applications, 3rd edn. Wiley, Hoboken
Martel A, Benmoussa A, Chouinard M, Klibi W, Kettani O (2013) Designing global supply networks for conflict or disaster support: the case of the Canadian Armed Forces. J Oper Res Soc 64:577–596
Montreuil B (2011) Towards a physical internet: meeting the global logistics sustainability grand challenge. Logistics Res 3(3):71–87
Muthukrishnan R, Shulman J (2006) Understanding supply chain risk: a McKinsey global survey. The McKinsey Quarterly
NASDAQ (2015) End of day commodity futures price quotes for crude oil. http://www.nasdaq.com/markets/crude-oil.aspx. Accessed 18 Feb 2015
Norrman A, Jansson U (2004) Ericsson's proactive supply chains risk management approach after a serious sub-supplier accident. Int J Phys Distrib Logistics Manage 34(5):434–456
Rowe G, Wright G (2001) Expert opinions in forecasting: the role of the Delphi technique. In: Scott Armstrong J (ed) Principles of forecasting: a handbook for researchers and practitioners. Kluwer Academic, Dordrecht
Scawthorn C, Shneider P, Shauer B (2006) Natural hazards: the multihazard approach. Nat Hazards Rev 7(2):39
Sheffi Y (2005) The resilient enterprise: overcoming vulnerability for competitive advantage. MIT Press, Cambridge
Shell (2011) Shell energy scenarios to 2050: signals & signposts. Shell International BV
Supply Chain Digest (2009) The greatest supply chain disasters of all time. Supply Chain Digest, May
Swiss Re (2004) Business interruption insurance. Swiss Re Publication 1501270_04_en
Taleb N (2007) The black swan: the impact of the highly improbable. Random House, New York
Taylor C (2013) Managing the value chain in turbulent times. Dynamic Markets Limited
Thomopoulos N (2013) Essentials of Monte Carlo simulation: statistical methods for building simulation models. Springer, Berlin
US Census Bureau (2015) http://www.census.gov/briefrm/esbr/www/esbr020.html. Accessed 17 Feb 2015
van der Heijden K (2005) Scenarios: the art of strategic conversation, 2nd edn. Wiley, Hoboken
van Opstal D (2007) The resilient economy: integrating competitiveness and security. Council on Competitiveness
WEF (2013) Building resilience in supply chains. World Economic Forum and Accenture
Wooldridge J (2008) Introductory econometrics: a modern approach, 4th edn. Cengage Learning, Boston
World Bank (2011) Global industrial production declined 1.1 % in April in the wake of the tsunami and earthquake in Japan. Prospects Weekly, 21 June

Chapter 11
Designing Robust SCNs Under Risk

We explained in Chap. 10 that the business environment in most companies is not deterministic and that several plausible futures need to be considered to design robust SCNs. Because the SCN design models proposed in Chaps. 7, 8 and 9 are deterministic, they show how design decisions can be optimized; but they are not sufficient to reach an adequate value-risk compromise. In this chapter, many of these models are revisited and extended to be able to cope with risk.[1] Initially, the structure of planning horizons under risk is examined. Then, the concepts of responsiveness, resilience, and robustness in an SCN design context are discussed. Subsequently, basic stochastic programming notions are introduced. The sample average approximation (SAA) method and coherent risk measures are explained and their application to SCN design is illustrated. In addition, we show how resilience strategies can be taken into account in the formulation of stochastic design models. Finally, a generic approach for the generation and evaluation of robust SCN designs is presented and illustrated with an example.

---

[1]In Chap. 10, the term *uncertainty* was defined as value neutral and the term *risk* was used to refer to the possibility that undesirable outcomes could occur, which is congruent with the terminology used in the risk analysis literature and with the English meaning of the word. This chapter draws heavily on the *stochastic programming* literature (Birge and Louveaux 2011). In this context, and in finance, the term *risk* refers to the volatility of possible outcomes associated with a decision (see Sect. 2.4) and, more specifically, to the chance that a decision's actual return will be different than expected. The expression *downside risk* is used to refer to possible undesirable (below average) results. We adopt this meaning in this chapter.

A. Martel and W. Klibi, *Designing Value-Creating Supply Chain Networks*,
DOI 10.1007/978-3-319-28146-9_11

11.1 Planning Structures Under Risk

SCN design problems deals with strategic capacity investment, resource deployment, and policy-making decisions that are the responsibility of top management. At that level, major preoccupations are long-term capital financing, expected return on investments, risk management, and, more generally, the impact of SCN design decisions on the value of the firm. As seen in Sect. 8.5.1, at the strategic level, the planning horizon considered covers a set of planning periods $t \in T$ (typically years or seasons depending on the context) amalgamated into reengineering cycles $h \in H$, a cycle h including a set of periods $t \in T_h$ (see Fig. 8.19). SCN design decisions are revised at the beginning of reengineering cycles. As explained in Sect. 3.1.4 on the nature of distributed planning systems, once an SCN design has been implemented, the resources deployed are used on a daily basis to perform operations such as sales, warehousing, transportation, production, procurement, emergency response, and so on, and the revenues and expenses generated by an SCN over a planning horizon are directly related to these *user* operations. Thus, alternative SCN designs cannot be evaluated without anticipating operational events and decisions. Operational events are usually stochastic (Chap. 10), and their occurrence over the operational periods $\tau \in T^{u}$ of a planning horizon is best anticipated through the generation of plausible future scenarios (see Fig. 10.20). Given the number of daily events and decisions occurring in an SCN, it is usually not possible to consider all of them explicitly in strategic planning models. To simplify the decision process, approximate anticipations involving planning-period event and decision aggregations generally are used (i.e., aggregations over periods $\tau \in T_t^{u}, t \in T$). The relationship between these different planning horizons is illustrated in Fig. 11.1.

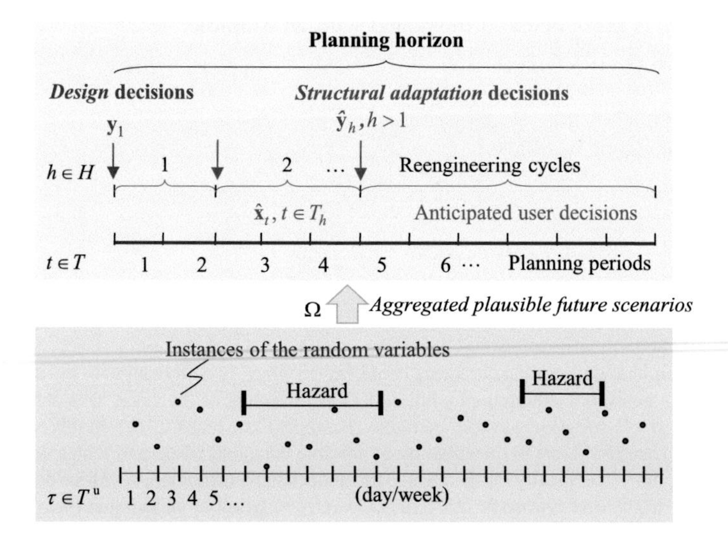

Fig. 11.1 Planning structure under risk

Under risk, the design decision and the anticipated user decision levels in Fig. 11.1 give rise to a complex multistage hierarchical decision problem. The information available at the beginning of the planning horizon takes the form of a set of plausible future scenarios, Ω, defined over planning periods $t \in T$ (see Sect. 10.5). Although the planning horizon covers several reengineering cycles, the only decisions implemented when the problem is solved are the first design decisions (location, platform, offer, and supplier selection decisions denoted here by the vector $\mathbf{y}_1$). Subsequent design decision vectors $\hat{\mathbf{y}}_h, h > 1$, are considered as future opportunities to adapt the network to its environment. Additional information, however, will be available at the beginning of following cycles, and it needs to be taken into account when reengineering the network. The '^' is used to make it clear that the vectors $\hat{\mathbf{y}}_h, h > 1$, are anticipations and that they will not be implemented. In practice, this leads to the solution of multi-period SCN design models on a rolling horizon basis. In order to avoid any ambiguity, in what follows we use the expression *design decisions* only for the decisions $\mathbf{y}_1$ to be implemented. Subsequent structural decisions are referred to as *structural adaptation decisions*. Anticipated user decisions vectors $\hat{\mathbf{x}}_t, t \in T$, corresponding to the flow, throughput, and inventory variables defined in previous design models, are also requited to adequately estimate expected returns. It should be clear that because the implementation of a new SCN design may take a long time, the design problem has to be solved several months before the reengineered SCN starts operating.

11.2 Fostering Robustness in SCN Design

It is widely recognized that SC systems (SCSs) and networks should be resilient and robust but the exact implications of these desirable attributes for SCS and SCN design are still debated. A system is *robust* if it is able to avoid or withstand a wide range of adverse conditions for a significant period of time. On the other hand, it is *resilient* if it can recover quickly from adverse conditions when hit. The first term captures the aim to design fail-proof systems but, admitting that complex systems are fallible, the second one expresses the aim to design systems that can recover quickly when they fail. In our context, adverse conditions are linked to the random, hazardous, and deeply uncertain events defined in Chap. 10. Mechanical or electronic systems fail when they stop functioning. However, SCSs are purposeful systems and the fact that they function does not mean that they perform well. A functioning SCS that is losing money would not be considered robust or resilient. Implicit in these concepts is the fact that the system must continue to accomplish its mission, which in our case is value creation. The time frame considered is also important. For a given planning horizon, it can be stated that a SCS is robust if it is capable of sustained value creation under any plausible future scenario defined over this horizon. For a plausible future defined over a planning horizon, several perturbations because of random, hazardous, and deeply uncertain events will occur. A SCS is resilient if it is designed to continue creating value when punctual perturbations occur. In other words, it is resilient if these perturbations do not generate deep and lengthy drops in value creation. From this point of view, resilience is necessary to achieve robustness.

The previous discussion applies to SCSs but not necessarily to SCNs. In Chap. 1, we saw that SCNs are the backbone of SCSs but that planning and execution processes form another vital part of these systems. These incorporate execution and emergency response processes that are crucial to achieve resilience. The *response policies* adopted by a company largely explain the efficiency of these processes. For example, the inventory management policies of a company determine how well they can cope with demand randomness; their customer reassignment policies govern how well they can react to capacity loss caused by hazards. When designing an SCN, these response policies are usually assumed to be predetermined. The operational decisions associated with these policies are part of the user decisions to anticipate in SCN design models. SCN structures, however, may facilitate some of these processes. For example, customers can be reassigned more efficiently (i.e., without generating a major drop in value creation) if there is a backup DC close to each of the reassigned customers. Such structures may be favored through the a priori adoption of adequate *prepositioning policies*. Finally, robustness may be improved by adopting SCN structures that help avoid or reduce risk. Before showing how to evaluate the robustness of an SCN design, we examine the impact of these policies and strategies more closely.

11.2.1 Resilience Strategies

The attributes of resilient enterprises are discussed in Sheffi (2005) and van Opstal (2007), among others, and on the Center for Resilience at Ohio State University website (www.resilience.osu.edu). The latter defines resilience as the ability of a system to survive, adapt, and develop in the face of unforeseen events, including catastrophic incidents, which is a broader definition than given previously. They developed a generic tool for SC resilience assessment and management (SCRAM). It includes a questionnaire based on a framework to reveal SC operations vulnerabilities and capabilities and helps managers develop appropriate resilience strategies (Pettit et al. 2013). Along the same lines, Blackhurst et al. (2011) developed a conceptual framework to identify resiliency enhancers and reducers and they show how it can be used to help develop resilience strategies. These frameworks highlight key elements of the problem, they help assess the resilience of a company, and they provide useful guidelines for strategy development, but they do not show how SC modeling can be used to help develop more resilient SCNs. Our previous discussion indicates that SCN resilience depends largely on the response and prepositioning policies of a company, and our main concern here is to show how to take these policies into account in SCN design models.

Response policies aim to provide an adequate reaction to short-term variations in supply, capacity, and demand. They provide means to reduce risk and to hedge against randomness and hazards in order to increase SCN expected value. On the supply side, response policies are typically associated with flexible sourcing contracts (Sheffi 2005; Tomlin 2006) and to resource flexibility mechanisms such as capacity buffers (Chopra and Sodhi 2004), production shifting (Graves and Tomlin 2003), and overtime and subcontracting (Bertrand 2003). On the demand side, they are often associated with safety stock pooling and placement tactics (Graves and Willems 2003), to the postponement of final production until demand is known, and to shortage response actions such as product substitution, lateral transfers, drawing products from insurance inventories, buying products from competitors, rerouting shipments, or delaying shipments (Tang and Tomlin 2008; Tomlin 2006). The type of response policies adopted by Home Depot to cope with hurricanes is described in the following box.

As mentioned, SCN design models usually assume that response policies are predetermined and some kind of response variable must be included in the model to anticipate their impact. These variables must be added to the user decision vectors $\hat{\mathbf{x}}_t, t \in T$, defined previously to represent the flow, throughput, and inventory variables found in deterministic design models. As we shall see, when using stochastic programming, all these surrogate decision variables are considered to be recourse variables whose value depends on the plausible future scenario considered. For example, if lateral transfers between DCs are permitted to help cope with demand surges because of natural disasters, then recourse variables corresponding to product flows between DCs are defined; if overtime is permitted within certain bounds to help circumvent capacity shortages, then recourse variables are added to reflect this policy; if dual sourcing is permitted to improve resilience, then flow variables from suppliers are defined accordingly; and so on.

Resilience may be enhanced by adopting SCN structures facilitating the application of response policies. These structures may be flexibility- or redundancy-based. Flexibility-based structures are developed by investing in SCN resources before they are needed. Examples of design decisions providing such capabilities include choosing geographically dispersed suppliers that are partially interchangeable and locating DCs to ensure that all customers can be supplied by a backup center with a reasonable service level if their primary supply DC fails. Redundancy-based structures involve a duplication of network resources in order to continue serving customers while rebuilding after a disruption. An important distinction between flexibility- and redundancy-based structures is that redundant resources may not be used. Examples of redundancy-based capabilities include insurance capacity, that is, maintaining production systems in excess of business-as-usual requirements, and insurance inventory dedicated to serve as buffers to prepare for critical situations. The Home Depot hurricane-specific DCs described in the following box illustrate this well. These structural tradeoffs need to be captured in SCN design models under risk. This usually requires the introduction of additional decision variables that must be added to the design and structural adaptation vectors $\mathbf{y}_1$ and $\hat{\mathbf{y}}_h, h > 1$, defined previously to represent location, platform, market offer, and supplier selection decisions.

Home Depot's Hurricane Strategy (Fortune 2010)

Home Depot (www.homedepot.com) is one of the most successful retail chains in North America. The company has more than 2250 retail stores in the United States and Canada and a typical store assortment consists of up to 40,000 different kinds of building materials, home improvement supplies, appliances, and lawn and garden products. However, as shown in Fig. 11.2, about 500 Home Depot stores are located in the path of potential hurricanes in the southeastern United States. Every year, the company uses its SCS to prepare for the hurricane season and respond quickly and efficiently when hit. The company's approach is based on a resilience strategy involving the setup of hurricane-specific DCs located in low-risk areas and of a command center located in Atlanta. It also involves response policies specifying how to react when hit and during the following recovery period. Home Depot collaborates with specialized organizations such as the American Red Cross (ARC) and the Federal Emergency Management Agency (FEMA) to mitigate the impact of hurricanes on coastal communities. It also helps shoreline populations prepare for the hurricane season by giving workshops on critical readiness and recovery topics in nearly 700 stores and by maintaining a website on these topics (www.homedepot.com/hurricane).

To prepare for the hurricane season, SC managers place orders in November, based on past storm data, so products such as generators, water, tarp, flashlights, and plywood are stocked by June in three hurricane-specific DCs. Before hurricane Gustav in 2008, 500 supply trucks were shipped to these DCs. When a serious storm is detected, the command center in Atlanta

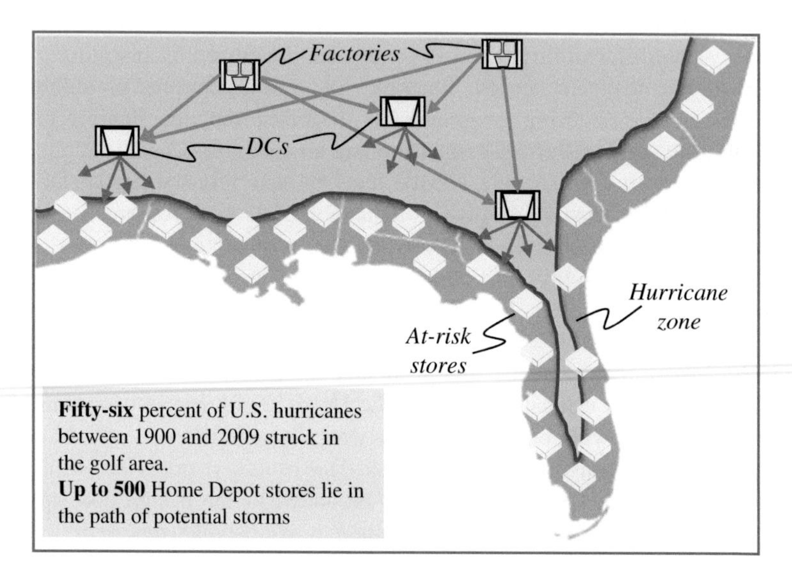

Fig. 11.2 Home depot SCN in Southeast States (*Source* Fortune 2010)

opens. Planners track the weather and handle everything from store needs to placing products near impact zones and finding shelter for personnel. Employees secure stores and when winds reach 45 mph the stores are closed. Home Depot tries to be the last retailer to close and the first to reopen after the storm. The command center takes the actions required to provide the additional supplies required. During Gustav, 300 trucks reached the 35 affected stores. Stores reopened as soon as field teams judged them safe, selling high-demand products such as tarps, lumber, batteries, gas cans, extension cords, chain saws, and trash bags. Batteries came in containers that could be opened directly on the floor and lumber could be loaded straight from a truck into a car. Temporary stores may also open in parking lots or elsewhere.

11.2.2 Selecting Robust SCN Designs

Resilience strategies enhance SCN robustness, but another more straightforward way of improving robustness is to design networks that avoid, or at least reduce, risks. Risk-avoidance strategies are used when the risk associated with potential product markets, suppliers, or facility locations is considered unacceptable, for example, because of the instability of the associated geographical area. This may involve closing some network facilities, delaying an implementation, or simply not selecting an opportunity. Another way to avoid risks may be through vertical integration (see Sect. 6.1.3). This may reduce risk through an improved control, but it converts variable costs into fixed costs. There is a tendency to produce internally for low-risk product markets and to outsource production for higher-risk product markets, thus transferring risks to contract manufacturers and suppliers. These are important issues that must be captured in SCN design models. The stochastic programming models introduced in the next sections provide a natural way to do this.

Now suppose that a number of SCN designs $\mathbf{y}_l, l = 1, 2, \ldots, L$, are considered.[2] Based on our previous discussion, which one should be selected? We stated that an SCN is robust if it is capable of sustained value creation under any plausible future scenario defined over a planning horizon $t \in T$. As before, we denote the set of plausible future scenarios over a planning horizon by Ω. Moreover, we indicated that to have *sustained* value creation, the design needs to be resilient, that is, to display relatively stable period-to-period value creation for any plausible scenario. Assume that the value $V^{\text{SCN}}(\mathbf{y}_l, \omega)$ of candidate designs $\mathbf{y}_l, l = 1, 2, \ldots, L$, can be calculated for each design and each plausible future $\omega \in \Omega$ (we show how to make

---

[2]We use $\mathbf{y}$ in what follows instead of $\mathbf{y}_1$ to simplify the notation but it should be clear that when the cycle index h is omitted, we are always concerned with the SCN design for the first reengineering cycle (h = 1).

these calculations later on). Then our previous definitions suggest that we should select the design $\mathbf{y}_l$ with the highest expected value $E_\Omega[V^{SCN}(\mathbf{y}_l, \omega)]$. This may not be sufficient, however, because even if a design's performance is good on average, its value may be pretty bad for some scenarios. Our definitions require that the design perform relatively well even for worst-case scenarios. This brings us back to our discussion of downside risk in Sect. 2.4 and to the need to consider some risk measures $\rho_\Omega[V^{SCN}(\mathbf{y}_l, \omega)]$ in the design evaluation process.

As shown in Fig. 2.17, when risk is explicitly considered, the search for a maximum expected value design is replaced by the search for efficient value-risk designs (the designs on the efficient frontier in Fig. 2.17) and the final decision depends on the attitude toward risk of the decision maker. Moreover, as discussed in Sect. 10.5.2, some executives may not be overly concerned by business-as-usual risks, but they may have serious concerns about the behavior of their SCN under highly hazardous scenarios. When this is the case, the set of scenarios Ω needs to be split into an acceptable-risk scenarios subset Ω^A and a serious-risk scenarios subset Ω^S, and the evaluation of candidate designs must be based on conditional expected value and risk measures. The design should then also be evaluated under worst-case scenarios and possibly even for some deep uncertainty scenarios. All the aforementioned performance measures are based on the cumulated design value $V^{SCN}(\mathbf{y}_l, \omega)$ over the whole planning horizon. Our definition of resilience, however, requires a relative stability of results within the planning horizon. Some additional performance measures thus may be used to evaluate resilience explicitly. These may be based on desirable resiliency attributes such as the average distance of backup depots to customers on some direct period-to-period stability measures or on indirect measures such as the impact of the number of hazardous events on the expected design value.

In Sect. 11.7, we will show how to generate good SCN designs and illustrate how the previous performance measures can be used to select a best design. It should be clear, however, that the designs obtained depend on the length of the planning horizon considered, on the attitude of the decision maker to risk, and on the resilience strategies adopted. The modeling approach proposed, however, can be used to investigate the impact of different risk tolerence levels and resilience strategies.

11.3 Modeling SCNs Under Risk

The modeling of uncertainty in an SCN design context was studied at length in Chap. 10. Recall that a plausible future scenario is the result of the juxtaposition of random, hazardous, and deeply uncertain events along a planning horizon. Random and hazardous events can be characterized by probability distributions or stochastic processes and, as shown in Sect. 10.5.2, different realizations of the underlying random variables over a planning horizon give rise to distinct plausible future scenarios. For the moment, deeply uncertain scenarios are neglected so that the set

Ω of all plausible scenarios is assumed to contain only probabilistic scenarios. Several approaches have been used to anticipate plausible futures in optimization models (see Fig. 10.19). The most common approach is no doubt to replace the set of plausible futures by a single scenario based on a recent past realization or on forecasts that results in deterministic SCN design models.

The static deterministic models studied in Chap. 7 provide a solid basis for the design of SCNs. However, any design obtained from these models is not robust enough to guarantee adequate performances for any plausible future. It may be reasonable to use these models on an annual basis when decisions are limited to the selection of public warehouses and contract manufacturers, and the future is expected to be a reproduction of the past. However, when supply, location, platform, and offer-selection decisions have long-lasting (several years) impacts these models are usually not sufficient. Under such conditions, the explicit consideration of representative plausible futures defined over a multi-year planning horizon is necessary to obtain robust designs. We therefore begin this section by an example that underlines the importance of considering multiple plausible futures in design models. Then, stochastic programming with recourse is introduced and we show how it can be used to address SCN design problems under risk.

Example 11.1 In order to show the limitations of deterministic models, we investigate the feasibility of solutions under risk using the distribution network design example given in Chap. 7 (Example 7.1). The design model solved in this example was MIP (7.1)–(7.4). We now assume that the demand in the five countries considered is a random variable and that the two equally probable demand scenarios given in Table 11.1 can occur.

To start with, let us solve MIP (7.1)–(7.4) for each of these two demand scenarios. The solution obtained when the MIP is solved with the Excel solver using scenario 2 is given in the upper part of Fig. 11.3, and the design obtained with scenario 1 is shown in its lower part. The two designs are different in terms of the number of DCs to open and the mission of the opened DC. Scenario 1 requires the opening of the three DCs whereas scenario 2 advocates opening one DC in Turin. Note that the solution produced for scenario 2 is not feasible for scenario 1 because the total demand for scenario 1 is much higher than the capacity of the Turin DC, which provides only 40 % of the required capacity. However, the design obtained with scenario 1

Table 11.1 Two demand scenarios for the distribution network design example

		Demand				
Scenario	Probability	Germany	France	Italy	Spain	Portugal
1	0.5	21,750	25,375	14,500	25,375	36,250
2	0.5	8250	9625	5500	9625	13,750
	Average	15,000	17,500	10,000	17,500	25,000

Optimal Design for Scenario 2

	Flows to demand zones					Open	Excess
Site	Germany	France	Italy	Spain	Portugal	DCs	capacity
Dusseldorf	-	-	-	-	-	0	-
Toulouse	-	-	-	-	-	0	-
Turin	8,250	9,625	5,500	9,625	13,750	1	3,250

Optimal Design for Scenario 1

	Flows to demand zones					Open	Excess
Site	Germany	France	Italy	Spain	Portugal	DCs	capacity
Dusseldorf	21,750	25,375	-	-	-	1	2,875
Toulouse	-	-	-	25,375	24,625	1	-
Turin	-	-	14,500	-	11,625	1	23,875

Fig. 11.3 Excel solver solutions for the two demand scenarios

would be very bad under scenario 2 because only 31 % of the capacity provided by the three opened DCs would be used.

As noted previously, one of the most popular approaches to anticipating the future is to use a single average scenario based on the expected value of the random variables involved. For this example, the average scenario is given on the last line of Table 11.1 and it corresponds to the demand used in Example 7.1. The design obtained when the model is solved with the average demand is thus found in Fig. 7.9. It involves the opening of two DCs; one in Düsseldorf and one in Turin. However, when inspecting this solution more closely, we see that the design obtained is not feasible under scenario 1 and that if scenario 2 prevails only 46.8 % of the capacity available will be used.

This example shows that because the demand is random, constraint (7.2) in the design model is not valid anymore because fixed shipments (left side) can never equal a random demand (right side). The only way to remove this flaw is to explicitly model the recourse actions that must be made in practice, once the demand has been observed, to match supply and demand. This is exactly what stochastic programming with recourse[3] is doing.

11.3.1 Stochastic Programming with Recourse

Stochastic programming is an optimization approach to solve decision problems under risk (Birge and Louveaux 2011; Shapiro et al. 2009), and we shall see how it can be applied to the design of SCNs. Basically, stochastic programs are

[3]Note that other modeling approaches such as chance-constrained programming and robust optimization can be used to tackle these issues but they are not studied in this book.

mathematical programs in which some of the data are random. An assumption made when using this modeling approach, however, is that the probability distribution of the random parameters is known. Consequently, it cannot cope with deeply uncertain events. Two main types of modeling constructs can be used to take risk into account. One explicitly models the recourses available in practice to bridge the gap between decisions made before the random variables are observed (e.g., designs, plans) and the real needs that develop when the random variables are observed, which gives rise to *stochastic programs with recourse* (SPRs). The other one imposes restrictions on the probability that a constraint is violated, which gives rise to *chance-constrained programs*. In what follows, we limit ourselves to the first class of models. To explain basic stochastic programming with recourse concepts, we start by looking at single-period SCN design models aiming to optimize the design expected value.

The stochastic programming with recourse approach assumes that decisions are made in two stages: initial plans are elaborated prior to the observation of random variables and, when the value of the random variables are observed, recourse decisions are made to adjust the plan to reality. In our context, referring to Fig. 11.1, when considering a single reengineering cycle including a single planning period, first-stage variables correspond to the design variables vector $\mathbf{y}$ and second-stage recourse variables correspond to the anticipated user decision vector $\hat{\mathbf{x}}$. To obtain better anticipations the reengineering cycle considered can cover several planning periods $t \in T$. The second-stage recourse variables then correspond to the decision vectors $\hat{\mathbf{x}}_t, t \in T$. When several reengineering cycles are considered a more complex multi-stage SPR (Shapiro et al. 2009) is obtained.

Recourse variables are required to ensure that a feasible solution can be found for any plausible future scenario. Some recourse variables anticipate primary operational activities such as the flow and platform activity decisions found in the Chap. 7 models; others anticipate short-term actions required to match supply and demand (as dictated by response policies) such as the use of overtime or subcontracting to provide additional capacity; and some are state variables such as end-of-period inventories or capacity not used during a period. Most recourse variables generate revenues or costs but some state variables may not (e.g., unused capacity). Some unit revenue and cost parameters, such as unit transportation costs or any unit revenue/cost affected by exchange rates, are typically random variables. In addition, the main random variables found in SCN design problems are linked to product-market demand and to platform capacity.

SPRs can be expressed directly in terms of the random variables present in the problem. However, because, as seen in Sect. 10.5, these random variables give rise to a set Ω of plausible future scenarios, we explain how to formulate SPRs directly in terms of these scenarios. To do this, we show how to transform the design models presented in Chap. 7 into two-stage SPRs. When using a matrix notation, the static deterministic design models presented in Chaps. 7, 8 and 9 can all be written as follows:

$$\max_{(\mathbf{x},\mathbf{y})} V^{SCN}(\mathbf{x},\mathbf{y}) = (\boldsymbol{\pi} - \mathbf{c})\mathbf{x} - \mathbf{f}\mathbf{y}$$
$$\text{s.t. } \mathbf{A}\mathbf{x} \le \mathbf{b}(\mathbf{y}), \quad \mathbf{W}\mathbf{x} = \mathbf{d}, \quad \mathbf{x} \ge 0, \ \mathbf{y} \in \mathcal{Y} \tag{11.1}$$

where $\mathbf{x}$ is a vector of continuous flow (F) and throughput (X) variables; $\mathbf{y}$ a vector of binary platform (Y), market offer (Y^{M}), and supplier (Y^{V}) selection variables; $\boldsymbol{\pi}$, $\mathbf{c}$, and $\mathbf{f}$ are, respectively, price, variable cost, and fixed design cost vectors; $\mathbf{A}$ and $\mathbf{W}$ are parameter matrices; $\mathbf{b(y)}$ a capacity and/or demand vector depending on design decisions; $\mathbf{d}$ a right-hand-side parameter vector independent of design decisions; and $\mathcal{Y}$ a design variables domain. In the generic SCN design model formulated in Sect. 7.5, for example, $\mathbf{A}\mathbf{x} \le \mathbf{b}(\mathbf{y})$ corresponds to vendor contract constraints (7.33) and (7.34) and to platform capacity constraints (7.39) and (7.40), $\mathbf{W}\mathbf{x} = \mathbf{d}$ to flow equilibrium constraints (7.36)–(7.38) and demand constraints (7.41), and $\mathcal{Y}$ is delimited by the platform selection constraints (7.35) and the binary value restrictions on $\mathbf{y}$.

Under risk, as discussed previously, the original user decisions vector $\mathbf{x}$, and the corresponding parameter vectors and matrices, must be expanded to accommodate additional recourse variables and constraints. A '^' is added over these vectors and matrices to distinguish them from the original ones. Because random variables can be present in the vectors $\hat{\boldsymbol{\pi}}$, $\hat{\mathbf{c}}$, $\mathbf{f}$, $\hat{\mathbf{d}}$, and $\mathbf{b(y)}$, their value depends on the scenario $\omega \in \Omega$ considered. This dependency is recognized by replacing the previous notation with $\hat{\boldsymbol{\pi}}(\omega)$, $\hat{\mathbf{c}}(\omega)$, $\mathbf{f}(\omega)$, $\hat{\mathbf{d}}(\omega)$, and $\mathbf{b}(\mathbf{y},\omega)$. Finally, in order to be able to capture the impact of alternative scenarios, a copy of the recourse vector $\hat{\mathbf{x}}$ must be made for each possible scenario, which leads to the definition of the vectors $\hat{\mathbf{x}}_\omega, \omega \in \Omega$. Remembering that each scenario $\omega \in \Omega$ has a probability of occurrence $\mathrm{p}(\omega)$, the SPR to solve to maximize the expected value of the SCN design is

$$\max_{\mathbf{y}\in\mathcal{Y}} \mathrm{E}_\Omega[V^{\mathrm{SCN}}(\mathbf{y},\omega)] = \sum_{\omega\in\Omega} \mathrm{p}(\omega)[(\hat{\boldsymbol{\pi}}(\omega) - \hat{\mathbf{c}}(\omega))\hat{\mathbf{x}}_\omega - \mathbf{f}(\omega)\mathbf{y}] \tag{11.2}$$

$$\text{s.t.} \quad \hat{\mathbf{A}}\hat{\mathbf{x}}_\omega \le \mathbf{b}(\mathbf{y},\omega), \quad \omega \in \Omega \tag{11.3}$$

$$\hat{\mathbf{W}}\hat{\mathbf{x}}_\omega = \hat{\mathbf{d}}(\omega), \quad \hat{\mathbf{x}}_\omega \ge \mathbf{0}, \quad \omega \in \Omega \tag{11.4}$$

Note that this mathematical program is still a MIP. It is much larger than the deterministic model, however, because most original constraints are now copied as many times as there are scenarios. Note also that the objective function (11.2) as it stands seeks to maximize the expected value $\mathrm{E}_\Omega[V^{\mathrm{SCN}}(\mathbf{y},\omega)]$ (or minimize expected costs when $\hat{\pi}(\omega) = \mathbf{0}$) and that it does not take risk into account explicitly (i.e., it assumes that the decision maker is risk neutral). If the decision maker is risk averse, then, as seen in Sect. 2.4, an adequate risk measure $\rho_\Omega[V^{\mathrm{SCN}}(\mathbf{y},\omega)]$ should be selected, and a value-risk efficient frontier (see Fig. 2.17) should be constructed to enable the decision maker to select an adequate value-risk tradeoff. This can be done by replacing objective function (11.2) with

$$\max_{\mathbf{y}\in\mathcal{Y}} \mathrm{E}_{\Omega}[V^{\mathrm{SCN}}(\mathbf{y},\omega)] - \lambda\rho_{\Omega}[V^{\mathrm{SCN}}(\mathbf{y},\omega)], \quad \lambda \geq 0 \tag{11.5}$$

The efficient frontier is then obtained by solving the design model for different weights λ, which could be seen as the cost of risk. An alternative would be to add a maximum risk constraint $\rho_{\Omega}[V^{\mathrm{SCN}}(\mathbf{y},\omega)] \leq \rho^{o}$ and to solve the model for several risk levels ρ^{o}. The choice of an adequate risk measure will be discussed in Sect. 11.3.3. In what follows, to illustrate how these concepts apply to SCN design, we show how the main models studied in Chap. 7 can be transformed into SPRs.

11.3.2 Distribution Network Design Model

Static distribution network design model (7.1)–(7.4) can be transformed into a two-stage SPR by considering a set of plausible future scenarios $\omega \in \Omega$. Assume that demand d_l for demand zone $l \in L^{\mathrm{D}}$ is a random variable with a known probability distribution and that transportation costs are estimated using a distance-based regression function as illustrated in Fig. 5.19. As a consequence, for a scenario ω, the demand is given by $d_l(\omega), l \in L^{\mathrm{D}}$ and the unit flow costs between DCs and demand zones by $f_{wl}(\omega), w \in L^{\mathrm{W}}, l \in L^{\mathrm{D}}$. As shown previously, model (7.1)–(7.4) is not adequate under risk. We showed that there is no guarantee that DC capacity and demand will match for all scenarios. To eliminate this problem, a copy of the flow variables must first be defined for each scenario leading to the following redefinition:

$F_{wl\omega}$ Quantity of products provided by DC w to demand zone l during the year considered under scenario ω

Also, as seen previously, for some scenarios, the DC selected may not have enough capacity. This problem can be solved by introducing the following recourse variables:

$B_{w\omega}$ Additional throughput capacity required during the year at DC w under scenario ω

If the DCs are public warehouses, the fact that some additional capacity may be required during the year can be included in the provider contract. Otherwise, this additional capacity may be obtained by renting temporary space and/or using overtime. In both cases, the additional capacity available is usually limited. In the following model, a parameter ζ_w is introduced to represents the maximum proportion of DC w capacity that may be added during the year considered. Also, this additional capacity is not free. An additional cost c_w^{+} is incurred for each capacity unit added at DC w. Using this additional notation, the design model to solve under risk can be formulated as follows.

$$\min \sum_{w \in L^{W}} y_w Y_w + \sum_{\omega \in \Omega} \mathrm{p}(\omega) [\sum_{w \in L^{W}} \sum_{l \in L^{D}} f_{wl}(\omega) F_{wl\omega} + \sum_{w \in L^{W}} c_w^+ B_{w\omega}] \tag{11.6}$$

subject to

– Demand constraints

$$\sum_{w \in L^{W}} F_{wl\omega} = d_l(\omega), \quad l \in L^{D}, \omega \in \Omega \tag{11.7}$$

– DC capacity constraints

$$\sum_{l \in L^{D}} F_{wl\omega} - B_{w\omega} \leq b_w Y_w, \quad w \in L^{W}, \omega \in \Omega \tag{11.8}$$

– DC local recourse constraints

$$B_{w\omega} \leq (\zeta_w b_w) Y_w, \quad w \in L^{W}, \omega \in \Omega \tag{11.9}$$

– Binary value and non-negativity constraints

$$Y_w \in \{0, 1\}, w \in L^{W}; \quad F_{wl\omega}, B_{w\omega} \geq 0, w \in L^{W}, l \in L^{D}, \omega \in \Omega$$

Demand constraints (11.7) are modeled as before but in this case they must be met for each scenario $\omega \in \Omega$. Capacity constraints (11.8) now include the recourse variables $B_{w\omega}$ on the left side, and they must also be satisfied for all scenarios $\omega \in \Omega$. New constraints (11.9) are added to limit the amount of additional capacity at a DC under any scenario. The objective function is also modified to calculate expected flow and capacity-adjustment costs. Constraints (11.8) and (11.9) correspond to constraint (11.3) in the generic SPR model formulated in Sect. 11.3.1, and constraint (11.7) corresponds to the generic constraint (11.4). The model is still a MIP, but instead of having $|L^{D}| + |L^{W}|$ constraints, we now have $(|L^{D}| + 2|L^{W}|) \times |\Omega|$ constraints. When the number of scenarios considered is large, the main difficulty encountered to solve these models is related to their size.

Example 11.2 Return to the design problem examined in Example 11.1. Suppose now, however, that the 3PLs managing the DCs can provide additional capacity during the year if required. The additional cost incurred per unit of capacity when this recourse is used is €100, €40 and €80 for the Düsseldorf, Toulouse, and Turin DCs, respectively. The maximum proportion of the contracted capacity that can be obtained in this manner is fixed at 40 % per year, which gives a potential 20,000 standard shipping loads of additional capacity for each DC. Using the two demand scenarios provided in Table 11.1,

	Scenario 1						
	Flows to demand zones					Excess capacity	Recourse
Site	Germany	France	Italy	Spain	Portugal		
Dusseldorf	0	0	0	0	0	0	0
Toulouse	0	8,375	0	25,375	36,250	0	20,000
Turin	21,750	17,000	14,500	0	0	0	3,250
Demand	21,750	25,375	14,500	25,375	36,250		

	Scenario 2						
	Flows to demand zones					Excess capacity	Recourse
Site	Germany	France	Italy	Spain	Portugal		
Dusseldorf	0	0	0	0	0	0	0
Toulouse	0	0	0	9,625	13,750	26,625	0
Turin	8,250	9,625	5,500	0	0	26,625	0
Demand	8,250	9,625	5,500	9,625	13,750		

Site	Opened DCs	Fixed costs	Flows scen. 1	Flows scen. 2	Local recourse
Dusseldorf	0	€ 0	€ 0	€ 0	€ 0
Toulouse	1	€ 3,500,000	€ 9,825,000	€ 3,313,750	€ 800,000
Turin	1	€ 2,300,000	€ 7,140,000	€ 3,121,250	€ 260,000
	Total:	€ 5,800,000	€ 16,965,000	€ 6,435,000	€ 1,060,000
	Expected costs:	€ 5,800,000	€ 11,700,000		€ 530,000
				Total expected costs:	€ 18,030,000

Fig. 11.4 Excel solver SPR solution with two demand scenarios

the optimal design can then be found with the Excel solver using model (11.6)–(11.9). The optimal distribution network design obtained is shown in Fig. 11.4. It suggests opening two DCs, one in Toulouse and one in Turin, and it gives their mission in terms of annual shipments to each demand zone under the two demand scenarios considered. Under scenario 1, to have a feasible solution, 20,000 units of recourse capacity are required in Toulouse and 3250 units in Turin. Under scenario 2, however, no recourse capacity is required and there is excess capacity at the two DCs. Expected recourse costs with this solution amount to about 3 % of the total expected design costs (€530,000). Note that the design obtained here is very different from the two designs in Fig. 11.3 found by solving the deterministic model with one scenario at a time, as well as from the design in Fig. 7.9 obtained using an average scenario.

11.3.3 Multi-period Distribution Network Design Model

The main limitation of the previous distribution network design model is that it considers a single planning period. In Sect. 10.4, we argued that basing an SCN design model on a recent year data is not sufficient and that plausible evolutionary paths should be explored. In Sect. 10.3, we also studied the perturbing impact that multi-hazards can have on SCN capacity and demand. Single-period models cannot capture these predicaments. Even if a single reengineering cycle is modeled, a multiple planning period $t \in T$ horizon should be used. These periods can be years, but they could also be months or quarters to better capture hazard impacts. Also, when considering alternative evolutionary paths, several model parameters are typically affected (see Table 10.5) when defining plausible future scenarios $\omega \in \Omega$. In the national distribution context studied in this section, these would include at least product costs, inventory-holding costs, and transportation costs, which are all captured here in the unit flow cost parameter $f_{wlt}(\omega)$. This also would include demand zone sales prices $\pi_{lt}(\omega)$. For this reason, the design model should seek to maximize value added, as done in Sect. 7.4.1, instead of minimizing costs.

This being said, the main modification required to the previous recourse variables ($F_{wlt\omega}$ and $B_{wt\omega}$) and parameters (c_{wt}^{+}, $d_{lt}(\omega)$, b_{wt}, and ζ_{wt}) notation is the addition of a time period index t. The design variables and fixed costs notation (Y_w and y_w) does not change, however, because we are still considering a single reengineering cycle. Because multi-hazards can occur, b_{wt} denotes the business-as-usual capacity, but, because DCs may be hit during planning period t, the capacity $b_{wt}(\omega)$ available during a period now depends on scenarios. We saw how to calculate them by using Monte Carlo methods in Sect. 10.5.2. Also, because we are now maximizing value added, demand constraints must be rewritten, as in Sect. 7.4. A market policy parameter $\underline{d}_{lt}$ is used, as before, to specify a minimum market penetration target for each period and demand zone. However, the demand upper bound $d_{lt}(\omega)$ now depends on scenarios. As before, the company may choose not to satisfy all customer demand. A recourse variable $D_{lt\omega}^{-}$ can be used in the model to represent the demand purposefully not satisfied in zone l during period t under scenario ω. Because, such actions may undermine the reputation of the company in the future, a penalty weight π_{wt}^{-} representing future opportunity loss is introduced in the objective function to discourage such actions. The multi-period two-stage SPR design model to use for this revised problem is the following:

$$\max \sum_{\omega \in \Omega} \mathrm{p}(\omega)\{\sum_{t \in T} [\sum_{w \in L^{\mathrm{W}}} \sum_{l \in L^{\mathrm{D}}} (\pi_{lt}(\omega) - f_{wlt}(\omega))F_{wlt\omega} - \sum_{w \in L^{\mathrm{W}}} c_{wt}^{+} B_{wt\omega} - \sum_{l \in L^{\mathrm{D}}} \pi_{wt}^{-} D_{lt\omega}^{-}]\} - \sum_{w \in L^{\mathrm{W}}} y_w Y_w \tag{11.10}$$

subject to

- Demand constraints

$$\sum_{w \in L^{\mathrm{W}}} F_{wlt\omega} \geq \underline{d}_{lt}, \quad l \in L^{\mathrm{D}},\ t \in T,\ \omega \in \Omega \tag{11.11}$$

$$\sum_{w \in L^{W}} F_{wlt\omega} + D^{-}_{lt\omega} = d_{lt}(\omega), \quad l \in L^{D}, t \in T, \omega \in \Omega \tag{11.12}$$

– DC capacity constraints

$$\sum_{l \in L^{D}} F_{wlt\omega} - B_{wt\omega} \leq b_{wt}(\omega) Y_w, \quad w \in L^{W},\ t \in T,\ \omega \in \Omega \tag{11.13}$$

– Local recourse constraints

$$B_{wt\omega} \leq (\zeta_{wt} b_{wt}) Y_w, \quad w \in L^{W},\ t \in T,\ \omega \in \Omega \tag{11.14}$$

– Binary value and non-negativity constraints

$$Y_w \in \{0,1\}, w \in L^{W}; F_{wlt\omega}, B_{wt\omega}, D^{-}_{lt\omega} \geq 0, \quad w \in L^{W},\ l \in L^{D},\ t \in T,\ \omega \in \Omega$$

If several years are considered, the operational cash flows in the objective function could be discounted, as we did, for example, in Sect. 8.5.

Example 11.3 Suppose that the distribution company described in Example 11.2 would now like to use a three-year planning horizon to be able to consider evolutionary paths and multi-hazards. Three equally probable ($p(\omega) = 1/3, \omega = 1, 2, 3$) plausible future scenarios are defined. The scenario capacities are generated using Monte Carlo methods as shown in Sect. 10.5.2.2. The three years of DC capacity used for each scenario are those presented in Fig. 10.28. The demand for each scenario is based on a distinct evolutionary path and is generated as shown in Example 10.4[4] (using demand expansion factors −0.2, 0.05 and 0.1 for scenarios $\omega = 1, 2, 3$, respectively). The demand zone prices for the three scenarios are based on the same evolutionary paths (using inflation factors −0.05, 0.05, and 0.1 for scenarios $\omega = 1, 2, 3$, respectively). To simplify the presentation, unit flow costs are assumed to be deterministic and constant, that is, they are the same for each period and scenario. All the scenarios' data are summarized in Fig. 11.5. The spreadsheet in the figure also gives the other data required by the model, that is, the demand lower bounds, the maximum local recourse capacity addition (10, 15 and 20 % of the business-as-usual capacity per year for Düsseldorf, Toulouse, and Turin, respectively), the unit flow costs, the local recourse capacity unit costs (assumed identical for each year), and the fixed DC location costs (three-year

[4] Usually, a much larger number of scenarios would be generated and their evolutionary paths would be selected randomly using the probabilities $p_\kappa, \kappa \in K$, as indicated in the Monte Carlo procedure found in Fig. 10.21. Here we define a single scenario for each evolutionary path to keep the size of the example to a minimum.

		Demand zones						Local recourse	
Flow costs		Germany	France	Italy	Spain	Portugal	Fixed costs	Maximum	Unit cost
DC Sites	Düsseldorf	€ 125	€ 120	€ 150	€ 160	€ 180	€ 8,100,000	5000	€ 120
	Toulouse	€ 160	€ 130	€ 145	€ 130	€ 150	€ 10,500,000	7500	€ 150
	Turin	€ 140	€ 130	€ 130	€ 170	€ 170	€ 6,900,000	10000	€ 100
Demand		Germany	France	Italy	Spain	Portugal			
Min deliveries	Year 1	11,500	14,000	7,500	13,000	18,500			
	Year 2	9,000	9,500	5,500	9,500	14,500	*DC Capacity*		
	Year 3	5,500	6,000	4,000	7,000	9,500	Düsseldorf	Toulouse	Turin
Scenario 1	Year 1	12,000	14,000	8,000	14,000	20,000	49,080	49,030	41,140
	Year 2	9,000	10,500	6,000	10,500	15,000	46,466	50,000	40,649
	Year 3	6,000	7,000	4,000	7,000	10,000	48,197	49,411	43,930
Scenario 2	Year 1	15,750	18,375	10,500	18,375	26,250	50,000	49,161	43,746
	Year 2	16,500	19,250	11,000	19,250	27,500	48,729	49,819	39,868
	Year 3	17,250	20,125	11,500	20,125	28,750	45,996	50,000	41,403
Scenario 3	Year 1	16,500	19,250	11,000	19,250	27,500	44,173	50,000	49,893
	Year 2	18,000	21,000	12,000	21,000	30,000	50,000	49,416	41,053
	Year 3	19,500	22,750	13,000	22,750	32,500	50,000	48,995	40,598
Prices		Germany	France	Italy	Spain	Portugal			
Scenario 1	Year 1	€ 257	€ 242	€ 233	€ 216	€ 208			
	Year 2	€ 247	€ 234	€ 229	€ 205	€ 199			
	Year 3	€ 224	€ 216	€ 212	€ 198	€ 182			
Scenario 2	Year 1	€ 283	€ 274	€ 261	€ 246	€ 236			
	Year 2	€ 297	€ 289	€ 273	€ 248	€ 241			
	Year 3	€ 309	€ 295	€ 288	€ 267	€ 257			
Scenario 3	Year 1	€ 295	€ 285	€ 271	€ 253	€ 244			
	Year 2	€ 320	€ 310	€ 304	€ 277	€ 262			
	Year 3	€ 353	€ 334	€ 324	€ 294	€ 285			

Fig. 11.5 Excel spreadsheet with Example 11.3 data

Site	Opened DCs	Fixed costs	Flows Scen. 1	Flows Scen. 2	Flows Scen. 3	Local recourse	Revenues
Dusseldorf	1	€ 8,100,000	€ 12,195,000	€ 19,113,500	€ 19,890,760	€ 600,000	€ 106,905,029
Toulouse	0	€ 0	€ 0	€ 0	€ 0	€ 0	€ 0
Turin	1	€ 6,900,000	€ 9,990,000	€ 19,168,570	€ 22,093,180	€ 1,000,000	€ 82,921,707
	Total:	€ 15,000,000	€ 22,185,000	€ 38,282,070	€ 41,983,940	€ 1,600,000	€ 189,826,736
	Expected value:	€ 15,000,000		€ 34,150,337		€ 533,333	€ 63,275,579
						Total expected profit:	€ 13,591,909

Fig. 11.6 Optimal design obtained with the multi-period stochastic program

fixed 3PL contract charge). The penalty weights for not satisfying demand are null.

The resolution of model (11.10)–(11.14) with Excel[5] produces the optimal solution partially shown in Figs. 11.6 and 11.7. The first figure indicates that the best design is to open DCs in Düsseldorf and Turin, and it gives an overview of the operational and fixed design costs involved over the

[5]The Premium Solver upgrade sold by Frontline Systems (www.solver.com) was used to solve the MIP obtained because it is too large for the default Excel solver.

Second-Stage Variables (Scenario 1)

Flows		Demand zones					Excess	Recourse
Period	Site	Germany	France	Italy	Spain	Portugal	capacity	capacity
1	Dusseldorf	12,000	14,000	-	14,000	-	9,080	0
2	Dusseldorf	9,000	10,500	-	10,500	-	16,466	0
3	Dusseldorf	6,000	7,000	-	7,000	-	28,197	0
1	Toulouse	-	-	-	-	-	-	0
2	Toulouse	-	-	-	-	-	-	0
3	Toulouse	-	-	-	-	-	-	0
1	Turin	-	-	8,000	-	20,000	13,140	0
2	Turin	-	-	6,000	-	15,000	19,649	0
3	Turin	-	-	4,000	-	10,000	29,930	0
1	Lost demand	-	-	-	-	-		
2	Lost demand	-	-	-	-	-		
3	Lost demand	-	-	-	-	-		

Second-Stage Variables (Scenario 3)

Flows		Demand zones					Excess	Recourse
Period	Site	Germany	France	Italy	Spain	Portugal	capacity	capacity
1	Dusseldorf	16,500	8,423	-	19,250	-	-	-
2	Dusseldorf	18,000	21,000	-	11,000	-	-	-
3	Dusseldorf	19,500	22,750	-	12,750	-	-	5,000
1	Toulouse	-	-	-	-	-	-	-
2	Toulouse	-	-	-	-	-	-	-
3	Toulouse	-	-	-	-	-	-	-
1	Turin	-	10,827	11,000	-	27,500	566	-
2	Turin	-	-	12,000	10,000	19,053	-	-
3	Turin	-	-	13,000	10,000	27,598	-	10,000
1	Lost demand	-	-	-	-	-		
2	Lost demand	-	-	-	-	10,947		
3	Lost demand	-	-	-	-	4,902		

Fig. 11.7 Second-stage variables for two scenarios

three-year planning horizon considered. The total expected value added of the design is €13,591,909. Note that this solution differs from the design obtained with a static stochastic model (see Fig. 11.4) in Example 11.2. Figure 11.7 provides the optimal values of second-stage recourse variables for two scenarios (1 and 3). It shows that, under scenario 1, all customer demands can be satisfied with the capacity provided by the two opened DCs and that, in fact, there is a substantial amount of excess capacity. Under scenario 3, however, even if some local recourse capacity is used in Düsseldorf and Turin during period 3, some of the demand from Portugal cannot be satisfied during periods 2 and 3. Note also that the flows between the DCs and the demand zones are not the same for the two scenarios. This shows how second-stage recourse variables adapt to scenarios at the user level.

Model (11.10)–(11.14) can be extended to cover two or three reengineering cycles. Strictly speaking, when this is done, the model becomes a multi-stage stochastic program with recourse. This complicates things significantly because structural adaptation decisions made for cycles $h > 1$ depend on the actual realization of the random variables up to the beginning of cycle *h,* which implies that conditional expectations must be used. When the design problem is formulated in terms of a set of multi-cycle scenarios, *nonanticipativity* constraints must be added to the model. This goes beyond the scope of this book and the interested reader is referred to Shapiro et al. (2009). Multi-cycle design models can be simplified, however, by using an approximate anticipation of structural adaptation decisions. We will show how to do this in Sect. 11.7.

11.3.4 Multi-period SCN Design Model

The previous sections are concerned with pure distribution networks. To complete our illustrations of how to proceed to transform deterministic design models into SPRs, we show in this section how the production–distribution network design model presented in Sect. 7.3.2 can be changed into a multi-period two-stage SPR when evolutionary paths and multi-hazards are considered. The basic notation used is the same as in Sect. 7.3.2. For design variables and costs, there is no change to make. Because we are now considering a planning horizon $t \in T$, most user-level parameters and variables now have an additional index t as in the previous section. In addition, all the random cost, demand, and capacity parameters value now depend on the scenario $\omega \in \Omega$ considered. Similarly, because all the flow variables are now recourse variables, depending on the scenario considered, they have an additional index ω. Finally, in the context considered, three additional types of local recourse variables must be introduced:

- We assume here that all customer demands must be satisfied even if they are random variables. When internal resources are not sufficient to satisfy the demand during a planning period, equivalent products can be shipped directly from an external recourse supplier to the demand zone. Under scenario ω, the quantity of product p that must be shipped to demand zone l during period t by the recourse supplier is denoted by $F^{+}_{plt\omega}$. The unit flow cost associated with each product shipped by the recourse supplier is f^{+}_{plt}.
- As in the previous section, because DCs and PDCs may be the victim of serious disruptions, the distribution capacity available depends on the scenario considered, and additional capacity can be provided during a year either by the facility owner, say a 3PL, or using local short-term storage. The recourse distribution capacity required on site $s \in L^{\mathrm{S}}$ during period t under scenario ω is denoted $B^{\mathrm{D}}_{st\omega}$ and the associated unit recourse cost is $c^{\mathrm{D}+}_{st}$.

- For the same reason, the production capacity available in a PDC also depends on the scenario considered. When production capacity is missing, overtime or subcontracting can be used to compensate. The number of recourse capacity units required at PDC $u \in L^{\mathrm{U}}$ during period t for product p under scenario ω is denoted $B^{\mathrm{P}}_{put\omega}$ and the associated unit recourse cost is $c^{\mathrm{P}+}_{put}$.

We assume in the following model that the external supply recourse $F^{+}_{plt\omega}$ is unlimited because of the presence of numerous suppliers and a wide offer on the market. However, production and distribution capacity recourses are bounded as in the previous sections. The maximum recourse capacity is limited to a proportion of regular capacity per site, and it is denoted by ζ_{pu} for PDCs and ζ_s, for DCs. Under these conditions, the required stochastic production–distribution network design model is formulated as follows:

$$\min \sum_{s \in L^{\mathrm{S}}} y_s Y_s + \sum_{\omega \in \Omega} p(\omega)\{\sum_{t \in T}[\sum_{p \in P}(\sum_{u \in L^{\mathrm{U}}}\sum_{w \in L^{\mathrm{W}}} f_{puw}(\omega)F_{puwt\omega} + \sum_{s \in L^{\mathrm{S}}}\sum_{l \in L^{\mathrm{D}}_p} f_{psl}(\omega)F_{pslt\omega}$$
$$+ \sum_{l \in L^{\mathrm{D}}_p} f^{+}_{plt}F^{+}_{plt\omega} + \sum_{u \in L^{\mathrm{U}}} c^{\mathrm{P}+}_{put}B^{\mathrm{P}}_{put\omega}) + \sum_{s \in L^{\mathrm{S}}} c^{\mathrm{D}+}_{st}B^{\mathrm{D}}_{st\omega}]\} \tag{11.15}$$

subject to

- Demand constraints

$$\sum_{s \in L^{S}} F_{pslt\omega} + F^{+}_{plt\omega} = d_{plt}(\omega), \quad p \in P,\ l \in L^{D}_p,\ t \in T,\ \omega \in \Omega \tag{11.16}$$

- Production capacity constraints

$$\sum_{l \in L^{\mathrm{D}}_p} F_{pult\omega} + \sum_{w \in L^{\mathrm{W}}} F_{puwt\omega} - B^{\mathrm{P}}_{put\omega} \le b_{put}(\omega)Y_u, \quad u \in L^{\mathrm{U}},\ p \in P_u,\ t \in T,\ \omega \in \Omega \tag{11.17}$$

- Distribution capacity constraints

$$\sum_{p \in P} e_p(\sum_{l \in L^{\mathrm{D}}_p} F_{pslt\omega}) - B^{\mathrm{D}}_{st\omega} \le b_{st}(\omega)Y_s, \quad s \in L^{\mathrm{S}}, t \in T, \omega \in \Omega \tag{11.18}$$

- Flow equilibrium constraints

$$\sum_{u \in L^{\mathrm{U}}} F_{puwt\omega} = \sum_{l \in L^{\mathrm{D}}_p} F_{pwlt\omega}, \quad p \in P,\ w \in L^{\mathrm{W}},\ t \in T,\ \omega \in \Omega \tag{11.19}$$

– Recourse capacity constraints

$$0 \leq B^{\mathrm{P}}_{put\omega} \leq (\zeta_{pu} b_{put}) Y_u, \quad u \in L^U,\ p \in P,\ t \in T,\ \omega \in \Omega \tag{11.20}$$

$$0 \leq B^{\mathrm{D}}_{st\omega} \leq (\zeta_s b_{st}) Y_s, \quad s \in L^S, t \in T, \omega \in \Omega \tag{11.21}$$

The usual binary value and non-negativity restrictions must also be included. Depending on the nature of the scenario set Ω, this MIP can be extremely large and, as seen in the next section, some scenario sampling methods may have to be used to be able to solve it.

11.4 SAA Method

In the previous sections, we implicitly assumed that the set of scenarios considered is of a manageable size. Unfortunately, as explained in Chap. 10, the set of plausible futures is usually extremely large. In fact, if any of the random variables involved have a continuous probability distribution, then the number of plausible scenarios is infinite. Clearly, when this is the case, the SPRs previously formulated cannot be solved. To circumvent this difficulty, these models are usually solved using only a representative subset of scenarios. When a sample $\Omega^N \subset \Omega$ of N scenarios is generated using Monte Carlo methods, as shown in Sect. 10.5.2, then all the scenarios considered are equiprobable, that is, $\mathrm{p}(\omega) = 1/N, \omega \in \Omega^N$. The generic SPR (11.2)–(11.4) defined in Sect. 11.3.1 can then be replaced by the smaller MIP

$$\bar{V}^N = \max_{\mathbf{y} \in \mathcal{Y}} \bar{V}^N(\mathbf{y}) = \frac{1}{N} \sum\nolimits_{\omega \in \Omega^N} [(\hat{\boldsymbol{\pi}}(\omega) - \hat{\mathbf{c}}(\omega))\hat{\mathbf{x}}_\omega - \mathbf{f}(\omega)\mathbf{y}] \tag{11.22}$$

$$\text{s.t.} \quad \hat{\mathbf{A}}\hat{\mathbf{x}}_\omega \leq \mathbf{b}(\mathbf{y}, \omega), \quad \omega \in \Omega^N \tag{11.23}$$

$$\hat{\mathbf{W}}\hat{\mathbf{x}}_\omega = \hat{\mathbf{d}}(\omega), \quad \hat{\mathbf{x}}_\omega \geq \mathbf{0}, \quad \omega \in \Omega^N \tag{11.24}$$

where $\bar{V}^N(\mathbf{y})$ is a *sample average approximation* (SAA) of $\mathrm{E}_\Omega[V^{\mathrm{SCN}}(\mathbf{y}, \omega)]$. For this reason, MIP (11.22)–(11.24) is usually referred to as a SAA program.

The main question now is how good is the design $\bar{\mathbf{y}}^N$, with objective function value $\bar{V}^N$, obtained when solving a SAA program with a scenario sample of size N. It has been shown that $\bar{V}^N(\mathbf{y})$ is an unbiased estimator of the SPR expected value $\mathrm{E}_\Omega[V^{\mathrm{SCN}}(\mathbf{y}, \omega)]$ and that it converges with probability one to $\mathrm{E}_\Omega[V^{\mathrm{SCN}}(\mathbf{y}, \omega)]$ as $N \to \infty$ (see Shapiro et al. 2009). This suggests that the quality of the approximation improves as the size of the scenario sample used grows. Unfortunately, even with the power of current solvers, the scenario samples which can be used for real design problems are still relatively small (in the hundreds at best). So, how good are the designs obtained with the resulting SAA models? It was also shown that valid statistical bounds for the true expected value V^* of an optimal design $\mathbf{y}^*$ of the original SPR are

$$\mathrm{E}_\Omega[V^{\mathrm{SCN}}(\bar{\mathbf{y}}^N,\omega)] \le V^* \le \mathrm{E}[\bar{V}^N]$$

A statistical optimality gap for a design $\bar{\mathbf{y}}^N$ obtained by solving a SAA model can be calculated by estimating these lower and upper bounds.

For a given design $\bar{\mathbf{y}}^N$, the lower bound can be estimated by solving the second-stage programs obtained by fixing $\mathbf{y}$ and ω, for a sample $\Omega^{N^+} \subset \Omega$ of N^+ scenarios with $N^+ \gg N$. More precisely, the second-stage program to solve for a given $\bar{\mathbf{y}}^N$ and ω is

$$Q(\bar{\mathbf{y}}^N,\omega) = \max_{\hat{\mathbf{x}}}(\hat{\boldsymbol{\pi}}(\omega) - \hat{\mathbf{c}}(\omega))\hat{\mathbf{x}} \tag{11.25}$$

$$\text{s.t.} \quad \hat{\mathbf{A}}\hat{\mathbf{x}} \le \mathbf{b}(\bar{\mathbf{y}}^N,\omega), \hat{\mathbf{W}}\hat{\mathbf{x}} = \hat{\mathbf{d}}(\omega), \hat{\mathbf{x}} \ge \mathbf{0} \tag{11.26}$$

For the SCN design problems studied, this linear program is usually easily solved. An estimate of $\mathrm{E}_\Omega[V^{\mathrm{SCN}}(\bar{\mathbf{y}}^N,\omega)]$ is then given by

$$\bar{V}^{N^+}(\bar{\mathbf{y}}^N) = \frac{1}{N^+}\sum_{\omega\in\Omega^{N^+}} V^{\mathrm{SCN}}(\bar{\mathbf{y}}^N,\omega) = \frac{1}{N^+}\sum_{\omega\in\Omega^{N^+}} Q(\bar{\mathbf{y}}^N,\omega) - \mathbf{f}(\omega)\bar{\mathbf{y}}^N \tag{11.27}$$

Note furthermore that $\bar{V}^{N^+}(\mathbf{y}^{\mathrm{o}})$ in fact provides an estimate of the true expected value of any proposed feasible design $\mathbf{y}^{\mathrm{o}}$. The upper bound $\mathrm{E}[\bar{V}^N]$ can be estimated by averaging. One can solve M SAA programs based on independently generated scenario samples each of size N. Let $\bar{\mathbf{y}}_j^N$ and $\bar{V}_j^N, j = 1,\ldots,M$, be, respectively, the optimal solutions and the optimal values of these SAA programs. Then an unbiased estimator of $\mathrm{E}[\bar{V}^N]$ is given by

$$\overline{V}^{N,M} = \frac{1}{M}\sum_{j=1}^{M} \bar{V}_j^N \tag{11.28}$$

From this, the SAA method summarized in Fig. 11.8 has been largely used to find near-optimal solutions for two-stage SPRs. Results are also available to calculate the variance of the estimated optimality gap and the scenario sample size required to obtain a predetermined precision. Importance sampling methods were also developed to improve the precision of SAA programs when small samples are used. The SAA method can also be extended to solve multi-stage SPRs. A detailed coverage of these subjects is found in Shapiro et al. (2009).

1. Generate M independent samples $\Omega_j^N, j=1,...,M$ of N scenarios and solve the SAA program (11.22) to (11.24) for each sample to get the solutions $(\overline{V}_j^N, \overline{\mathbf{y}}_j^N)_{j=1,...,M}$.
2. Compute the statistical upper bound $\overline{V}^{N,M}$ with (11.28).
3. Let $\{\overline{\mathbf{y}}_j^N\}_{j=1,...,L\leq M}$ be the set of distinct solutions found in step 1. For each $j=1,...,L$, using an independently generated sample of N^+ scenarios, estimate the true expected value of solution $\overline{\mathbf{y}}_j^N$ using (11.27).
4. Select the solution j^* with the best expected value $\overline{V}^{N^+}(\overline{\mathbf{y}}_{j^*}^N)$.
5. Estimate the optimality gap of solution $\overline{\mathbf{y}}_{j^*}^N$ with

   $$gap^{N,M,N^+}(\overline{\mathbf{y}}_{j^*}^N) = \overline{V}^{N,M} - \overline{V}^{N^+}(\overline{\mathbf{y}}_{j^*}^N)$$

   If this gap is acceptable, stop.
   Otherwise, return to step 1 using larger N and/or M.

Fig. 11.8 SAA method for two-stage SPRs

Example 11.4 To illustrate the SAA method, let's go back to the multi-period distribution network design problem solved in Example 11.3. This example was based on three scenarios generated in Chap. 10. Capacities were taken from the example in Sect. 10.5.2.2 based on natural disasters with exponential inter-arrival times and uniformly distributed impacts. Because these two probability distributions are continuous, the set of plausible futures $\omega \in \Omega$ contains an infinite number of scenarios. Using three scenarios as we did is far from sufficient, and the SAA method should be used to solve this example. When a sample $\Omega^N \subset \Omega$ of N scenarios generated using Monte Carlo methods is used, the SAA program to solve is

$$\max \frac{1}{N} \sum_{\omega \in \Omega^N} \{ \sum_{t \in T} [\sum_{w \in L^W} \sum_{l \in L^D} (\pi_{lt}(\omega) - f_{wlt}(\omega)) F_{wlt\omega} - \sum_{w \in L^W} c_{wt}^+ B_{wt\omega} - \sum_{l \in L^D} \pi_{wt}^- D_{lt\omega}^-] \} - \sum_{w \in L^W} y_w Y_w$$

with Ω replaced by Ω^N in constraints (11.11)–(11.14).

Now, if four samples ($M = 4$) of three scenarios[6] ($N = 3$) are generated using Monte Carlo methods, four SAA programs must be solved. The four designs thus obtained are given in Fig. 11.9. When inspecting these designs, we see that two distinct designs ($L = 2$) were found from the four samples used. The same design (one DC in Düsseldorf and one in Turin) was found when solving with samples 1 and 3, but another one (one DC in Toulouse and

[6]A much larger number of scenarios should be generated to obtain a robust design. We kept N and M deliberately small in this example to be able to solve the SAA models with Excel and to illustrate the approach succinctly.

	Opening decisions			
	Sample 1	*Sample 3*	*Sample 2*	*Sample 4*
	Design 1		*Design 2*	
Düsseldorf DC	1	1	0	0
Toulouse DC	0	0	1	1
Turin DC	1	1	1	1
SAA value (N = 3)	€ 13,546,909	€ 13,326,879	€ 12,766,827	€ 13,267,173
Expected value (N^+ = 20)	€ 13,508,410		€ 12,808,736	

Fig. 11.9 Spreadsheet of the designs obtained with four three-scenario samples

one in Turin) was obtained with samples 2 and 4. The next line in the table gives the optimal SAA value $\bar{V}_j^N$ for the four designs (for example, $\bar{V}_1^N$ = €13,546,909). From these values, the statistical upper bound $\overline{V}^{3,4}$ = €13,226,947 can be calculated with (11.28). To be able to compare the two distinct designs obtained, we estimate their true expected value $E_\Omega[V^{SCN}(\bar{\mathbf{y}}_j^3, \omega)], j = 1, 2$, with (11.27) using an independently generated sample of 20 scenarios ($N^+ = 20$). The values $\bar{V}^{20}(\bar{\mathbf{y}}_j^3), j = 1, 2$, obtained are given on the last table row. This indicates that design 1 (obtained with samples 1 and 3) is the best design. The size and number of scenario samples used are too small, however, to be able to estimate a valid statistical gap and to make any conclusive decision. The gap calculated is −€281,463, that is, -2 % of $\bar{V}^{20}(\bar{\mathbf{y}}_1^3)$. Usually, the gap value should be positive. To obtain conclusive results, much larger values of M, N, and N^+ should be used.

11.5 Risk Measures

The generic SPR formulation (11.2)–(11.4) as well as the stochastic SCN design models presented previously all include an objective function aiming to maximize the expected value $E_\Omega[V^{SCN}(\mathbf{y}, \omega)]$ of the SCN designed. However, as pointed out in Sect. 11.2.2, even if a design is good on average, it may be pretty bad for some scenarios. To obtain more robust designs, risk-averse decision makers generally want to consider some risk measures $\rho_\Omega[V^{SCN}(\mathbf{y}, \omega)]$ in the SCN design process and eventually select an efficient value-risk design. For instance, designs 1 and 2 in Fig. 11.9 have relatively close expected values but one of them may be less risky and it is important to evaluate this to make a final selection. As explained in Sect. 11.3.1, this leads to replacing objective function (11.2) by the weighted value-risk function (11.5) or to the addition of a maximum risk constraint $\rho_\Omega[V^{SCN}(\mathbf{y}, \omega)] \leq \rho^\circ$. In Sect. 2.4, we indicated that several risk measures were proposed in the literature but that downside risk measures are more appropriate in our context. More specifically, the most commonly used measures are

- The mean absolute lower semi-deviation from the mean or from a target

$$\rho_\Omega[V^{SCN}(\mathbf{y},\omega)] = \mathrm{E}_\Omega[(\mathrm{E}_\Omega[V^{SCN}(\mathbf{y},\omega)] - V^{SCN}(\mathbf{y},\omega))_+] \tag{11.29}$$

$$\rho_\Omega[V^{SCN}(\mathbf{y},\omega)] = \mathrm{E}_\Omega[(V^{T} - V^{SCN}(\mathbf{y},\omega))_+] \tag{11.30}$$

  where $(v)_+ = \max(v, 0)$ and V^T is a predetermined SCN value target.
- The mean lower semi-standard deviation

$$\rho_\Omega[V^{SCN}(\mathbf{y},\omega)] = \left(\mathrm{E}_\Omega[(\mathrm{E}_\Omega[V^{SCN}(\mathbf{y},\omega)] - V^{SCN}(\mathbf{y},\omega))_+^2]\right)^{1/2} \tag{11.31}$$

- Weighted mean deviations from an α-quantile

$$\begin{aligned}&\rho_\Omega[V^{SCN}(\mathbf{y},\omega)]\\ &= \min\{\mathrm{E}_\Omega[\max\{(1-\alpha)(z - V^{SCN}(\mathbf{y},\omega)), \alpha(V^{SCN}(\mathbf{y},\omega) - z)\}]\}\end{aligned} \tag{11.32}$$

  where $\alpha \in (0, 1)$. This measure can be shown to be equivalent to the conditional value-at-risk used in finance (Shapiro et al. 2009).
- For finite scenario sets, the maximum lower semi-deviation (worst case)

$$\rho_\Omega[V^{SCN}(\mathbf{y},\omega)] = \max_\Omega[(\mathrm{E}_\Omega[V^{SCN}(\mathbf{y},\omega)] - V^{SCN}(\mathbf{y},\omega))_+] \tag{11.33}$$

A weighted sum of the previous measures could also be used. Recent theoretical work in stochastic programming and financial engineering led to the identification of properties a risk measure should have to be a good quantifier of risk. A number of desirable convexity, monotonicity, translation equivalence, and positive homogeneity properties were identified, and measures satisfying these properties are known as *coherent risk measures* (Shapiro et al. 2009). It was shown that when (11.29), (11.31), or (11.32) are introduced in objective function (11.5) with $\lambda \in [0, 1]$, coherent risk measures are obtained. Measure (11.33) is not computable for continuous random variables but it can be estimated for a scenario sample. However, it is often considered to be too conservative. All the risk measures are easily calculated for a given feasible design $\mathbf{y}^o$. However, when added to the objective function or the constraints of an SCN network design model, (11.31) and (11.32) introduce significant additional complications. However, the use of (11.29) is relatively straightforward. For these reasons, the absolute downside semi-deviation from the mean is the more convenient risk measure to use in SCN design models. Also, with this measure, the SAA method applies as before.

To be more precise, when (11.29) is introduced in the objective function of generic design model (11.2)–(11.4), the SAA program to solve becomes

$$\max \bar{V}^N - \lambda \frac{1}{N} \sum_{\omega \in \Omega^N} S_\omega^- \tag{11.34}$$

$$\text{s.t.} \quad \hat{\mathbf{A}}\hat{\mathbf{x}}_\omega \leq \mathbf{b}(\mathbf{y}, \omega), \quad \omega \in \Omega^N \tag{11.35}$$

$$\hat{\mathbf{W}}\hat{\mathbf{x}}_\omega = \hat{\mathbf{d}}(\omega), \quad \hat{\mathbf{x}}_\omega \geq \mathbf{0}, \quad \omega \in \Omega^N \tag{11.36}$$

$$V_\omega^{\text{SCN}} = (\hat{\pi}(\omega) - \hat{\mathbf{c}}(\omega))\hat{\mathbf{x}}_\omega - \mathbf{f}(\omega)\mathbf{y}, \omega \in \Omega^N \tag{11.37}$$

$$\bar{V}^N = \frac{1}{N} \sum\nolimits_{\omega \in \Omega^N} V_\omega^{\text{SCN}} \tag{11.38}$$

$$\bar{V}^N - V_\omega^{\text{SCN}} - S_\omega^- \leq 0, S_\omega^- \geq 0, \quad \omega \in \Omega^N \tag{11.39}$$

Points on the value-risk efficient frontier (see Fig. 2.17) can then be found by applying the SAA method, using model (11.34)–(11.39), with different weights λ. The same approach can be used to introduce a risk constraint in the model instead of a weighted mean-risk objective function.

Example 11.5 To illustrate this, let us go back to Example 11.4, which was solved with the SAA method. Suppose also that we now want to include a maximum risk constraint in the model expressed in terms of a maximum mean absolute lower semi-deviation from the mean. The new SAA model to solve would then be

$$\bar{V}^N = \max ACF - \sum_{w \in L^{\text{W}}} y_w Y_w$$

with Ω replaced by Ω^N in (11.11)–(11.14) as before, and with the following additional cash flow and risk definition relations:

- Scenarios' operational cash flow definitions:

$$OCF_\omega = \sum_{t \in T} [\sum_{w \in L^{\text{W}}} \sum_{l \in L^{\text{D}}} (\pi_{lt}(\omega) - f_{wlt}(\omega))F_{wlt\omega} - \sum_{w \in L^{\text{W}}} c_{wt}^+ B_{wt\omega} - \sum_{l \in L^{\text{D}}} \pi_{wt}^- D_{lt\omega}^-], \omega \in \Omega^N$$

- Sample average operational cash flow definition:

$$ACF = \frac{1}{N} \sum_{\omega \in \Omega^N} OCF_\omega$$

– Scenarios' absolute lower semi-deviation definitions:

$$ACF - OCF_\omega - S_\omega^- \leq 0, S_\omega^- \geq 0, \ \omega \in \Omega^N$$

– Maximum downside risk constraint:

$$\frac{1}{N} \sum_{\omega \in \Omega^N} S_\omega^- \leq \rho^o$$

As before, by solving this SAA program with M independent samples $\Omega_j^N, j = 1, \ldots, M$, of N scenarios, $L \leq M$ distinct designs $(\bar{V}_j^N, \bar{\mathbf{y}}_j^N)_{j=1,\ldots,L}$ are obtained, where $\bar{\mathbf{y}}_j^N$ in this case is the vector of optimal binary DC selection variables (Y_w) defining design j. The best design $j*$ is determined by evaluating $\bar{V}^{N^+}(\bar{\mathbf{y}}_j^N)$ with (11.27) using an independent sample of $N^+ \gg N$ scenarios. The sample mean absolute lower semi-deviation of design $j*$ can then be estimated using

$$\begin{aligned}\bar{\rho}^{N^+}(\bar{\mathbf{y}}_{j*}^N) &= \frac{1}{N^+} \sum\nolimits_{\omega \in \Omega^{N^+}} \left(\bar{V}^{N^+}(\bar{\mathbf{y}}_{j*}^N) - V^{\text{SCN}}(\bar{\mathbf{y}}_{j*}^N, \omega) \right)_+, \ V^{\text{SCN}}(\bar{\mathbf{y}}_{j*}^N, \omega) \\ &= Q(\bar{\mathbf{y}}_{j*}^N, \omega) - \sum\nolimits_{w \in L^{\text{W}}} y_w Y_w\end{aligned}$$

The point $(\bar{V}^{N^+}(\bar{\mathbf{y}}_{j*}^N), \ \bar{\rho}^{N^+}(\bar{\mathbf{y}}_{j*}^N))$ thus obtained can be used to plot a value-risk efficient frontier. Other points are obtained by repeating the process with other ρ^o values. The easiest way to do this is first to solve the SAA program without the downside risk constraint and then to evaluate $\bar{\rho}^{N^+}(\bar{\mathbf{y}}_{j*}^N)$ for the solution obtained. If the risk involved is acceptable, we can stop the process and use the design found. Otherwise, a maximum risk value $\rho^o < \bar{\rho}^{N^+}(\bar{\mathbf{y}}_{j*}^N)$ is set and the SAA method is reapplied. If the estimated risk $\bar{\rho}^{N^+}(\bar{\mathbf{y}}_{j*}^N)$ of the new design found is acceptable, we can stop. Otherwise, we continue in this way until an acceptable value-risk tradeoff is found.

For our small example, when the SAA method is applied without a risk constraint, design 1 in Fig. 11.9 is obtained. Its value, estimated with a sample of 20 scenarios, is $\bar{V}^{20}(\bar{\mathbf{y}}_1^3) = €\ 13{,}508{,}410$, and its risk, estimated with the same scenario sample, is $\bar{\rho}^{20}(\bar{\mathbf{y}}_1^3) = €\ 233{,}160$, that is, only 1.7 % of $\bar{V}^{20}(\bar{\mathbf{y}}_1^3)$. This is relatively small and most executives would probably be satisfied with this design. To find an alternative, however, the design model can be solved with a maximum downside risk constraint of € 200,000. When this new model is solved with the SAA method, design 2 in Fig. 11.9 is obtained. The value and risk of design 2, estimated with the same 20 scenarios, are $(\bar{V}^{20}(\bar{\mathbf{y}}_2^3)$ = €12,808,736, $\bar{\rho}^{20}(\bar{\mathbf{y}}_2^3)$ = €128,408). The mean absolute lower semi-deviation estimated is 1 % of the design value. Knowing this, the decision maker can make a final decision.

11.6 Modeling Resilience Strategies

In Sect. 11.2, we stressed that to design resilient SCNs, one needs to incorporate an anticipation of user responses to random and hazardous events in design models and to include facility prepositioning constructs that lead to network structures facilitating the application of response policies. As seen in previous sections, when using stochastic programming, the anticipation of response policies leads to the inclusion of additional underage (e.g., missing capacity, unsatisfied demand, external supply) or overage (e.g., excess capacity) recourse variables, along with standard flow and state variables in the second-stage vectors $\hat{\mathbf{x}}_\omega, \omega \in \Omega^N$. Because resilience is concerned with short-term response to events, a better anticipation is obtained when short planning periods (months or quarters) are used. This is not always possible, however, because it increases the size of the design models considerably. As we shall see in Sect. 11.7, a compromise may be to use a more detailed user model when evaluating the candidate solutions found with design models.

As explained in Chap. 3 (see Fig. 3.10), SCN design models provide instructions to network users. These include the network structure but also precisions on facility missions. In a distribution context, for example, the mission of a DC may take the form of a list of products to stock and of ship-to points to serve. These instructions are not necessarily all explicitly modeled in the vector $\mathbf{y}$ of design variables. In Example 11.3, for example, the DCs' mission is implicit. When examining the solution of the second-stage program in Fig. 11.7, we can conclude from scenario 1 flows that the primary mission of the Düsseldorf DC should be to serve German, French, and Spanish ship-to points and for the Turin DC to serve Italy and Portugal. Also, from scenario 3 flows, we can conclude that Turin should serve as backup DC for France and Spain when the demand is stronger. In what follows, we use the notation $\mathbf{y}^+$ to represent all the instructions to network users derived from design variables $\mathbf{y}$ and second-stage variables $\hat{\mathbf{x}}_\omega, \omega \in \Omega^N$. To obtain more precise instructions, however, one may want to include explicit primary and backup customer assignment variables in vector $\mathbf{y}$. Additional first- and/or second-stage constraints based on response policy rules may also be added to enforce some required behavior.

To clarify this, the modeling of alternative response policies can be examined in a distribution network design context. Consider first a *single-sourcing first-come first-serve* response policy. Under this policy, each ship-to point is assigned to a single DC, and the orders received by a DC are queued and shipped as soon as capacity permits; that is, on a given day, if demand exceeds capacity, the unsatisfied orders are simply moved to the next day. When yearly planning periods are used, an approximate anticipation of this policy is provided by a multi-period variant of model (11.6)–(11.8) with no local capacity recourse variable. Because, under this policy, there is no need to transfer customer orders to other DCs or to an external supplier, the annual DC demand $d_{lt}(\omega)$ is simply the sum of all orders received during the year. The yearly capacity b_w must be interpreted as a maximum annual throughput when complying with this response policy. This revised model does not guarantee single sourcing, however. To do this, additional binary assignment variables $Z^{\mathrm{S}}_{wl} \in \{0,1\}, w \in L^{\mathrm{W}}, l \in L^{\mathrm{D}}$, must be added to the design vector $\mathbf{y}$, as explained in Sect. 7.3.3.6, and demand constraints (11.7) must be replaced by

$$\sum_{w \in L^{\mathrm{W}}} Z^{\mathrm{S}}_{wl} = 1, \quad l \in L^{\mathrm{D}} \tag{11.40}$$

$$F_{wl\omega} - d_l(\omega) Z^{\mathrm{S}}_{wl} = 0, \quad w \in L^{\mathrm{W}},\ l \in L^{\mathrm{D}},\ \omega \in \Omega \tag{11.41}$$

This response policy may lead to serious order shipment delays and to poor resiliency. An alternative is a *dual-sourcing next-day-delivery* policy. Under this policy, orders received on a given day must be shipped to customers at the beginning of the next day from a DC sufficiently close from the ship-to point to be able to make the delivery during that day under usual conditions. On a given day, the orders received from a ship-to point in demand zone $l \in L^{\mathrm{D}}$ are assigned to its primary DC, denoted $w(l)$, while capacity permits (including available local recourse capacity). When demand exceeds capacity, excess demand is transferred to the backup DC,[7] denoted $w'(l)$. To be able to do this, each ship-to point is pre-assigned to a *primary* DC and to a *backup* DC. The additional difficulty here is to determine $w(l)$ and $w'(l)$ for all $l \in L^{\mathrm{D}}$. Incorporating these decisions explicitly in the design model and providing an exact anticipation of user response is almost impossible. An approximate design model capturing the essence of the problem must be used. The multi-period design model (11.10)–(11.14) provides a good starting point to do this. Note that to capture the impact of this response policy precisely, daily time periods should be used. This would lead to extremely large models, and a first approximation that needs to be made is the use of longer time periods[8] (say, months). Four alternative approaches, of increasing complexity, to determine primary and backup DCs will now be examined. To illustrate their impact on network structures, we use a case dealing with the distribution of soap products in the northeast United States studied in detail in Klibi and Martel (2012). This case is presented graphically in Fig. 11.10, which shows the 15 potential DC

[7] We assume here to simplify that the backup DC always has sufficient capacity (including local recourse capacity) to cover transferred orders. The policy, however, can be extended to involve a third DC or an external supplier if required.

[8] This can be done by aggregating or sampling the daily period. Klibi et al. (2015) show that period sampling gives better results.

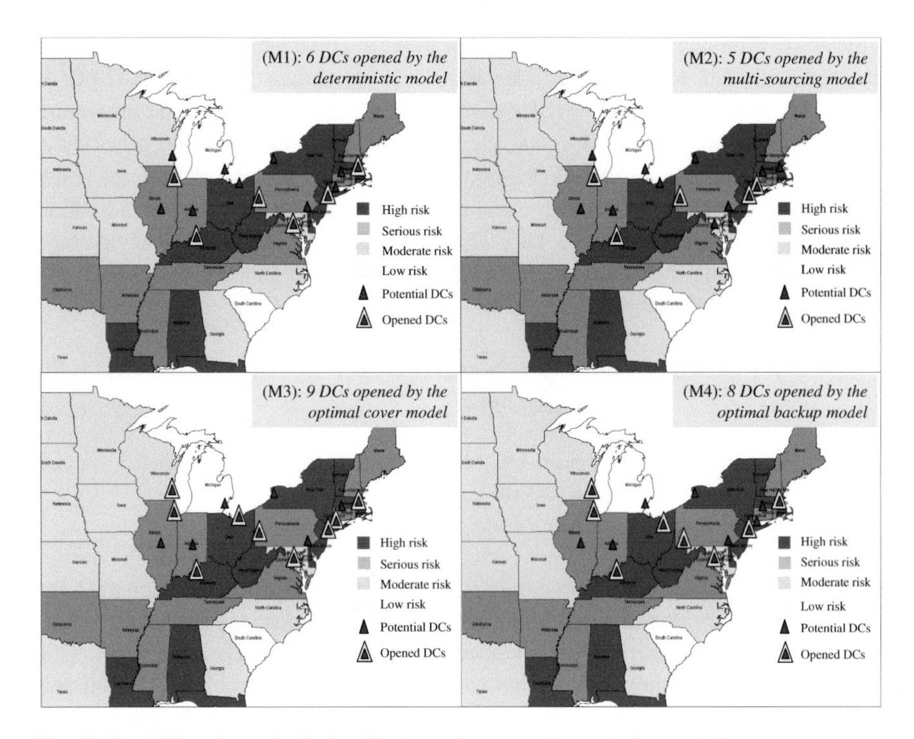

Fig. 11.10 SCN designs obtained with alternative approximate design models

locations considered as well as the states where the 706 demand zones serve are located and their risk exposure level.

Average Scenario Deterministic Model (M1)

The simplest way to find an SCN design is to solve model (11.10)–(11.14) with a single average scenario. As explained in Sect. 7.3.3.1, all network arcs that are too long to be able to provide next-day delivery are eliminated from the model a priori. The solution of the model gives the network structure. For the case considered, the six DCs shown in quadrant (M1) of Fig. 11.10 were selected. The primary DC $w(l)$ for demand zone l is the one with the maximum horizon inflow ($\max_w \Sigma_{t \in T} F_{wlt}$). The backup DC $w'(l)$ is the one with the second largest horizon inflow. If all the zone l demand is satisfied by a single DC in the optimal solution, then $w'(l)$ is the closest DC, excluding $w(l)$. Note that when this model is used, the generic design vector $\mathbf{y}$ corresponds to binary location variables. However, the generic instruction vector $\mathbf{y}^+$ includes these location decisions as well as the DC mission decisions derived, as indicated previously, from the flow variables.

Multi-sourcing Stochastic Model (M2)

A first alternative is to solve model (11.10)–(11.14) with the SAA method as shown previously. Again, arcs that are too long to be able to provide next-day delivery are eliminated from the model. For the case considered, the optimal design obtained with this approach includes five DCs, as shown in quadrant (M2) of Fig. 11.10. The primary DC $w(l)$ for demand zone l is the one with the maximum sample average horizon inflow ($\max_w \frac{1}{N'}\Sigma_{\omega\in\Omega^{N'}}\Sigma_{t\in T}F_{wlt\omega}$). As discussed previously, for some demand zones l, it is possible that the demand is satisfied from the same DC for all scenarios ω considered, meaning that the solution provides no indication of how to help select a backup DC. To avoid this, one can add a second-stage constraint requiring that the flows from a DC to a demand zone never exceed, say, 90 % of the zone demand:

$$\sum_{\omega\in\Omega^{N'}}\sum_{t\in T}F_{wlt\omega}\leq 90\%\sum_{\omega\in\Omega^{N'}}\sum_{t\in T}d_{lt}(\omega),\quad w\in L^{\mathrm{W}}, l\in L^{\mathrm{D}} \tag{11.42}$$

When this is done, the backup DC $w'(l)$ selected can be one with the second largest sample average horizon inflow.

Maximum Covering Stochastic Model (M3)

The two previous models do not include any construct to try to force the design to include good backups. The previous models were constructed, however, to ensure that primary depots would all be located within a distance sufficiently short to permit next-day delivery. The same idea can be applied to the backup depot. This gives rise to what is known as a maximal covering location problem (Church and Revelle 1974). Let $L_l^1\subseteq L^{\mathrm{W}}$ be the set of DCs located within the next-day delivery distance required for ship-to point l, and $L_l^2\subseteq L^{\mathrm{W}}$ be the set of DCs located within the backup distance specified, the latter being greater than the required primary DC distance so that $L_l^1\subset L_l^2$. The idea is to impose for each ship-to point $l\in L^{\mathrm{D}}$ that at least one depot is in L_l^1 and at least two depots are in L_l^2. This is done by adding the following constraints to the first stage program:

$$\sum_{w\in L_l^1}Y_w\geq 1,\quad l\in L^{\mathrm{D}} \tag{11.43}$$

$$\sum_{w\in L_l^2}Y_w\geq 2,\quad l\in L^{\mathrm{D}} \tag{11.44}$$

This tends to increase the number of opened DCs and to spread them more evenly in the territory. When applying this approach to our illustrative case, nine DCs are selected as shown in quadrant (M3) of Fig. 11.10. As expected, this model opens more DCs than the two previous ones. As for the previous case, the primary DC $w(l)$ for demand zone l is the one with the maximum sample average horizon inflow ($\max_w \frac{1}{N'}\Sigma_{\omega\in\Omega^{N'}}\Sigma_{t\in T}F_{wlt\omega}$). The backup DC $w'(l)$ is the closest one excluding the primary depot.

Stochastic Backup DC Selection Model (M4)

A fourth alternative is to optimize the backup location explicitly. This requires, as for the single-sourcing policy previously examined, the explicit introduction of primary DC assignment variables $Z_{wl}^{S} \in \{0,1\}, w \in L^{W}, l \in L^{D}$. It also requires the definition of explicit backup DC assignment variables $Z_{wl}^{B} \in \{0,1\}, w \in L^{W}$, $l \in L^{D}$. The variables, $Z_{wl}^{r}, r = S, B$, take the value 1 if demand zone l is allocated to DC w as level r (primary or backup) DC and 0 otherwise. In the model, these mission specification variables are first-stage variables and thus they remain the same for all the planning horizon. However, for each period t of scenario ω, second-stage binary variables also need to be introduced. The resulting stochastic program is thus much more difficult to solve than the previous ones. It can be solved only if the number of periods and scenarios considered are relatively small, which trades an improvement in mission assignment precision for a reduction in user anticipation precision. The advantage of the approach is that the primary and backup depots are directly given by the binary variables $Z_{wl}^{r}, r = S, B$. When applied to our case example, this approach gives the eight DCs shown in quadrant (M4) of Fig. 11.10. Note that when this design model is used, the generic design and instruction vectors $\mathbf{y}$ and $\mathbf{y}^{+}$ are identical.

As can be seen, the formulation of the design model has a significant impact on the network structure and DC missions obtained. This stresses the importance of evaluating candidate designs with adequate value, risk, and resilience measures before making a final decision, as discussed in Sect. 11.2.2. For the case studied, a value and a risk measure were estimated based on a more precise user anticipation than embedded in each individual design model (more on this in the next section). These two measures are

- The SAA estimate of the expected value added by the design (11.27)
- The SAA estimate of the absolute lower deviation from the mean (11.29)

They are plotted in Fig. 11.11 for the four designs obtained with the previous modeling approaches. The graph show that from a value-risk point of view, model (M4) is very conservative: it provides a design with a lower expected value but with less risk. At the other extreme, models (M1) and (M2) are more aggressive: they provide higher value designs but with more risk. Model (M3) is a compromise between these two extremes. Note, however, that for all models the downside risk of the designs obtained is relatively low in comparison with the expected value. In other words, for the case considered, risk does not stand out as a strong discriminating factor.

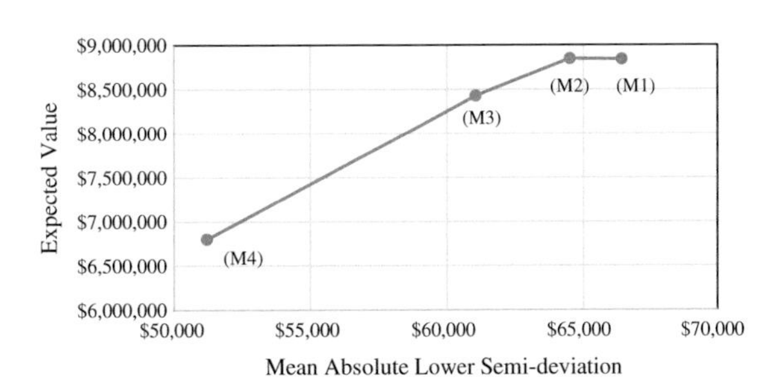

Fig. 11.11 Approximate design model results comparison

11.7 Generation and Selection of Robust Designs

Our previous discussions led us to believe that SCN design problems ideally should be formulated as multi-stage stochastic programs based on explicit anticipations of adaptation-response decisions and a multi-objective reward function. Tremendous progress has been made in recent years for the solution of multi-stage stochastic mixed-integer programs using exact and heuristic methods. Unfortunately, for realistic SCN design problems including continuous random variables, the stochastic programs to solve are intractable. Complexity-reduction methods must thus be used to obtain simpler models capturing the essence of the problem but solvable with current optimization software or heuristic methods. Our purpose in this concluding section is to look at the SCN design problem in general and to propose accuracy-solvability tradeoffs likely to yield robust value-creating SCNs. The design methodology proposed is summarized in Fig. 11.12. It includes three phases: scenario generation, design generation, and design evaluation and selection. Each phase is discussed in the following subsections, but it should be clear at the outset that the aim of the design generation phase (2) is to obtain good candidate designs $\mathbf{y}_l^+, l = 1, \ldots, L$, using simplified approximate design models that can be solved relatively easily and not to try to find the best design using a monolithic all-encompassing model. It should also be clear that in the evaluation and selection phase (3) the candidate designs considered are evaluated using more exhaustive and precise anticipations and evaluation criteria than in the design models.

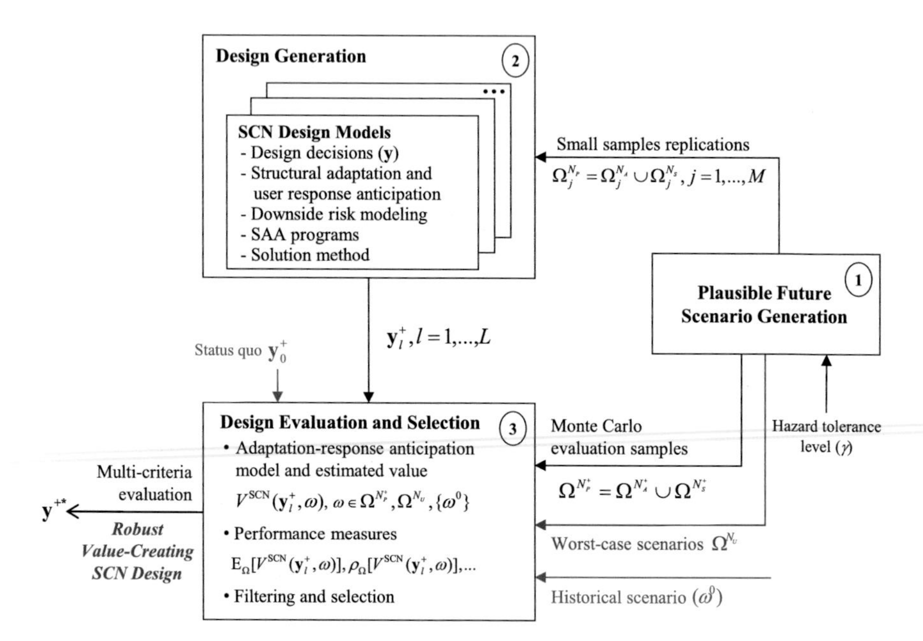

Fig. 11.12 Robust value-creating SCN design methodology

11.7.1 Design Generation Phase

Four complexity reduction avenues can be used to formulate solvable SCN design models. The first one involves the use of reduced design vectors ($\mathbf{y}$ instead of $\mathbf{y}^+$) and the incorporation of approximate user response anticipations in the design model. In the models studied previously, user decisions are anticipated using continuous throughput, flow, state, and/or local recourse variables. These anticipation variables are typically aggregates over several products, customers, transportation means, and operational periods. In some context they may correspond to tactical planning variables. They may also be used as explained in Sect. 11.6 to derive instructions on the mission of facilities. Klibi et al. (2015) studied various approximate anticipations for stochastic location-transportation problems based on different representations of demand (average demand instead of multiple scenarios), transportation decisions (origin–destination flows instead of delivery routes), customers (demand zones instead of ship-to points), and time (planning periods instead of operational periods). Such approximations can transform highly combinatorial stochastic models into stochastic programs with continuous recourse variables such as those studied in previous sections or even into deterministic design models such as those studied in previous chapters. The Klibi et al. (2015) results show that, although significant gains can be made by using more precise anticipations, the modeling and computation complexity they add is significant and some tradeoffs must be made. Ideally, the approximate user response submodel used should be a linear program.

A second complexity-reduction avenue is to disregard some SCN design objectives and to consider only primordial expected value and risk aversion criteria associated with probabilistic scenarios. Because the recourses included in the anticipation submodel are usually very expensive, stochastic programs tend to eliminate any extreme behavior, which naturally reduces variability even if risk measures are not included in the objective function. For this reason, it is often sufficient to assume that the decision makers are risk neutral and to formulate the objective function only in terms of expected value. If risk needs to be considered explicitly, one should limit this to the inclusion of a risk measure based on mean absolute downside deviations such as (11.29) and, when multi-hazards are considered, to the distinction between acceptable and serious risk scenarios, as discussed in Sect. 10.5.2.

The dynamics of the multi-stage decision structure described in Fig. 11.1 is another important source of complexity. When several reengineering cycles are considered, first-stage design decisions $\mathbf{y}_1$ are made here and now but subsequent structural adaptation decisions, $\hat{\mathbf{y}}_h, h > 1$, take the form of decision rules depending on the history up to the beginning of reengineering cycle h. When this is taken into account explicitly, the size of the problem tends to blow up. However, because design decisions are made on a rolling horizon basis, decision rules $\hat{\mathbf{y}}_h, h > 1$, are never used. In our context, they are essentially an anticipation mechanism. Moreover, because the planning period residual cash flows are discounted, the weight of structural adaptation decisions decreases as h increases. Under these conditions, a third complexity-reduction mechanism is to assume that the decisions $\hat{\mathbf{y}}_h, h > 1$, must be made at the beginning of the planning horizon, and thus that they cannot depend on the history up to cycle h. This transforms multi-stage design models into multi-cycle two-stage SPRs, the structural adaptation variables $\hat{\mathbf{y}}_h, h > 1$, being all considered as first-stage decision variables. Despite this simplification, adding reengineering cycles increases the size of the problem significantly. For this reason, one generally considers only two or three reengineering cycles. The second and third cycles can include several planning periods to provide a better anticipation of structural adaptation impacts.

Another difficulty is that when some random variables in the model have normal, exponential, or Poisson distributions, which is usually the case in practice, the number of plausible future scenarios is infinite and the probabilities $\mathrm{p}(\omega), \omega \in \Omega^{\mathrm{P}}$, cannot be estimated. In order to obtain solvable stochastic programs, one needs to limit the number of scenarios considered and to find a way to calculate their probability. This can be done by replacing the population set Ω^{P} in the design model by a representative Monte Carlo sample Ω^{N_P} of N_P equiprobable scenarios. As seen for the SAA method in Sect. 11.4, to get better designs, the model can also be solved for M independent scenario samples $\Omega_j^{N_P}, j = 1, \ldots, M$.

When all the previous complexity-reduction recommendations are applied, the approximate multi-cycle SCN design models to solve, assuming that yearly planning periods $t \in T$ are used, have the following form:

$$\max \bar{V}^{N_P} = \frac{1}{N_P} \sum_{\omega \in \Omega^{N_P}} V_\omega^{\mathrm{SCN}} \tag{11.45}$$

$$\text{s.t.} \quad V_\omega^{\mathrm{SCN}} = \sum_{t \in T} [(\hat{\pi}_t(\omega) - \hat{\mathbf{c}}_t(\omega))\hat{\mathbf{x}}_{t\omega} - \mathbf{f}_t(\omega)\mathbf{y}_{h(t)}]/(1+r)^t, \quad \omega \in \Omega^{N_P} \tag{11.46}$$

$$(\mathbf{y}_0, \mathbf{y}_1) \in \mathcal{Y}_1, \tag{11.47}$$

$$(\hat{\mathbf{y}}_{h-1}, \hat{\mathbf{y}}_h) \in \mathcal{Y}_h, h > 1 \tag{11.48}$$

$$\hat{\mathbf{A}}_t \hat{\mathbf{x}}_{t\omega} \leq \mathbf{b}_t(\mathbf{y}_{h(t)}, \omega), \quad t \in T, \omega \in \Omega^{N_P} \tag{11.49}$$

$$(\hat{\mathbf{x}}_{t-1,\omega}, \hat{\mathbf{x}}_{t\omega}) \in \mathcal{X}_t(\omega), \quad t \in T, \ \omega \in \Omega^{N_P} \tag{11.50}$$

where $h(t)$ is the index of the reengineering cycle containing period t, and r is the weighted average cost of capital of the company. As seen in previous chapters, the set of feasible designs $\mathcal{Y}_1$ is defined by platform, offer, and supplier-selection constraints and by cycle-to-cycle change of state relationships, and so is the feasible structural adaptation decisions set $\mathcal{Y}_h$ for $h > 1$. Constraints (11.49) specify relationships between the first-stage variables $(\mathbf{y}_1, \hat{\mathbf{y}}_2, \ldots, \hat{\mathbf{y}}_{|H|})$ and the second-stage recourse variables $\hat{\mathbf{x}}_{t\omega}, t \in T, \omega \in \Omega^{N_P}$. Finally, $\mathcal{X}_t(\omega)$ is defined for scenario ω by period-to-period inventory accounting relations and by flow equilibrium and recourse restriction constraints.

If the decision maker is risk averse, then objective function (11.45) can be replaced by

$$\max \bar{V}^{N_P} - \lambda \frac{1}{N_P} \sum\nolimits_{\omega \in \Omega^{N_P}} \left(\bar{V}^{N_P} - V_\omega^{\mathrm{SCN}} \right)_+, \ \lambda \in [0,1] \tag{11.51}$$

where, $\left(\bar{V}^{N_P} - V_\omega^{\mathrm{SCN}} \right)_+$, the downside semi-deviation from the sample mean $\bar{V}^{N_P}$ for scenario ω, can be calculated using a relation similar to (11.39). A maximum risk constraint could be used instead, as explained in Sect. 11.5. When multi-hazards are considered, if the decision maker is not overly concerned by business-as-usual variations but would like the SCN design to be resilient when catastrophic events occur, then the sample of scenarios generated Ω^{N_P} can be split into a subset of acceptable-risk scenarios Ω^{N_A} and a subset of serious-risk scenarios Ω^{N_S}, using a predetermined hazard tolerance level γ, as explained in Sect. 10.5.2.3. Objective (11.45) can then be replaced by an objective function based on weighted conditional expected values $\mathrm{E}_{\Omega^{N_A}|A}[V^{\mathrm{SCN}}(\mathbf{y}_1, \omega)]$ and $\mathrm{E}_{\Omega^{N_S}|S}[V^{\mathrm{SCN}}(\mathbf{y}_1, \omega)]$. When sample average estimates are used, this objective function is written

$$\max \frac{w_A}{N_A} \sum_{\omega \in \Omega^{N_A}} V_\omega^{\mathrm{SCN}} + \frac{w_S}{N_S} \sum_{\omega \in \Omega^{N_S}} V_\omega^{\mathrm{SCN}} \tag{11.52}$$

where $0 \le w_A \le 1$, $w_S = 1 - w_A$, are subjective weights specified by the decision maker. Note that when $w_A = \mathrm{p}_A = \Sigma_{\omega \in \Omega^A} \mathrm{p}(\omega)$, the probability that an acceptable-risk scenario will occur, (11.52) is equivalent to (11.45). The probability p_A is usually not known, but it can be estimated using Monte Carlo methods (Klibi and Martel 2013). A decision maker averse to extreme events should then use (11.52) with a weight $w_A < \mathrm{p}_A$. Changing risk-attitude weights λ or (w_A, w_S) is easy and, because these weights are not hard data, varying them is, as shown in Sect. 11.5, an adequate approach to generate alternative candidate designs.

Example 11.6 To illustrate how candidate designs are generated, let's go back to the North American soap distribution case introduced in Sect. 11.6 and exemplified in Fig. 11.10 for different designs under a dual-sourcing next-day delivery response policy. Suppose that to generate candidate designs we decide to use two versions of a multi-cycle multi-sourcing stochastic design model with a conditional expected value objective function akin to (11.52) and monthly planning periods. A 5-year planning horizon with two reengineering cycles (the first covering 24 months, and the second 36) is considered. One version of the model includes a constraint similar to (11.42) to ensure that all customers are served by at least two DCs (with maximum flows of 60 % of the zone demand), and the other does not. When these two

Table 11.2 Candidate designs generated in phase 2 of the design methodology

	Design 1	Design 2	Design 3	Design 4	Design 5	Design 6	Design 7	Design 8
No. of DCs	4	4	4	4	5	5	5	5
DC	% of demand zones assigned to DC							
1					12.1	11.7	8.7	12.6
2				33.5	41.3	32.0	32.0	49.0
3	20.9	18.9	21.4		19.4	30.1	24.3	18.0
4	17.0	16.5	17.0	17.0	8.3	8.7	11.7	11.2
5	34.5	33.0	34.0	22.8	18.9	17.5	23.3	9.2
6								
7	27.7	31.6	27.7	26.7				
% single-sourcing	1.94	3.88	8.25	10.19	0	0	0	0

models are each solved for four distinct samples of 30 scenarios and with weights $w_A = 0.44$ and $w_S = 0.56$, the eight candidate designs described in Table 11.2 are obtained. For each design, the table gives the number of opened DCs, the percentage of demand zones assigned to opened DCs, and finally the percentage of demand zones served by a single DC. Three distinct network structures (**y**) are obtained (two involving four DCs and one involving five DCs) and DCs 3, 4 and 5 are present in all of them. However, as shown by the assignment percentages, for the designs with the same network structure, the opened DCs' mission (and thus $\mathbf{y}^+$ under a dual-sourcing next-day-delivery policy) is not the same.

11.7.2 Design Evaluation and Selection Phase

The aim of the design evaluation phase is to select the best SCN design among those generated ($\mathbf{y}_l^+, l = 1, \ldots, L$) and to compare them to the status quo, $\mathbf{y}_0^+$, first using a sample of Monte Carlo scenarios $\Omega^{N_P^+} = \Omega^{N_A^+} \cup \Omega^{N_S^+}$ much larger than the samples used to generate candidate designs. In order to test the robustness of the designs, executives also typically want to see how they would perform for a selected set of extreme scenarios Ω^{N_U} possibly including deeply uncertain events. Finally, they typically also want to compare candidate designs for a recent historical scenario ω^0. These evaluations need to be made for the instruction vectors $\mathbf{y}_l^+, l = 1, \ldots, L$, derived from the solutions **y** of the approximate design models solved for scenario samples $\Omega_j^{N_P}, j = 1, \ldots, M$ and not for the **y**'s directly as in the SAA method. Also, this evaluation should be based on an operational response model as close as possible to the real user decision processes and not on a second-stage SPR model as in the SAA method.

When a single reengineering cycle is considered, it may be possible to evaluate designs with the real operational decision processes of the company, using its ERP/APS systems (see Sect. 3.2.3), for example. When $|H| > 1$, however, the fact that the network structure may change at the beginning of subsequent reengineering cycles should be taken into account. Using the operational response systems of the company is then difficult because the resulting joint adaptation-response decision process is too complex. A feasible fallback position is to make the evaluation using an approximate adaptation-response model similar to those embedded in SAA programs. However, these models are much simpler than SAA programs because they are solved separately for each scenario and because binary design variables are predetermined. For this reason, to improve the evaluation process, it is usually possible to formulate and solve much more precise adaptation-response models than those embedded in SAA programs. For example, to evaluate transportation costs more precisely, daily periods can be used instead of years, and daily routing decision variables can be included in the model instead of approximate flow variables. Independent of the approach selected to evaluate the value added by a design $\mathbf{y}^+$ under a plausible future scenario ω, the estimated value added is denoted $VA(\mathbf{y}^+,\omega)$, as opposed to $V^{\text{SCN}}(\mathbf{y},\omega)$ in the SAA method [see Eq. (11.27) and second-stage program (11.25)–(11.26)] to reflect the fact that in the evaluation phase adaptation-response decisions can be anticipated more precisely than in the design models, even if an exact anticipation is usually not possible.

The essence of the evaluation phase is to compare the candidate designs $(\mathbf{y}_0^+,\mathbf{y}_1^+,\ldots,\mathbf{y}_L^+)$ using appropriate performance measures. As indicated previously, in order to evaluate the value, risk, resilience, and robustness of the designs adequately, the battery of performance measures required is much more elaborate than those embedded in the objective function of any of the design models solved. These measures are based on those discussed in the previous sections but, for a given design $\mathbf{y}^+$, they are estimated using large samples of probabilistic and worst-case scenarios. The following expected value and risk measures based on the Monte Carlo scenarios $\Omega^{N_P^+} = \Omega^{N_A^+} \cup \Omega^{N_S^+}$ are commonly used:

- The expected value

$$\overline{VA}_P(\mathbf{y}^+) = \frac{1}{N_P^+} \sum_{\omega \in \Omega^{N_P^+}} VA(\mathbf{y}^+,\omega) \tag{11.53}$$

- The mean absolute lower semi-deviation from the mean

$$MSD_P(\mathbf{y}^+) = \frac{1}{N_P^+} \sum_{\omega \in \Omega^{N_P^+}} \left(\overline{VA}_P(\mathbf{y}^+) - VA(\mathbf{y}^+,\omega)\right)_+ \tag{11.54}$$

- The combined value-risk measure

$$VR_P(\mathbf{y}^+) = \overline{VA}_P(\mathbf{y}^+) - \lambda MSD_P(\mathbf{y}^+), \quad \lambda \in [0, 1] \tag{11.55}$$

When multi-hazards are considered, these measures can be replaced by the following weighted expected value, semi-deviation, and value-risk measures:

$$WV(\mathbf{y}^+) = w_A\overline{VA}_A(\mathbf{y}^+) + w_S\overline{VA}_S(\mathbf{y}^+) \tag{11.56}$$

$$WD(\mathbf{y}^+) = w_A MSD_A(\mathbf{y}^+) + w_S MSD_S(\mathbf{y}^+) \tag{11.57}$$

$$WVR(\mathbf{y}^+) = w_A VR_A(\mathbf{y}^+) + w_S VR_S(\mathbf{y}^+) \tag{11.58}$$

where the conditional expected values and $\overline{VA}_S(\mathbf{y}^+)$, the semi-deviations $MSD_A(\mathbf{y}^+)$ and $MSD_S(\mathbf{y}^+)$, and the value-risk measures $VR_A(\mathbf{y}^+)$ and $VR_S(\mathbf{y}^+)$ are estimated by replacing sample $\Omega^{N_P^+}$ by $\Omega^{N_A^+}$ or $\Omega^{N_S^+}$ in (11.53), (11.54) and (11.55), respectively.

To assess the robustness of the designs in difficult conditions, an extreme scenarios sample Ω^{N_U} is used. One then wants to determine the lowest design value (worst-case) under all the disastrous plausible futures considered, that is,

$$LV(\mathbf{y}^+) = \min_{\omega \in \Omega^{N_U}} VA(\mathbf{y}^+, \omega) \tag{11.59}$$

The previous measures can also be combined to obtain a single multi-criteria measure. A natural combined performance measure is given by

$$CM(\mathbf{y}^+) = (1-\theta)WVR(\mathbf{y}^+) + \theta LV(\mathbf{y}^+) \tag{11.60}$$

where $0 \le \theta \le 1$ is the weight given to performances under worst-case scenarios.

Example 11.6 (continued) Return to the previous US soap company case for which the eight designs in Table 11.2 were obtained. Which of these designs should be selected? Although only three distinct network structures were found, the assignment of ship-to points to depots (and thus the DCs' mission) is very sensitive to the sample of scenarios used, and a more extensive evaluation is required to make a final decision. Let's say that it was decided to use 100 Monte Carlo scenarios and five worst-case scenarios in the evaluation phase to reach a final selection. It was also decided to base the evaluation of the candidate designs on a more accurate user response anticipation model than included in the SAA programs solved in the design-generation phase. Daily periods are used instead of months and a detailed user-response procedure (based on Klibi and Martel 2012) is applied to solve daily location-transportation problems under disruptions. It combines a method to reassign customer orders under disruptions and a heuristic to solve transportation problems. This procedure is employed to calculate the value added over the planning horizon, $VA(\mathbf{y}_l^+, \omega)$, for candidate designs $(\mathbf{y}_l^+, l = 1, \ldots, 8)$ under all 105 scenarios.

The values $VA(\mathbf{y}_l^+, \omega)$ estimated can then be used to compute several performance measures as shown in Table 11.3. The two first columns provide acceptable and serious risk sample averages, $\overline{VA}_A(\mathbf{y}^+)$ and $\overline{VA}_S(\mathbf{y}^+)$, and they show that *design 4* perform better. From an expected value point of view, it is clear that *designs 1 to 4* are more efficient than the others, probably because the model used to generate them is less constrained (*multiple-sourcing* is optional, not mandatory). However, the two following columns provide mean semi-deviations $MSD_A(\mathbf{y}^+)$ and $MSD_S(\mathbf{y}^+)$, and they show that *designs 5 to 8* provide more stable returns. The value-risk plot (based on $WV(\mathbf{y}^+)$ and $WD(\mathbf{y}^+)$) on the left of Fig. 11.13 makes it clear that *design 4* provides the best value but with more risk, and that *design 8* has the smallest risk but with the lowest value. The next column in Table 11.3 gives the

Table 11.3 Performance measures for the eight candidate designs

Design	$\overline{VA}_A$	$\overline{VA}_S$	MSD_A	MSD_S	LV	WVR	CM
1	28,270,972	27,452,966	1,261,509	1,585,055	16,414,150	27,957,158	26,807,100
2	28,027,940	27,243,454	1,246,384	1,568,418	16,287,382	27,731,300	26,590,972
3	28,342,152	27,522,026	1,264,443	1,590,121	16,458,371	28,027,564	26,874,898
4	28,797,070	28,204,780	1,316,189	1,614,736	16,760,006	28,613,725	27,431,390
5	25,280,400	24,787,532	1,212,189	1,496,112	14,092,717	25,141,513	24,039,142
6	25,639,664	25,119,228	1,218,849	1,513,225	14,307,558	25,486,590	24,371,338
7	24,891,280	24,404,660	1,206,993	1,488,181	13,802,253	24,755,219	23,662,398
8	25,081,406	24,594,560	1,150,338	1,422,519	14,134,633	24,939,048	23,861,088

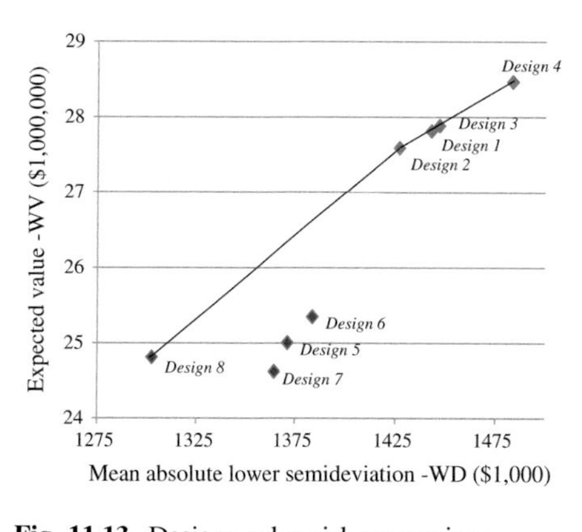

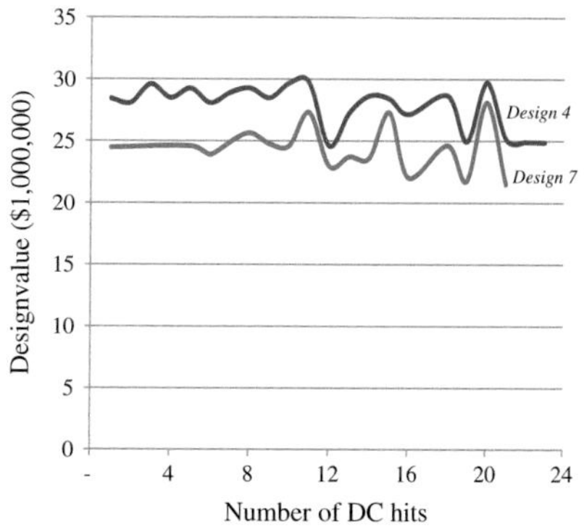

Fig. 11.13 Designs value-risk comparison

lowest value $LV(\mathbf{y}^+)$ obtained for the five worst-case scenarios generated. For this criterion, *design 7* is the best and *designs 5 to 8* dominate *designs 1 to 4*. The two last columns of the table give combined measures $WVR(\mathbf{y}^+)$ and $CM(\mathbf{y}^+)$ calculated with weights $\lambda = \theta = 0.1$. They both favor *design 4,* which thus seems to be a good compromise.

The analysis can be pushed further first by examining what happens when the risk-aversion weights (λ, θ) are increased. When this is done, *design 4* still has the best $CM(\mathbf{y}^+)$ value. The resilience of the designs obtained can be analyzed by examining their behavior when hit by natural disasters. The right plot in Fig. 11.13 presents a value-hit graph for *design 4* and *design 7*: it gives their average value for all scenarios including $0, 1, 2, \ldots$ hits. In terms of average value creation under a given number of hits, *design 4* dominates *design 7*. However, *design 7* performs better than *design 4* in terms of value variability (*MSD*). Robust designs should not only perform well when hit but also should minimize downside risk. Taking all this into account, a very conservative decision maker may prefer *design 8,* but, noting that the *MSD*s are relatively small when compared to expected values (about 5 %), most executives would select *design 4*.

Example 11.6 showed how to use expected value, downside semi-deviation, and worst-case performance measures to select a final SCN design. In some cases other attributes of the solutions can provide additional guidance. For example, to assess resilience, one could calculate the weighted sum of the average distance between clients and their backup DC for the response policy considered. Other attributes, such as structural connectivity (Dong and Chen 2007), operational flexibility (Tang and Tomlin 2008), or network fragility (Lim et al. 2010), can be captured through the calculation of additional performance indicators. Following this evaluation, a multi-criteria decision making (Figueira et al. 2005) or filtering approach can be used to select a best robust value-creating SCN design $\mathbf{y}^{+*}$. Filtering eliminates dominated solutions a priori and then the selection can be done through further analysis of remaining designs, including the status quo design $\mathbf{y}_0^+$. In practice, a final selection among remaining designs is often based on caparisons for a historical scenario ω^0 and a sample of extreme scenarios.

11.7.3 Scenario Generation Phase

Plausible future scenario samples are required in phases (2) and (3) of the design methodology presented in Fig. 11.12. As explained in Chap. 10, scenarios are juxtapositions of random, hazardous, and deeply uncertain events over the planning horizon, and they are shaped by evolutionary paths. In Sect. 10.5.2, we saw how

Monte Carlo methods can be used to generate probabilistic scenarios including random and hazardous events. It is important to use sufficiently large scenario samples to ensure that each evolutionary path is well represented. If the scenario samples embedded in phase 2 design models are small, to obtain an adequate representatively it may be necessary to use hierarchical sampling methods (Ducapova et al. 2000). We also saw how a hazard tolerance level γ can be used to partition scenario samples into acceptable and serious risk subsamples (see Fig. 10.29). The probability p_A of having an acceptable risk scenario can be estimated simply by counting the proportion of acceptable risk scenarios in the generated sample. This proportion then provides guidelines for the subjective specification of extreme events aversion weights (w_A, w_S).

We saw that, in the evaluation phase, we may want to use deep uncertainty scenarios to assess the robustness of the designs under extreme events. As pointed out at the end of Sect. 10.5.2, decision makers interested in deep uncertainty scenarios are driven mainly by their desire to examine worst-case scenarios including unusual extreme events. Such scenarios can be elaborated first by identifying potential new SCN threats through structured brainstorming sessions and/or expert interviews (van der Heijden 2005) and then by expressing them quantitatively in terms of capacity, demand, and other SCN design parameters. Because these scenarios would necessarily include probabilistic events, they can be constructed by perturbing worst-case scenarios in the tail of the distribution of the number of hits (see Fig. 10.29) using deep uncertainty events imagined by experts. More specifically, in order to obtain a set Ω^{N_U} of worst-case scenarios, a large sample $\Omega^{N_P^+}$ of probabilistic scenarios is generated and a subset of tail scenarios is selected. These are then taken as is or modified manually by adding extraordinary imaginative events.

Review Questions

11.1. Why is SCN design considered to be a decision problem under risk?
11.2. How would you define the concepts of robustness and resilience? Explain the links between them.
11.3. What is the difference between a response policy and a resilience strategy?
11.4. How should we measure risk in SCN design?
11.5. How would you explain the role of each of the three phases of the SCN design methodology proposed in Sect. 11.7? Describe the benefits of such an approach.
11.6. How would you explain the differences between the multi-sourcing and the coverage models proposed to foster resilience in Sect. 11.6? Discuss their advantages and disadvantages.

Exercises

Exercise 11.1 Consider the distribution network presented in Example 11.1 but assume now that the yearly demand in each country follows a Normal distribution

Table 11.4 Exercise 11.1 data

	Demand				
	Germany	France	Italy	Spain	Portugal
Mean	15,000	17,500	10,000	17,500	25,000
Standard deviation (%)	10	5	12	15	8

Table 11.5 Value added (in €) by three candidate designs for 20 scenarios

	Design 1	Design 2	Design 3
Scenario 1	12,935,400	7,493,300	12,007,640
Scenario 2	−2,134,500	−8,766,000	−1,366,000
Scenario 3	11,670,475	5,414,500	11,902,680
Scenario 4	12,786,355	7,681,160	12,975,680
Scenario 5	−1,940,500	−8,579,500	−1,214,000
Scenario 6	11,042,440	4,867,040	10,966,640
Scenario 7	13,298,365	8,058,960	13,090,680
Scenario 8	−1,888,500	−8,535,000	−1,164,000
Scenario 9	12,002,250	5,862,000	12,127,560
Scenario 10	12,007,490	8,594,180	12,197,940
Scenario 11	−1,874,500	−8,510,500	−1,134,000
Scenario 12	11,080,060	4,952,580	11,273,580
Scenario 13	12,185,160	7,871,840	12,463,360
Scenario 14	−2,411,500	−9,047,500	−1,663,000
Scenario 15	12,211,420	6,143,760	11,687,280
Scenario 16	12,070,460	7,066,240	12,245,320
Scenario 17	−1,967,000	−8,594,000	−1,221,000
Scenario 18	11,320,300	5,085,480	11,582,480
Scenario 19	12,141,405	7,262,000	12,389,480
Scenario 20	13,116,925	7,551,000	12,144,000

with the parameters given in Table 11.4. The standard deviation is expressed as a percentage of the average demand. Using the Monte Carlo procedure presented in Chap. 10, generate five samples of four demand scenarios. Then optimize the distribution network for each sample, compare the solutions obtained, and propose a process to choose the most robust design.

Exercise 11.2 At the end of Sect. 11.3.2, we stated that the single-cycle multi-period distribution network design model (11.10)–(11.14) can be extended to cover several reengineering cycles. Assuming that all structural adaptation decisions must be made at the beginning of the planning horizon, propose a two-stage stochastic programming extension of model (11.10)–(11.14) covering several reengineering cycles.

Exercise 11.3 Table 11.5 gives the value of three distribution network designs related to Example 11.3. These values were estimated with a detailed user model for a sample of 20 scenarios. Propose adequate performance measures to compare these three designs and choose the best. Display the designs on a value-risk plot to help justify your decision.

Exercise 11.4 Return to model (11.6)–(11.9) in Sect. 11.3.2, which assumes that a local recourse can be used when additional capacity is required at a DC. Suppose now that no such local recourse is available but that an external supplier can be used to make direct deliveries to customer when DC capacity is missing. Modify model (11.6)–(11.9) to take this into account. Then, find the optimal design obtained with this new model if a cost of €200 is added every time the external supplier is used to deliver a product to a demand zone. Compare the design thus obtained with the solution previously found.

Exercise 11.5 Go back to multi-period stochastic SCN design model (11.15)–(11.21), and propose multi-cycle version of this model (assume that all structural adaptation decisions must be made at the beginning of the planning horizon). Then adapt the model obtained to a *maximum covering* resilience strategy for distribution activities.

Bibliography

Bertrand J (2003) Supply chain design: flexibility considerations. In: de Kok A, Graves S (eds) Supply chain management: design, coordination and operation. Handbooks in OR & MS, Vol. 11. Elsevier

Birge J, Louveaux F (2011) Introduction to stochastic programming, 2nd edn. Springer, Berlin

Blackhurst J, Dunn K, Craighead C (2011) An empirically derived framework of global supply resiliency. J Bus Logistics 32(4):374–391

Chopra S, Sodhi M (2004) Managing risk to avoid supply-chain breakdown. MIT Sloan Manag Rev 46:52–61

Church R, Revelle C (1974) The maximal covering location problem. Papers Reg Sci Assoc 32:101–118

Dong M, Chen F (2007) Quantitative robustness index design for supply chain networks. In: Jung H et al (eds) Springer Series in advanced manufacturing, Part II: 369–391

Ducapova J, Consigli G, Wallace S (2000) Scenarios for multistage stochastic programs. Ann Oper Res 100:25–53

Figueira J, Greco S, Ehrgott M (2005) Multiple criteria decision analysis: state of the art surveys. Springer, New York

Fortune (2010) Home depot's hurricane plan. http://archive.fortune.com/galleries/2010/fortune/1008/gallery.home_depot_hurricane.fortune/index.html. Accessed 27 June 2015

Graves S, Tomlin B (2003) Process flexibility in supply chains. Manag Sci 49:907–919

Graves S, Willems S (2003) Supply chain design: safety stock placement and supply chain configuration. In: de Kok A, Graves S (eds) Supply chain management: design, coordination and operation. Handbooks in OR & MS, Vol. 11. Elsevier

Klibi W, Martel A (2012) Modeling approaches for the design of resilient supply networks under disruptions. Int J Prod Econ 135:882–898

Klibi W, Martel A (2013) The design of robust value-creating supply chain networks. OR Spectr 35(4):867–903

Klibi W, Lasalle F, Martel A, Ichoua S (2010) The stochastic multi-period location-transportation problem. Transp Sci 44(2):221–237

Klibi W, Martel A, Guitouni A (2015) The impact of operations anticipations on the quality of supply network design models. Omega, forthcoming

Kouvelis P, Yu G (1997) Robust discrete optimization and its applications. Kluwer Academic Publishers

Lim M, Bassamboo A, Chopra S, Daskin M (2010) Flexibility and fragility: use of chaining strategies in the presence of disruption risks. Working paper, University of Illinois, Urbana-Champaign

Mansini R, Ogryczak W, Speranza M (2014) Twenty years of linear programming based portfolio optimization. Eur J Oper Res 234:518–535

Pettit T, Croxton K-L, Fiksel J (2013) Ensuring supply chain resilience: development and implementation of an assessment tool. J Bus Logistics 34(1):46–76

Shapiro A, Dentcheva D, Ruszczynski A (2009) Lectures on stochastic programming. Society of industrial and applied mathematics and mathematical programming society

Sheffi Y (2005) The resilient enterprise: overcoming vulnerability for competitive advantage. MIT Press

Tang C, Tomlin B (2008) The power of flexibility for mitigating supply chain risks. Int J Prod Econ 116:12–27

Tomlin B (2006) On the value of mitigation and contingency strategies for managing supply chain disruption risks. Manag Sci 52(5):639–657

van der Heijden K (2005) Scenarios: the art of strategic conversation, 2nd edn. Wiley

van Opstal D (2007) The resilient economy: integrating competitiveness and security. Council on Competitiveness

Chapter 12
SCNs for Sustainable Development

The previous chapters have concentrated on the design of robust value-creating SCNs. It was stated at the outset that, from an economic point of view, an SCN creates value if it contributes to the long-term sustainable improvement of its company's market value, the latter being in principle measured as the discounted sum of all yearly residual cash flows (RCFs) or economic value added (EVA) over the company's lifetime. To improve RCFs or EVA, sales revenues must be maximized. This can be done only if the company is persistently better at winning orders than its competitors. Inasmuch as customers value the well-being of the environment and society, striving to improve value creation thus indirectly motivates companies to favor environmental and social sustainability. Not doing so endangers their long-term survivability. This vision amounts to pursuing shareholder and customer value creation with value for other stakeholders as a by-product. However, as discussed in Sect. 2.4, good corporate citizens strive to do more than that. They adopt sustainability goals based on a triple bottom line: economic, environmental, and social. According to the CEO and VP of sustainability for Walmart Stores, "every healthy, high-performing company has an obligation to use its strengths to help society, and each can do so in ways that enhance the viability of the business" (McLaughlin and McMillon 2015). Or, in the words of the chief executive of Unilever, "we can choose to be givers and not takers from the system that gives us life in the first place" (Polman 2015). The aim of this chapter is to show how SCNs can be designed to enhance sustainable development.

We start by presenting a generic framework and approach for sustainable SCN design and then we examine two particular cases in more detail. The first one deals with the design of eco-efficient SCNs. We show how the SCN design models presented in previous chapters can be extended to take the environmental bottom line into account. Finally, we study the design of reverse and closed-loop SCNs, that is, networks incorporating used product collection and revalorization activities and resources.

A. Martel and W. Klibi, *Designing Value-Creating Supply Chain Networks*,
DOI 10.1007/978-3-319-28146-9_12

12.1 Environmentally and Socially Sustainable SCs

12.1.1 Sustainable SCN Development Framework

The stakeholders of a company do not all have the same perception of its raison d'être. Individual members of society tend to perceive companies as a means for social development. At the other extreme, investors increasingly consider companies as commodities to be traded. Both groups in a sense perceive firms as instruments to enhance their well-being, despite the fact that they usually do not entertain a close relationship with them and do not necessarily understand what will make them successful. Both, however, exert pressures that influence company governance. Public corporations are particularly sensitive to quarterly earnings targets fixed by their board in response to financial market expectations. This, coupled with the need to solve arriving crises, tends to make them concentrate on the short term. Society would like companies to pay more taxes to be redistributed to the poor, preserve the environment, and provide an equitable and ethical workplace. Achieving these expectations requires a long-term perspective. Private corporations are less sensitive to external pressures and they are driven more by the vision and leadership of their owners. In the last part of the twentieth century, pressures from the market led to widespread "quarterly capitalism," which resulted in remarkable failures such as the Enron scandal revealed in 2001. Increased pressures from environmental and social activists, media, and governments have since change the perceptions of company managers who, as evidenced by the top executives' statement cited previously, now believe a sustainability strategy is a competitive necessity (Haanaes et al. 2012).

The proactive improvement of business, environmental, and societal networks and systems requires a long-term holistic vision. As indicated at the beginning of this book, SCs can be seen as the cardiovascular system of the economy, and SCNs are the backbone of individual companies. SCNs consume and transform the planet resources and they produce wastes and emissions that pollute it. The planet is thus altered by their activity on both the input and output sides. Similarly, people are an essential ingredient of any SC system. They are simultaneously the ultimate consumers of their products and services and the workforce required to perform their activities. SCNs are thus inseparable from the planet and from people. This holistic view of SCNs is represented in Fig. 12.1. In the long term, to survive and create value, companies must contribute to the preservation of the planet and the emancipation of society, which is congruent with the definition of sustainable development. A closer examination of the framework in Fig. 12.1 will enable us to identify the role that SCN design can play to enhance sustainable development.

Let us start by looking at planet–SCN interactions. Currently, the demand for natural resources (water, fossil fuels, wood, metals, land, etc.) continues to increase, but the supply of natural resources continues to diminish. Clearly, this is not sustainable in the long term. To survive, companies must act to preserve and renew

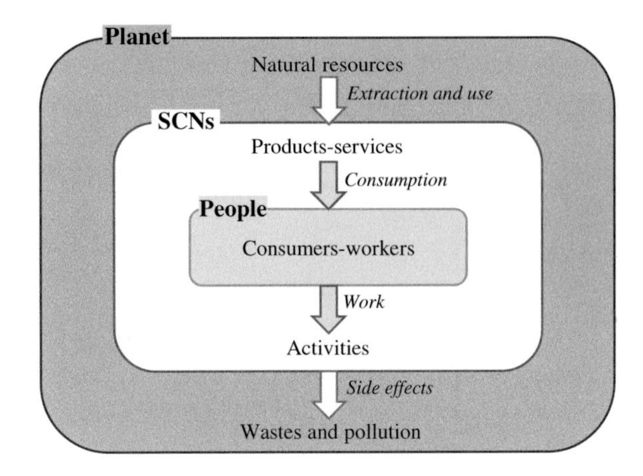

Fig. 12.1 Sustainable SCN development framework

natural resources. This can be done in two ways: they can consume fewer resources; additionally, they can help improve the supply of natural resources. Some natural resources are renewable and others are not. Whenever possible, the latter should be substituted by the former. For renewable natural resources or energy sources, companies should make all efforts possible to improve the renewal process of the resource they consume. This can take several forms. For example, in the pulp and paper industry, companies must plant as many trees as they harvest. In the food industry, producers and retailers can engage in R&D, certification, and training programs to improve the yield, water consumption, and reach of smaller farmers.

The resources consumed can be reduced in several ways. Recall from Chap. 4 that SCNs are technological systems transforming input raw materials into output products and services using production-distribution processes. Improvements can be obtained by acting at the output, processor, and input levels. At the output level, companies can design more reliable products and improve maintenance processes to increase the useful life of finished products and thus reduce demand. At the process level, more efficient transformation and distribution technologies can be adopted to reduce material and energy consumption. For example, wastes generated during transformation can be burned to produce a part of the energy required by the process. At the input level, transformed resources can be reintegrated into the SCN through several revalorization paths to reduce material needs. Possible revalorization paths are illustrated in Fig. 12.2, which depicts a generic OEM closed-loop SCN (CLSCN), that is, an SCN incorporating a forward production–distribution network of the type studied previously but also a reverse-logistics network for used product recuperation, restoring, refurbishing, cannibalization, and/or recycling. To simplify, no warehousing-storage activities are shown in the revalorization network but depots are usually present at some level.

Revalorization paths all start by a collection, inspection, and dispatching activity. Used products return and collection takes several forms depending mainly

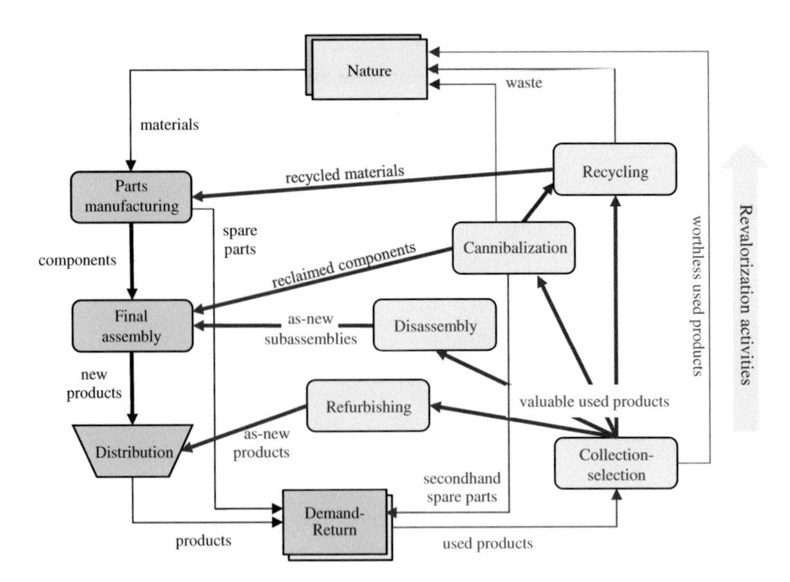

Fig. 12.2 Activity graph of a generic OEM closed-loop SCN

on whether the OEM has property rights on the product and the used product has a significant residual value. High-value proprietary products typically give rise to product return, recall, and inventory repatriation processes, often through reverse distribution channels. At the other extreme, valueless nonproprietary products are typically recovered through third-party garbage collection processes either at consumption or sales points (SPs). All sorts of intermediate situations are possible. For example, in Germany, used electrical and electronic equipment is recovered through a network of public collection points and subsequently reclaimed by independent agents for reselling or disposal. Incentives may also be used to encourage disposal methods. These may take the form of a financial compensation (buyback, rebates on new products, cost-free take back, etc.), deposit fee, or contractual or legal obligation.

Collected used products are inspected to assess their state. Worthless products are shipped directly to a waste disposal site where they are either land-filled or incinerated. Valuable used products are dispatched to a refurbishing, disassembly, cannibalization, or recycling process, depending on their state and on the context. A used product is valuable if there is a demand for revalorized parts, subassemblies, or products and if revalorization costs do not surpass savings or if society is prepared to pay any loss incurred in order to help preserve the planet resources. Returned products that are still usable can be refurbished (minor repairs or maintenance, functional tests, repackaging) and reinserted as new in the product distribution network or sold on a secondary market. Alternatively, the product could be disassembled and its subassemblies recertified and reinserted into the product assembly process. If this is not possible, another alternative is cannibalization, that is, a selective disassembly of the product to remove functional components, which can then be reconditioned and sold as secondhand spare parts or reinserted into the production process. Residual nonfunctional components are sent to recycling or to waste disposal depending on their nature. Recycling is typically a mechanical process involving the shredding of residual product parts and the sorting of the resulting material. Valuable recycled material can be reintroduced as parts manufacturing inputs and non-valuable residuals are sent to waste disposal.

Returning to the planet–SCN interactions depicted in Fig. 12.1, it is clear that production–distribution activities generate waste and pollution that may harm the planet. We just saw that useless products usually end up as waste, but extraction, production, and distribution processes also generate waste, such as leftover crops, livestock manure, sawdust, used water, scrap metal, chemical residues, used tires, used oils, and so on. Some wastes, such as biomass (forest residues, wood chips, animal matter, etc.), can be used to produce energy (electricity, steam, ethanol) and some wastes may be recycled (e.g., used water). Some wastes are toxic and their disposal then presents serious problems. SCN activities may also pollute air, water, and land. Greenhouse gas (GHG) emissions and global warming have become a serious concern. Spillages in surface water, infiltration in ground water, and soil pollution also have disastrous effects. Other undesirable industrial side effects include noise, smells, visual pollution, vibrations, and radiations. Clearly, it is the responsibility of industrial and commercial organizations to reduce waste and pollution as much as possible.

In addition to SCN–planet interactions, Fig. 12.1 also highlights SCN–people interactions. As mentioned, individuals simultaneously consume products and contribute to their production and distribution. As consumers, through their buying behavior, people can influence the decisions of companies. This brings us back to our previous discussions of order winners. As workers, people expect to find jobs with fair labor conditions (wages, benefits, working time, rest allowances, disciplinary practices, etc.), personal development opportunities, and a secure work environment (accident prevention, physical and mental health, etc.). This precludes any abuse to human rights (civil liberties, child and forced labor, freedom of association, discrimination related to sex, race, or other, etc.). Collectively, people have larger social concerns related to product responsibility (safe and healthy products), community development, education, R&D investments, infrastructure development, healthcare, culture, standard of living, reducing poverty, equitable trade, business ethics, and so on. Responsible companies feel that they must help society at all these levels.

It is usually not possible to make improvements on all fronts without compromise. For example, burning residues to save energy produces GHG emissions; collecting used wine bottles for reuse requires additional transportation, which increases emissions; using high-tech ecological technologies may cut jobs; favoring regional development may increase transport emissions because of the increased distance between producers and markets. This raises the necessity to elaborate adequate performance measures to assess various impacts and to find ways to make adequate tradeoffs, particularly when designing SCNs.

Although not explicitly displayed in Fig. 12.1, governments are present everywhere behind the scene because, through laws and regulations, they specify the rules to be followed by companies and consumers. Regulatory policies elaborated to influence the environmental and social behaviors of companies can vary widely between states. For example, in order to reduce GHG emissions, governments can implement carbon emission pricing or trading schemes. The former is a tax applied to carbon emissions; in 2012 for instance Australia enforced a $23 tax per ton of carbon pollution (Zakeri et al. 2015). The latter is usually a cap-and-trade program creating a market and a price for emission reductions. The government sets an emission target (cap) for covered companies and allows these companies to sell part of their allowances on a carbon market (trade). On April 30, 2015, the price of one ton of CO_2-equivalent pollution on the California carbon market was $12.72 (calcarbondash.org).

In Chap. 1, we stressed that the mission of a company's SCN is sustainable value creation for the company and its stakeholders. With the previous discussion in mind, it should be clear that SCNs must be designed and managed to create value, "while protecting, sustaining, and enhancing the human and natural resources that will be needed in the future" (IISD 1992, p. 11).

12.1.2 Sustainable SCN Design Approach

The previous chapters examine key decisions required when designing value-creating SCNs and propose a modeling approach to help make these decisions. In this section, we show how these decisions relate to sustainable development and how our modeling approach can be applied to design sustainable SCNs. Clearly, not all sustainable development issues are related to SCN design. For example, the fact that some companies produce harmful products such as cigarettes or weapons is a larger societal concern. Regulating or prohibiting the production and distribution of these products has nothing to do with SCN design. Also, the development of adequate working conditions, the adoption of ethical business practices, and the respect of human rights depends on company policies but it does not rest directly on the SCN design decisions studied in this book.

The value offer of a company is directly related to its product design. The design of a product determines how easily and ecologically it can be made, distributed, used, revalorized, and disposed at the end of its life. It also determines how much nonrenewable resources it consumes and how safely it can be used. Greener, safer, and more equitable products may be rewarded by a price premium. Product-market offer decisions thus clearly have a marked impact on the triple bottom line. Technology selection also clearly affects the three pillars of sustainability. High-tech platforms may be more efficient and reliable but may cut jobs. Some processes generate less waste and are safer than others but they are typically more costly. Some energy sources are cleaner and more easily renewable (e.g., biofuels) than others but they are often more expensive.

Site location and transportation means selection decisions also involve clear tradeoffs. Transportation is an important source of pollution, and as seen in Chap. 5, some transportation modes are more eco-efficient than others. The location of production–distribution sites determines transportation needs. A small number of sites requires more transportation but enables the use of more efficient production-warehousing technologies because of scale, and vice versa. The location of production sites in developing countries favors the creation of jobs in poor regions and the reduction of some costs, but it involves more transportation. Sourcing decisions are particularly important when designing for sustainability. It has been shown (Clift 2003) that resource extraction, processing, and refining associated with the first stages of an SC incur disproportionally high environmental impact, that is, the ecological footprint of an SC is much sharper for upstream than downstream activities. Hence, a large part of the life cycle wastes and emissions generated by a company's products typically comes from its sourcing decisions. Similarly, unfair trade practices and poor working conditions are more frequently observed for extraction and second or third tier manufacturing stages. At the other end of the SC, used products collection, revalorization, and disposal processes also have a significant impact on sustainability and, whenever possible, they should be considered explicitly in SCN design projects.

Companies often take a piecemeal approach to sustainability (Lee 2010). They select greener or more socially responsible suppliers, for example, or they modify their processes to be more energy efficient. "Although these changes often seem worthwhile individually, they may in the grand scheme generate unintended consequences, such as higher financial, social, or environmental costs" (Lee 2010, p. 65). We saw in Chaps. 4, 5, and 6 that multi-criteria decision-making approaches can be used to select business sites, transportation means, and suppliers and that the criteria applied typically cover environmental and social factors. However, as discussed in Chap. 6, these local decision approaches are not capable of making the holistic tradeoffs required to get major breakthroughs. The SCN design models presented in Chaps. 7–11, however, are, and they can help reinvent sourcing, manufacturing, and distribution processes, even collaborating with competitors, to become more eco-efficient and socially responsible while improving value creation.

How can this be done? In a nutshell, adjustments to the SCN design model objective function and solution space must be made. In the previous chapters, we saw that value-creation objectives can be expressed in terms of cash flows. This leads to the formulation of SCN design models of the following form:

$$\max_{(\mathbf{x},\mathbf{y})} V^{\mathrm{SCN}}(\mathbf{x},\mathbf{y}) \quad \text{s.t.} \quad \mathbf{x} \in \mathcal{X}, \mathbf{y} \in \mathcal{Y}, \mathbf{A}\mathbf{x} \leq \mathbf{b}(\mathbf{y}) \tag{12.1}$$

where $\mathbf{x}$ is a vector of continuous flow, throughput, and inventory variables and $\mathbf{y}$ a vector of binary design variables. This matrix formulation is the same as that presented in Sect. 11.3.1 with $\mathcal{X} = \{\mathbf{x} | \mathbf{W}\mathbf{x} \leq \mathbf{h}, \ \mathbf{x} \geq \mathbf{0}\}$. Some environmental and social impacts, such as resource renewal costs, carbon taxes, and waste disposal costs, can be costed, but these are particular cases. Environmental and social objectives are multidimensional and it is difficult to find a single measure to express them.

The dominant approach for the estimation of environmental impacts involves a life cycle assessment (LCA) from extraction of raw materials (cradle) to disposal of end products (grave) (ISO 1997). To assess an SC, an inventory of all flows from and to the planet is required, and an evaluation of all extractions, land use, and emissions (in air, water, and soil) must be made. This evaluation is based on a hierarchy of indicators that can be integrated into a single score. Several distinct life cycle impact assessment hierarchies were proposed. Two that can readily be used to construct an environmental impact objective function $EI^{SCN}(\mathbf{x}, \mathbf{y})$ are as follows:

- *Eco-indicator 99* (Goedkoop and Spriensma 2000), which aggregates 11 midpoint impact indicators into 3 endpoint categories (human health, ecosystem quality, and resources) combined to obtain a single score
- *ReCiPe 2008* (Goedkoop et al. 2013), which, as shown in Fig. 12.3, aggregates 18 midpoint indicators first into 3 endpoint damage indicators and then into a final score

Because compiling all these indicators may require significant efforts and expertise, several authors base their environmental impact objectives on a partial surrogate measure such as CO_2-equivalent emissions. Also, because forward and closed-loop SCNs usually cover a subset of several SCs, boundary activities such as supply and sales should include any relevant external upstream and downstream impacts in their evaluations.

Efforts have been made to develop a societal LCA approach (Jorgensen et al. 2008); however, characterizing and measuring social well-being has proved very

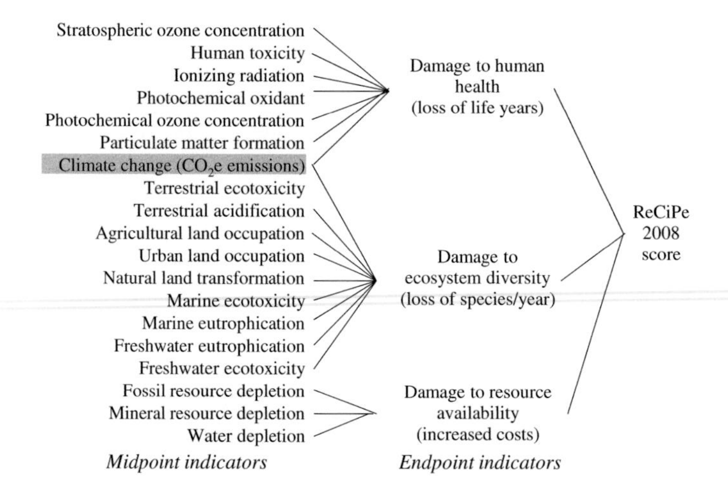

Fig. 12.3 ReCiPe 2008 environmental damage indicators hierarchy

difficult. The definition of social sustainability itself is still under elaboration and the factors to consider depend largely on the context. The explanation of a social impact objective function $SI^{\mathrm{SCN}}(\mathbf{x}, \mathbf{y})$ therefore depends largely on the application. Employment is the main social indicator used. Employment level objectives are easily formulated by associating a number of jobs to platform use variables (Y) or work hours to throughput variables (X) and by weighing them, for example, to favor regional development (Mota et al. 2014).

Social objectives also often take the form of an equity concern related to space, time, returns, or production factors. These typically can be expressed using summary inequality constraints. International or regional (space) equity concerns can be expressed by introducing local content constraints similar to (9.7), but where the proportion β_k is an objective fixed by the company for region k instead of a lower bound imposed by government. Intergenerational (time) equity concerns can take the form of constraints to ensure that nonrenewable resource use and pollution gradually decrease in the successive time periods of the planning horizon. Production factor equity concerns could take the form of constraints to ensure that yearly labor costs are at least as large as a proportion of fixed asset annuities. Equity constraints can be expressed in terms of lower and/or upper bounds, deviation from a mean, absolute differences, and so forth (Karsu and Morton 2015).

Another important issue when modeling for sustainability is the breadth of the SCN design options $\mathbf{y}$ considered. Instead of examining only classical options, one should include potential product-market offers for greener products, consider technologies and platforms that consume less energy and nonrenewable resources and produce less waste, consider labor-intensive technologies to help create jobs, and so on. Conversely, highly destructive options should be eliminated a priori. One can also include environmental and social taxes in the value-creation objective function. These taxes may be real, as for carbon taxes, or they may be artificially used to reflect a company policy, for example, a natural resource renewal tax, a waste disposal tax, or a facility shutdown tax. The value-creation objective also usually depends on the discount factor α used by a company. As explained in Chap. 2, α is typically based on weighted average cost of capital (WACC) and financial risks, and discounting gives more weight to short-term decisions. By selecting a discount factor greater (that is closer to 1) than suggested by the WACC and risks, a company gives more weight to long-term decisions, which, as explained previously, is necessary for sustainable development.

Given all this, SCN design model (12.1) can be rewritten as

$$\begin{aligned} &\max_{(\mathbf{x},\mathbf{y})}\left(V^{\mathrm{SCN}+}(\mathbf{x},\mathbf{y}), SI^{\mathrm{SCN}}(\mathbf{x},\mathbf{y})\right) \text{ and } \min_{(\mathbf{x},\mathbf{y})} EI^{\mathrm{SCN}}(\mathbf{x},\mathbf{y}) \\ &\text{s.t.}\quad \mathbf{x}\in\mathcal{X}^{+},\ \mathbf{y}\in\mathcal{Y}^{+},\ \mathbf{A}^{+}\mathbf{x}\leq\mathbf{b}^{+}(\mathbf{y}) \end{aligned} \tag{12.2}$$

where the superscript "+" is added to the value-creation function and solution space notations to reflect the model enhancements discussed. Model (12.2) is a multi-objective mixed-integer program (MOMIP), which may be quite difficult to solve. Furthermore, as discussed in Chaps. 2 and 11, under risk, recourse variables must be added, the value-creation objective $V^{SCN+}(\mathbf{x}, \mathbf{y})$ must be expressed in terms of expected value, and a risk minimization objective function based on an appropriate risk measure must be added.

In general, there is no single optimal solution that simultaneously optimizes all the objectives of an MOMIP. In a sustainable SCN design context, the designer is looking for a *most preferred* design instead of an optimal solution. The concept of optimality is then replaced with that of efficiency (Pareto optimality). Efficient designs are non-dominated or non-improvable solutions, that is, they cannot be improved in one objective function without deteriorating performances in another. Solving MOMIP (12.2), therefore, involves finding all efficient SCN designs (or at least a representative subset) so that the decision maker can make a final selection, possibly based on additional soft criteria not included in the formulation. Several methods were proposed to solve MOMIPs. For our purposes, the *weighting method* or the *ε-constraint method* can be used to find representative efficient solutions. The former associates a weight with each objective and combines them into a single objective function. The latter optimizes one objective function and considers the others as constraints.

The ε-constraint method, however, has several advantages over the weighting method (Mavrotas 2009) and it is particularly appropriate in our context. To apply this method, model (12.2) must be rewritten as follows:

$$\max_{(\mathbf{x},\mathbf{y})} V^{SCN+}(\mathbf{x}, \mathbf{y}) \tag{12.3}$$

subject to

$$EI^{SCN}(\mathbf{x}, \mathbf{y}) \leq EI \tag{12.4}$$

$$SI^{SCN}(\mathbf{x}, \mathbf{y}) \geq SI \tag{12.5}$$

$$\mathbf{x} \in \mathcal{X}^+, \ \mathbf{y} \in \mathcal{Y}^+, \ \mathbf{A}^+\mathbf{x} \leq \mathbf{b}^+(\mathbf{y}) \tag{12.6}$$

where *EI* and *SI* are respectively an upper bound on environmental impacts and a lower bound on social impacts. Efficient SCN designs are obtained by varying the bounds *EI* and *SI* in discrete steps in a predetermined or calculated interval. Guidelines for the implementation of the method are found in Mavrotas (2009). For the reasons previously discussed, in our context it is often convenient to consider *SI* to be a corporate social impact goal based on current achievements and to fix its value. When this is done, constraints (12.5) blend with (12.6) and efficient designs can be examined by plotting a (value, environmental impact) efficient frontier of the type shown in Fig. 2.17.

To summarize, sustainable SCN designs can be obtained by proceeding as follows:

1. Plot the activity graph of the current extended SCN. In addition to all the activities that could be internalized (Chap. 6), depict major flows from (natural

resources, energy) and to (wastes, emissions) the planet and, when relevant, include used products collection and revalorization activities.
2. Map current SCN suppliers, assets (land, platforms, equipment), and product markets.
3. Use the activity graph and resource maps produced to identify where potential environmental and social responsibility problems and opportunities lie.
4. Imagine more sustainable ways (new locations, technologies, suppliers, market offers, partnerships, etc.) of performing activities and deploying the SCN (i.e., enrich the solution space).
5. Select performance measures for the evaluation of value creation, environmental impacts, and social impacts.
6. Formulate the SCN design problem as an MOMIP and solve it to find a set of representative efficient designs.
7. Perform a detailed evaluation of the efficient designs found, possibly using additional performance criteria not explicitly included in the MOMIP, to select a sustainable SCN design congruent with the company's SC strategy.

This SCN design approach is embedded in the more complete SCN design methodology presented and illustrated in Chap. 13.

12.2 Eco-Efficient SCNs

Following the adoption of the Kyoto Protocol on climate change in 1997, initiatives around the world were launched to regulate GHG emissions, and the need for companies to develop GHG accounting and reporting procedures became obvious. This led to the publication by the World Business Council for Sustainable Development (WBCSD, www.wbcsd.org) and the World Resource Institute (WRI, www.wri.org) of corporate, product life cycle, and value chain GHG accounting and reporting standards (WBCSD-WRI 2004, 2011a, b). It also led to the development of several emission databases by organizations, such as the Intergovernmental Panel on Climate Change (IPCC, www.ipcc.ch) and the Ecoinvent Center (www.ecoinvent.org), and to the elaboration of software to facilitate the evaluation of the carbon footprint of companies. GHGs include several natural and artificial gases that trap heat near the earth surface and their emission is generally reported in tons of carbon dioxide equivalent (CO_2e). Emissions for a given gas are converted into CO_2e by multiplying them by their global warming potential (GWP) for a given time period (20, 100, or 500 years). The GWPs of the six GHGs covered by the Kyoto Protocol are given in Table 12.1.

The development of these standards and tools makes the evaluation of the impact of GHG emissions on climate change much easier than the assessment of other environmental damage indicators (see Fig. 12.3). For this reason, CO_2e emissions are often used as a surrogate measure for the evaluation of environmental damages because of SC activities. As for environmental damages in general, GHG emissions

Table 12.1 Global warming potentials for Kyoto protocol gases (IPCC 2007)

Gas	Formula	20-year GWP	100-year GWP	500-year GWP
Carbon dioxide	CO_2	1	1	1
Methane	CH_4	72	25	7.6
Nitrous oxide	N_2O	289	298	153
Hydrofluorocarbons	HFCs	437–12,000	124–14,800	38–12,200
Perfluorocarbons	PFCs	5210–8630	7390–12,200	11,200–18,200
Sulfur hexafluoride	SF_6	16,300	22,800	32,600

are not uniformly distributed across the SC. For extraction, agricultural, and primary transformation activities, emissions are much higher than for secondary and tertiary production and distribution activities. Also, the majority of GHG emissions come from energy consumption. Most activities produce GHGs directly through the use of fossil fuels on-site and during transport and indirectly through the use of electricity and heat. The technology used to produce electricity has a marked impact on emissions. Electricity generated using coal, oil, or gas produces much more emissions than hydro or wind energy. This incidentally means that the location of activities has a significant impact. Locating an activity in a region where hydro-electricity dominates leads to much less emission than for regions where electricity is produced using fossil fuels. The main sources of GHG emissions in the United States are summarized in Fig. 12.4.

The first implication of the preceding discussion for eco-efficient SCN design is that it is important to consider energy-intensive activities explicitly in design models and to develop options to reduce pollutant energy consumption. In a recent study, the Word Economic Forum (WEF 2009) identified a series of strategic opportunities to decarbonize SCs. Table 12.2 relates the main focus of these opportunities to the SCN design decisions studied in this book. It reveals that all the

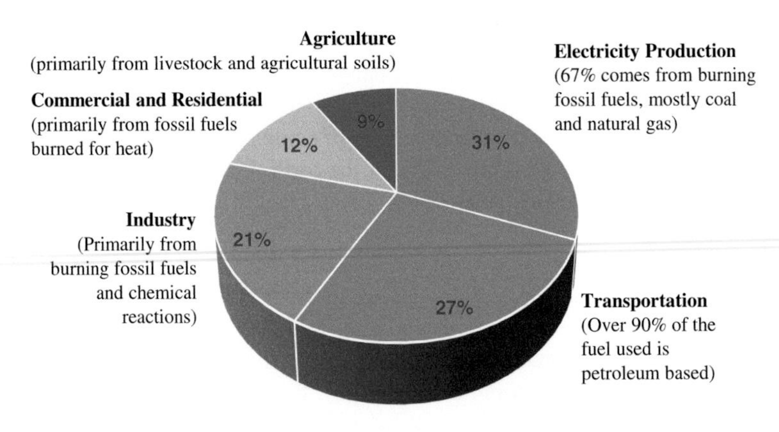

Fig. 12.4 United States GHG Emissions by Sector (U.S. EPA 2015)

Table 12.2 Focus areas in SCN design for GHG emission reduction

Focus area	Decarbonization opportunities	Design decisions
Clean vehicle technologies	• Improved vehicle-driver efficiency • Alternative fuels	Transportation options
Modal switches	• Long haul road to train or waterways • Short haul air to road • Intercontinental air to ocean	Transportation options
Shipper collaboration	• Shipments consolidation • Carbon-offsetting solutions	Transportation options
Despeeding the SC	• Decrease shipment frequency (i.e., increase lot sizes) • Reduce vehicle speed	Transportation options Market offer
Products and packaging design initiatives	• Green product design • Green packaging design • Packaging elimination	Market offer
Low carbon raw material and component sourcing	• Select suppliers with clean extraction-transformation processes • Select low energy consumption sources • Nearshoring	Supplier selection
Low carbon manufacturing	• Energy consumption reduction • Process optimization • Economies of scale • Change localization	Site and platform selection
Energy efficient buildings	• Improved building specifications • Improved management systems • Better lighting, heating, and cooling	Platform selection
Optimized network structure	• Eliminate intermediaries • Reduce network flows	Site selection
Reverse logistics	• Improved used products collection • Better recycling solutions • Improved waste processing	Site and platform selection

design decisions studied in previous chapters have at least some impact on GHG emissions. However, our previous discussion highlights the fact that sourcing and transportation-means selection decisions are particularly important. Another implication is that in order to formulate a cradle-to-grave (CTG) carbon emission minimization objective function, unit CO_2e emissions must be estimated for internal platform selection, activity, and flow variables, as well as for upstream and downstream emissions related to supply contracts and product-market offers. This is illustrated in Fig. 12.5 for a simplified multi-tier car industry SC.

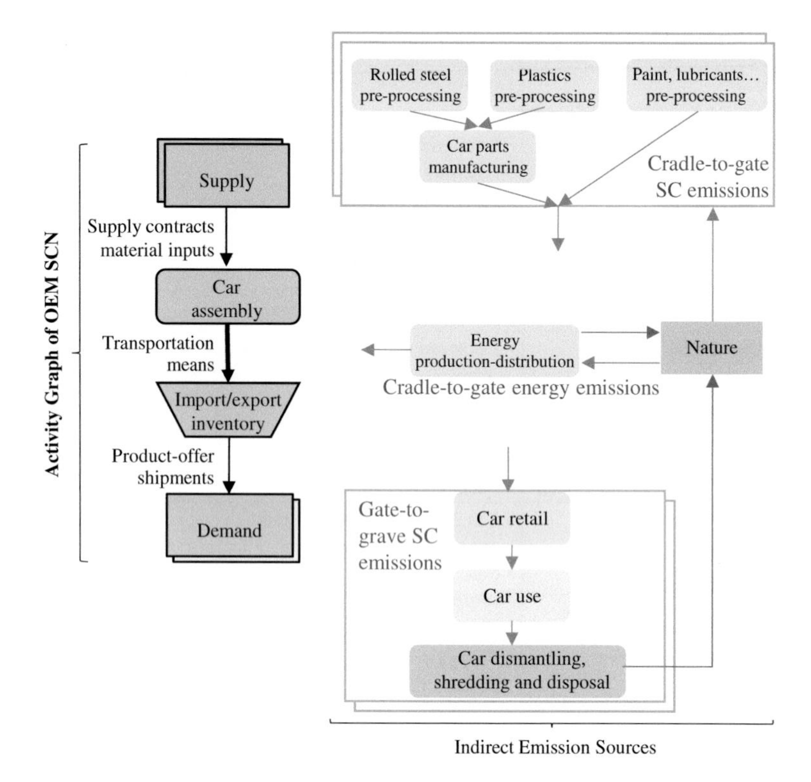

Fig. 12.5 Cradle-to-Grave emission sources for a car industry SCN

12.2.1 SCN Arcs GHG Emissions

The arcs of SCNs correspond to transportation or handling activities. The approach used to compute associated GHG emissions should comply with applicable standards (e.g., European standard EN 16258 for the calculation of GHG emissions of transport services) and with the GHG Protocol (WBCSD-WRI 2004). Several software tools such as EcoTransIT World (www.ecotransit.org) are available to facilitate these calculations. When evaluating transportation emissions, two basic steps are involved. One deals with the calculation of emissions associated with the consumption of energy when a vehicle is operated (TTW: tank-to-wheels) and the other with the estimation of upstream emissions associated with the extraction, production (in refineries or power plants), and provision of the energy used by the vehicle (WTT: well-to-tank). The sum of the two provides total well-to-wheels (WTW) emissions. However, default conversion factors provided by standards such as EN 16258 enable the calculation of WTW emissions directly.

Consider a transportation means m (internal fleet, annual 3PL contract, pooled transportation alliance, common carrier, etc.) for which the type(s) of vehicle used (truck, train, ship, barge, aircraft, etc.), the type(s) of energy consumed (gasoline, diesel, ethanol, biodiesel, liquefied petroleum gas, electricity, etc.), the topography of the shipping lanes (highway vs. local roads, gradient [flat, hilly, mountains], pavement, etc.), the average speed of vehicles, and the type of goods transported (see Fig. 5.15) are known. Then, for an arc (l, l') with an origin-destination distance $D_{ll'}$ and an annual flow of product p of $F_{pll'}$ unit loads (pallets, cwt, ton, TEU, etc.), CO_2e emissions are given by the following relation:

$$\mathrm{E}_{m}^{\mathrm{T}}\left(F_{pll'}\right) = \bar{\varepsilon}_{pm}^{\mathrm{T}}(D_{ll'})F_{pll'} \tag{12.7}$$

where $\bar{\varepsilon}_{pm}^{\mathrm{T}}(D)$ is a function giving average CO_2e emissions per unit load of product p shipped on an arc of length D when transportation means m is used. This function can be derived from the CO_2e emissions $\varepsilon_{m}^{\mathrm{T}}(D, W)$ of a load of weight W shipped using means m over a distance D.

The emissions $\varepsilon_{m}^{\mathrm{T}}(D, W)$ calculated depend on the standard used (European, US EPA [United States Environmental Protection Agency], or Japanese). In what follows, to comply with standard EN 16258, we assume without loss of generality that load weights are measured in tons (t), fuel consumption in liters (l), and distances in kilometers (km). Let

E_m Fuel (energy) consumption in liters per ton-kilometer (l/tkm) under the settings (vehicle type, fuel type, speed, traffic, topology, type of goods, etc.) of transportation means m. This is usually estimated by the carrier providing the transportation service. Typical diesel consumption values for motorways under average speed and traffic conditions are provided in Table 12.3 for diverse situations

g_{m}^{WTW} WTW GHG emissions in kgCO_2e per liter of fuel used for transportation means m. This factor is obtained from published standards or estimated by the carrier using standard directives. The factors specified in EN 16258 for different types of fuel are given in Table 12.4

W_{m}^{empty} Empty weight in tons of the vehicles used with transportation means m. Typical empty vehicle weights are given in Table 12.3

Table 12.3 Typical empty weights and diesel consumptions for diverse vehicles, goods, and topologies

		Diesel consumption in l/tkm					
		Hilly			Level ground		
Vehicle	Empty weight	Volume goods	Average goods	Bulk goods	Volume goods	Average goods	Bulk goods
7.5–12 t truck	6 t	0.108	0.061	0.050	0.105	0.059	0.048
12–24 t truck	9 t	0.063	0.036	0.029	0.060	0.034	0.027
24–40 t van	14 t	0.038	0.023	0.020	0.033	0.020	0.016

Source Schmied and Knorr (2012)

Table 12.4 GHG emissions in $kgCO_2e/l$ for diverse fuels

Fuel	Tank-to-wheels (TTW)	Well-to-wheels (WTW)
Petrol	2.42	2.88
Ethanol	0.00	1.24
Petrol E 10 (10 vol. % ethanol)	2.18	2.72
Diesel	2.67	3.24
Biodiesel	0.00	1.92
Diesel D7 (7 vol. % biofuel)	2.48	3.15
Liquefied petroleum gas	1.7	1.9
Jet kerosene	2.54	3.10
Marine diesel oil	2.92	3.53
Marine gas oil	2.88	3.49

Source European Standard EN 16258

LTF_m Loaded trip factor for transportation means m, that is, (total km/km loaded) ratio for the transportation solution proposed ($1 \leq LTF_m \leq 2$). This efficiency factor is estimated by the carrier

With this data, the CO_2e emissions of a load of weight W shipped using means m over a distance D can be estimated as follows:

$$\varepsilon_m^T(D, W) = g_m^{WTW} E_m \left(LTF_m W_m^{empty} + W \right) D \tag{12.8}$$

The expression in parenthesis gives the weight in tons of the vehicle and its load on the origin–destination lane plus the additional weight moved because of empty vehicle repositioning. Now, suppose that the average payload of the shipments made with means m is given as $W_m = \Sigma_p \bar{w}_p \bar{Q}_{pm}$, where $\bar{w}_p$ is the average weight in ton of a unit load of product family p, and $\bar{Q}_{pm}$ is the average number unit loads of product p in a transportation means m shipment. Then the CO_2e emissions per ton are given by $\bar{\varepsilon}_m^T(D) = \varepsilon_m^T(D, W_m)/W_m$ and the unit load emission function $\bar{\varepsilon}_{pm}^T(D)$ required in (12.7) is

$$\bar{\varepsilon}_{pm}^T(D) = \bar{w}_p \bar{\varepsilon}_m^T(D) = \bar{w}_p g_m^{WTW} E_m \left(LTF_m W_m^{empty} + W_m \right) D / W_m \tag{12.9}$$

Example 12.1 Suppose that GHG emissions must be estimated for shipments between The Hague and Paris in Europe for a transportation means with the following characteristics:

- Vehicles: 40 t vans (tractor-semitrailer) with a 14 t empty weight (W_m^{empty})
- Fuel: diesel
- Speed: normal speed and traffic conditions
- Products: family of wood briquettes classified as average goods

- Unit loads: 0.75 ton ($\bar{w}$) pallets (on average)
- Topography: highways on level ground
- Loaded trip factor: 1.5 (LTF_m)
- Average number of unit loads per shipment: 25 pallets ($\bar{Q}_m$)

From Table 12.3, we can see that the diesel consumption for 40 t vans on flat highways when shipping average goods is 0.02 l/tkm (E_m). Also, Table 12.4 indicates that the standard WTW emissions per liter of diesel are 3.24 kgCO$_2$e (g_m^{WTW}). Furthermore, the average payload of the shipments made on The Hague–Paris lane is $W_m = \bar{w}\bar{Q}_m = 0.75(25) = 18.75$ tons. Knowing that the distance between The Hague and Paris is D = 500 km, based on (12.8), the CO$_2$e emissions per shipment can be estimated as follows:

$$\begin{aligned}\varepsilon_m^{\mathrm{T}}(500, 18.75) &= 3.24\,\mathrm{kgCO_2e/l}(0.02\,\mathrm{l/tkm})[1.5(14)t + 18.75\mathrm{t}]500\,\mathrm{km}\\ &= 1287.9\,\mathrm{kgCO_2e}\end{aligned}$$

Now, based on (12.9), the GHG emissions per unit load on The Hague–Paris lane are calculated as follows:

$$\begin{aligned}\bar{\varepsilon}_m^{\mathrm{T}}(500) &= 0.75\varepsilon_m^{\mathrm{T}}(500, 18.75)/18.75 = 0.75(1287.9)/18.75\\ &= 51.5\,\mathrm{kg\ CO_2e/pallet}\end{aligned}$$

Note that the function $\bar{\varepsilon}_m^{\mathrm{T}}(D) = 0.103D$ obtained from (12.9) can be used to calculate emissions per wood briquette pallet as a function of distance for any other lane using this transportation means.

Expression (12.9) is appropriate for the calculation of GHG emissions in most road transportation contexts. For other modes, the approach must be adjusted but the basic principle is the same (Schmied and Knorr 2012). When negotiating or signing transportation contracts, companies should ask 3PLs or carriers to provide information on the GHG emissions associated with their offers. Most transportation service suppliers are well equipped to do this nowadays. For handling arcs, GHG emissions depend on the energy consumption of handling equipment, and their calculations per unit load follows the principles used for any type of equipment. These are examined in the next section.

12.2.2 SCN Nodes GHG Emissions

As illustrated in Fig. 12.6, an SCN node corresponds to an activity $a \in A_{lo}$ performed in a platform $o \in O_l$ using a specific dedicated or flexible technology on a

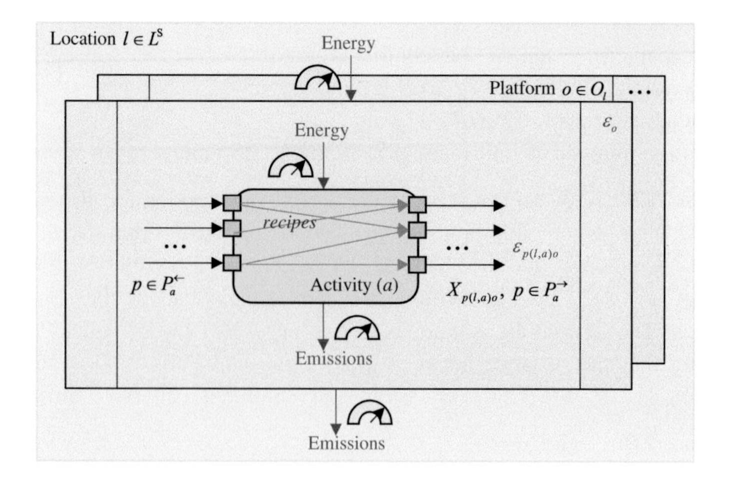

Fig. 12.6 Generic production activity node

given site location $l \in L^S$. Companies should collect emissions data for all the internal activities and facilities of their SCN. For a specific activity, emissions should be measured directly when possible and allocated to output products unit loads. The energy consumed by buildings (for lighting, heating, cooling, etc.) should also be monitored and used to estimate platform emissions. This may not be easy, however, and each industrial context requires adapted estimation procedures (WBCSD-WRI 2011b). For production activities, for example, the appropriate approach for continuous processes (e.g., refinery, chemical) and discrete processes (e.g., assembly line, batch production, job shop) is not the same. When products are made in parallel or when coproducts are obtained, the process emissions must be allocated to output products to avoid double counting. For example, for an automated car assembly line, if 10 cars are moving on the line at any time, then the line emissions must not be allocated to each car; they must be divided by 10.

Emissions can be calculated directly or indirectly. For some processes, emissions may be measured using a continuous monitoring system, stoichiometric equation balancing (for chemical reactions), mass balance (for refrigerants), or similar methods. In most cases, however, emissions must be based on process activity measures, facility dimensions, and published emission factors, that is, GHG emissions per unit of activity or space. Emissions factors may include only one type of GHG (say CO_2 or CH_4) or they may be expressed in CO_2e. They may cover only the process emissions or they may include cradle-to-gate emissions. Examples of emissions factor sources include LCA databases, industry association reports, governmental agencies, company-developed factors, and scientific publications (see www.ghgprotocol.org for a list of available databases).

In several sectors, the majority of emissions come from the use of energy (fuel, electricity, etc.). For buildings, electricity consumption depends on construction specifications and lighting, heating, and cooling technologies. Electricity emission factors, however, are strongly influenced by the facility location because some electricity production technologies pollute much more than others and emissions depend on regional electricity production-distribution technologies. For wind, hydro, and nuclear power, emissions are less than 100 gCO_2e/kWh. However, for electricity produced with natural gas, diesel, and coal they are about 450, 850, and 1100 gCO_2e/kWh, respectively (Raadal et al. 2011). To illustrate the impact of this, average electricity emissions factors for selected countries are given in Table 12.5. Buildings also may be heated using oil. Handling equipment such as forklift trucks and conveyer belts used for sorting, warehousing, and staging activities are usually powered by diesel, liquefied petroleum gas, or electricity. Energy consumption is easily evaluated using meters installed by energy providers or energy bills. However, when a single meter is used for a site, it is difficult to distinguish between platform and activities intakes. An allocation process must then be elaborated.

Independently of the context and of the approach used, the following node emission parameters must be estimated:

$\varepsilon^{A}_{p(l,a)o}$ GHG emissions (in $kgCO_2e$) per unit load of product p flowing out of activity $a \in A_{lo}$ performed in platform $o \in O_l$ on-site location $l \in L^S$

ε_{ol} Yearly GHG emissions (in $kgCO_2e$) associated with the use of platform $o \in O_l$ on-site location $l \in L^S$

In order to avoid double counting, activity and platform emissions calculations should include cradle-to-gate energy emissions but not upstream and downstream SC emissions. Additional emissions parameters could be derived to account for the construction, renovation, and disposal of facilities and equipment associated with platforms but overall their impact is usually relatively small. If the platform considered is owned by a third party (e.g., a public warehouse), then the emissions parameters can be provided by this third party.

Table 12.5 Emissions per kWh of electricity consumed

Country	$kgCO_2$/kWh	$kgCH_4$/kWh	kgN_2O/kWh	$kgCO_2e$/kWh[a]
Brazil	0.109907407	0.00000211589	0.00000064114	0.110151364
China	1.037486940	0.00001114415	0.00001618993	1.042590143
France	0.075693212	0.00000107817	0.00000069265	0.075926576
Germany	0.714637591	0.00000767552	0.00000967384	0.717712283
India	1.800805423	0.00002096774	0.00002716280	1.809424131
Japan	0.465951477	0.00000746038	0.00000416633	0.467379553
United Kingdom	0.548402315	0.00000728401	0.00000552340	0.550230388
United States	0.586666503	0.00000702729	0.00000776512	0.589156191

[a]Computed using the 100-yr GWPs in Fig. 12.1
Source Brander et al. (2011)

Example 12.2 Consider a company operating a set of DCs in Europe. The DCs have similar building construction; electricity-based heating, cooling, and lighting technologies; and electrically powered handling equipment; but they have different dimensions and annual throughputs. All the company's products are stored and shipped in pallets (or pallet-equivalent shipping loads). For each year of operation the DCs compile their annual electricity consumption in kWh based on the electricity bills received. Using this information, they would like to estimate the GHG emissions of individual DC platforms as well as emissions per pallet shipped. The data cumulated in recent years for all the DCs can be used to estimate an explanatory energy consumption model in terms of DC annual throughput and cubic space (volume), by regression.

For a given DC and year, let

E^{kWh} Electricity consumed in kWh during the year by the DC

V Volume of the DC in m^3

X Total annual throughput of the DC in pallets, that is, $X = \Sigma_p e_p X_p$ where $e_p \leq 1$ is the number of pallets occupied by a unit load of product p and X_p is the annual throughput of product p

Then the model estimated by regression with the (E^{kWh}, V, X) observed would take the form

$$E^{kWh} = \beta_0 + \beta^{\mathrm{V}} V + \beta^{\mathrm{X}} X$$

where $(\beta_0, \beta^{\mathrm{V}}, \beta^{\mathrm{X}})$ are the estimated regression coefficients. Based on this, for a platform o with a volume V_o and using the same technology as the other DCs to perform warehousing-storage activity a, the required GHG emission parameters would be given as

$$\varepsilon^{\mathrm{A}}_{p(l,a)o} = g_l^{\mathrm{CTG}} e_p \beta^{\mathrm{X}} \text{ and } \varepsilon_o = g_{l(o)}^{\mathrm{CTG}} (\beta_0 + \beta^{\mathrm{X}} V_o)$$

where g_l^{CTG} is the CTG electricity emission factor for location l.

To illustrate this, suppose that the regression model estimated is

$$E^{\mathrm{kWh}} = 400,000 + 1.2V + 5X \text{ (kWh/year)}$$

Suppose also that we want to estimate the emission parameters for a potential platform with a 2,000,000 m^3 volume located in Germany and that all product unit loads are pallets (i.e., $e_p = 1$ for all p). Because the DC considered is in Germany, from Table 12.5 we see that $g_l^{\mathrm{CTG}} = 0.717712283$ kgCO$_2$e/kWh. Based on this, we have

$$\varepsilon^{\mathrm{A}}_{p(l,a)o} = 0.717712283(5) = 3.5885614\,\mathrm{kgCO_2e/pallet\ for\ all}\ p$$

$$\varepsilon_o = 0.717712283(400,000 + 1.2V_o) = 2,009,594.4\,\mathrm{kgCO_2e/year}$$

12.2.3 Upstream and Downstream GHG Emissions

The estimation of cradle-to-gate and gate-to-grave (GTG) GHG emissions may be difficult because, as shown in Fig. 12.5, they cover upstream and downstream SC activities that are not under the control of the company. From an SCN design perspective, CTG emission parameters $\varepsilon^{\mathrm{V}}_{pll'}$ in kgCO$_2$e/unit load must be estimated for inbound flows $F_{pll'}$ associated with supply contracts (vendors) $l \in L^{\mathrm{V}}$. When transportation means are selected by the supplier, the parameter value must include transportation emissions between source l and destination l'. The easiest way to obtain emission parameters values is to negotiate supply contracts requiring that vendors provide estimates of CO$_2$e emissions per unit load based on GHG protocol standards. When this is not reliable or possible, an alternative is to take steps to collect primary emission or activity data from vendor premises and then to use the methods described in previous sections to estimate the required parameters. For example, PepsiCo, whose brands include Tropicana juices, collects its own primary data for the growing of oranges in its vendor farms to capture differences in fertilizers, processes, and transportation means (WBCSD-WRI 2011b).

GTG emission parameters $\varepsilon^{\mathrm{D}}_{jpl'lm}$ in kgCO$_2$e/unit load must also be estimated for outbound flows $F_{jpl'lm}$ to demand zones $l \in L^{\mathrm{D}}$. These depend on the transportation means m used, and we saw how to calculate their emissions in Sect. 12.2.1, but they may also depend on the product-market policy j selected (see Sect. 7.4.3). Because market policies can be used to differentiate green products from standard products, to enable demand fulfillment despeeding options, or to capture the effect of programs to encourage the return of used products, they may have an impact on emissions ensuing from product deliveries, product use during service life, and end-of-life recovery. However, for SCNs involved in primary or intermediate transformations, it may be very difficult to determine in which end goods their products are used and a large number of end goods may be involved. In such contexts, performing a complete CTG emission analysis including all postprocessing activities may be impossible and the study may have to be limited to immediate customers.

Calculating a product's lifetime use emissions (from its energy consumption and direct release of GHGs) typically requires the product specifications and assumptions about how consumers use it. National average statistics may then be used to support calculations. Calculating emissions for the end-of-life treatment of sold products requires assumptions about the disposal methods used by consumers, and data collected by waste management providers may be helpful. Corporate GHG protocol standards (WBCSD-WRI 2011a) provide guidance on how to obtain primary and secondary emission data from SC partners and make downstream emissions calculations. Note finally that from an SCN design point of view some downstream emissions may not be relevant. As for costs (see Sect. 2.2.1), they are relevant only if they are affected by some design decisions. Otherwise, they do not have to be included explicitly in design models.

12.2.4 Optimization Model

All the SCN design models presented in previous chapters can be adapted to minimize GHG emissions relatively easily. The main difficulties encountered come from the estimation of emission parameters, as seen in the previous sections, and from the requirement to find efficient designs under multiple conflicting objectives. To illustrate the modeling impacts of our previous discussion on eco-efficiency without overly complicating the presentation, we show in this section how to extend the platform selection model presented in Sect. 8.4 to take GHG emissions into account. Because, as discussed previously (Table 12.2), transportation means selection and product-market offers may have a marked impact on life cycle GHG emissions, we extend the model to take these aspects into account. Usually, supply contract decisions should also be considered but this would require the explicit modeling of production recipes, as done in Sects. 7.5 and 8.5, which would complicate the presentation needlessly. We saw that when considering market policies it is more appropriate to maximize value added than to minimize costs (Sect. 7.4.3). The SCN design model proposed therefore seeks, as in (12.2), to maximize net operating profits (NOP) and minimize GHG emissions.

The mathematical notation used and the assumptions made in the design model are essentially the same as in Sects. 7.4.3, 8.4, and 12.2.1–12.2.3. However, in order to take into account transportation options and GHG emissions, some of the previous notations must be altered and a few additional parameters, variables, and assumptions are required. The SCN of the company considered has the following characteristics:

- All the transportation in the SCN is made by external contract carriers (or 3PLs). The company has to choose between a set M of potential annual transportation contracts. Because a contract does not necessarily cover all the arcs of the SCN, the company may need or want to sign several contracts simultaneously. A transportation contract $m \in M$ has the following features:

- The nature of the transportation means involved (type of vehicle and fuel, average speed, average shipment payload, etc.) is specified, and a function $\bar{\varepsilon}_m^{\mathrm{T}}(D)$ is given to evaluate CO_2e emissions per ton for given distances.
- The sets of SCN internal arcs $UW_{pm} \subset L^{\mathrm{U}} \times L^{\mathrm{W}}$ and demand arcs $SD_{pm} \subset L^{\mathrm{S}} \times L^{\mathrm{D}}$ over which each product $p \in P$ can be shipped using transportation means m is specified.
- The transportation rates c_{pslm}^{T} charged per unit load of product p are specified for all lanes $(s, l) \in UW_{pm} \cup SD_{pm}$.
- The minimum annual billing required $\underline{b}_m^{\mathrm{T}}$ to sign the contract is specified.
- The maximum network transportation capacity (assumed here to take the form of a maximum annual billing amount) available $\bar{b}_m^{\mathrm{T}}$ is specified (for the formulation to be valid, an upper bound must be defined even if no capacity limits are specified in the contract).
- Annual contract management costs, if any, are specified.

- Transportation means for raw materials are selected by the vendor and upstream supply costs and emissions are all associated with production activities.
- GHG emissions related to the construction, upgrade, or disposal of platforms are not included but this assumption is easily relaxed.

The following modified or additional decision variables are required to formulate the model:

Y_m^{T} Binary variable with value 1 if transportation contract $m \in M$ is selected and 0 otherwise

F_{puwm} Annual flow of product p on internal lane (u, w) when using transportation means m

F_{jpslm} Annual flow of product p on demand lane (s, l) when using transportation means m under market policy j

The following additional sets and data are also needed:

SM_{jpl} Set of (site, transportation means) pairs (s, m) that can be used to ship product p to demand zone l under market policy j

y_m^{T} Fixed annual costs incurred to manage transportation contract m (see Sect. 6.5)

c_{pslm} Unit flow cost of product p on lane (s, l) when using transportation means m (c_{pslm}^{T} plus any relevant picking, staging, and in-transit inventory holding costs)

$\varepsilon_{puo}^{\mathrm{VP}}$ GHG emissions (in kgCO_2e) per unit load of product p produced in plant u when platform o is implemented (includes upstream emissions for all raw material required to produce a unit load as well as the production process emissions)

$\varepsilon_{pso}^{\mathrm{S}}$ GHG emissions (in kgCO_2e) per unit load of product p stored and shipped from site s when platform o is implemented

$\varepsilon^{\mathrm{T}}_{puwm}$ GHG emissions (in $kgCO_2e$) per unit load of product p shipped on lane (u, w) using transportation means m ($\varepsilon^{\mathrm{T}}_{puwm} = \bar{w}_p \bar{\varepsilon}^{\mathrm{T}}_m(D_{uw})$)

Using this notation, the required bi-criterion SCN design model is formulated as follows:

$$\max NOP \text{ and } \min GHG \tag{12.10}$$

subject to

– Net operating profit definition

$$\begin{aligned}
NOP = & \sum_{p\in P}\sum_{l\in L_p^{\mathrm{D}}}\sum_{j\in J_{k(l)}}\sum_{(s,m)\in SM_{jpl}} \pi_{jp}F_{jpslm} && \text{Revenues}\\
& -\sum_{j\in J} y_j^{\mathrm{M}} Y_j^{\mathrm{M}} && \text{Market policy fixed costs}\\
& -\sum_{s\in L^{\mathrm{S}}}\sum_{o\in O_s} y_{so}^{+}Y_{so}^{+} + y_{so}Y_{so} + y_{so}^{-}Y_{so}^{-} && \text{Platform fixed costs}\\
& -\sum_{m\in M} y_m^{\mathrm{T}} Y_m^{\mathrm{T}} && \text{Carrier contract fixed costs}\\
& -\sum_{u\in L^{\mathrm{U}}}\sum_{o\in O_u}\sum_{p\in P_u} c_{puo}^{\mathrm{P}} X_{puo}^{P} && \text{Production costs}\\
& -\sum_{s\in L^{\mathrm{S}}}\sum_{o\in O_s}\sum_{p\in P} c_{pso}^{\mathrm{S}} X_{pso}^{S} && \text{Storage costs}\\
& -\sum_{p\in P}\sum_{m\in M}\sum_{(u,w)\in UW_{pm}} c_{puwm}F_{puwm} && \text{Internal flow costs}\\
& -\sum_{p\in P}\sum_{l\in L_p^{\mathrm{D}}}\sum_{j\in J_{k(l)}}\sum_{(s,m)\in SM_{jpl}} c_{pslm}F_{jpslm} && \text{Demand flow costs}
\end{aligned} \tag{12.11}$$

– GHG emissions definition

$$\begin{aligned}
GHG = & \sum_{s\in L^{\mathrm{S}}}\sum_{o\in O_s} \varepsilon_{so}Y_{so} && \text{Platform emissions}\\
& + \sum_{u\in L^{\mathrm{U}}}\sum_{o\in O_u}\sum_{p\in P_u} \varepsilon_{puo}^{\mathrm{VP}} X_{puo}^{P} && \text{Upstream emissions}\\
& + \sum_{s\in L^{\mathrm{S}}}\sum_{o\in O_s}\sum_{p\in P} \varepsilon_{pso}^{\mathrm{S}} X_{pso}^{S} && \text{Shipping emissions}\\
& + \sum_{p\in P}\sum_{m\in M}\sum_{(u,w)\in UW_{pm}} \varepsilon_{puwm}^{\mathrm{T}} F_{puwm} && \text{Internal flow emissions}\\
& + \sum_{p\in P}\sum_{l\in L_p^{\mathrm{D}}}\sum_{j\in J_{k(l)}}\sum_{(s,m)\in SM_{jpl}} \varepsilon_{jpslm}^{\mathrm{D}} F_{jpslm} && \text{Downstream emissions}
\end{aligned} \tag{12.12}$$

– Demand constraints

$$Y_j^{\mathrm{M}} \underline{d}_{pl} \le \sum_{(s,m)\in SM_{jpl}} F_{jpslm} \le Y_j^{\mathrm{M}} \bar{d}_{jpl}, \quad p\in P, l\in L_p^{\mathrm{D}}, j\in J_{k(l)} \tag{12.13}$$

$$\sum_{j \in J_k} Y_j^{\mathrm{M}} \leq 1, \quad k \in K \tag{12.14}$$

– Platform selection constraints

$$\sum_{o \in O_s} Y_{so} \leq 1, \quad s \in L^{\mathrm{S}} \tag{12.15}$$

$$\left.\begin{array}{ll} Y_{so} + Y_{so}^{-} + Y_{s,re(o)}^{+} = 1 & \text{if } Y_{so}^{0} = 1 \\ Y_{so} - Y_{so}^{+} = 0 & \text{if } Y_{so}^{0} = 0 \end{array}\right\}, \quad s \in L^{\mathrm{S}}, o \in O_s \tag{12.16}$$

– Platform activity-level definition and flow equilibrium constraints

$$\sum_{o \in O_u} X_{puo}^{\mathrm{P}} = \sum_{l \in L_p^{\mathrm{D}}} \sum_{j \in J_{k(l)}} \sum_{m | (u,m) \in SM_{jpl}} F_{jpulm} + \sum_{m \in M} \sum_{w | (u,w) \in UW_{pm}} F_{puwm}, \quad u \in L^{\mathrm{U}}, p \in P_u \tag{12.17}$$

$$\sum_{o \in O_s} X_{pso}^{\mathrm{S}} = \sum_{l \in L_p^{\mathrm{D}}} \sum_{j \in J_{k(l)}} \sum_{m | (s,m) \in SM_{jpl}} F_{jpslm}, \quad s \in L^{\mathrm{S}}, p \in P \tag{12.18}$$

$$\sum_{m \in M} \sum_{u | (u,w) \in UW_{pm}} F_{puwm} = \sum_{o \in O_w} X_{pwo}^{\mathrm{S}}, \quad w \in L^{\mathrm{W}}, p \in P \tag{12.19}$$

– Production and distribution capacity constraints

$$X_{puo}^{\mathrm{P}} \leq \bar{b}_{puo} Y_{uo}, \quad u \in L^{\mathrm{U}}, o \in O_u, p \in P_u \tag{12.20}$$

$$\sum_{p \in P} e_p X_{pso}^{\mathrm{S}} \leq \bar{b}_{so} Y_{so}, \quad s \in L^{\mathrm{S}}, o \in O_s \tag{12.21}$$

– Transportation contract restrictions

$$\begin{aligned} \underline{b}_m^{\mathrm{T}} Y_m^{\mathrm{T}} \leq & \sum_{p \in P} \sum_{(u,w) \in UW_{pm}} c_{puwm}^{\mathrm{T}} F_{puwm} \\ & + \sum_{p \in P} \sum_{l \in L_p^{\mathrm{D}}} \sum_{j \in J_{k(l)}} \sum_{s | (s,m) \in SM_{jpl}} c_{pslm}^{\mathrm{T}} F_{jpslm} \leq \bar{b}_m^{\mathrm{T}} Y_m^{\mathrm{T}}, \quad m \in M \end{aligned} \tag{12.22}$$

The usual restrictions on the binary or nonnegative value of the decision variables must also be included. This model is a bi-objective MIP and it can be solved to find efficient SCNs by proceeding as indicated in Sect. 12.1.2.

Example 12.3 Let us go back to the SCN design problem solved in Ex. 8.3, which involves a choice between alternative platforms at current and potential sites and seeks to minimize costs under a fixed demand. We now assume that the company wants to maximize NOP, considers alternative platforms engineered to reduce GHG emissions, and examines the possible use of an additional, less polluting transportation means. Also, unit flow costs are equal to carrier transportation rates, that is, picking, staging, and in-transit inventory holding costs are neglected. The modified cost, revenue, and capacity data required under these revised conditions are given in Fig. 12.7.

To simplify, we assume that product-market policies are predetermined, that markets correspond to demand zones, and that all sites can use any transportation means to deliver products to demand zones. Consequently,

Platforms data:

			Fixed costs			Production costs		Storage costs		Production capacity		Distrib.
Site	Plateform	State	Open	Use	Close	$p=1$	$p=2$	$p=1$	$p=2$	$p=1$	$p=2$	capacity
The Hague	1 (status quo)	1		€ 21,000,000	€ 10,000,000	€ 80.00	€ 110.00	€ 5	€ 7	90000	210000	100000
The Hague	2 (expansion)	0	€ 1,500,000	€ 25,000,000		€ 70.00	€ 100.00	€ 5	€ 7	100000	225000	100000
Barcelona	1 (addition)	0	€ 2,000,000	€ 12,000,000		€ 70.00	€ 100.00	€ 5	€ 6	50000	100000	130000
Dusseldorf	1 (status quo)	1		€ 5,500,000	€ 500,000			€ 6	€ 7			100000
Bordeaux	1 (contract)	0	€ 1,000,000	€ 5,000,000				€ 9	€ 10			120000
Bordeaux	2 (contract)	0	€ 1,000,000	€ 4,000,000				€ 11	€ 12			100000
Turin	1 (status quo)	1		€ 4,500,000	€ 500,000			€ 7	€ 8			100000
Turin	2 (contract)	0	€ 500,000	€ 2,500,000				€ 7	€ 8			50000

Unit transportation costs to markets with transportation means $m=1$:

	Product $p=1$ markets						Product $p=2$ markets					
Site	Germany	France	Italy	Spain	Portugal	Switzerl.	Germany	France	Italy	Spain	Portugal	Switzerl.
The Hague	€ 23	€ 16	€ 53	€ 56	€ 73	€ 26	€ 25	€ 17	€ 58	€ 61	€ 80	€ 28
Barcelona	€ 60	€ 33	€ 43	€ 20	€ 40	€ 33	€ 66	€ 36	€ 47	€ 22	€ 44	€ 36
Dusseldorf	€ 18	€ 16	€ 50	€ 56	€ 73	€ 20	€ 19	€ 17	€ 55	€ 61	€ 80	€ 22
Bordeaux	€ 53	€ 20	€ 50	€ 23	€ 36	€ 33	€ 58	€ 22	€ 55	€ 25	€ 39	€ 36
Turin	€ 36	€ 26	€ 23	€ 50	€ 66	€ 13	€ 39	€ 28	€ 25	€ 55	€ 72	€ 14
Min demand	14000	16000	10000	17000	25000	10000	56000	45000	30000	30000	45000	28000
Max demand	17500	20000	12000	20000	30000	12000	68000	50000	35000	32000	50000	32000
Sales price	€ 270	€ 265	€ 265	€ 255	€ 260	€ 275	€ 310	€ 295	€ 305	€ 290	€ 300	€ 305

Unit transportation costs to DCs with transportation means $m=1$:

	DC (product $p=1$)			DC (product $p=2$)		
Site	Dusseldorf	Bordeaux	Turin	Dusseldorf	Bordeaux	Turin
The Hague	€ 6	€ 36	€ 40	€ 6	€ 39	€ 44
Barcelona	€ 46	€ 20	€ 30	€ 50	€ 22	€ 33

Trans. means	Fixed cost	Lower bound	Upper bound
1	€ 25,000	€ 1,000,000	€ 15,000,000
2	€ 25,000	€ 1,000,000	€ 15,000,000

Unit transportation costs to markets with transportation means $m=2$:

	Product $p=1$ markets						Product $p=2$ markets					
Site	Germany	France	Italy	Spain	Portugal	Switzerl.	Germany	France	Italy	Spain	Portugal	Switzerl.
The Hague	€ 25	€ 17	€ 58	€ 61	€ 80	€ 28	€ 27	€ 19	€ 63	€ 67	€ 87	€ 31
Barcelona	€ 66	€ 36	€ 47	€ 22	€ 44	€ 36	€ 72	€ 39	€ 51	€ 24	€ 48	€ 39
Dusseldorf	€ 19	€ 17	€ 55	€ 61	€ 80	€ 22	€ 21	€ 19	€ 60	€ 67	€ 87	€ 24
Bordeaux	€ 58	€ 22	€ 55	€ 25	€ 39	€ 36	€ 63	€ 24	€ 60	€ 27	€ 43	€ 39
Turin	€ 39	€ 28	€ 25	€ 55	€ 72	€ 14	€ 43	€ 31	€ 27	€ 60	€ 79	€ 15

Unit transportation costs to DCs with transportation means $m=2$:

	DC (product $p=1$)			DC (product $p=2$)		
Site	Dusseldorf	Bordeaux	Turin	Dusseldorf	Bordeaux	Turin
The Hague	€ 6	€ 39	€ 44	€ 7	€ 43	€ 48
Barcelona	€ 50	€ 22	€ 33	€ 55	€ 24	€ 36

Unit loads/product	
e_1	e_2
1	1

Fig. 12.7 Cost, revenue, and capacity data spreadsheet for Example 12.3

index j and binary policy selection variables can be dropped, and the revenue term in (12.11) reduces to

$$\sum_{p \in P} \sum_{l \in L_p^D} \pi_{pl} (\sum_{s \in L^S} \sum_{m \in M} F_{pslm})$$

and demand constraints (12.13) and (12.14) reduce to

$$\underline{d}_{pl} \leq \sum_{s \in L^S} \sum_{m \in M} F_{pslm} \leq \bar{d}_{pl}, \quad p \in P, l \in L_p^D$$

The bounds on the demand are given in Fig. 12.7.

The current SCN of the company generates about 90,000 tons of CO_2e emissions per year and the company would like to reengineer its network to decrease these emissions. The plant upgrade now considered for The Hague is based on top green technologies. All the GHG emissions data required are given in Fig. 12.8.

In order to obtain efficient SCN designs, we use the ε-constraint method introduced in Sect. 12.1.2. The models solved therefore include all the constraints defined previously, but the bi-criteria objective function (12.10) is replaced by the following NOP maximization objective and GHG emissions bound:

$$\max NOP \text{ subject to } GHG \leq GHG^{\max}$$

where $GHG^{\max}$ is an upper bound on annual emissions expressed in $kgCO_2e$. The MIPs thus obtained include 204 variables (26 binary) and 65 constraints and they cannot be solved with the default solver included in Excel because it is limited to 200 variables. They are easily solved however with the Premium Solver upgrade sold by Frontline Systems (www.solver.com).

To initialize the solution process, the model is solved with an emissions bound equal to the current emissions level of the company, that is, $GHG^{\max} = 90,000,000$ $kgCO_2e$. The SCN design thus obtained is essentially the same as that found in Ex. 8.3 and it generates emissions of 84,334,000 $kgCO_2e$. Thus, simply by maximizing *NOP*, the GHG emissions are reduced. This is not surprising because transportation emissions are usually highly correlated with transportation costs (because they both depend on distance). In order to explore additional reduction possibilities, the model is solved several times with gradually reduced emissions bounds in order to detect changes in design leading to lower emissions. The efficient frontier shown in Fig. 12.9 results. Alternative designs on the frontier are given in Table 12.6 with corresponding network costs and emissions.

The solutions in Table 12.6 provide indications on how to proceed to improve the eco-efficiency of the company. Design 2 indicates that the first

Platform emissions in $kgCO_2e$:

Site	Plateform	Platform emissions	Upstream emissions $p=1$	Upstream emissions $p=2$	Shipping emissions $p=1$	Shipping emissions $p=2$
The Hague	1 (status quo)	2,000,000	80.0	110.0	5.0	5.0
The Hague	2 (expansion)	1,500,000	60.0	80.0	3.0	3.0
Barcelona	1 (addition)	1,500,000	80.0	110.0	5.0	5.0
Dusseldorf	1 (status quo)	750,000			5.0	5.0
Bordeaux	1 (contract)	250,000			2.0	2.0
Bordeaux	2 (contract)	500,000			5.0	5.0
Turin	1 (status quo)	600,000			6.0	6.0
Turin	2 (contract)	300,000			6.0	6.0

Downstream emissions in $kgCO_2e$/unit load with transportation means $m=1$:

	Product $p=1$ markets						Product $p=2$ markets					
Site	Germany	France	Italy	Spain	Portugal	Switzerl.	Germany	France	Italy	Spain	Portugal	Switzerl.
The Hague	70	50	160	170	220	80	84	60	192	204	264	96
Barcelona	180	100	130	60	120	100	216	120	156	72	144	120
Dusseldorf	55	50	150	170	220	60	66	60	180	204	264	72
Bordeaux	160	60	150	70	110	100	192	72	180	84	132	120
Turin	110	80	70	150	200	40	132	96	84	180	240	48

Internal flow emissions in $kgCO_2e$/unit load with transportation means $m=1$:

	DC (product $p=1$)			DC (product $p=2$)		
Site	Dusseldorf	Bordeaux	Turin	Dusseldorf	Bordeaux	Turin
The Hague	20	110	120	24	132	144
Barcelona	140	60	90	168	72	108

Downstream emissions in $kgCO_2e$/unit load with transportation means $m=2$:

	Product $p=1$ markets						Product $p=2$ markets					
Site	Germany	France	Italy	Spain	Portugal	Switzerl.	Germany	France	Italy	Spain	Portugal	Switzerl.
The Hague	56	40	128	136	176	64	70	50	160	170	220	80
Barcelona	144	80	104	48	96	80	180	100	130	60	120	100
Dusseldorf	44	40	120	136	176	48	55	50	150	170	220	60
Bordeaux	128	48	120	56	88	80	160	60	150	70	110	100
Turin	88	64	56	120	160	32	110	80	70	150	200	40

Internal flow emissions in $kgCO_2e$/unit load with transportation means $m=2$:

	DC (product $p=1$)			DC (product $p=2$)		
Site	Dusseldorf	Bordeaux	Turin	Dusseldorf	Bordeaux	Turin
The Hague	16	88	96	20	110	120
Barcelona	112	48	72	140	60	90

Fig. 12.8 GHG emission data spreadsheet for Example 12.3

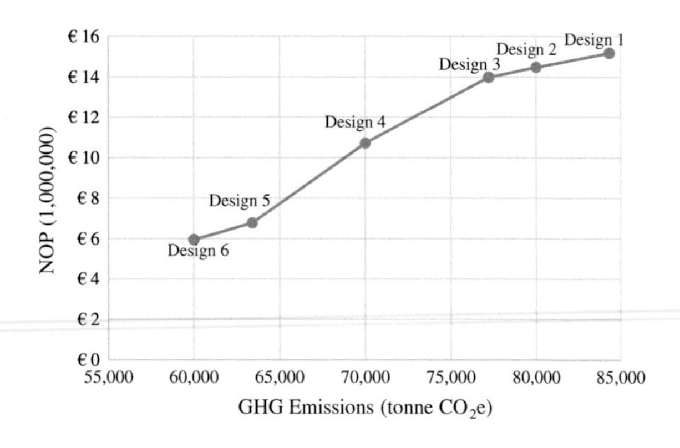

Fig. 12.9 SCN designs efficient frontier

Table 12.6 Efficient SCN designs

Design decisions	Design 1	Design 2	Design 3	Design 4	Design 5	Design 6
The Hague—status quo	1	1	1	**0**	0	0
The Hague 2—upgrade	0	0	0	**1**	1	1
Barcelona—new plant	1	1	1	1	1	1
Dusseldorf—status quo	1	1	1	1	1	1
Bordeaux—contract 1	0	0	0	0	0	0
Bordeaux—contract 2	0	0	0	0	0	0
Turin—status quo	0	0	0	0	0	0
Turin—new contract	1	1	1	1	**0**	0
Transportation means 1	1	1	**0**	0	**1**	**0**
Transportation means 2	0	**1**	**1**	1	1	1
Revenues	€ 109,950,000	€ 109,950,000	€ 109,857,161	€ 109,633,380	**€ 95,840,000**	€ 95,444,909
Platform fixed costs	€ 43,500,000	€ 43,500,000	€ 43,500,000	€ 49,000,000	€ 46,500,000	€ 46,500,000
Carrier fixed costs	€ 25,000	€ 50,000	€ 25,000	€ 25,000	€ 50,000	€ 25,000
Production costs	€ 36,790,000	€ 36,790,000	€ 36,755,959	€ 34,399,460	€ 30,120,000	€ 30,015,636
Storage costs	€ 2,406,500	€ 2,417,000	€ 2,432,334	€ 2,427,112	€ 2,046,000	€ 2,054,545
Transportation costs	€ 12,044,000	€ 12,709,333	€ 13,155,602	€ 13,069,073	€ 10,344,000	€ 10,902,164
Net Operating Profit	*€ 15,184,500*	*€ 14,483,667*	*€ 13,988,266*	*€ 10,712,734*	*€ 6,780,000*	*€ 5,947,564*
Platform emissions	4,550,000	4,550,000	4,550,000	4,050,000	3,750,000	3,750,000
Upstream emissions	38,290,000	38,290,000	38,255,959	31,965,568	27,960,000	27,790,545
Shipping emissions	1,941,000	1,941,000	1,939,143	1,734,668	1,450,000	1,442,545
Internal flow emissions	8,352,000	7,584,000	6,763,197	6,684,127	2,272,000	1,874,182
Downstream emissions	31,201,000	27,635,000	25,691,701	25,565,637	27,968,000	25,142,727
Total GHG ($kgCO_2e$)	*84,334,000*	*80,000,000*	*77,200,000*	*70,000,000*	*63,400,000*	*60,000,000*

action to take is to gradually switch from the current transportation means to the greener alternative. A comparison of Design 3 with Design 1 shows that using green transportation means 2 for all shipments reduces emissions by 8.5 %, at the cost of a 7.9 % reduction in profits (1.16 % increase in costs). Design 4 goes further by upgrading the platform at The Hague. This reduces emissions by 17 % but it also reduces profits by 29.5 %. Designs 5 and 6 go even further by closing the Turin DC, which can be done only if sales are reduced in some product market. It is by studying all these possibilities with regard to their sustainability strategy that top management would make a final decision.

As discussed in Sect. 12.1.2, several countries and states are implementing emissions trading schemes, such as the EU Emission Trading System (EU ETS) launched in 2005, which currently covers more than 11,000 power stations and industrial plants in 31 countries (European Commission 2013). These usually take the form of cap-and-trade programs. Under such schemes, allowances for GHG emissions are allocated to companies or auctioned and subsequently can be traded. Prices are determined by supply and demand and trading is done between buyers and sellers through organized exchanges or intermediaries. The model presented in this section is easily adapted to take such schemes into account.

To illustrate this, consider the following simplified trading context. First assume that traded emissions are measured in ton CO_2e (currently, EU ETS covers only CO_2, N_2O, and PFCs emissions). Assume also that under the system considered, the company must comply with a maximum annual emissions target GHG^{caps} in ton CO_2e. Within specified limits, the company, however, can buy or sell carbon credits, one credit corresponding to the right to emit one ton of CO_2e. Let

- CC^+ Credits purchased during the year on the carbon market
- CC^- Credits sold during the year on the carbon market
- $\overline{CC}^+$ Maximum number of credits that can be purchased during the year
- $\overline{CC}^-$ Maximum number of credits that can be sold during the year
- π^{GHG} Average price obtained per credit sold during the year
- c^{GHG} Average price paid per credit purchased during the year

where CC^+ and CC^- are decision variables. To take this trading scheme into account, the following constraints must be added to the previous model:

$$0.001GHG + CC^+ - CC^- \leq GHG^{\text{caps}}, \quad 0 \leq CC^+ \leq \overline{CC}^+, \; 0 \leq CC^- \leq \overline{CC}^-$$

The definition of the *NOP* must also be revised to include credit sale revenues and credit purchase expenses, that is, the cash flows $\pi^{\text{GHG}}CC^- - c^{\text{GHG}}CC^+$ must be added to the right-hand side of (12.11).

12.3 Reverse and Closed-Loop SCNs

As discussed in Sect. 12.1.1, the recuperation and revalorization of used products can take several forms, depending on the industrial context. It should also be clear that not all used products or components can or should be revalorized. For instance, worn-out or technically obsolete products are worthless unless some of the material they include can be recycled. Also, products are worth remanufacturing and functional components disassembling and reconditioning only if a market exist for the revalorized items. This could be a secondhand market independent of the OEM or an internal demand for parts, subassemblies, or used products by OEM manufacturing or remanufacturing activities. Even if these possibilities exist, however, the opportunity is interesting only if the efforts made to remanufacture, disassemble, or recycle used units generate some savings (or significantly reduce environmental impacts). In some contexts, however, OEMs may, by law, have to collect and ecologically dispose used products even if this is not economically viable.

From a network reengineering point of view, the approach used to design reverse or closed-loop SCNs is pretty much the same as for forward SCNs. The strategic decisions involved include the location of facility sites; the choice of facility platforms among a set of potential platforms with distinct forward and/or reverse activity set, technologies, and capacity; the elaboration of demand and return shaping policies; the allocation of flows to facilities; and so on. One important additional decision for OEMs is whether they should get involved in the first place in used products collection and revalorization activities or not. The complexity of design models also pretty much depends on the same factors, that is, whether they are deterministic or stochastic, static or dynamic (multi-period), two-echelon or multi-echelon, single objective or multi-objective, and so forth. Two classes of problems can be distinguished: pure reverse logistics network (RLN) and closed-loop SCN (CLSCN) design problems.

In some contexts, EOMs are not involved in reverse flows. Used product collection, sorting, disposal, disassembly, and remanufacturing activities are performed by independent parties operating RLNs. The items collected are typically low-value nonproprietary products that were originally made by several brand manufacturers. The products, components, and materials remanufactured or recycled are sold on a secondhand market. Metal, glass, and sand recycling businesses are typical examples. Cellular phone and truck tire refurbishing companies also run RLNs. The optimization models required to design these networks are very similar to the ones studied in previous chapters. The main difference is that upstream flows are not decision variables but they result from the used product return behavior of users. Another difference is that waste disposal is usually modeled explicitly.

EOMs need to set up reverse product flows mainly to cope with commercial returns, warranty replacements, buyback programs, or leased equipment returns and upgrades, that is, to comply with some kind of contractual engagement. In such contexts, the resources used to operate reverse flows are typically the same as those used for forward flows so that forward and reverse activities are tightly coupled in the resulting CLSCN. In other words, defective or used products are returned by customers to sales or service points, or collected at customer premises, and then sent back using reversed distribution channels. This typically occurs, for example, for leased copiers, several types of electronic equipment, and car parts or subassemblies. In this context, inspection, sorting, forwarding, and disposal processes are typically implemented in DCs and disassembly-remanufacturing processes in plants. A typical case is illustrated in Fig. 12.10. We show how to formulate a CLSCN design model for this case later on in this section.

As explained previously, in order to help reduce and renew the natural resources consumed or to obtain rare resources at a reduced cost, OEMs may want to reintegrate transformed resources through several revalorization paths, which brings us back to the generic CLSCN activity graph presented in Fig. 12.2. In such a context, there is no reason to impose a tight coupling between forward and reverse activities. If revalorized products are sold in a secondhand market in developing countries, for example, it may be much better to remanufacture products in these countries than in the original plants. Similarly, if a secondhand spare parts market exists, it is often more effective to set up independent collection, disassembly, or cannibalization centers close to these markets. To achieve economies of scale, it may also be more efficient to set up independent product remanufacturing or material recycling plants. When considering the situation this way, designers must formulate multi-platform multi-period models similar to the one presented in Sect. 8.5 for forward SCNs (or in Sect. 9.4 for multinational SCNs). Clearly, some adaptations are required but the approach presented in this book involving a characterization of the design problem in terms of an activity graph and of alternative network resources (platforms) still applies.

In order to illustrate how our modeling approach can be adapted to the design of CLSCNs, we show in what way the model presented in Sect. 7.3.2 to solve two-echelon production–distribution network design problems (see Fig. 7.10) can be extended to solve the coupled CLSCN design problem depicted in Fig. 12.10.

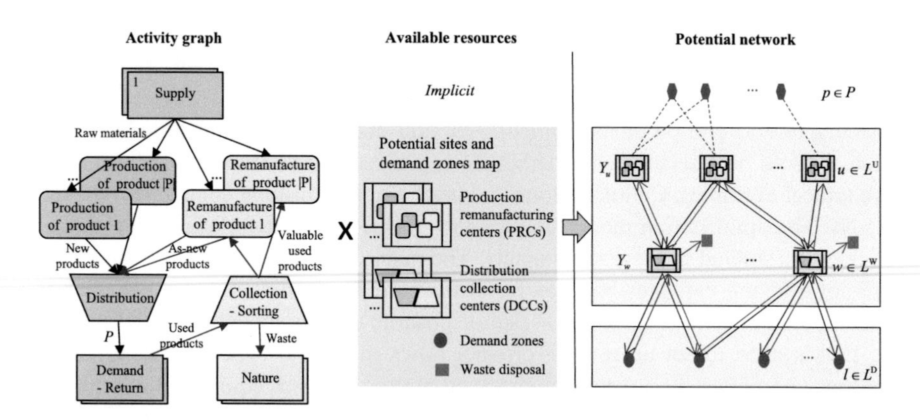

Fig. 12.10 Two-Echelon coupled closed-loop SCN design problem

We assume that the problem features described in Sect. 7.3.2 for the forward SCN still apply. The main features of the reverse SCN are the following:

- Products are made-to-stock by an OEM and distributed to external SPs grouped into demand zones. The SPs sell the products to final customers. Used products can be returned by customers to the SPs and the OEM must recover the returned used products. However, customers do not necessarily return all the products they buy at the end of their service life. The products returned can be remanufactured if they are in a sufficiently good state.
- The collected products are inspected at the distribution centers to determine their state. If they are in a good state, they can be shipped to the plants to be remanufactured. If not, the products are prepared for disposal and shipped to a waste disposal site. The resources used in distribution–collection centers (DCCs) to process used products are not the same as those required to store and ship new products, that is, they constitute distinct activity centers.
- Remanufacturing activities are performed in the same activity centers as production activities and they share the same resources. When a product is remanufactured, however, some operations can be skipped and little new raw material is required. As a consequence, the capacity used is just a fraction of what is needed for new products, and considerable savings in raw material, labor, and other production costs are made. Also, the products remanufactured can be considered to be new. The platform of production–remanufacturing centers (PRCs) may be designed to process only a subset of the product families manufactured by the OEM.

As in Sect. 7.3.2, the OEM wants to design its CLSCN to minimize total relevant costs for a base year. To formulate the CLSCN design model, all the sets, data, and decision variables defined in Sect. 7.3.2 for the forward SCN are still required. However, the following additional data related to the reverse SCN are also needed:

c^{P}_{pw} Average unit processing (inspection and handling) cost of products p shipped to PRCs from DCC w

c^{R}_{pl} Average amount paid to SPs in demand zone l for the recuperation of product p

c^{-}_{pu} Unit cost savings realized at PRC u when product p is remanufactured from a used product

c^{R}_{pls} Unit picking, shipping, transportation, and in-transit holding cost associated with the reverse shipment of product p from node $l \in L^{\mathrm{W}} \cup L^{\mathrm{D}}$ to site $s \in L^{\mathrm{S}}$

f^{R}_{plw} $= c^{\mathrm{R}}_{pl} + c^{\mathrm{R}}_{plw}, \quad l \in L^{\mathrm{D}}, w \in L^{\mathrm{W}}$

f^{R}_{pwu} $= c^{\mathrm{P}}_{pw} + c^{\mathrm{R}}_{pwu} - c^{-}_{pu}, \quad w \in L^{\mathrm{W}}, u \in L^{\mathrm{U}}$

f^{D}_{pw} Average unit disposal (inspection, preparation, handling, and transportation) cost of products p shipped to a waste disposal site at DCC w

b^{F}_{w} Maximum annual outflow in standard unit loads from DCC $w \in L^{\mathrm{W}}$ to demand zones imposed by available storage space

b_w^{R} Annual DCC $w \in L^{\mathrm{W}}$ reverse flow processing capacity in a standard unit (e.g., processing time)
q_p^{R} Number of standard capacity units required at a DCC to process products p shipped to PRCs to be remanufactured
q_p^{D} Number of standard capacity units required at a DCC to process products p sent to waste disposal
λ_{pl} Customer return rate for product p in demand zone l
γ_p Minimum waste disposal proportion for product p
χ_p Fraction of the capacity used to make a new product p saved when the product is remanufactured

The following reverse flow decision variables are also needed:

F_{plw}^{R} Annual flow of used product p collected by DCC w from demand zone l
F_{pwu}^{R} Annual flow of used product p delivered to PRC u by DCC w
F_{pw}^{D} Annual flow of used product p sent to a waste disposal site by DCC w

Using this additional notation, the two-echelon coupled CLSCN design problem considered can be formulated as follows:

$$C^{\mathrm{SCN}} = \min \sum_{s \in L^{\mathrm{S}}} y_s Y_s + \sum_{p \in P} \sum_{u \in L^{\mathrm{U}}} \sum_{w \in L^{\mathrm{W}}} f_{puw} F_{puw} + \sum_{p \in P} \sum_{w \in L^{\mathrm{W}}} \sum_{l \in L_p^{\mathrm{D}}} f_{pwl} F_{pwl} + \sum_{p \in P} \sum_{l \in L_p^{\mathrm{D}}} \sum_{w \in L^{\mathrm{W}}} f_{plw}^{\mathrm{R}} F_{plw}^{\mathrm{R}} + \sum_{p \in P} \sum_{w \in L^{\mathrm{W}}} \sum_{u \in L^{\mathrm{U}}} f_{pwu}^{\mathrm{R}} F_{pwu}^{\mathrm{R}} + \sum_{p \in P} \sum_{w \in L^{\mathrm{W}}} f_{pw}^{\mathrm{D}} F_{pw}^{\mathrm{D}} \tag{12.23}$$

subject to

- Demand constraints

$$\sum_{w \in L^{\mathrm{W}}} F_{pwl} = d_{pl}, \quad p \in P, l \in L_p^{\mathrm{D}} \tag{12.24}$$

- Product returns recovery constraints

$$\sum_{w \in L^{\mathrm{W}}} F_{plw}^{\mathrm{R}} = \lambda_{pl} d_{pl}, \quad p \in P, l \in L_p^{\mathrm{D}} \tag{12.25}$$

– Production capacity constraints

$$\sum_{w \in L^{\mathrm{W}}} F_{puw} - \chi_p \sum_{w \in L^{\mathrm{W}}} F^{\mathrm{R}}_{pwu} \leq b_{pu} Y_u, \quad u \in L^{\mathrm{U}}, p \in P_u \tag{12.26}$$

– DCC forward distribution capacity constraints

$$\sum_{p \in P} e_p \Big(\sum_{l \in L^{\mathrm{D}}_p} F_{pwl}\Big) \leq b_w Y_w, \quad w \in L^{\mathrm{W}} \tag{12.27}$$

– DCC reverse flow processing capacity constraints

$$\sum_{p \in P} \Big(q^{\mathrm{R}}_p \sum_{u \in L^U} F^{\mathrm{R}}_{pwu} + q^{\mathrm{D}}_p F^{\mathrm{D}}_{pw}\Big) \leq b_w Y_w, \quad w \in L^{\mathrm{W}} \tag{12.28}$$

– Minimum waste disposal constraints

$$F^{\mathrm{D}}_{pw} \geq \gamma_p \sum_{l \in L^{\mathrm{D}}_p} F^{\mathrm{R}}_{pwl}, \quad p \in P, w \in L^{\mathrm{W}} \tag{12.29}$$

– Flow equilibrium constraints

$$\sum_{w \in L^{\mathrm{W}}} F_{puw} \geq \sum_{w \in L^{\mathrm{W}}} F^{\mathrm{R}}_{pwu}, \quad u \in L^{\mathrm{U}}, p \in P_u \tag{12.30}$$

$$\sum_{u \in L^{\mathrm{U}}} F_{puw} = \sum_{l \in L^{\mathrm{D}}_p} F_{pwl}, \quad p \in P, w \in L^{\mathrm{W}} \tag{12.31}$$

$$\sum_{l \in L^{\mathrm{D}}_p} F^{\mathrm{R}}_{pwl} = \sum_{u \in L^{\mathrm{U}}} F^{\mathrm{R}}_{pwu} + F^{\mathrm{D}}_{pw}, \quad p \in P, w \in L^{\mathrm{W}} \tag{12.32}$$

The usual restrictions on the binary or nonnegative value of the decision variables must also be added.

Production capacity constraints (12.26) account for the fact that the capacity available for each product at a PRC is shared by manufacturing and remanufacturing activities. The second term on the left is the capacity saved when a product is remanufactured from a used product. Constraints (12.27) and (12.28) account for the fact that distribution and reverse processing activities at the DCC use distinct resources. The resulting MIP can be solved using standard solvers. In practice, customer return rates and minimum waste disposal proportions are clearly random variables. The stochastic programming approach studied in Chap. 11 can be used to take this into account. Also, for CLSCN design models, the impact of the decisions made on the environment should be taken into account. A second objective aiming to minimize GHG emissions is easily added to the model by proceeding as indicated in Sect. 12.2.

Review Questions

12.1 What is the aim of sustainable development?
12.2 Should businesses be concerned by sustainable development?
12.3 How does the planet interact with SCNs?
12.4 How do people interact with SCNs?
12.5 How can the natural resources consumed by SCNs be reduced?
12.6 What is the difference between a reverse logistics network and a closed-loop logistics network?
12.7 Are all wastes produced by SCNs equally damaging to the environment?
12.8 How can businesses be socially responsible?
12.9 How can the triple bottom line required for sustainable development be taken into account in SCN design?
12.10 Why is a holistic approach required to design sustainable SCNs?
12.11 What kind of adjustments to design model solution spaces are required to formulate sustainable development design models?
12.12 What is the life cycle assessment (LCA) approach?
12.13 Give examples of available life cycle impact measurement hierarchies. Are these hierarchies all equivalent?
12.14 How can we proceed to solve multi-objective MIPs?
12.15 How are GHG emissions typically measured?
12.16 What is the global warming potential of greenhouse gases?
12.17 What are the main SCN decarbonization opportunities?
12.18 What is meant by cradle-to-gate and GHG emissions?
12.19 Why is energy consumption important to calculate GHG emissions?
12.20 How can we proceed to calculate the CO_2e emissions of a transportation means?
12.21 What kinds of CO_2e emissions are associated with SCN facilities?
12.22 How can we proceed to calculate the cradle-to-gate and GHG CO_2e emissions of a SCN?
12.23 Are eco-efficient SCN design models much more difficult to solve than standard SCN design models? Why?
12.24 What are the main types of reverse and closed-loop SCNs found in practice?
12.25 Can the approach proposed in previous chapters to formulate forward SCN design models be used to formulate reverse or closed-loop SCNs? Explain.

Exercises

Exercise 12.1 Suppose that GHG emissions must be estimated for palletized product shipments made in Europe with 24 t trucks. The trucks and shipments made have the following characteristics:

- Empty truck weight: 9 ton
- Fuel: diesel
- Speed: normal speed and traffic conditions
- Products type: Volume goods
- Unit loads: 0.25 ton pallets (on average)
- Topography: highways on hilly ground
- Loaded trip factor: 1.4
- Average number of unit loads per shipment: 20 pallets

Derive a formula to calculate CO_2e emissions per pallet as a function of distance. Use this formula to calculate the emissions of a 20-pallet shipment from Geneva to Madrid.

Exercise 12.2 A company is currently operating six DCs in Europe and it intends to open a new 1,000,000 m^3 DC in France. All the company products are stored and shipped in standard pallets. Based on its electricity bills, the company has made the observations listed in Table 12.7 for each of its current DCs. Using the data in Table 12.7, estimate a regression model relating energy consumption to DC cubic space and annual throughput. Using this function:

- Estimate CO_2e emissions per pallet shipped from the new DC in France
- Estimate the yearly GHG emissions associated with the use of this new DC

Exercise 12.3 Return to Ex. 12.3 in Sect. 12.2.4. As indicated in the problem statement, this example is based on Ex. 8.3, which was solved using the Excel spreadsheet and solver model shown in Fig. 8.18. Starting from this Excel spreadsheet and using the additional data given in Figs. 12.7 and 12.8, construct an Excel solver model to solve the problem discussed in Ex. 12.3. Verify that the solutions you obtain correspond to the solutions given in Fig. 12.9 and Table 12.6.

Table 12.7 Exercise 12.2 current DCs data

DC	Cubic space (m^3)	Annual throughput (pallets)	Electricity consumption (kWh)
1	1,200,000	140,000	3,000,000
2	600,000	50,000	1,500,000
3	1,000,000	240,000	4,000,000
4	500,000	120,000	2,000,000
5	1,500,000	280,000	5,000,000
6	800,000	130,000	2,500,000

Exercise 12.4 A cellphone remanufacturer is collecting used cellphones from SPs using regional collection, inspection, and disposal centers. The phones that are in a good state and not obsolete are then remanufactured in national plants. The activity graph of its SCN is similar to the reverse logistics network depicted in the CLSCN of Fig. 12.10. Propose a RLN design model to optimize the SCN of the company.

Exercise 12.5 Model (12.23)–(12.32) was proposed to help design two-echelon coupled CLSCNs that have a cost minimization objective. However, in this context, GHG emissions usually are also an important issue. Propose a second objective function that could be added to the model to minimize CO_2e emissions. How would you proceed to solve the resulting model?

Exercise 12.6 Reconsider the situation of the OEM described in Fig. 12.10. Suppose now that the company believes it could improve its situation by considering the possibility of implementing inspection, sorting, and shipping activities in dedicated facilities independent of DCs and remanufacturing activities in dedicated facilities independent of plants. At the same time, the company is not certain that its reverse logistics operations are bringing some economies and it wonders whether it should not drop them altogether. How should model (12.23)–(12.32) be modified to be able to answer these interrogations?

Bibliography

Akçalı E, Çetinkaya S, Uster H (2009) Network design for reverse and closed-loop supply chains: an annotated bibliography of models and solution approaches. Networks 53(3):231–248

Brander M, Sood A, Wylie C, Haughton A, Lovell J (2011) Electricity-specific emission factors for grid electricity. Econometrica Technical Paper

Chabane A, Ramudhin A, Paquet M (2012) Design of sustainable supply chains under the emission trading scheme. Int J Prod Econ 135:37–49

Chardine-Baumann E, Botta-Genoulaz V (2014) A framework for sustainable performance assessment of supply chain management practices. Comput Ind Eng 76:138–147

Chen L, Olhager J, Tang O (2014) Manufacturing facility location and sustainability: a literature review. Int J Prod Econ 149:154–163

Clift R (2003) Metrics for supply chain sustainability. Clean Techn Environ Policy 5:240–247

Dekker R, Bloemhof J, Mallidis I (2012) Operations research for green logistics—an overview of aspects, issues, contributions and challenges. Eur J Oper Res 219:671–679

Easwaran G, Uster H (2010) A closed-loop supply chain network design problem with integrated forward and reverse channel decisions. IIE Trans 42:779–792

Eskandarpour M, Dejax P, Miemczyk J, Péton O (2015) Sustainable supply chain network design: an optimization-oriented review. Omega 54:11–32

European Commission (2013) The EU emissions trading system (EU ETS). European Union

Fleischmann M, Beullens P, Bloemhof-Ruwaard J, Van Wassenhove L (2001) The impact of product recovery on logistics network design. Prod Oper Manage 10(2):156–173

Goedkoop M, Heigungs R, Huijbregts M, De Schryver A, Struijs J, Van Zelm R (2013) ReCiPe 2008: a life cycle impact assessment method which comprises harmonized category indicators at the midpoint and the endpoint level (v. 1.08), Report I. Rijksinstituut voor Volksgezondheid en Milieu, The Netherlands

Goedkoop M, Spriensma R (2000) The Eco-Indicator 99: a damage oriented method for life cycle impact assessment: methodology report, 3rd edn, Technical Report. PRé Consultants

Guillen-Gosalbez G, Grossmann I (2009) Optimal design and planning of sustainable chemical supply chains under uncertainty. AIChE J 55(1):99–121

Haanaes K, Reeves M, Strengvelken I, Audretsch M, Kiron D, Kruschwitz N (2012) Sustainability nears a tipping point, Research Report, MIT Sloan Management Review

Hugo A, Pistikopoulos E (2005) Environmentally conscious long-range planning and design of supply chain networks. J Cleaner Prod 13:1471–1491

IISD (1992) Business strategy for sustainable development: leadership and accountability for the 90's. DIANE

IPCC (2007) IPCC fourth assessment report: climate change 2007. Intergovernmental Panel on Climate Change

ISO (1997) ISO 14040: environmental management life cycle assessment principles and framework

Jorgensen A, Le Bocq A, Nazarkina L, Hauschild M (2008) Methodologies for social life cycle assessment. Int J Life Cycle Assess 13(2):96–103

Karsu O, Morton A (2015) Inequity averse optimization in operational research. Eur J Oper Res 245:343–359

Lebreton B (2007) Strategic closed-loop supply chain management. Springer, Berlin

Lee H (2010) Don't tweak your supply chain: rethinking it end to end. Harvard Bus Rev 88:63–69

Mavrotas G (2009) Effective implementation of the ε-constraint method in multi-objective mathematical programming problems. Appl Math Comput 213:455–465

McLaughlin K, McMillon D (2015) Business and society in the coming decades. In: Kirkland J, Kuntz M (eds) Perspectives on the long term: building a stronger foundation for tomorrow, focusing capital on the long term, pp 57–61

Mota B, Gomes M, Carvalho A, Barbosa-Povoa A (2014) Towards supply chain sustainability: economic, environmental and social design and planning. J Cleaner Prod. doi:10.1016/j.jclepro.2014.07.052

Polman P (2015) Solving problems that matter. In: Kirkland J, Kuntz M (eds) Perspectives on the long term: building a stronger foundation for tomorrow. Focusing capital on the long term, pp 42–47

Raadal H, Gagnon L, Modahl I, Hanssen O (2011) Life cycle greenhouse gas (GHG) emissions from the generation of wind and hydro power. Renew Sustain Energy Rev 15(7):3417–3422

Schmied M, Knorr W (2012) Calculating GHG emissions for freight forwarding and logistics services. European Association for Forwarding, Transport, Logistics, and Customs Services

Tang C, Zhou S (2012) Research advances in environmentally and socially sustainable operations. Eur J Oper Res 223:585–594

U.S. EPA (2015) Inventory of U.S. Greenhouse gas emissions and sinks: 1990–2013. U.S. Environmental Protection Agency Report EPA 430-R-15-004, Washington

Ward P, Kiron D (2012) The four organizational factors that built Kimberly-Clark's remarkable sustainability goals. MIT Sloan Management Review Reprint 53420

WBCSD-WRI (2004) A corporate accounting and reporting standard, rev edn. World Business Council for Sustainable Development and World Resource Institute, USA

WBCSD-WRI (2011a) Corporate value chain (scope 3) accounting and reporting standard. World Business Council for Sustainable Development and World Resource Institute, USA

WBCSD-WRI (2011b) Product life cycle accounting and reporting standard. World Business Council for Sustainable Development and World Resource Institute, USA

WEF (2009) Supply chain decarbonization. World Economic Forum, Genova

Zakeri A, Dehghanian F, Fahimnia B, Sarkis J (2015) Carbon pricing versus emission trading: a supply chain planning perspective. Int J Prod Econ 164:197–205

Chapter 13
SCN Reengineering Methodology

Previous chapters have progressively introduced SC and SCN design concepts, issues, and models. The didactic approach adopted was to gradually introduce different aspects of the SCN design problem, examine related practical issues, and then show how various decision models and methods could be exploited to help solve the generic subproblem considered. In practice, design problems do not necessarily fit with any of the subproblems studied. Every SCN has its particularities, and tailor-made models and methods may be required to improve them. Supply chain systems are very complex entities and are rarely engineered holistically on a green field basis. Rather, major projects are commissioned at discrete points in time to analyze and improve a portion of an existing SCN. SCN reengineering can be seen as a cyclical improvement process in which parts of the network are periodically restructured. This chapter examines the nature of SCN reengineering projects and it proposes a methodology that can be followed in practice to structure, manage, and realize such projects. Its content cuts across all previous chapters, showing how all the pieces can be assembled together, but it also covers a number of system analysis and design tasks that were not addressed explicitly before.

Chapter 1 defined SCNs as a major component of SC systems (SCSs) concerned with the use of internal and/or external resources to perform primary sourcing, production, distribution, and recovery activities (see Figs. 1.2 and 1.3). While designing SCNs, the key concern is the acquisition, disposal, and deployment of long-term resources to achieve value creation. SCSs, however, also incorporate P&C processes (see Chap. 3) used to manage the flow of material, information, and cash in the network. In reality, these two subsystems are inseparable and interdependent. While elaborating a SC strategy, both subsystems must be considered and, ideally, as discussed in Sect. 1.3, they should be reengineered simultaneously. However, the SCSs of large corporations are extremely complex entities, and it is rarely possible to redesign the two subsystems simultaneously. Most reengineering projects concentrate on structural issues, under the assumption that processes will not change, or on functional issues, under the assumption that structures will not

A. Martel and W. Klibi, *Designing Value-Creating Supply Chain Networks*,
DOI 10.1007/978-3-319-28146-9_13

change. In this chapter, we adopt the first of these two points of view. This does not mean, however, that P&C processes can be neglected. We insisted repeatedly on the fact that SCN design models must incorporate an anticipation of short-term user plans and actions. Good anticipations can be elaborated only if processes are well understood. Consequently, although process reengineering is not our main focus, a good knowledge of processes must be developed in SCN design projects.

SCNs can have different scopes depending on who owns them. In the simplest case, the SCN to reengineer is completely under the control of a single company. The SCN could also be owned by a large multinational with foreign divisions or subsidiaries, an OEM orchestrating a multi-tier supply structure (see Fig. 6.5), or even a network of collaborating companies. In what follows, we assume that independently of who owns the SCN, the reengineering project is under the responsibility of a steering committee representing all stakeholders and that the committee members can agree on a common design objective related to SCN value creation. In other words, we assume that the steering committee acts as a team and that no antagonistic or opportunistic behaviors are present. In the next section, we start by presenting the rationale behind the proposed design methodology and the major tasks to perform. The following sections explain all these tasks.

13.1 The Methodology in Perspective

SCNs are complex socio-technical systems. Their internal resources (personnel, equipment, facilities, and knowledge) and processes (procurement, production, distribution, selling, recovery, planning, and control methods) as well as their external environment (suppliers, markets, investors, governments, unions, competitors, technology sources, etc.) continually evolve along with the goals of their stakeholders. These changes are unavoidable, and to create value for stakeholders, internal structures and processes must be adapted to cope with economic, ecological, and social changes. Of course, it is neither possible nor desirable to modify the formal structures and processes of SCSs whenever a change occurs in the environment. Consequently, there is always a gap between implemented structures and processes and those needed to perform best. Small *maintenance* alterations are continuously made to try to bridge that gap. However, when the gap becomes too large, it is necessary to rethink and reengineer the system. These “reorganizations” are expensive and often painful and it is wise to keep them few and far between. One of the main goals of reengineering projects is thus to redesign and implement new robust system structures and processes that will last as long as possible, with the least possible effort.

These continuous improvement changes can be viewed as mutations of the SCS owing to environmental pressures. The system life cycle that results from these changes is illustrated in Fig. 13.1a. As with all life cycles, there is a gestation period and an active life period. When the system is to be reengineered, an analysis

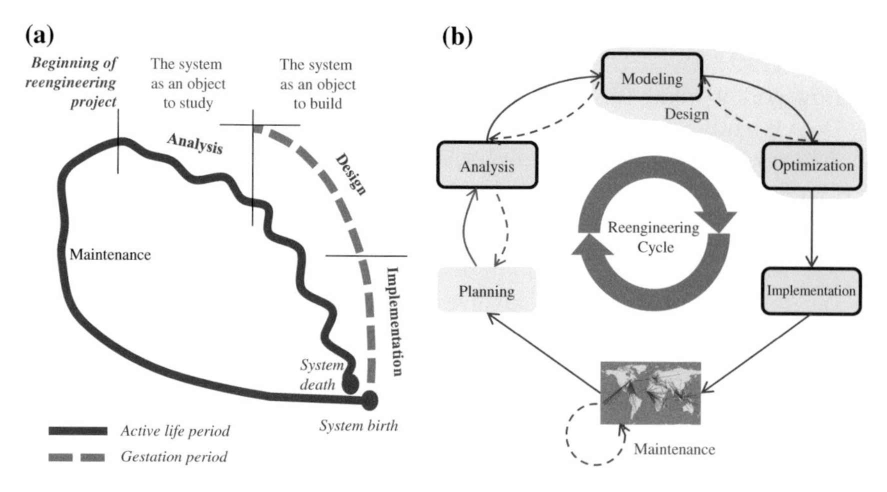

Fig. 13.1 Socio-technical system life cycle and reengineering process. **a** System life cycle. **b** Continuous improvement process

(intelligence) phase is undertaken during which it is viewed as a socio-technical object to study and understand. During this phase the system and its environment is observed, which requires the collection and analysis of a lot of historical data, the mapping of processes and structures, and the evaluation of performances in order to diagnose strengths and weaknesses. It also involves the prospective analysis of evolutionary trends to identify future opportunities and threats as well as the elaboration of strategic directions on how the company wants to position its SCS in the future. The next phase is one of design in which the goal is to produce the specifications of the new system. During this phase, the system is viewed as an object to revalorize. In our context, it involves significant modeling and optimization activities. Finally, the new designed system is implemented. This requires the acquisition and installation of new equipment, the building of new facilities, the negotiation of long-term contracts, the testing of installed subsystems, and so on. During the design and implementation phases, the SC must continue to operate under the old (current) system. It is at the end of the implementation phase that the current system is replaced with the new one. Parts of the two systems may also work in parallel for a while.

The main activities of the system reengineering process are congruent with the phases of the system life cycle just described and they are portrayed in Fig. 13.1b. A reengineering cycle starts with a planning activity to set up the reengineering project. Companies do not necessarily have the internal expertise required to complete such projects successfully, which may involve the selection and hiring of external consultants. This is followed by the system analysis. The analysis starts with the observation of the physical system, and it is essentially an abstraction process involving the preparation of conceptual multifaceted system representations to facilitate the elaboration of a diagnostic and of strategic revalorization directions. Six fundamental questions must be asked during this system analysis activity (Hare 1967):

1. What is it? (the study of the system structures and environment, and of user needs)
2. How does it work? (the study of processes and methods)
3. Does it work according to needs? (the measurement of performances)
4. Why not? (the diagnosis of deficiencies)
5. Can it be fixed and how? (the search for possible revalorization directions)
6. What are the impacts of the proposed changes? (the costs and benefits of possible revalorization directions)

The following design phase involves modeling and optimization activities. During the modeling activity, the conceptual system representations and design directions provided by the analysis are partly converted into symbolic representations that can be mathematically manipulated. Modeling constructs such as product families, demand zones, hazard zones, supply sources, potential facility locations, demand processes, transportation means, and so on are identified; design options as well as value-creation, socioecological, and risk-impact frameworks are elaborated; and finally all these constructs are embedded into normative design and user response (process) models. The following optimization activity employs the formulated models to generate plausible future scenarios (mostly using Monte Carlo methods) and candidate designs (mostly by solving mathematical programs, which in some cases may require the development of specialized solution algorithms). Candidate designs are then evaluated with a user model for different scenarios, and various performance measures are estimated to enable the final selection of a robust revised system design. The reengineering cycle is completed by a final implementation activity. During this activity, one goes back from the symbolic level to the physical level. This involves the conversion of the optimal design into concrete implementation plans, taking cultural and organizational issues into account, and the execution of these plans.

During reengineering projects, analyst-designers need to be supported by data mining, statistical, simulation, optimization, system engineering, and project management techniques and tools to improve their productivity. Analysis techniques and tools fall into four main categories: variety reduction, mapping, data gathering and measurement, and statistical inference. Variety reduction tools such as Pareto (ABC) analysis and statistical sampling help identify a small number of things that, if observed in detail, provide a good understanding of the system. Mapping involves the use of visual formalisms to represent system structures and processes. Some formalisms provide a holistic (macro) view of the system (SCN maps, facility layout plans, activity graphs, value stream maps, data flow diagrams, entity–relationship diagrams, organization chart, etc.) and others provide a detailed (micro) view of some system elements (flowcharts, process charts, routings, BOMs, workflow diagrams, action diagrams, etc.). Some specialize in structural system facets and others on processes. Some are more appropriate for material resources and flows and others for information structures and flows (Martin and McClure 1985). System analysis tools are now largely computerized. They must be selected with care to provide relevant complementary views of the system. Several system representation examples will be presented in the following sections.

The following sections discuss the activities performed during a reengineering project sequentially. However, in practice, the process is not linear; it is rather a gradual refinement process, which is why back arrows are shown between major activities in Fig. 13.1b. As one progresses, it is common to have to go back to get additional information or to refine previous system representations and design directions. Also, in what follows, we cover most of the elements that may be relevant to redesign a system completely. As mentioned, in real-life projects, depending on the specific mandate given to the design team, only a subset of these elements are usually covered. One can never be exhaustive; choices must be made about what to observe/model and the point of view to adopt. The depth and precision of analysis and design activities may vary widely depending on the project objectives. Unfortunately, companies often wait until they have serious problems before engaging in reengineering projects. In such contexts, they usually need to react quickly and the analysis and design must be based on readily available information and rough-cut models. On the other end, world class companies may engage in reengineering projects on a regular basis and have the necessary time and resources to seek much better precision. In any case, one must beware of the paralysis by analysis syndrome. Note, finally, that not everything can be quantified and modeled. Soft cultural, organizational, and strategic issues are often neglected in SCN modeling, but they cannot be neglected when selecting the design to implement.

13.2 Planning the Reengineering Project

The resources and time required to complete SCN reengineering projects can be very significant depending on the scope of the SCN redesigned and on the depth of the analysis performed. They usually take a few months, and they may require several person-years of work and generate significant expenses. Also, their results generally have an impact on a large number of people at all levels in the organization. For the project to be successful, that is, to improve value creation, it must have top management's support and it must be perceived as necessary by all affected personnel. For all these reasons, the project must be planned and controlled very carefully, and key actors must be continually informed of the progress made and of coming activities. Such projects, when poorly managed, provoke organizational resistance and their chance of success is low. Also, the final implementation activity in Fig. 13.1b may require the acquisition or disposal of resources, the building or reorganization of facilities, the negotiation of long-term contracts with 3PLs and contract manufacturers, and so on. These are major undertakings and they usually give rise to separate implementation projects. The reengineering project discussed here does not include these activities. It stops with the elaboration of detailed implementation plans.

The first responsibilities of the top executive or steering committee commissioning the project is to define its specific mandate (deliverables), select the design team members, and set up a project reporting structure. The selection of the design team is a critical success factor in all reengineering projects. As seen in previous chapters, in addition to an excellent knowledge of the current SCN and the industry, the design team must have strong data mining, analysis, forecasting, modeling, and optimization competencies. It must also be headed by a motivated and respected leader. Some enterprises may have staff capable of providing all these competencies, mainly when they are structured to reengineer parts of their SCN on a periodic basis. Most companies, however, lack the technical knowledge required for SCN design projects, and it is often difficult for them to assign several key resources to the project for a few months. Also, it is usually desirable to include someone in the design team with no prior knowledge of the system to provide a fresh look and avoid any entrenched bias. For these reasons, external expertise is often required to form a strong design team. When this is the case, a request for proposal (RFP) may be prepared explaining the project objectives, target subsystems, and competencies required as well as service provider selection criteria. After studying the proposals received, the external consultant selection is usually based on a multi-criteria procedure, as explained in Chap. 6.

There are different ways to staff a design team, depending on the context and resources available, but some key roles must be covered:

- *Project leader*: He is responsible for the overall planning and coordination of the project. He has a deep understanding of the reengineering methodology and a holistic view of the system designed. He is involved, if required, in the formulation of tailor-made SCN design models and solution algorithms. He sees to the allocation of resources to activities, ensures that work progresses according to plan, prepares and presents progress reports to the steering committee, and is responsible for the timely preparation of deliverables. He follows the work closely, is directly involved in critical tasks, and ensures quality control.
- *Project champion*: She is the interface between the design team and the rest of the organization. She is well known and respected in the company, has an extensive contact network, and knows the current system structures and processes inside–out. She organizes facility visits, meetings between team members and users, and obtains relevant company reports and documentation. She is also responsible for informing company personnel on progress made and coming activities requiring their involvement. It is through her actions that the project is perceived positively and that organizational resistance is avoided.
- *Methodological expert(s)*: They provide the expertise in data mining, modeling, and optimization required to adapt the methodology to the specific needs of the company. They intervene only if need be.
- *Data wizard*: He knows the company information systems inside out. He has the authority to access databases (DBs) and is familiar with the data query tools (SQL and variants) used by the company. Because team members cannot

usually access company DBs directly for security reasons, he is responsible for getting all the primary data required and transferring it to a project DB. He is also responsible for the validation of the primary data thus obtained.

- *User/partner delegates*: They represent the main SCN users and partners (suppliers, customers, etc.). They are critical to obtaining information on the current system and validating candidate designs. They are used as needed and they may be structured as an advisory committee. Their participation is also crucial to ensure the seamless implementation of the proposed design.
- *Analyst-designers*: These are the persons doing most of the analysis, modeling, and optimization work. They must have in-depth knowledge of the techniques and tools used. They are often junior consultants with a good technical background.

Depending on the size of the project, several roles can be played by the same person. When internal resources are involved in modeling activities, some a priori training on the tools used may be required.

When the design team leader has been appointed, several initial project planning tasks must be completed:

- Clarify who the client is (specific top executive, company board, steering committee, etc.), the client expectations (i.e., the mandate), and the responsibility structure.
- Adapt the reengineering methodology selected to any specific needs of the client.
- Select the tools to support the methodology. These may include a project management tool (e.g., MS Project), visual modeling tools (e.g., Visio, Smartdraw, etc.), a project data depository software (e.g., Excel, Access, FileMaker Pro, MySQL, etc.), a GIS (e.g., MapInfo, ArcGIS, etc.), a statistical analysis package (e.g., Excel, SPSS, Minitab, SAS, etc.), a model generation software (e.g., AMPL, AIMMS, GAMS, OPL, etc.), an optimization software (e.g., Excel with Frontline's solver, IBM CPLEX, Gurobi, FICO Xpress, etc.), Monte Carlo simulation software (e.g., @RISK, Crystal Ball, etc.), and an SCN design package (SAILS, Supply Chain Guru, LogicNet, etc.).
- Set up a project war room and office space where all the project information can be centralized and the team can meet and work.
- Complete the selection and installation of team members.
- Organize training with project tools for team members as needed.
- Prepare detailed plans for the analysis activity and present them to the client for approbation.
- Set up communication mechanisms to inform the client and the personnel affected of the progress made and obtain user feedback. Nowadays, this is best done by setting up a project blog and/or wiki.

Reengineering projects include major milestones when progress reports are presented, decisions about the orientation of the project are made, and plans (including budgets) for the next phase are approved. Milestones are usually placed at the end of the planning, analysis, and design phases. When the planning activity is completed, a precise project mandate, detailed plans for the analysis phase, and rough-cut plans for the design phase must be approved. At the end of the analysis phase, the diagnostic made and proposed strategic SCN development directions must be approved. A detailed plan for the realization of the design phase must also be accepted. Finally, at the end of the design phase, the new SCN design proposed and plans for its implementation must be approved. Each of these milestones may give rise to adjustments to the analysis, modeling, or optimization work previously done. Depending on the context, the reports produced may be formal syntheses of the work done, the results obtained, and recommendations for the future, or they may simply take the form of a detailed PowerPoint presentation.

13.3 Strategic SCN Analysis

The aim of the analysis phase is to understand the SCN and the business environment in which it is evolving and specify the design phase objectives, deliverables, and plans. The analysis must be done from several points of view. The SCN is considered first as an actor performing on the industrial scene, which leads to markets, competitors, and industry structures investigations. It is then viewed as a complex socio-technical system to understand, which requires structural, functional, and performance analyses. The SCN must also be regarded as a system interacting with the planet and with society (see Fig. 12.1), which give rise to sustainability analyses. The SCN must finally be viewed as a vulnerable system under threats, which gives rise to some risk analyses. These analyses require a large amount of data, and a significant part of reengineering projects time is usually devoted to data collection and validation. These analyses provide the information required to perform a SCN diagnostic, thus identifying strengths, weaknesses, opportunities, threats, and critical success factors. Strategic SCN development directions can then be elaborated to improve the competitive position of the company, and a precise design phase mandate can be specified. The various tasks to perform during the analysis phase are summarized in Fig. 13.2.

13.3.1 Data Collection and Validation

The analysis is important and necessary to understand the current system and design objectives but it should be performed as quickly as possible. To keep this phase short, one must use any strategic plans, business models, and performance reports already available in the company as well as external industry sources and concentrate on missing information and updates. In some case, part of the data accumulated during previous reengineering projects may be recuperated. Much of this information, however, may be obsolete. The information found in available

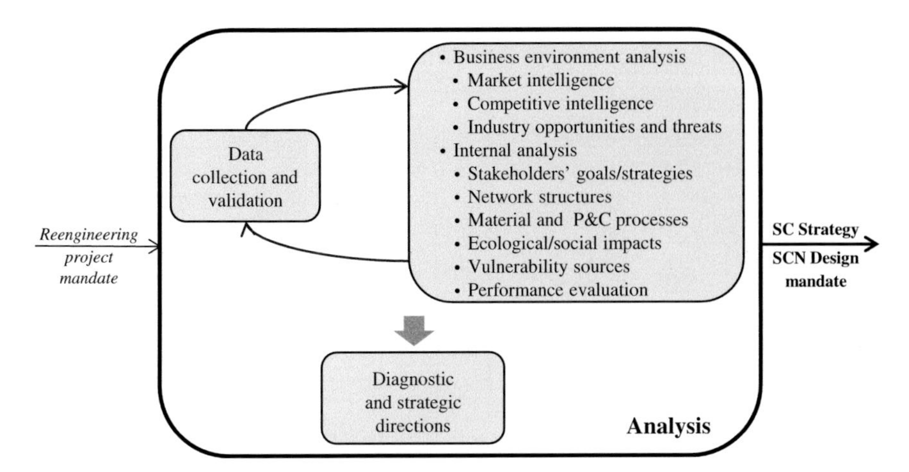

Fig. 13.2 Tasks performed during the analysis

strategic and financial reports is already aggregated and is not necessarily in the form required to make detailed statistical analysis and eventually to estimate model parameters. These analyses typically must be based on primary transactional data extracted from the company ERP/APS (see Chap. 3) systems DB, from direct observation of a sample of SCN entities or processes, and/or from SC partners. These data must then be transformed as needed to perform the analysis required (e.g., aggregation of sales transactions into weekly sales, apportionment of transportation costs of multi-destination shipments to ship-to points, calculation of distances between network locations, etc.). Typical data required include the following:

- The list of products, their characteristics, and their BOM when production facilities are involved
- The address of ship-to points, DCs, plants, suppliers, and so on
- Distance and transit times between network nodes
- Product demands and prices by customers and sales region
- The service level required by product-markets
- Transportation, warehousing, inventory holding, production, supply, and capital cost functions (per capacity type, product type, etc.). As explained in Chap. 2, these cost functions usually cannot be obtained directly from the company accounting systems. Cost models must be derived from primary cost data.
- The capacity and prices of supply sources
- Exchange rates, tariffs, and income tax rates by country (state) and/or product types
- The capacity provided by the various technological options considered and the rate of consumption of capacity per product
- CO_2e emission and waste data per process, source, and market
- Hazard occurrences and impact data

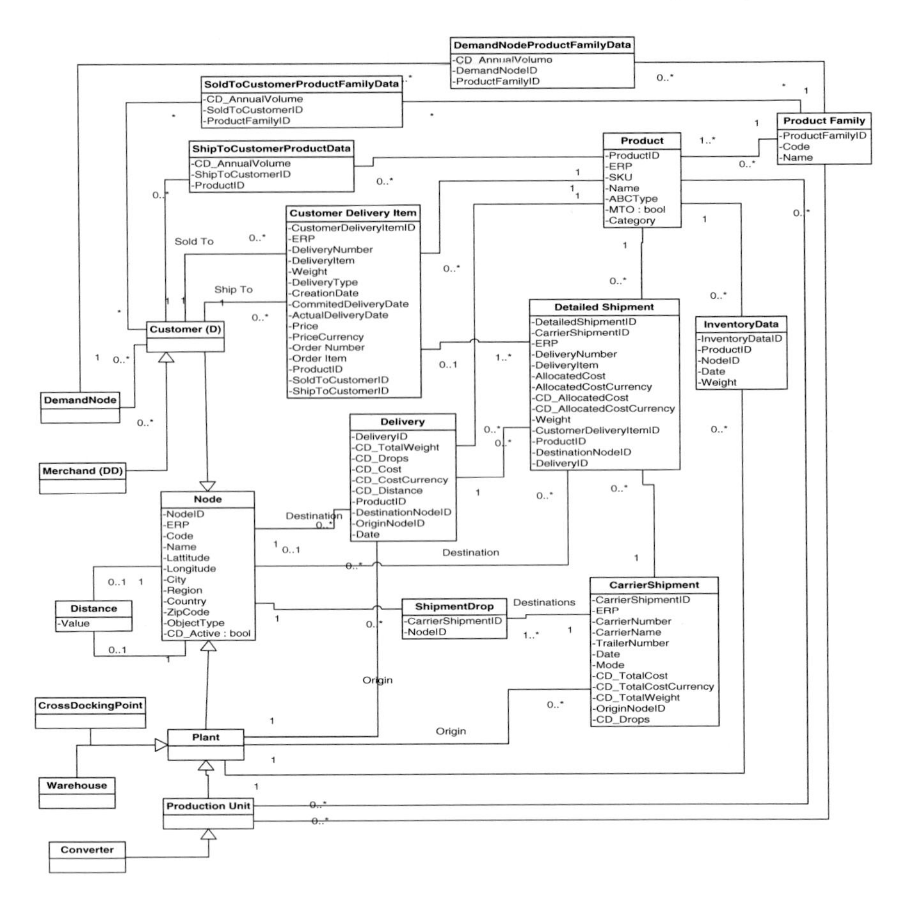

Fig. 13.3 Database schema example for a distribution network design project

In order to collect and manipulate these data efficiently, it is usually necessary to build a project DB. Large companies have thousands, if not millions, of products, customers, and transactions, and it is not possible to handle all related information manually. The relational DB schema of the project DB used to support the analysis of a North American distribution network in the pulp and paper industry is shown in Fig. 13.3. Most of the primary data required to populate this DB were obtained through queries and transfers from the company's multiple ERP systems (multiple systems were in use because the project followed a number of acquisitions and each original company had its own ERP system). When this is done, it is important to validate the data captured in the project DB. For all sorts of reasons, even large companies operating ERP/APSs purchased from leading software vendors can suffer from data integrity problems. The types of problems often encountered include customer/suppliers/shippers being present in the system under different names (say ABC vs. ABC inc. vs. ABC Inc.), invalid zip (postal) codes, inventory/flow imbalances, missing data, and so on. Also, clearly, when the company has several ERP/APS systems with different coding structures, equivalence conversion must be made. Validations, and the actions required to correct errors or convert data, are usually very time-consuming.

Another important issue to address at the beginning of a reengineering project is the selection of standard units for data collection and analysis. The units of measure used in the project have a direct impact on the value and interpretation of data. They should be selected carefully and, above all, respected rigorously during data entry. Otherwise, the data available will be impossible to interpret. The first decision to make is the choice of a base period for data acquisition as well as the timing of the planning horizon to consider. Usually, the base period is the most recent year for which the company is able to obtain all required data. For data transformed into time series, however, several years should be examined. Because all data are usually not updated instantaneously, and because the reengineering project takes time, there is necessarily a lag between the base period used and the beginning of the planning horizon. The length of the operational periods and planning periods used in the project do not have to be the same as the length of the base period. They are often shorter (say, weekly operational periods, three-month planning periods, and a base year). Appropriate weight, distance, area, volume, unit loads, and currency units must be selected.

13.3.2 Understanding the Business Environment

Thorough business environment analyses are usually performed when a company prepares its strategic plans (Rothaermel 2014). The first question to address in strategic planning is "What business are we in?" and the strategic plans of the company should give a clear answer to that question, thus providing clear boundaries of the product-markets, industry, and competitors to consider in the SCN reengineering project. If this information is not readily available, however, one must develop at least an understanding of existing and potential markets, emerging technologies (production, storage, transportation, etc.), potential partners (suppliers, subcontractors, public warehouses, carriers, 3PLs, etc.), and competitors (potential acquisitions, market shares, etc.) that may contribute to improving the actual SCN. The nature of contracts with partners, capital and labor market constraints, environmental regulations, and inherent industry risks must also be understood.

Several frameworks have been developed for analyzing industries and competitors. A popular approach to understand the rules of competition in an industry is the five forces framework (competitive rivalry, threat of new entrants, threat of substitutes, bargaining power of suppliers, and bargaining power of customers) proposed by Porter (1980), based on industrial organization (economics) concepts (Scherer and Ross 1990). The position and strengths of competitors in an industry (see Fig. 2.16) value chain must be understood to establish a profitable and sustainable competitive strategy. Another somewhat complementary approach that has evolved over the years is a PESTLE analysis (Team FME 2013). It identifies six categories of external factors to investigate:

- **P**olitical: Political factors to examine include government stability, trade agreements and restrictions, labor protection, corruption, bureaucracy, security threats, and so on. As discussed in Chap. 9, these factors are particularly important for multinational companies.
- **E**conomic: Economic factors to investigate include inflation levels, interest rates, availability of capital, exchange rates, cost of living, consumer revenues, tax levels, skill levels, wage levels, labor costs trends, absenteeism, and so on.
- **S**ociocultural: Sociocultural factors to consider include demographic variables (age distribution, population growth rate, etc.), employment levels, lifestyle, consumer behavior, education, religion, ethics, historical issues, cross-cultural attitudes, and so on.
- **T**echnological: Technological factors to review include product innovation, new production technology development and adoption, outsourcing possibilities, industrial clusters, quality standards, productivity, information technology advances, transportation, and communication infrastructures.
- **L**egal: Legal factors involve the study of legal systems and regulatory bodies (labor, consumer, health, safety, environmental, trade, competition, accounting, investment, and tax regulations).
- **E**cological: Ecological factors relate to issues discussed in Chap. 12, such as natural resources availability, waste disposal, used products recuperation systems, ecological impact measurement regulations, and packaging restrictions. It also involves the identification of the hazard types that may affect the company as discussed in Sect. 10.3 and the evaluation of multi-hazard exposure levels using publicly available data, such as the maps shown in Fig. 10.11.

Mind mapping or PESTLE diagramming tools (e.g., PESTLEWeb) can be used to facilitate PESTLE analyses. These diagramming tools help to show how critical external factors may combine to affect the company (opportunities and threats) and they provide good discussion, synthesis, and communication means. They help formalize environment analysis results and provide a good starting point for the study of evolutionary paths and eventually the elaboration of plausible future scenarios (see Sect. 10.4).

The industry market structure and development opportunities are particularly important inputs for SCN design studies. These should usually be documented in the strategic marketing plan of the company (Lambin and Schuiling 2012). It is necessary to identify the products to be manufactured and sold, the customers to be served, and eventually the order winners to develop. In most industries, there is great diversity of needs stemming from basic variations in customer behavior. To cope with this, companies focus on well-defined groups of customers, thus dividing the industry market into relatively homogenous segments (product-markets) in order to adapt their offer to specific segment needs. Customers are not seeking a product or service as such but rather the solution to a problem. This solution may be provided by existing products or by products under development. One must thus identify existing as well as potential (latent) product-markets. Segments are defined in terms of products, geographical regions, benefits, and so on. Once identified, they represent value-creation opportunities whose attractiveness must be evaluated. The attributes and desirability of product-markets need to be characterized in qualitative (needs, buying habits, sensitivity to value drivers, strength of competition, accessibility, etc.) and quantitative (size, potential, market share, growth, etc.) terms. All these attributes are out of the control of the firm but they must be well understood. The relationships among demand, PESTLE variables, and value drivers such as prices, response times, and quality must also be grasped.

13.3.3 Understanding Internal SC Systems

This task is concerned with the study of business structures, processes, and factors that are under the control of a company (network of company) or of its stakeholders, that is, elements that are shaped by strategic business decisions. Stakeholders must be identified and their goals clarified. As indicated previously, we assume that stakeholders work as a team and that they share a common vision of the SCN mission. Clearly, this mission must be well understood because it drives all SCN design decisions. The SCN mission is defined mainly in terms of the product portfolio and product-markets targeted by the company, the value proposition (offer) it wants to make to market segments, and sustainable value-creation objectives. This brings us back to the discussion of SCN mission and SC strategy in Chap. 1 and in particular to Fig. 1.12. Understanding the company offer and doctrine is a critical success factor in a SCN design project. Information on the current company value proposition and directions is usually found in strategic marketing and operations plans and in related documentation (product classification and catalogs, sales regions, etc.). Much of this is now often available on the company website. Meeting with key stakeholders should be organized, however, to update and validate the documented vision. It should be clear also that SCN reengineering projects are strategic planning ventures and that the current offer, doctrine, and SCN mission of the company may be altered during the project.

The structure and behavior of the current SCN usually is studied using data mining tools, geographical information systems (GISs), mapping formalisms, and statistical inference methods. These analyses require large amounts of historical data on products, customers, facilities, suppliers, subcontractors, carriers, sales, shipments, inventories, production quantities, and so on, obtained mainly from the company DBs. At the structural level, the analyses performed provide a classification of products, ship-to points, facilities, suppliers, partners, transportation means and lanes, and so on. These classifications are often based on Pareto analyses, geospatial analyses, and cluster analyses. Results from typical structural analyses are shown in Fig. 13.4 for a pulp and paper company. The top illustrations display the dispersion of ship-to points in North America and their relative importance in terms of annual shipment volumes (Pareto plot). The lower

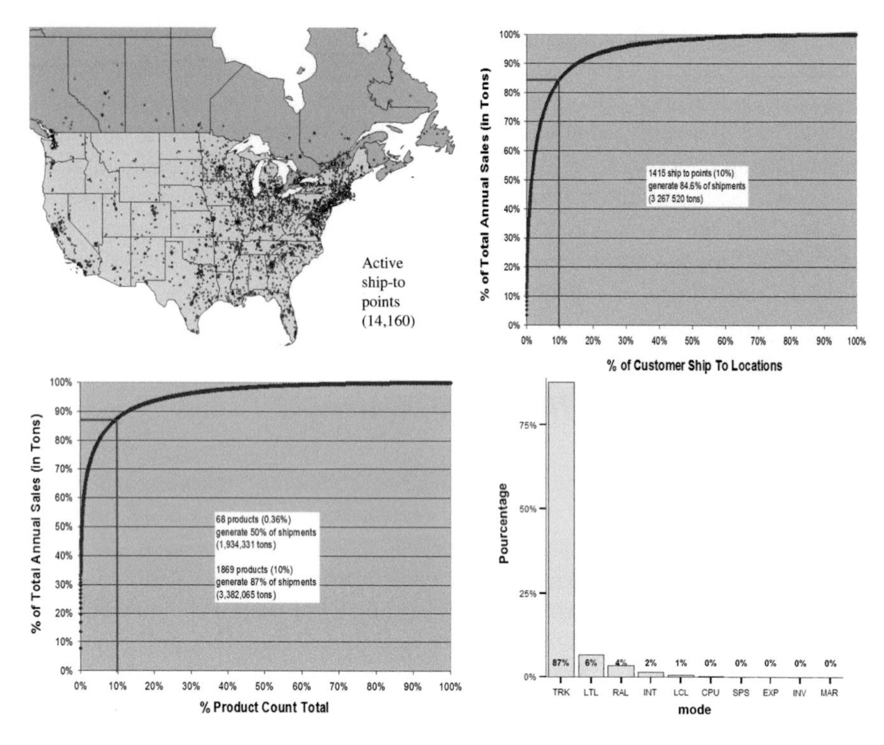

Fig. 13.4 Structural analyses for a pulp and paper company

illustrations pinpoint the relative prominence of products and transportation modes in the current SCN. These illustrations show that the company customers are concentrated in the eastern part of North America, a very small proportion of customers and products generate most of the flows in the network (10 % of products ordered by 10 % of customers generate about 75 % of annual flows), and, although the company uses LTL, rail, and intermodal transportation, most of its deliveries are made by external carriers under multi-stop TL shipment contracts. This sheds light on where to concentrate in the study to capture the essence of the business without examining everything in detail.

A general SCN schema, such as the one in Fig. 1.7 for the supply system of an electric utility, usually is useful to illustrate the echelons-stages structure of the network. Maps of current flow patterns for product classes among suppliers (or supplier groupings), facilities, and ship-to point groupings can also be produced to portray the current SCN. Such a map, created using a GIS for a pulp and paper company, is illustrated in Fig. 13.5. This type of representation is useful to formalize the status quo SCN design that will be compared to other candidate designs at the end of the optimization activity. It is also required for model validation purposes. Note that when the flows are compiled for a base year from the company DB, facilities input and output flows do not necessarily balance. This may be

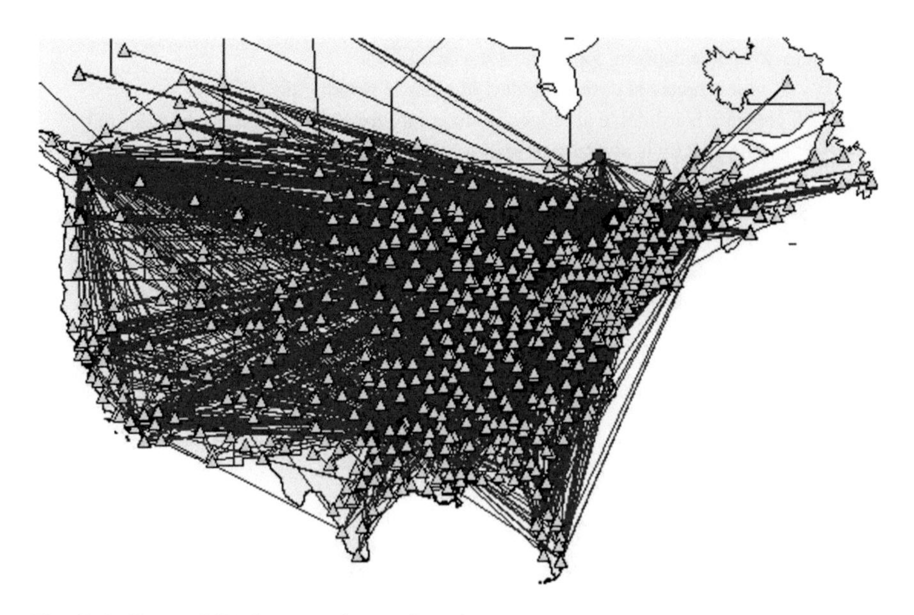

Fig. 13.5 Current SCN flow map for a pulp and paper company

caused by data capture errors, but it is mostly because inventory levels at the beginning and at the end of the base year are not the same. A corrected status quo design can be produced using a linear programming model including flow equilibrium constraints similar to those formulated in Chap. 7 but with an objective function defined to minimize the deviations from observed flows. Snapshots of the resources associated with current facilities (e.g., floor plans as in Fig. 4.16 or internal flow maps as in Fig. 1.5) and current partners (based on typical vendor, 3PL, and contract manufacturer contracts) may also be elaborated. The capacity of available resources must also be assessed (see the Sect. 8.2 discussion).

At the functional level, the analyses yield process diagrams describing the logic of the company supply, production, and distribution activities and its planning and control methods. Activity graphs, such as the one showed in Fig. 4.2 for a sawing company or in Fig. 6.3 for a fine paper manufacturer, are used to describe physical SC processes. The use of activity graphs to model SCNs was discussed in Sect. 7.2. Planning, control, and response policies and methods can be described using data flow diagrams, flow charts, or pseudocode. It is generally useful to elaborate a high-level diagram describing the main policies and decisions needed to manage the SCN (e.g., Fig. 1.18 for a grocery retail network) or its main P&C subsystems (e.g., Fig. 3.9 for a pulp and paper company). In order to be able to formulate design models incorporating adequate user response anticipations and to evaluate candidate designs accurately, it is necessary to understand SCN operational methods (related to the assignment of orders to DCs, the elaboration of shipping routes, the management of inventories, the planning of production lot sizes, etc.) relatively well. These methods

```
Compute available capacity for the next day at all DCs
    For all orders received during the day, in order of priority, do:
        If capacity is available at order ship-to point primary DC, then assign order to it
        Else, if capacity is available at ship-to point secondary DC, then assign order to it
        Else, transfer the order to an external supplier
        Update available capacity at used DC
    End do
```

Fig. 13.6 End-of-day order assignment procedure

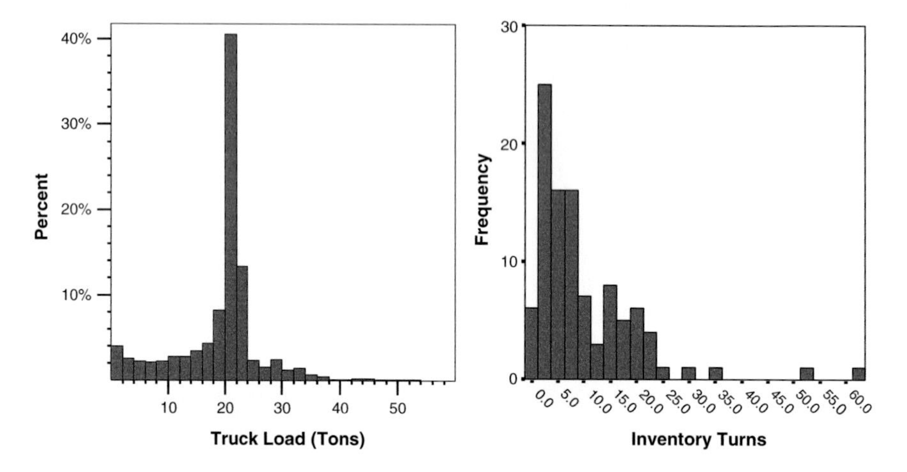

Fig. 13.7 Truck load and inventory turn profiles for a pulp and paper company

often can be described using pseudocode, as shown in Fig. 13.6 for an order assignment procedure.

During the functional analysis, the supply, demand, and resource consumption behaviors resulting from these processes, and associated costs and revenues, are also examined. This is done using data visualization methods, such as histograms and resource usage profiles, as well as statistical methods, such as time series and regression analyses. Critical system parameters, probability distributions, and cost functions are also estimated. For example, Fig. 13.7 shows truck loads (number of tons loaded per TL shipment) and inventory turns profiles, Fig. 2.10 shows an inventory-throughput function obtained for the power poles used by an electric utility company, and Fig. 5.20 includes a plot of ($ per ton, distance) points for all the shipments made from a DC during the considered base year as well as the estimated regression function. The characterization of demand behaviors is a particularly important aspect of the analysis. Sales[1] time series should be examined for

[1]Note that sales are not necessarily equal to demand, but true demand data is rarely available in practice.

Demand Processes	*Product Families* 10.6% Measure equip-ment	6.4% Hard-ware (VD)	4.2% Hard-ware (vD)	11.3% Hard-ware (d)	5.3% Poles	6.8% Heavy equip-ment	25.8% Cable reels	19.1% Trans-formers	2.8% Line material
Nonstationary Normal (VD)							12.1%	4.0%	
Stationary Normal (VD)	4.4%	5.1%			2.1%		6.2%		
Stationary Normal (vD)			4.2%						0.3%
Sporadic Poisson (V)	6.2%			3.2%	2.0%	4.4%	2.8%	9.0%	1.2%
Sporadic Poisson (v)				6.3%					
Planned requirements		1.3%		1.8%	1.2%	2.4%	4.7%	6.1%	1.3%

D: Fast movers d: Slow movers V: High value products v: Low value products

Fig. 13.8 Demand processes characterization for an electric utility company

a representative sample of products. The result of such an analysis for an electric utility is presented in Fig. 13.8. The identification of demand processes is essential to generate plausible future scenarios and determine how they are affected by evolutionary paths.

The analysis should also include a study of the SCN ecological and social impacts. Ecological aspects essentially correspond to the inventory analysis and impact assessment phases of classical life cycle assessment (LCA) studies, as discussed in Sect. 12.1. In our context, the activity graphs produced should be extended to show all closed-loop SCN components (see Fig. 12.2). Also, major flows of natural resources and energy from the planet as well as wastes and emissions should be quantified. In several countries, ecological impacts must now be assessed and reported by law. The methodology used by the company to do this should be adopted. As indicated in Sect. 12.2, however, impacts must not be quantified only for internal resources (land, platform, equipment) and processes but also for cradle-to-gate material and energy suppliers and for gate-to-grave consumption and disposal. Social impacts are more difficult to quantify; however, an effort should be made to identify any social problem resulting from, or possibly exacerbated by, company practices.

SCN vulnerabilities and threats arising from hazardous and deeply uncertain events must also be identified. More specifically, as discussed in Sect. 10.3.1, this involves the classification of SCN locations into vulnerability sources. As illustrated in Fig. 13.9, when considering potential risks arising from natural, accidental, and willful hazards, a large set of vulnerability sources can be identified. However, the impact of hazards on these vulnerability sources can vary greatly. In SCN reengineering projects, the number of vulnerability sources considered explicitly should be reduced to a minimum. A filtering process based on a subjective evaluation of the vulnerability identified should be used to select the sources with

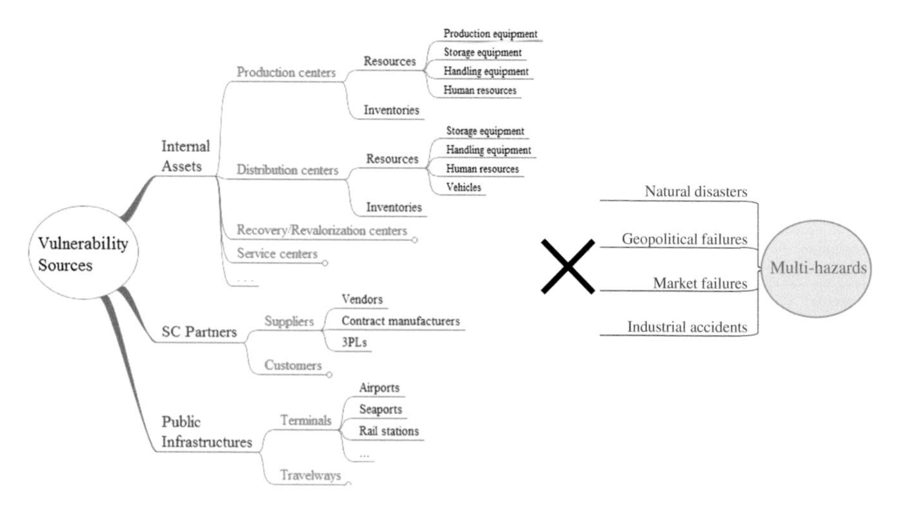

Fig. 13.9 Examples of vulnerability sources and multi-hazards

potential strategic consequences, and only those should be considered. The vulnerability sources retained usually include major internal production and distribution resources (plants, DCs), product-markets, and raw material and energy supply sources.

A final important analysis dimension is the evaluation of SCN performances. Performance evaluation was discussed in detail in Chap. 2. Most companies follow their financial performances closely but they do not necessarily have a framework directly relating SC activities and resources to value creation (see Fig. 2.8). In order to foster sustainable value creation, one must not monitor performances only in terms of expected value; as discussed in Sect. 11.2, robustness measures and aversion to risk must also be taken into account. The objective here is not to contest the performance evaluation system of the company but rather to elaborate a performance evaluation framework that will enable the design team to formulate adequate SCN design models and subsequently evaluate candidate designs in order to identify the best one. This framework also is used to evaluate the current SCN design, thus providing an improvement baseline.

13.3.4 Diagnostic and Strategic Positioning

The environmental, structural, functional, ecological, social, risk, and performance analyses discussed previously provide the information required to perform a SCN diagnostic. This is often done by synthesizing strengths, weaknesses, opportunities, and threats (SWOTs), and then by delineating critical success factors. Strategic SCN development directions can then be elaborated to improve the competitive position of the company. As indicated at the beginning of the section, the first analysis task to accomplish is to understand the offer and doctrine of the company and the mission of the SCN. Having completed the SCN diagnostic, it is possible that modifications to the offer, doctrine, and SCN mission are recommended to improve value creation. Once approved, these changes become the foundation for the design of the new SCN. Before the design phase can be started, the following strategic positioning elements must be clear:

- Company product-market penetration targets and value proposition variants
- An extended SCN activity graph clarifying production stages and distribution echelons prepositioning requirements as well as SCN decoupling points
- Planning, response, and control policies to support
- Potential new production, distribution, and transportation technologies (capacity options) and partners to consider
- Activity insourcing–outsourcing opportunities and restrictions
- International deployment opportunities and restrictions
- Current facilities, suppliers, subcontractors, and 3PLs to preserve
- Sustainable development objectives and constraints
- Risk mitigation imperatives
- Performance framework for the evaluation of candidate designs

These strategic directions delimit the SCN boundaries to consider in the design phase, and they enable the detailed planning of its activities.

13.4 SCN Modeling

Throughout the book we have seen how to use descriptive and normative models to help formalize SCN design problems and to facilitate the search for superior designs. A design model is a simplified symbolic representation of reality that can be mathematically optimized. It suggests good candidate designs only if it captures the essence of the SCN business context. This requires the reduction of millions of interrelated factors to a manageable number of variables and parameters that are linked together using appropriate mathematical functions. The role of the modeling activity is to define these variables, parameters, and functions. If a SCN design software package is used, this may simply involve reducing the SCN context to predefined tool constructs. This may be dangerous, however, because it requires bending the real problem to fit a predetermined generic structure. At this point in the project, it is much better to formulate a personalized model and use it at the beginning of the optimization activity to validate and justify the use of specific tools offered on the market. In what follows, we look at how the design model variables, parameters, and functions can be specified based on the analyses made during the previous activity.

13.4.1 Design Variables Definition

SCN design models are typically formulated as network optimization problems composed of internal and external nodes, as well as of arcs representing the movement of products between nodes. The first task when formulating a design model is thus to define the sets of products, demand nodes, supply nodes, intermodal transfer points, facility sites, and supply or transportation arcs to consider. Clearly, the number of products, suppliers, and ship-to points involved can be huge (usually in the millions for large corporations), and they cannot all be represented explicitly in the model. To reduce complexity, they must be clustered into product families, demand zones, and supply sources. Much of the design options considered in the network optimization model depend on how products, nodes, and arcs are defined. The defined clusters are often based on Pareto analysis, geospatial analysis, or cluster analysis methods (Ernst and Cohen 1990). The defined zones also must be congruent with any other geographical areas, such as countries or states, product-market regions, and hazard zones, represented in the model. When a node corresponds to a physical location, it has an address. When it is a geographic zone, its centroid latitude and longitude must be calculated. This is required to position nodes on GIS maps and calculate distances between them.

13.4.1.1 Product Families

Products must be classified in families with similar operational characteristics. Marketing or accounting product classifications are often readily available in the company, but, in general, they are not adequate for the formulation of SCN design models. To be useful in our context, all the products in a given family must have similar demand processes (fast or slow movers, stationary or not, explanatory variables, etc.); production, storage, and handling technologies; distribution channels and market offers; and supply sources. They must also all be transported using the same type of equipment. The number of defined product families should be kept to the minimum required to capture relevant design nuances (ideally less than 100). The classification should take account of Pareto's law, that is, pay much more attention to high demand, high-value products. Products with very low demand can generally be excluded without bias. An example of a product family classification used in an electric utility supply network design project is found in Fig. 13.8. Note that poles, for instance, are distinguished from other products because they require different storage (they are stored in an external yard) and transportation (they require specially equipped flatbed trucks) technologies.

One should be able to associate the defined product families to the movements of an activity graph without ambiguity. This is illustrated in Fig. 13.10 for the closed-loop activity graph elaborated during a project to engineer a global supply network to support humanitarian and military missions (Martel et al. 2013). In production–distribution network design projects, intermediate products, and raw

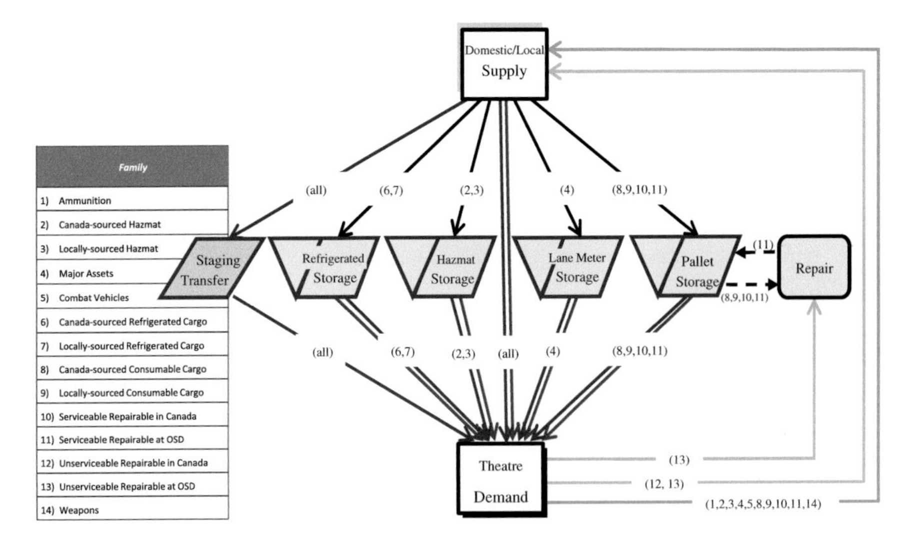

Fig. 13.10 Product families displayed on a closed-loop activity graph

Table 13.1 Product families defined for a pulp and paper SCN design project

Product type	Number of families	Outsourced	Insourced	Sold
Logs	3	3		
Chips	7	4	3	
Chemicals	4	4		
Pulp	7	2	5	1
Paper m/c jumbo roll grades	5		5	
Paper roll grades	6		6	6
Paper sheets	4	4	4	4

materials also must be aggregated into families to get an aggregated BOM. This may not be easy. Again, one should be able to assign coherently intermediate products and raw material families to the activity graph. For a project in the pulp and paper industry with an activity graph similar to Fig. 6.3, the product families summarized in Table 13.1 were defined.

13.4.1.2 External Nodes

In most SCN design projects, ship-to points must be grouped into demand zones, usually based on their postal codes or geographic coordinates (longitude and latitude) and on proximity and demand density criteria. Country postal codes are now generally available on the web (see www.geonames.org for a geographical database including the postal codes of most countries). Recall, however, that demand zones

1. Start with one demand zone per postal code (e.g., 900 three-digit zip codes in the United States) and take the demand-weighted average of the longitude and latitude of each ship-to point in the zone as its centroid.
2. Calculate the distance between the centroids of each zone pair and retain the pair with the smallest distance.
3. Merge the two zones retained in a single zone (thus decreasing the number of zones by one) and take the demand-weighted average of the longitude and latitude of each ship-to point in the new zone as its centroid.
4. Return to step 2 until the desired number of zones is obtained.

Fig. 13.11 Ship-to point clustering procedure

must be defined by product-markets (see Sect. 7.4.3), that is, that a given postal code could belong to more than one demand zone. When demand zones are used, the demand of all the ship-to points in the zone is associated with its centroid, and transportation lanes are assumed to link node centroids. This means that if demand zones are too large, serious errors in the calculation of transportation costs may result. Some studies were performed to determine the best number of demand zones to use to reduce transportation cost errors. In the United States, common practice has been to use between 100 and 200 zones in design projects. Ballou (1994) shows that this is not always appropriate. He suggests that the number of zones used should be 5–10 times larger than the number of potential internal sites considered in the study.

Figure 13.11 presents a basic ship-to point clustering procedure suggested by Geoffrion (1976) and based on postal code proximity. This clustering approach, however, does not take demand density into account. Ideally, to minimize transportation errors, smaller zones should be used when demand density is high and larger zones when demand density is low. Geoffrion's procedure can be modified to weigh the proximity measure used by zone demands when selecting the next two zones to merge. When several product-markets are considered, the procedure can be applied to each market segment separately using only applicable demand points. More sophisticated clustering approaches can be applied when using GISs. Demand polygons then can be constructed based on ship-to point geocodes and demands. A partial map of demand polygons obtained this way for a North American reengineering project is shown in Fig. 13.12.

When required, the same type of approach can be employed to cluster vendor ship-from points into supply sources, although, as seen in Chap. 7, when supplier selection decisions are not modeled, suppliers usually are considered only implicitly by including material supply costs in the calculation of production costs. When product-markets are considered, they are associated with geographical regions incorporating a set of demand zones. The delimitation of these sales regions usually is based on marketing criteria specified by the company but they must be geographically congruent with demand zone definitions. Also, when multi-hazards are modeled, zones delineating geographical areas with similar hazard characteristics

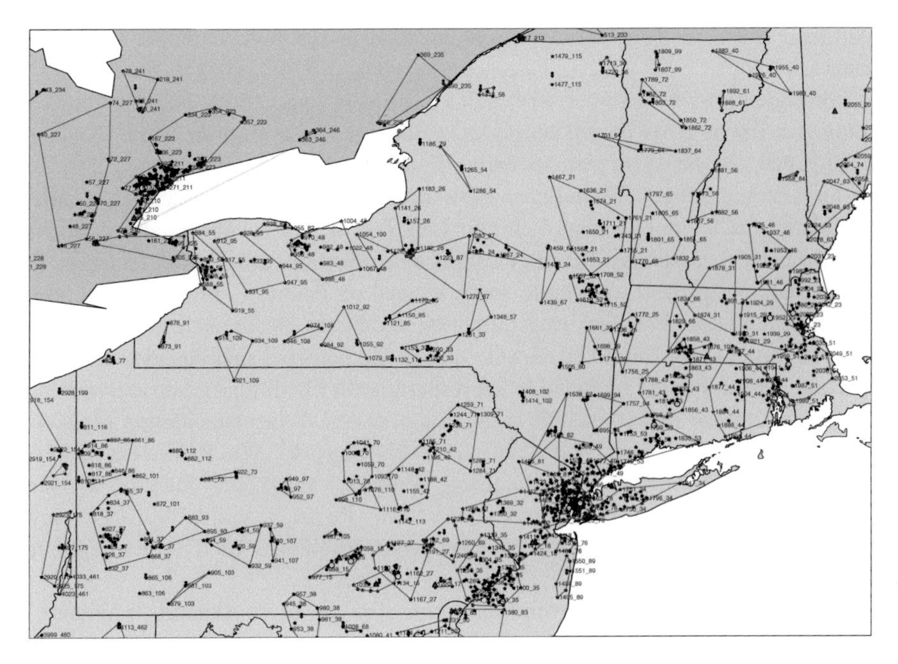

Fig. 13.12 Demand polygons example for a North American project

must be specified. These hazard zones need to be demand zone aggregates. When the modeling approach used assumes that extreme events in different hazard zones occur independently (see Sect. 10.3.1), the defined zones typically correspond to states (provinces, departments, etc.) or to countries, depending on the level of precision desired and the data available.

13.4.1.3 Potential Design Nodes

Most binary design variables in SCN optimization models are associated with network nodes (company, 3PL and subcontractor facility sites, supply sources (contracts), intermodal transfer points, etc.). Some of them correspond to the entities present in the current SCN. However, to develop superior SCNs, the designer must, as we discuss in Sect. 12.1.2, enrich the solution space, that is, add new potential nodes that are susceptible to lead to more robust value-creating SCN designs. The question here is how do we find these additional design options (facility sites, subcontractors, suppliers, etc.) and how many should be added to the design model? The answer to this question was provided in good part in Sect. 4.4 on site selection and in Sect. 6.4 on partner selection. These sections describe first how to make geographical explorations to find potential sites (partners) and second how to use multi-criteria filtering methods to restrict the list of potential sites (partners) found to those satisfying critical criteria and highly ranked subjective criteria.

For nodes providing direct services to customers (e.g., DCs), one can start by building a list of areas (based on postal codes, for example) with a high demand density that are far from current network service facilities. If potential sites or for-hire resources can be found near these areas, they can provide natural potential candidates. Industrial or logistics clusters (see Sect. 6.3) favored by industry actors are also natural candidates for the location of production–distribution facilities. In an international context, free trade zones located on major SC flow paths are also natural candidates. For first transformation facilities, locations near major raw material sources also should be considered. All the potential sites found using these different approaches are included in the multi-criteria filtering process performed to select the potential sites to include in the design model. When considering suppliers or subcontractors, network nodes are not purely physical. When different supply, component manufacturing, or warehousing contracts can be considered for a given potential partner, then these alternative contracts are associated with distinct network nodes. In that context, contractual possibilities also need to be explored attentively.

As discussed in Chap. 8, several platforms also can be considered for potential facility sites. For current sites, they include the existing platform but they may also include reorganizations or expansions of the facility in place. For new sites, they can be alternative layouts of existing facilities that could be purchased and renovated or alternative construction plans for vacant sites providing distinct activity capacity levels. Alternative technologies for performing activities should also be considered when elaborating the set of potential platforms to include in the model. This is particularly important when one wants to design a more sustainable SCN, as discussed in Sect. 12.1.2. As noted in Sect. 4.3.2, the preparation of detailed facility plans is a demanding engineering task. For this reason, only a few alternatives typically are considered in SCN design projects. For potential platforms, it is not necessary to finalize all platform design details. One needs to go far enough to be able to calculate the capacity that would be provided for all the activities performed in the platform. A block layout (see Fig. 4.15) of the considered platforms should at least be elaborated.

13.4.1.4 Network Arcs

Network arcs also are a source of important design options. These are mainly associated with the choice of transportation means. The nature of contemporary transportation options was discussed in Chap. 5. The consideration of alternative transportation means usually gives rise to parallel arcs in the design model. However, in a context in which most companies now use third-party transportation services, design options are frequently related to the choice of service providers who can accommodate a subset of network arcs (see Sect. 12.2.4). The selection of service providers for predetermined network subsets is then the main design issue.

Alternative contracts with potential partners also may then be considered, as discussed previously.

Independent of the options considered, as seen in Sect. 5.4, the costs associated with a network arc usually depend on the distance between its origin and destination, which may itself rest on the mode selected because they do not all use the same transport infrastructures. Also, the service level provided to customers usually depends on the transit time (or distance) required to reach them. An important issue in SCN design projects is thus the calculation of distances and delays associated with network arcs. Several methods are available to calculate distances with different precision and effort involved. Distances and transit times generally can be calculated with GISs. However, when the network is large, processing time can be relatively long. Because the level of precision provided by these tools is not always necessary, approximate methods are often used to calculate distances and times.

When the node coordinates are defined on a grid, Love and Morris (1988) show that the following function provides a good approximation to calculate the distance D between two points a and b with coordinates (x_a, y_a) and (x_b, y_b):

$$D = k\left[|x_b - x_a|^q + |y_b - y_a|^q\right]^{1/q}$$

where for a given network, the scale conversion factor k and the parameter q are estimated empirically. When $q = 2$, the formula gives straight-line distances and when $q = 1$, it provides rectilinear distances. For most road networks, a value of q between these two extremes offers the best approximation.

When distances are required for a large SCN including arcs between remote sites, it is convenient to use latitude-longitude coordinates because they are readily available. A good approximation is then obtained by calculating the distance, in miles, between points a and b with the following great-circle trigonometry formula (Ballou et al. 2002):

$$D = 3959\{\arccos[\sin(lat_a)\sin(lat_b) + \cos(lat_a)\cos(lat_b)\cos|long_b - long_a|]\}$$

where $(lat_a, long_a)$ and $(lat_b, long_b)$ are the latitude and longitude of a and b in radians. This formula accounts for the curvature of the earth, which is preferable when long distances are involved. However, the formula gives crow-fly distances and it does not consider the fact that vehicles cannot travel in a straight line because of road network meandering, variations in altitude, and broken connectivity. To cope with this, straight-line (crow-fly) distances must be multiplied by a circuitry factor. Ballou et al. (2002) provide empirically estimated circuitry factors for several countries and regions. Some of them are given in Table 13.2.

When the distance D is available, it is not too difficult to calculate transit times under certain conditions (mode, equipment type, road type, etc.). Long distances can be converted to transit times using the average speed S associated with the transportation means, that is, by calculating D/S. This approximation can be very rough, however. More accurate estimates for a given mode are provided by an empirical relationship of the type illustrated in Fig. 5.13. When distances are short,

Table 13.2 Circuitry factors for selected countries

Country	Factor	Country	Factor
Brazil	1.23	Italy	1.18
Canada	1.3	India	1.31
China	1.33	Japan	1.41
England	1.4	Russia	1.37
France	1.65	Spain	1.58
Germany	1.32	United States	1.2

Source Ballou et al. (2002)

a tailor-made empirical model may have to be developed. Kolesar et al. (1975), for example, propose a model of this kind to estimate the time firefighters take to reach a fire scene in New York City.

Web-based atlases such as Google Maps can now be used to calculate relatively precise distances and travel times, and everyone is familiar with their functionality. These systems represent road networks as a set of nodes and arcs of known length. To calculate the distance between two addresses, they simply find the shortest path between them on the road network. Each arc has a road type (e.g., interstate highway, state road, residential street, etc.) and an average speed is associated with each road type. Using this information, once the shortest path has been found, travel times can be calculated for each arc on that path and summed to get a total travel time estimate. These applications are excellent to find itineraries one at a time but they are not very good when thousands of distances must be quickly calculated for an entire SCN. Distance and time calculations can also be made with GISs (e.g., *MapInfo* [www.pitneybowes.com], *ArcGIS* [www.esri.com]). GISs are very flexible and they incorporate their own programming language, which enables the development of custom applications. They are, however, very costly and require specialized knowledge. A prevailing alternative in SCN design projects is the use of route planning software developed for the transportation industry, such as PC*Miler (www.pcmiler.com). These tools take many more details into account (e.g., traffic density, toll costs, hazmat restrictions, etc.), they very quickly can calculate thousands of distances in batches, and they can be embedded in tailor-made SCN design applications. Similar tools are also available for rail transportation. Commercial SCN design packages typically incorporate these types of tools.

13.4.2 Models Formulation

Once basic modeling variables have been specified, functional relationships must be established between them to formulate the objective function and constraints of the design model. Thousands of cost and revenue parameters must be associated with the decision variables defined, and it is not possible to evaluate each of them individually. Instead, costs and revenue models must be developed for classes of variables and then applied to calculate specific parameters. For example, a TL transportation cost model may be developed for a country and product family, as shown in Sect. 5.4, and then applied to calculate unit flow costs for all the network arcs requiring this kind of transportation. Chap. 2 discussed the formulation of such descriptive cost models. One generally has to start by the elaboration of a value-creation framework such as the one given in Fig. 2.8. Then, the cost elements associated with the model variables must be determined. Figure 2.9, for example, shows some of the costs and revenues typically associated with the arcs and nodes in a design model. Some of the costs involved, such as order, reception, and shipping costs, are at the interface between network arcs and nodes. In the design model, depending on the context, these costs could be associated with flow variables (F) or with throughput variables (X). This must be elucidated to finalize the cost function used to evaluate specific cost parameters.

The main types of cost models required in SCN design projects were discussed in Sect. 2.2.3. The data used to estimate these models typically come from the company DBs, potential partner proposals, or public sources. In some cases, raw transactional data accumulated during the base year can be employed (e.g., transportation bills for the year), but one may also have to use preprocessed data from the company accounting systems. In the latter case, procedures (e.g., SQL DB queries) must be elaborated to convert accounting data into required cost parameters. In some cases, the elaborated cost models may be embedded directly into the design model. For example, if a capacity cost function of the type studied in Sect. 8.1.3 is elaborated, and one wants to optimize capacity levels, then this function may be embedded directly in the design model instead of using it to calculate predetermined capacity options costs.

Usually, the estimation of cost models starts during the analysis phase because they provide useful means to study the current system behavior. However, it is only during the modeling phase that they can be finalized because the exact cost parameters to estimate are not known beforehand. Also, the potential SCN incorporates several new options and their costs may have to be evaluated from data not previously available. If ecological impacts are taken into account in the design project, emission functions also must be derived to be able to calculate the value of related design model parameters. The approach to take was presented in Sect. 12.2.

The model constraints establish relationships between decision variables and product demand, supply, production, storage and shipping capacity, and so on. Demand processes must be formalized and associated probability distributions estimated. The approach to do this was discussed in Sect. 10.2. Other relevant stochastic processes associated, for example, with exchange rates and hazard arrival and impact processes must also be estimated. The capacity provided by modeled resources and capacity consumption factors must also be assessed. As discussed in Sect. 8.2, this must be done with care, especially in contexts in which the need for capacity is random, thus involving idle periods and congestion periods, that is, an irregular need for capacity. Recipes providing material consumption factors and inventory-throughput functions must also be elaborated in terms of the product families defined. Again, this may involve thousands of parameters to estimate and the idea is to elaborate generic relationships that can be applied to evaluate as many specific parameters as possible.

The elaboration of generic relationships is particularly important when a multi-period model is used. This multiplies the number of parameters to estimate by the number of periods in the model. The idea is then to relate the generic functions developed to time so that they can be used to estimate parameter values for all periods. When a stochastic programming model is used, this amounts to relating parameters to evolutionary paths (see Sect. 10.4) to be able to generate plausible multi-period scenarios. Another important issue is the aggregation level of the user anticipation embedded in the design model. Aggregated user response anticipations simplify the solution of resulting design models, but they may generate serious aggregate parameter estimation problems. Also, recall that the user response model used to evaluate candidate designs (see Sect. 11.7.2) may be more detailed than the one embedded in the design model. It is thus usually better to start by formulating the detailed user response model to use for the evaluation of candidate designs and then to derive a condensation of this model to be included in the design model. This generally helps develop adequate procedures for the estimation of the aggregated user anticipation parameters in the design model.

13.5 SCN Design Optimization

The design phase is completed by an optimization activity, as shown in Fig. 13.1. The modeling activity described in the previous section provides the descriptive (generic cost functions, stochastic processes, etc.) and normative (design objects [product families, nodes and arcs], decision variables, and functional relationships) models formulations required to initiate the optimization activity. Most SCN design models cannot be solved analytically; rather, they are mathematical programs that must be solved numerically using some kind of iterative method (e.g., the simplex method or a metaheuristic). This requires the use of a commercial or tailor-made solver implemented on a computer. To solve a mathematical program, its formulation must be converted into input data files formatted as required by the solver used, and the solution files provided by the solver must be converted back into meaningful design variable reports. The first objective of the optimization activity is thus to convert formulated design models into data files that can be interpreted by a solver to obtain candidate designs.

As discussed in Sect. 11.7., candidate designs generation is just one of three basic tasks needed to design robust value-creating SCNs. These tasks and their interrelations are shown in Fig. 13.13, which is based on the stochastic design approach proposed in Fig. 11.12. Although the generic SCN design process in Fig. 13.13 is based on stochastic programming concepts, it is quite general and most SCN design methods used in practice can be cast as particular cases of this framework. This is shown in Fig. 13.14, which highlights the characteristics of three commonly used design approaches (a, b, and c), as well as those of the

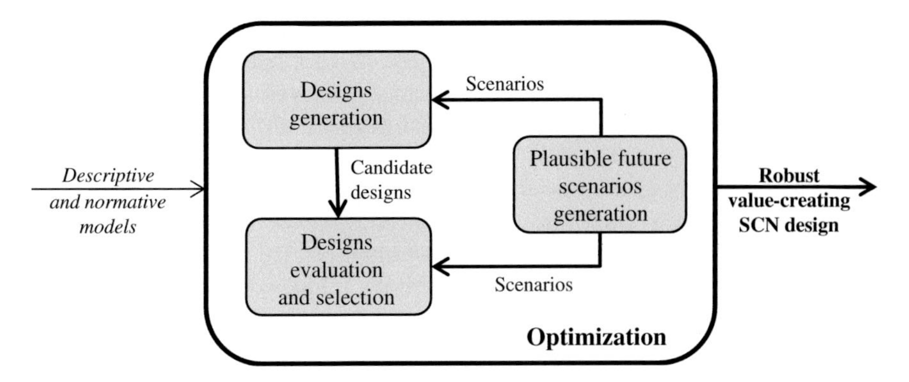

Fig. 13.13 Tasks performed during the optimization activity

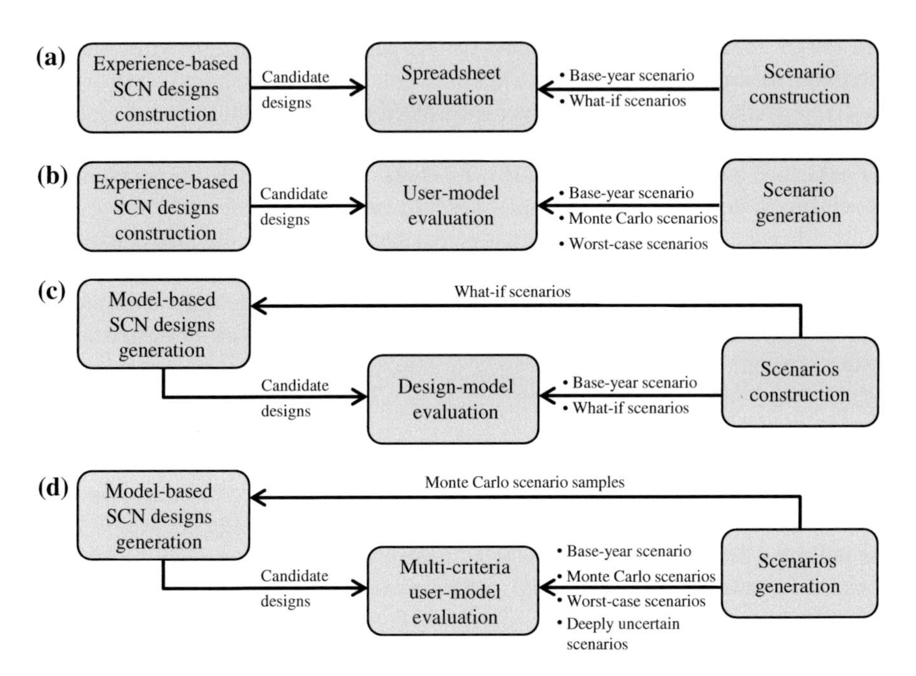

Fig. 13.14 Alternative SCN design approaches. **a** Spreadsheet approach. **b** Simulation approach. **c** Classical deterministic optimization approach. **d** Proposed stochastic optimization approach

approach proposed in this book (d). The two first approaches (a, b) do not use optimization models to obtain candidate designs. Instead, alternative designs are constructed manually by experienced managers or analysts. Manually constructing good feasible candidate designs for large SCNs is not an easy task. For this reason, SCN designs elaborated this way are often perturbations of the current network structure. There is no guarantee, however, that manually constructed designs are near-optimal. For approach (a) and (c), the plausible futures used to compare candidate designs are typically variants of a base-year scenario. One assumes that the future will be almost identical to the recent past. What-if scenarios, that is, perturbations of the base-year scenario, are frequently employed to try to evaluate the behavior of candidate designs under marginal plausible changes. When approach (a) or (c) is used, the evaluation of candidate designs for the considered scenarios is usually based on projected financial statements, often produced using a spreadsheet.

Despite its obvious limitations, the *spreadsheet approach* (a) is probably the most widely used in practice. The *simulation approach* (b) provides much better evaluations of candidate designs based on Monte Carlo scenarios and a detailed implementation of operational decision processes (user model). In fact, the evaluation is then often made using the ERP or APS of the company with detailed scenario data. The only problem with this approach is that the candidate designs evaluated are constructed manually instead of being optimized. Alternative (c) is the *classical deterministic optimization approach* used in practice to design SCNs. The design models are typically static (single-period) deterministic MIPs such as those introduced in Chap. 7, and the plausible futures considered are variants of the base-year scenario as in (a). Moreover, the objective function of the design model is often used to compare candidate designs to the status quo, which is a serious flaw. As explained previously, the objective function of these models are approximate anticipations, and a more detailed performance evaluation framework is needed to make adequate comparisons. Alternative (d) corresponds to the *stochastic optimization approach* proposed in Chap. 11. It corrects the defects of the other approaches. Clearly, all sorts of variants of the four SCN approaches illustrated in Fig. 13.14 are also possible. Various implementation issues are discussed in the next sections.

13.5.1 Model-Based SCN Design Generation

As indicated previously, to solve a SCN design model, the mathematical program formulated must be converted into machine-readable data specifying all model variables and functional relations (objective functions and constraints). Because millions of variables and relations may have to be defined, an efficient way of doing this must be found. This usually involves the definition of a *SCN design problem database* (not to be confused with the project DB) containing all the data required to specify design model structures and parameters, inspect the candidate designs obtained by solving mathematical programs, and evaluate these designs. In static design models, as shown by generic formulation (11.1), decision variables can be split into design variables $\mathbf{y}$ (platform, market offer, and supplier selection variables) and user response variables $\mathbf{x}$ (flow and throughput variables), and relations related to $\mathbf{x}$, $\mathbf{y}$, or a combination of both must be defined. In dynamic models, copies of the design variables $\mathbf{y}_h$ typically must be made for each reengineering cycle h, copies of the user response variables $\mathbf{x}_t$ for each planning period t, and associated relations must be extended accordingly. Also, in deterministic models a single plausible future is defined, but stochastic programming models cover a sample of plausible future scenarios. User response variable $\mathbf{x}_{t\omega}$ and associated constraints then must also be recopied for each scenario ω. However, although the underlying structure is similar for each t, h, and ω, specific parameter values differ. The basic idea to generate complex multi-period stochastic design models is first to identify the underlying *potential network structure,* second to define a *template* of parameter values and projection functions, and then to use them to project parameter values into time and scenario spaces.

Defining the underlying potential structure relates to the specification of an activity graph and potential network resources, as shown for example in Fig. 7.18, and to use this information to automatically generate potential network nodes and arcs, as shown in Fig. 7.19. Recall that a network node is an association between an activity and a site location and that the latter can be associated with supply contracts, site platforms, or demand zone offers. These associations first must be defined to generate network nodes. The activity graph provides definitions of activities, product families, and potential movements of products between activities. Transportation means can also be associated with activity graph movements, thus providing crucial information for the generation of network arcs. Going back to the example in Fig. 7.18, one could specify that movements of raw material 2 (say steel rods) between suppliers and plants performing activity 2 (machining) can be made on flatbed trucks or railcars, intermediate products are moved in full load van or railcar shipments, and final products are shipped to customers using LTL road transportation. Using this information, parallel arcs between node pairs can be automatically generated for all relevant product families and transportation means. Note, however, that some of the arcs generated this way are usually not feasible because, for example, they may not respect some service conditions specified by product-market offers. Additional rules must thus be added to ensure that all arcs created are feasible. Manual fine-tuning of potential nodes and arcs also should be possible. This permits, for example, the elimination of arcs that, although conceptually valid, would never be used in practice. Several thousand potential lanes generated using this approach for an international SCN are shown in Fig. 13.15.

The potential network structure illustrated in Fig. 13.15 reflects the main design and user response variables of the design model, but it does not specify the model objective function and constraints parameters. Parameter values must be superimposed on network structure variables. For static models, parameter values are either captured directly in the problem DB or calculated using predefined functions. For example, suppose that the transportation cost function shown in Fig. 5.20 was fitted to the TL shipment costs of the company for all its lanes in a given country. Then the unit flow cost on all network arcs in that country involving TL-shipments can be calculated using this function based on arc lengths in miles. Now if the model is defined over a multi-period planning horizon, then copies of user response variables must be made for each period t, and base-period costs must be defined and projected in time based on evolutionary path explanatory variables (see Table 10.5). Suppose for example that the transportation cost function in Fig. 5.20 was based on data taken from a complete year of operation starting in mid-2013 and that a ten-year

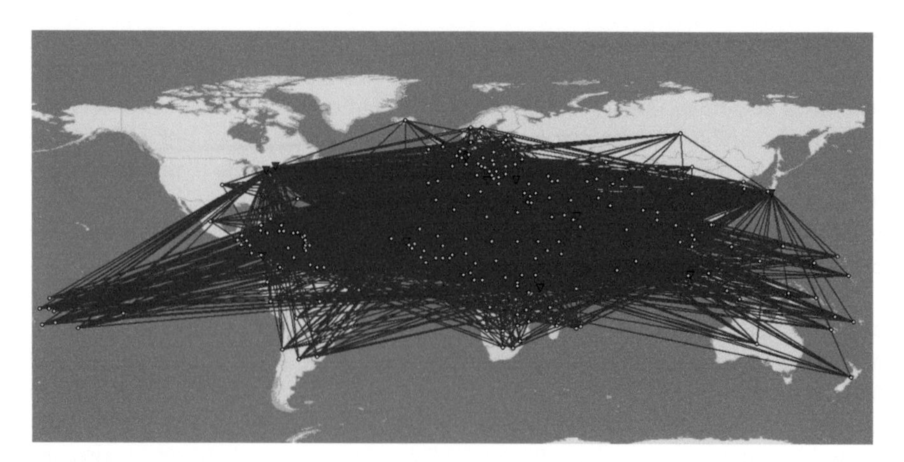

Fig. 13.15 Potential nodes and lanes for a global network example

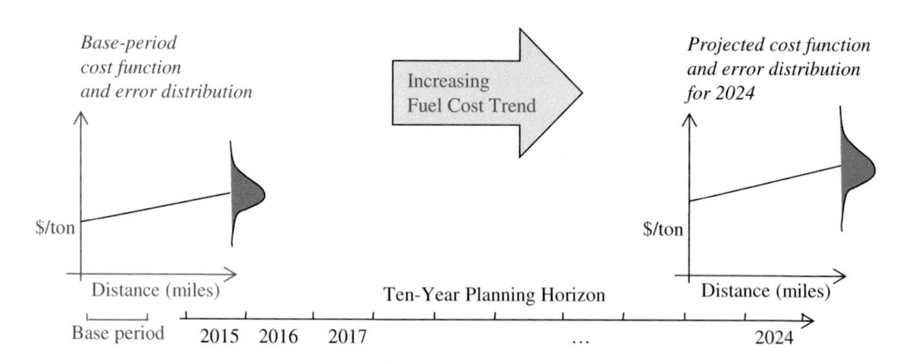

Fig. 13.16 Projection of transportation costs in time based on a fuel cost trend

planning horizon starting at the beginning of 2015 is used. Suppose also that fuel costs are expected to increase by 6 % per year during the planning horizon. Then the function used to compute unit flow costs during each year of the horizon would be modified to consider fuel cost increases as shown in Fig. 13.16. If the model is deterministic, a single evolutionary path is considered and expected costs are used. If the model is stochastic, a sample of plausible futures is defined using Monte Carlo methods (see Sect. 10.5.2) based on a set of evolutionary paths and taking transportation cost error distributions into account. Saving the model parameter values for all scenarios generated in the problem DB enables their inspection for validation purposes. It also facilitates the generation of the model data files required by the solver used. The generation of model files from the problem DB can be done with any programming environment including embedded SQL instructions (e.g., MS Visual Studio) or using specialized model generation tools developed to facilitate this process such as OPL (from IBM ILOG CPLEX Optimization Studio),

AMPL (ampl.com), AIMMS (www.aimms.com), or GAMS (www.gams.com). Once a solution has been found with a solver, it is usually desirable to reimport it back into the problem DB. Inspecting a solution is greatly facilitated by the use of a DB query language and displaying some of its features using a GIS.

Independent of the type of model employed to generate candidate designs, an important task is the validation of the data and model used. This is typically done by trying to reproduce the current financial and operational performance of the company by applying the design model to the status quo SCN for the base year considered. In other words, if the design variables in the model are fixed to their current value (current platforms, supply sources, and market offers) and the user response variables (flows and throughputs) are fixed to the value observed for the base year, then the revenues, costs, and service levels calculated with the model should match those found in the accounting and information system of the company. If the user model formulated to evaluate candidate designs is more detailed than the one embedded in the design model, the evaluations it provides should also be validated. Current flows and throughputs should be checked to ensure that flow equilibrium relations are satisfied. If not, adjustments must be made. If any discrepancy between actual results and calculated results are observed, they should be explainable. Without this essential exercise, the client will not trust any subsequent design recommendations based on the results provided by the design and evaluation models.

13.5.2 Candidate Designs Evaluation

The evaluation of candidate designs and the selection of a final robust value-creating SCN structure was discussed in general terms in Sect. 11.2.2 and studied more closely in Sect. 11.7.2. Although the evaluation approach presented was proposed for design models under risk, it applies whatever the method used to generate or construct candidate SCN designs. Recall that to evaluate a candidate design $\mathbf{y}$, a user response model must be solved for each scenario ω considered to get corresponding design values $V^{\mathrm{SCN}}(\mathbf{y}, \omega)$. As discussed in Sect. 12.1.2, when designing for sustainable development, other environmental and social impact measures, $EI^{\mathrm{SCN}}(\mathbf{y}, \omega)$ and $SI^{\mathrm{SCN}}(\mathbf{y}, \omega)$, may also be assessed. In the simplest case, for a given scenario, optimal user response variables are obtained by solving the design model with fixed $\mathbf{y}$ values. However, whenever possible, a more detailed and realistic user model should be developed to obtain better evaluations. When this is done, a solver or specialized algorithm must be implemented to solve the user model, and this requires the generation of input solver data files as discussed in the previous section. The plausible futures assessed may include historical, Monte Carlo, worst-case, and deep-uncertainty scenarios, as discussed previously, as well as any what-if scenarios devised by the designer.

Having done this, a set of performance indicators must be selected and estimated. For Monte Carlo scenarios, sample expected value and risk estimators are computed (see Sect. 11.5). In addition to financial, environmental, and social impact expected value and risk measures, other specialized indicators may be used to assess network flexibility, resilience under catastrophic events, and so on. Based on all these performance indicators, a multi-criteria filtering and ranking approach must be applied to make a final selection. We have seen that plotting efficient frontiers is an excellent way to make bi-criterion design comparisons. Typical efficient frontiers were illustrated in Fig. 2.15 for cost-service tradeoffs, in Fig. 11.3 for value-risk tradeoffs, and in Fig. 12.9 for value-GHG emissions tradeoffs.

The previous evaluations are based on quantifiable measures. However, in any SCN design project organizational and strategic factors difficult to quantify must also be taken into account. Each company has specific roots, culture, and public image; existing employees have particular skills; and so on so that some SCN designs may fit much better with these soft characteristics than others. One cannot recommend the implementation of a design that upsets all this and expect to get good results. Candidate designs must be evaluated not only from an economic and sustainable development point of view but also from an organizational point of view. Finally, when implementation plans are prepared based on the design selected, a gradual modular implementation ideally should be proposed. This facilitates the organizational insertion of new systems, and it permits subsequent adaptations if unforeseen events occur.

13.5.3 Commercial SCN Design Tools

The realization of the optimization tasks summarized in Fig. 13.13, as we just saw, may require substantial database definition and computer programming efforts. These developments are typically time-consuming, and they require expertise that often is not available in companies. In SCN design projects, in order to save time and avoid costly tailor-made SCN optimization tool developments, most companies adopt commercial SCN design tools. Several specialized packages are available on the market to facilitate the design of SCNs. They typically incorporate tools for the definition of potential SCN structures, the construction or generation of plausible future scenarios, the generation and solution of design models, as well as the exploration and evaluation of generated designs. They are often based on commercial DBMS, GIS, and optimization software, which are embedded in a transparent manner in the package. The products currently available on the market include the following:

- *Supply Chain Designer* originally developed by CAPS Logistics and now offered by Infor (www.infor.com)
- *Strategic Analysis for Integrated Logistics Systems* (SAILS) offered by Insight (www.insight-mss.com)

- *LogicNet* originally developed by Logic Tools and now offered by LLamasoft (www.llamasoft.com)
- *Supply Chain Guru* offered by LLamasoft (www.llamasoft.com)
- *Supply Chain Strategist* offered by JDA (www.jda.com)
- *LOPTIS* offered by Ketron Optimization (www.ketronms.com/loptis.shtml)
- *CAST* originally developed by Barloworld SC Software and now offered by LLamasoft (www.llamasoft.com)
- *SC-Studio* offered by Modellium (www.modellium.ca)

The main advantage of these tools is that they greatly facilitate the model generation, validation, and solution processes, as well as the candidate design evaluation process. They provide near-optimal solutions when the problem under consideration is part of the class of problems covered by the software. Unfortunately, they usually cover only a subset of the design issues discussed in this book and they may be difficult to apply in some industrial contexts. Most of them support only deterministic models. They usually facilitate the examination of alternative what-if scenarios, but they rarely support the formulation of stochastic programs. They may also be quite expensive. Some of them have a relatively open architecture and can be adapted to specific cases. However, this is a fairly demanding task. When the time comes to choose an SCN design tool, the following factors should be considered:

- The design and user response variables taken into account by the tool and the objectives and constraints it can accommodate. As indicated previously, the tool should be able to cope with all the aspects of the design problem captured by the models formulated during the modeling activity. The problem should not be twisted to fit the tool. If equivalent formulations are available and the discrepancies found are not critical, then the tool should be flexible enough to make the required adjustments.
- The costs and revenue modeling assumptions made and, in particular, the ability to account for economies of scale
- Activity graph and recipes (BOM) definition features. Some products limit themselves to the optimization of distribution networks and cannot cope with multi-stage production–distribution networks.
- Geocoding and automatic arc generation and edition capabilities
- Multiple platform, transportation means, and market offer selection features
- Multi-period parameter value projection capabilities. Some products limit themselves to static models.
- Monte Carlo scenarios and stochastic programming model generation capabilities
- What-if scenario construction and validation capabilities
- Optimization engine capabilities and size of models that can be solved
- Candidate design evaluation capabilities and in particular the ability to add additional performance metrics
- Multi-criteria candidate design screening and selection capabilities
- Automatic efficient-frontier analysis capabilities
- Quality of graphical user interfaces and network mapping capabilities

- Model and solution query capabilities
- Interfaces with main stream ERP/APS systems (e.g., mySAP) and standard spreadsheet and DBMS (e.g., Excel, Access)
- Report builder flexibility and ability to export reports to other applications
- Cloud-based processing and collaboration capabilities
- Tool acquisition, customization, and use prices
- Quality of the support provided by the company (training, documentation, troubleshooting, etc.)
- The financial health and experience of the company

The candidate designs generated and scenario analysis made with design tools should provide all the information required to elaborate final SCN design recommendations and prepare detailed implementation plans.

Review Questions

1. What is the link between a system's life cycle and the methodology adopted for systems reengineering?
2. What are the main roles of the planning, analysis, modeling, and optimization activities of a SCN reengineering project?
3. Why is it important to plan reengineering projects carefully?
4. What types of tools are required to support a SCN reengineering project?
5. What types of expertise should SCN reengineering project team members have and what are the key functions team members need to perform?
6. What are the main milestones of a typical SCN reengineering project?
7. What fundamental questions should be asked when performing a system analysis?
8. Why must a system analysis be performed from several points of view and what are the main points of view to adopt?
9. What type of data must be collected during a SCN reengineering project?
10. Why is it useful to set up a design project database?
11. What external factors should be investigated when studying the SCN environment?
12. What type of diagramming tools can be useful when doing a SCN analysis?
13. What are the main facets to examine when studying an existing SC system?
14. Why are Pareto analyses useful in SC system analysis?
15. What is the difference between an existing SCN and a potential SCN?
16. Why is the characterization of demand processes particularly important in SCN design projects?
17. What form does a SCN diagnostic usually takes?
18. What strategic positioning elements should be clear at the end of the SCN analysis activity?
19. What are the main variables to define to formulate a SCN design model?
20. Is it desirable to explicitly consider all the products manufactured or sold by a company when formulating a SCN design model? Why?

21. What types of approach can be used to define product families and demand zones?
22. Why is the calculation of distances important during the design phase and how can distances be calculated?
23. How can we proceed to identify potential design options when formulating a SCN design model?
24. What is the difference between a descriptive and a normative model? Give examples of how they can be used in a SCN design project.
25. Is it always necessary to use SCN design models in a SCN reengineering project?
26. When a design model is generated, is it necessary to specify the value of the thousands of parameters involved individually?
27. When using a multi-period model, how can model parameter values be projected in time?
28. What is the difference between a design model and a user response model and how are both used?
29. What are the advantages and disadvantages of commercial SCN design tools?
30. What specific tasks in a SCN reengineering project can be supported by commercial SCN design tools?

Bibliography

Ackermann F, Howick S, Quigley J, Walls L, Houghton T (2014) Systemic risk elicitation: using causal maps to engage stakeholders and build a comprehensive view of risks. Eur J Oper Res 238(1):290–299

Ballou R (1992) Business logistics management, 3rd edn. Prentice-Hall, Upper Saddle River

Ballou R (1994) Measuring transport costing error in customer aggregation for facility location. Transp J 49–59

Ballou R (1995) Logistics network design: modelling and informational considerations. Int J Logistics Manage 6(2):39–54

Ballou R, Rahardja H, Sakai N (2002) Selected country circuitry factors for road travel distance estimation. Transp Res Part A Policy Pract 36(9):843–848

Bender P (1985) Logistic system design. In: Robeson J, House R (eds) The distribution handbook. Free Press, New York, pp 143–224

Cadle J, Paul D, Turner P (2014) Business analysis techniques: 99 essential tools for success, Revised edn. BCS The Chartered Institute for IT

D'Amboise G, Martel A, Oral M, St-Pierre J, Côté M (1995) L'élaboration de projets stratégiques. In: Martel A, Oral M (eds) Les défis de la compétitivité. Publi-Relais, France

Ernst R, Cohen M (1990) Operations related groups (ORGs): a clustering procedure for production/inventory systems. J Oper Manage 9(4):574–598

Geoffrion A (1976) Customer aggregation in distribution modeling. Working paper 259, Western Management Science Institute, UCLA

Hare VC (1967) Systems analysis: a diagnostic approach. Harcourt, Brace & World

Kolesar P, Walker W, Hausner J (1975) Determining the relation between fire engine travel times and travel distances in New York City. Oper Res 23(4):614–627

Lambin JJ, Schuiling I (2012) Market-driven management: strategic and operational marketing, 3rd edn. Palgrave Macmillan

Love R, Morris J (1988) On estimating road distances by mathematical functions. Eur J Oper Res 36:251–253

Martel A, Klibi W (2012) A reengineering methodology for supply chain networks operating under disruptions. In: Gurnani H, Mehrotra A, Ray S (eds) Supply chain disruptions: theory and practice of managing risk. Springer, Berlin, pp 241–273

Martel A, Mantha R, Pascot D (1990) A conceptual framework for the development of computer integrated organizational systems. Working paper 90-06. FSA, Université Laval

Martel A, Benmoussa A, Chouinard M, Klibi W, Kettani O (2013) Designing global supply networks for conflict or disaster support: the case of the Canadian Armed Forces. J Oper Res Soc 64:577–596

Martin J, McClure C (1985) Diagramming techniques for analysts and programmers. Prentice-Hall, Englewood Cliffs

Martin K, Osterling M (2014) Value stream mapping. McGraw-Hill, New York

Porter M (1980) Competitive strategy: techniques for analyzing industries and competitors. Free Press, New York

Rothaermel F (2014) Strategic management: concepts, 2nd edn. McGraw-Hill, New York

Scherer F, Ross D (1990) Industrial market structure and economic performance, 3rd edn. Houghton Mifflin, Boston

Shapiro J (2001) Modeling the supply chain. Duxbury, North Scituate

Team FME (2013) PESTLE analysis. www.free-management-ebooks.com

Watson M, Lewis S, Cacioppi P, Jayaraman J (2013) Supply chain network design. FT Press

Index

A. Martel and W. Klibi, *Designing Value-Creating Supply Chain Networks*,
DOI 10.1007/978-3-319-28146-9

Printed in the United States
By Bookmasters